CLICK CHEMISTRY IN GLYCOSCIENCE

CLICK CHEMISTRY IN GLYCOSCIENCE

New Developments and Strategies

Edited by

ZBIGNIEW J. WITCZAK
ROMAN BIELSKI

WILEY

Published by John Wiley & Sons, Inc., Hoboken, New Jersey.
Published simultaneously in Canada.

Library of Congress Cataloging-in-Publication Data:

Click chemistry in glycoscience : new developments and strategies / edited by Zbigniew J. Witczak, Roman Bielski.
p. ; cm.
Includes bibliographical references and index.
ISBN 978-1-118-27533-7 (cloth)
I. Witczak, Zbigniew J., 1947- II. Bielski, Roman, 1946-
[DNLM: 1. Glycoconjugates–chemistry. 2. Click Chemistry–methods.
3. Glycoconjugates–physiology. QU 75]
572′.567–dc23

2012040254

Printed in the United States of America

ISBN: 9781118275337

10 9 8 7 6 5 4 3 2 1

CONTENTS

III CARBOHYDRATE CLICK CHEMISTRY FOR NOVEL SYNTHETIC TARGETS

IV CARBOHYDRATE CLICK CHEMISTRY IN BIOMEDICAL SCIENCES

FOREWORD

This book, compiled by Zbigniew Witczak and Roman Bielski, brings together contributions from authors around the world in addressing the impact of a single chemical reaction that permits the covalent connection of two complex precursor molecules under mild conditions. The reaction has found wide application in the fields of carbohydrate chemistry and glycobiology for building complex glycoconjugate target structures of interest in many biomedical areas.

The particular reaction is the (3 + 2) 1,3-dipolar cycloaddition of an alkyne to an azide at room temperature under copper(I) catalysis to generate a 1,2,3-triazole. Since its initial discovery by Huisgen in 1963, there have been numerous publications where it has been employed for a multitude of purposes, including applications in the carbohydrate field that date back to the 1970s.

However, it was not until the beginning of the present millennium that an "explosion" of research on the scope of this reaction began after Sharpless, and independently, Meldal, employed this procedure for the rapid synthesis, through heteroatom links, of many useful new compounds, peptide conjugates, and combinatorial libraries. At that point in time, the awkward formal name of Huisgen's excellent reaction was whimsically dubbed "click" chemistry, presumably because the simplicity of the reaction could be likened to the ends of a bracelet being "clicked" together.

The same concept of covalently connecting two complex molecules under very mild conditions has more recently led to valuable new procedures that also meet the "click" reaction criteria. These include the photoinduced reaction between a thiol and either an alkene or an alkyne, and the coupling between a thiol and a Michael enone acceptor.

The 13 chapters in the Witczak–Bielski book bring together a wide range of applications of these ligation strategies directed toward carbohydrate-based targets, including various types of glycoconjugates, such as neoglycoproteins, glycoclusters, glycodendrimers, and cyclodextrin conjugates. The reaction offers potential in diverse biomedical areas, including synthetic antigens, analogs of cell-surface receptors, immobilized enzymes, targeted drug delivery systems, multivalent cancer vaccines, and many others.

Beyond the original Huisgen reaction, there now have evolved several variants, some involving modifications of the original alkyne and azide reactants, along with such adaptations as novel catalysts for effecting the reaction under the mildest conditions, and performing the ligation reaction *in vivo*. Many of these extensions are detailed in different chapters in the book, together with the thiol–alkene, thiol–alkyne, and thiol–enone conjugation reactions that also merit the "click" designation.

This book derives from presentations made in a 2011 symposium at a meeting of the American Chemical Society. Not all relevant literatures on the alkyne–azide cycloaddition are documented. With new work being published almost daily, the coverage, even in the carbohydrate field alone, can never be complete. Nevertheless, the compilers of this volume have made a valuable contribution by bringing together the collective efforts of more than 30 researchers working on the applications of "click" chemistry to numerous targets in the carbohydrate and glycoconjugate area. The book provides a valuable resource for both the specialist researcher and the general reader.

Derek Horton
Professor of Chemistry Emeritus
Ohio State University

PREFACE

Synthesis of compounds designed to fulfill special requirements or exhibit specific properties has belonged and will belong to the most important targets of organic chemistry. This area of synthesis, particularly when applied to synthesis of constructs containing carbohydrates, has experienced a dramatic acceleration in recent years. One factor explaining the observed acceleration has been a better understanding of the function and structure of glycoproteins and other naturally occurring sugar derivatives. Another factor, perhaps even more significant, is the introduction of the concept of click chemistry by Finn, Kolb, and Sharpless, which dramatically facilitated the formation of various constructs. Chemistry of carbohydrates containing molecules has benefited tremendously from the introduction of this concept. The presented book attempts to offer an insight into the new developments created by marrying click and carbohydrate chemistries.

Carbohydrates represent a unique family of polyfunctional compounds that can be chemically or enzymatically manipulated in a multitude of ways. They have been extensively used as starting materials in enantioselective synthesis of many complex natural products with a plurality of chirality centers. Synthetic organic chemistry that utilizes these carbohydrate building blocks continues to spawn revolutionary discoveries in medicinal chemistry, pharmacology, molecular biology, glycobiology, and medicine simply by providing not only the raw material but also the mechanistic insight of modem molecular sciences.

Click chemistry was introduced over 10 years ago. Its founders offered the following description:

> *A click reaction must be modular, wide in scope, high yielding, create only inoffensive by-products (that can be removed without chromatography), are stereospecific, simple to perform, and that require benign or easily removed solvent.*

Since then, Sharpless' concept of click chemistry was quickly transplanted to carbohydrate chemistry and the number of publications in the field has been steadily growing.

This book originates from the symposium "Click Chemistry Approaches in Carbohydrate Chemistry," which we organized during the 242nd Meeting of the American Chemical Society in Anaheim in Spring 2011. It attracted several prominent speakers, had a relatively large attendance, and was met with a lot of interest. Some of the chapters in this book are based on the presentations delivered at the symposium. Other contributions are also from leading experts in the field of carbohydrate chemistry. Some of the chapters are reviews of the recent literature; some describe recent experiments performed at the authors' laboratories. We believe that all the articles are of very high standard and offer a novel perspective on the discussed subjects.

The medical and biomedical applications of synthetic organic chemistry were probably more affected than any other area of research (perhaps with the exception of polymer chemistry) by the development of click chemistry. Thus, it is not surprising that almost all of the chapters in the book are to some extent concerned with such applications. It confronts the editors with a dilemma that is impossible to address satisfactorily – how to divide the book into consistent segments. By no means are we satisfied with our choice but some kind of division had to be introduced. Thus, the reader should not be surprised finding biomedical applications described in segments carrying a title suggesting that the emphasis was put on an entirely different topic.

Each chapter in the book covers issues related to click chemistry and glycoconjugation, and discusses synthetic methodologies and potential applications of the synthesized constructs. In the last few years, it has been documented that the addition of thiols to unsaturated compounds is a legitimate click process. Thus, we made sure that this type of click reactions is represented in the book together with the most popular click reaction, that is, 1,3-dipolar addition of azides to alkynes.

The topics covered in the book are grouped into four categories:

1. Click chemistry strategies
2. Thio-click chemistry of carbohydrates
3. Click chemistry related to life science and glycobiology
4. Click chemistry related to medicinal chemistry

The introductory chapter written by both editors of the volume discusses the important aspects of carbohydrate click chemistry methodologies and proposes a novel strategy applicable to synthesizing certain glycoconjugates.

Chapter 2 authored by Witczak deals with thio-click strategies employed to the synthesis of thiodisaccharides and other sulfur-bridged oligosaccharide scaffolds. Interestingly, the most efficient methodologies compiled in this review were developed before the official birth of thio-click chemistry.

Chapter 3 authored by Dondoni and Marra describes intriguing results of the experiments employing various thio-click glycoconjugation processes.

The development and application of clickable lipid analogs for elucidating and harnessing lipid function were thoroughly explored by Best in Chapter 4. The chapter discusses most types of natural molecules, including nucleic acids, proteins, various lipids, and many substituted glycerols and, of course, carbohydrates.

In Chapter 5, Uhrig and Kovensky review the important topic of syntheses of oligosaccharide analogs and glycoclusters on carbohydrate scaffolds.

Mellet and coworkers offer in Chapter 6 a different perspective on the closely related subject of click multivalent glycomaterials, including glycosurfaces, glycodendrimers, and glycopolymers.

Chapter 7 by Chapleur and coworkers addresses clickable formation of carbohydrates labeled with radiotracers for molecular imaging.

Chapter 8 by Friscourt and Boons is a very broad review of bio-orthogonal reactions for labeling glycoconjugates. The review discusses the processes belonging to click chemistry as well as other important coupling reactions, such as Staudinger ligation and labeling with photoactivatable sugars.

Potopnyk and Jarosz explore in Chapter 9 quite a new topic of click functionalization of sucrose in the synthesis of interesting sucrose-based macrocycles.

Brimble and coworkers contribute in Chapter 10 with a discussion of novel syntheses of neoglycoproteins via copper-assisted azide–alkyne click reaction and native chemical ligation.

Chapter 11 by Wang and coworkers describes the formation and new fascinating biomedical applications of "click-modified" cyclodextrins.

In Chapter 12, Tripathi, Tiwari, and coworkers explore the important applications of triazolyl glycoconjugates in medicinal chemistry. The review highlights the synthesis of prototypes of drug molecules with high chemotherapeutic potential.

Finally, Chapter 13 by Campo and Carvalho provides a general overview of the potential applications of click reactions in the synthesis of highly valuable, bioactive, carbohydrate-based ligands for lectins, antitumor vaccines, and various enzyme inhibitors.

With the increasing complexity of modern sciences in the twenty-first century, a need to educate industrial leaders, public, and governmental funding agencies about the intellectual and technical potential and economic importance of specific areas of life sciences has become more and more crucial. One such area is the part of glycoscience emerging as a result of a marriage between the concepts of click and carbohydrate chemistries. We hope that this book will fill this need, at least to some extent.

In conclusion, we believe the presented collection of articles offers an insight into the present stage of click-based glycosciences and will help steer future discoveries to fulfill the enormous potential in the area of click carbohydrate chemistry.

Acknowledgment

We wish to thank all the authors for excellent contributions to this volume. We also wish to thank the peer reviewers of the chapters for their expertise and helpful efforts to improve the quality of the manuscripts. We dedicate this book to our wives Wanda and Barbara.

Zbigniew J. Witczak, Ph.D
Department of Pharmaceutical Sciences
Nesbitt School of Pharmacy
Wilkes University, PA, USA

Roman Bielski
Value Recovery, Inc., NJ
Department of Pharmaceutical Sciences
Wilkes University, PA, USA

CONTRIBUTORS

Michael D. Best, Department of Chemistry, University of Tennessee, TN USA

Roman Bielski, Value Recovery, Inc., Bridgeport, NJ, USA; Department of Pharmaceutical Sciences, Wilkes University, PA, USA

Geert-Jan Boons, Complex Carbohydrate Research Center, University of Georgia, GA, USA

Margaret A. Brimble, School of Chemical Sciences, University of Auckland, Auckland, New Zealand; School of Biological Sciences, University of Auckland, Auckland, New Zealand

Vanessa Leiria Campo, Faculdade de Ciências Farmacêuticas de Ribeirão Preto, São Paulo, Brazil

Ivone Carvalho, Faculdade de Ciências Farmacêuticas de Ribeirão Preto, São Paulo, Brazil

Yves Chapleur, Laboratoire Structure et Réactivité des Systèmes Moléculaires Complexes, Université de Nancy, Groupe SUCRES, Vandoeuvre les Nancy, France

Françoise Chrétien, Laboratoire Structure et Réactivité des Systèmes Moléculaires Complexes, Université de Nancy, Groupe SUCRES, Vandoeuvre les Nancy, France

Alessandro Dondoni, Dipartimento di Chimica, Laboratorio di Chimica Organica, Università di Ferrara, Ferrara, Italy

Pratibha Dwivedi, Department of Chemistry, Faculty of Science, Banaras Hindu University, Varanasi, India

Clive W. Evans, School of Biological Sciences, University of Auckland, Auckland, New Zealand

José Manuel García Fernández, Instituto de Investigaciones Químicas, CSIC – Universidad de Sevilla, Sevilla, Spain

Frédéric Friscourt, Complex Carbohydrate Research Center, University of Georgia, GA

Sławomir Jarosz, Institute of Organic Chemistry, Polish Academy of Sciences, Warsaw, Poland

Zhenshan Jia, Department of Pharmaceutical Sciences, College of Pharmacy, University of Nebraska Medical Center, NE, USA

Stephen B.H. Kent, Department of Chemistry, Institute for Biophysical Dynamics, Center for Integrative Science, University of Chicago, IL, USA

José Kovensky, Laboratoire des Glucides-CNRS FRE 3517, Université de Picardie Jules Verne, Amiens, France

Divya Kushwaha, Department of Chemistry, Faculty of Science, Banaras Hindu University, Varanasi, India

Sandrine Lamandé-Langle, Laboratoire Structure et Réactivité des Systèmes Moléculaires Complexes, Université de Nancy, Groupe SUCRES, Vandoeuvre les Nancy, France

Dong Jun Lee, School of Chemical Sciences, University of Auckland, Auckland, New Zealand; Department of Chemistry, Institute for Biophysical Dynamics, Center for Integrative Science, University of Chicago, IL, USA

Kalyaneswar Mandal, Department of Chemistry, Institute for Biophysical Dynamics, Center for Integrative Science, University of Chicago, IL, USA

Alberto Marra, Dipartimento di Chimica, Laboratorio di Chimica Organica, Università di Ferrara, Ferrara, Italy

Carmen Ortiz Mellet, Departamento de Química Orgánica, Facultad de Química, Universidad de Sevilla, Sevilla, Spain

Alejandro Méndez-Ardoy, Departamento de Química Orgánica, Facultad de Química, Universidad de Sevilla, Sevilla, Spain

Mykhaylo A. Potopnyk, Institute of Organic Chemistry, Polish Academy of Sciences, Warsaw, Poland

Anindra Sharma, Medicinal and Process Chemistry Division, Central Drug Research Institute, Lucknow, India

Rakesh K. Singh, Department of Pathology and Microbiology, College of Medicine, University of Nebraska Medical Center, NE, USA

Vinod Kumar Tiwari, Department of Chemistry, Faculty of Science, Banaras Hindu University, Varanasi, India

Rama Pati Tripathi, Medicinal and Process Chemistry Division, Central Drug Research Institute, Lucknow, India

Maria Laura Uhrig, CIHIDECAR-CONICET, Departamento de Química Orgánica, Facultad de Ciencias Exactas y Naturales, Universidad de Buenos Aires, Ciudad Universitaria, Buenos Aires, Argentina

Christine Vala, Laboratoire Structure et Réactivité des Systèmes Moléculaires Complexes, Université de Nancy, Groupe SUCRES, Vandoeuvre les Nancy, France

Dong Wang, Department of Pharmaceutical Sciences, College of Pharmacy, University of Nebraska Medical Center, NE, USA

Zbigniew J. Witczak, Department of Pharmaceutical Sciences, Nesbitt School of Pharmacy, Wilkes University, Wilkes-Barre, PA, USA

Joanna M. Wojnar, School of Chemical Sciences, University of Auckland, Auckland, New Zealand

LIST OF ABBREVIATIONS

[18F]FC	[18F]fluorocholine
[18F]FLT	3′-deoxy-3′-[18F]fluorothymidine
α-MSH	α-melanocyte-stimulating hormone
Ac	acetyl
ADIBO	azodibenzocyclooctyne
AFM	atom force microscopy
AIBN	azobisisobutyronitrile
AIDS	acquired immune deficiency syndrome
All	allyl
AMEs	aminoglycoside-modifying enzymes
ANL	*Aspergillus niger* lipase
ATR-FTIR	attenuated total reflectance-Fourier transform infrared spectroscopy
ATRP	atom transfer radical polymerization
BLG	bovine liver galactosidase
Bn	benzyl
Boc	*t*-butyloxycarbonyl
BrBz	bromobenzyl
BSA	bovine serum albumin
BTMSS	bis(trimethylsilyl)sulfide
BTTES	tris(triazolylmethyl)amine
CAD	coupling and decoupling
CAL B	Candida antarctica B lipase
Cbz	benzyloxycarbonyl
CCCTP	cobalt catalyzed chain transfer polymerization
CCTP	catalytic chain transfer polymerization

CD	cyclodextrin
CMP	cytidine monophosphate
CNBr	cyanogen bromide
CNTs	carbon nanotubes
ConA	concanavalin A
CSA	10-camphorsulfonic
CT	computerized tomography
CuAAC	copper-catalyzed azide–alkyne cycloaddition
DABCO	1,4-diazobicyclo[2,2,2]octane
DAG	diacylglycerol
DBU	1,8-diazabicyclo[5.4.0]undec-7-ene
DCM	dichloromethane
DCR	1,3-dipolar cycloaddition
DDQ	dichlorodicyanoquinone
DEGMA	di(ethylene glycol) methyl ether methacrylate
DF	diclofenac
DFT	density functional theory
DIAD	diisopropyl azidodicarboxylate
DIPEA	*N*,*N*-diisopropylethylamine
DLS	dynamic light scattering
DMAc	*N*,*N*-dimethylacetamide
DMAP	*N*,*N*-dimethylaminopyridine
DMF	*N*,*N*-dimethylformamide
DMPA	2,2-dimethoxy-2-phenylacetophenone
DMPP	dimethylphenylphosphine
DMSO	dimethyl sulfoxide
DMTr	4,4′-dimethoxytrityl
DNA	deoxyribonucleic acid
DNJ	deoxynojirimycin
ECG	*Escherichia coli* galactosidase
EDC	ethylene dichloride
ELISA	enzyme-linked immunosorbent assay
ELLA	enzyme-linked lectin assay
EMC	*N*-(ε-maleimidocaproyl)
ER	endoplasmic reticulum
Et	ethyl
FDG	2-deoxy-2-fluoro-D-glucose
FRET	Förster resonance energy transfer
GalNAz	*N*-azidoacetylgalactosamine
GDP-fucose	guanosine diphosphate β-L-fucose
GlcNAz	*N*-azidoacetylglucosamine
GM3	monosialo ganglioside 3
HA	hyaluronan
HEMA	2-hydroxyethyl methacrylate
HNE	4-hydroxynonenal

HPGs	hyperbranched polyglycerols
HPLC	high-performance liquid chromatography
K222	Kryptofix® 2.2.2 (4,7,13,16,21,24-hexaoxa-1,10-diazabicyclo[8.8.8]-hexacosane)
KLH	keyhole limpet hemocyanin
LAS	lipase AS Amano (from *Aspergillus niger*)
LCST	lower critical solution temperature
LDI-MS	laser desorption ionization mass spectrometry
LPC	lysophosphatidylcholine
MBL	mannose-binding lectin
m-CPBA	*m*-chloroperoxybenzoic acid
Me	methyl
MEM	methoxyethoxymethyl
MF	mefenamic acid
MHC	minimal hemolytic concentration index
MIC	minimal inhibitory concentration
MOM	methoxymethyl
MRI	magnetic resonance imaging
MRSA	methicillin-resistant *Staphylococcus aureus*
MUNANA	2'-(4-methylumbelliferyl)-α-D-*N*-acetylneuraminic acid
NaAsc	sodium ascorbate
NAP	naphthylmethyl
NBD	7-nitrobenz-2-oxa-1,3-diazol-4-yl
NCL	native chemical ligation
Neu5Ac	*N*-acetylneuraminic acid
NIS	*N*-iodo succinimide
NMR	nuclear magnetic resonance
NP	(*S*)-naproxen
NT	neurotensin
NTR	neurotensin receptor
OEG	oligoethylene glycol
PA	phosphatidic acid
PABTC	2-propanoic acid butyltrithiocarbonate
PBS	phosphate buffered saline
PC	phosphatidylcholine
PC	photocleavable
PDT	photodynamic therapy
PE	phosphatidylethanolamine
PEG	polyethylene glycol
PET	positron emission tomography
PG	phosphatidylglycerol
Ph	phenyl
PIPn	phosphatidylinositol polyphosphates
Piv	pivaloyl
PKC	protein kinase C

PLE	porcine liver esterase
PMAM	polyamidoamine
PMB	4-methoxybenzyl
PMP	*p*-methoxyphenyl
PNA	peanut agglutinin
Poc	propargyloxycarbonyl
PPI	poly(propylene imine)
PPTS	pyridinium toluene-4-sulfonate
Pr	propyl
PS	phosphatidylserine
p-TSA	*p*-toluenesulfonic acid
Py	pyridine
QCM	quartz crystal microbalance
RAFT	reversible addition–fragmentation chain transfer
RGD	arginine-glycine-aspartic acid peptidic sequence
RGD	arginylglycylaspartic acid
RGDC	H-Arg-Gly-Asp-Cys-OH
RNA	ribonucleic acid
ROP	ring opening polymerization
RuAAC	Ru-catalyzed azide alkyne cycloaddition
SAG	sweet almond glucosidase
SAM	self-assembled monolayer
SAR	structure-activity relationship
SATE	*S*-acetylthioethyl
SBS	sequencing by synthesis
sec-But	*sec*-butanol
SiRNA	small interfering ribonucleic acid
SLB	supported lipid bilayer
SM	sphingomyelin
SPECT	single-photon emission computerized tomography
SPPS	solid phase peptide synthesis
SPR	surface plasmon resonance
SST	somatostatin
TAMRA	tetramethylrhodamine
TBAF	tetrabutylammonium fluoride
TBAHS	tetrabutylammonium hydrogen sulfate
TBDMS	*t*-butyldimethylsilyl
TBDPS	*t*-butyldiphenylsilyl
TBTA	tris-(triazolyl)benzylamine
TCEP	tris(2-carboxyethyl)phosphine
TcTS	*Trypanosoma cruzi* trans-sialidase
TEA	triethylamine
TEAB	tetraethylammonium bromide
TEC	thiol-ene coupling
TEG	tetraethylene glycol

TFA	trifluoroacetic acid
TfOH	trifluoromethanesulfonic acid
THF	tetrahydrofuran
TLL	*Thermomyces lanuginosa* lipase
TMEA	thiol Michael enone addition
TMS	trimethylsilyl
TMSOTf	trimethylsilyl trifluoromethanesulfonate
Tol	tolyl
Ts	*p*-toluenesulfonyl
TTMA	triethyl-2,2′,2″-(4,4′,4″-nitrilotris(methylene)tris(1H-1,2,3-triazole-4,1-diyl))-triacetate
TYC	thiol-yne coupling
Tyc	tyrocidine
UV	ultraviolet
VRE	vancomycin resistant enterococci

PART I

CLICK CHEMISTRY STRATEGIES AND DECOUPLING

1

PARADIGM AND ADVANTAGE OF CARBOHYDRATE CLICK CHEMISTRY STRATEGY FOR FUTURE DECOUPLING

Roman Bielski and Zbigniew J. Witczak

1.1 INTRODUCTION

When discussing click chemistry, the exceptionally successful methodology of connecting molecules, it seems natural to look at ways of disconnecting them. Thus, while preparing a symposium on applications of the click chemistry in carbohydrates, we began thinking about effective methods of disconnecting molecular units from each other. It turns out that the number of options is rather limited. We reasoned that the need for such reactions must be rather rare. However, it is easy to list several situations that require decoupling after certain experiments or procedures had been completed. Recently, we discussed circumstances calling for a design of coupling of the molecular units that takes into account a future necessity for the decoupling of these units [1]. The review lists almost a dozen such situations. Examples include releasing of the radio part from the radiotherapeutic or the chemo part from the chemotherapeutic after the treatment had been completed; decoupling of molecular units to enable or facilitate analytical procedures; decoupling various constructs from the surface either to produce a specific structure on the surface or to release a product from the surface after its special properties had been taken advantage of; and cleavage of the synthesis's product from the solid support.

Click Chemistry in Glycoscience: New Developments and Strategies, First Edition.
Edited by Zbigniew J. Witczak and Roman Bielski.

For a variety of reasons, coupling is performed much more often than decoupling. The language gives strong support to this notion. While there are at least five words describing various types of coupling processes leading to the increase in the molecular weight (bioconjugation, derivatization, labeling, ligation, tagging), there are only two words describing decoupling processes (cleavage, scission). There are, additionally, at least three words whose negation is a proper word (coupling, protection, connection).

Let us consider possible protocols of coupling molecular units when it is known that the decoupling will be necessary later on. Let the molecules of interest potentially include large biomolecules. One obvious option is to couple molecular units by taking advantage of the click chemistry reactions, and later, utilize processes that are the reverse of the click reactions. Unfortunately, most click processes are not reversible or the retro reactions require conditions that cannot be applied to most (bio)macromolecular structures. Recently, Bielawski and coworkers [2] published an interesting paper showing that in the presence of ultrasound the popular click products, cyclic triazoles, can be transformed back to an azide and an alkyne. However, at this point, it seems unlikely that the ultrasound-generated reverse Huisgen cycloaddition process can find a broad application.

Since click reactions are usually not reversible, one has to develop other strategies. At least, two potentially successful strategies can be devised:

- One strategy asks for the use of reversible reactions (not belonging to click chemistry) such as formation of esters, amides, benzyl ethers during the coupling process and hydrolysis or debenzylation during the decoupling process (Scheme 1.1a). While such an option is often useful and effective, it is worth noting that hydrolysis of a specific ester or amide group from a construct containing protein(s) with many peptide bonds or polysaccharide with many acetal and/or ester groups may be problematic. The same applies to debenzylation, which may not be sufficiently regio- or chemoselective for various constructs.
- An alternative strategy asks for the introduction of a unit that is coupled (preferably via the click chemistry forming XZ and WB connections) to two (or more) molecular units and contains an easy cleavable functionality (AB) somewhere in the middle (Scheme 1.1b). We call such a unit a "sacrificial unit" containing a "sacrificial functionality." We coined the term for such chemistry—coupling and decoupling chemistry (CAD) [1]. The advantage of this option is that the sacrificial functionality (AB) can be tailor-made for specific molecular units; that is, it can be a functionality that is not found in the molecular units and thus can be safely decoupled when needed.

A construct **5** containing a sacrificial functionality can be synthesized on a step-by-step basis that comprises the introduction of appropriate linkers and other necessary components of sacrificial units. However, taking advantage of the pre-synthesized sacrificial units **4** equipped with all the required linkers, as shown in Scheme 1.1b, seems to be a better approach.

First, let us discuss the requirements of the reactive groups Z and W which are to connect the sacrificial unit to (macro)molecules of interest. Of course, the choice

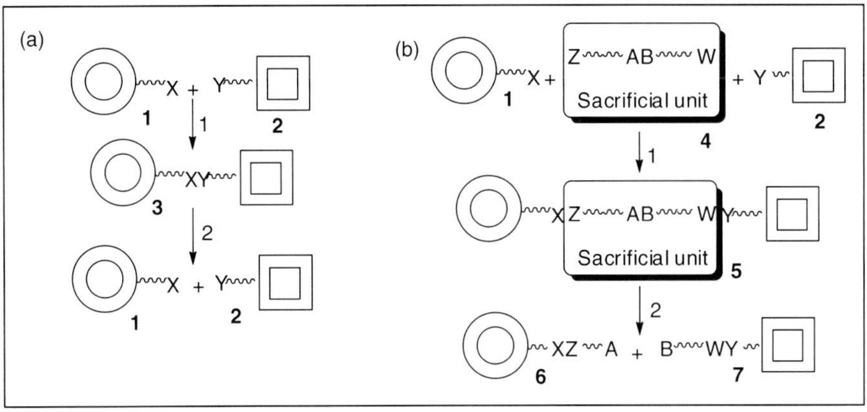

SCHEME 1.1 Possible approaches to the issue of future decoupling of connected molecular units: (a) use of reversible reactions; (b) use of sacrificial units.

of Z and W depends very strongly on the type of available X and Y groups and on reactivity of the various other functional groups present in molecular units **1** and **2**. It must be emphasized that the click chemistry should be used to connect the molecular units whenever possible. Besides all the virtues of the click chemistry reactions, the products of these reactions are, as a rule, very stable and we may be sure that unexpected and undesired disconnections will not happen.

While experts disagree on the exact scope of the click chemistry and what reactions truly belong to the click chemistry domain [3], it is probably worth at least listing the chemical reactions considered to belong there. Carvalho, Field, and coworkers recently reviewed applications of Cu(I)-catalyzed azide–alkyne cycloaddition (CuAAC) "click chemistry" in carbohydrate drug and neoglycopolymer synthesis [4]. In the review, they list four categories of "click reactions" (Scheme 1.2).

The Huisgen addition of azides to terminal alkynes (1A) is, by far, the most popular click reaction. The second most often employed click reactions is the addition of thiols to double (and triple) bonds. Depending on the substituents of the double bond, the reaction mechanism can be free radical or nucleo or electrophilic. Only a few of other reactions listed by Carvalho and Field [4] fulfill the conditions that open a door to a "distinguished" club of click processes. There are two more reviews of click-related carbohydrate chemistry. One of them discusses the impact of click chemistry on the carbohydrate-based drug development and glycobiology [5]. The other one, recently published by Lucas and coworkers [6], describes novel developments of click chemistry in polysaccharides. The authors are mainly focused on the catalyzed version of Huisgen 1,3-dipolar cycloaddition between terminal alkynes and azides.

Interestingly, the list of click reactions utilized in polymer chemistry [7] (Scheme 1.3) is slightly different from the one outlined in Reference 4.

Luckily, there are more than one click processes. It is important because taking advantage of a single click process (to form XZ and WY connections—see Fig. 1.1b) is rather problematic. Assume that a square is a functionality forming the

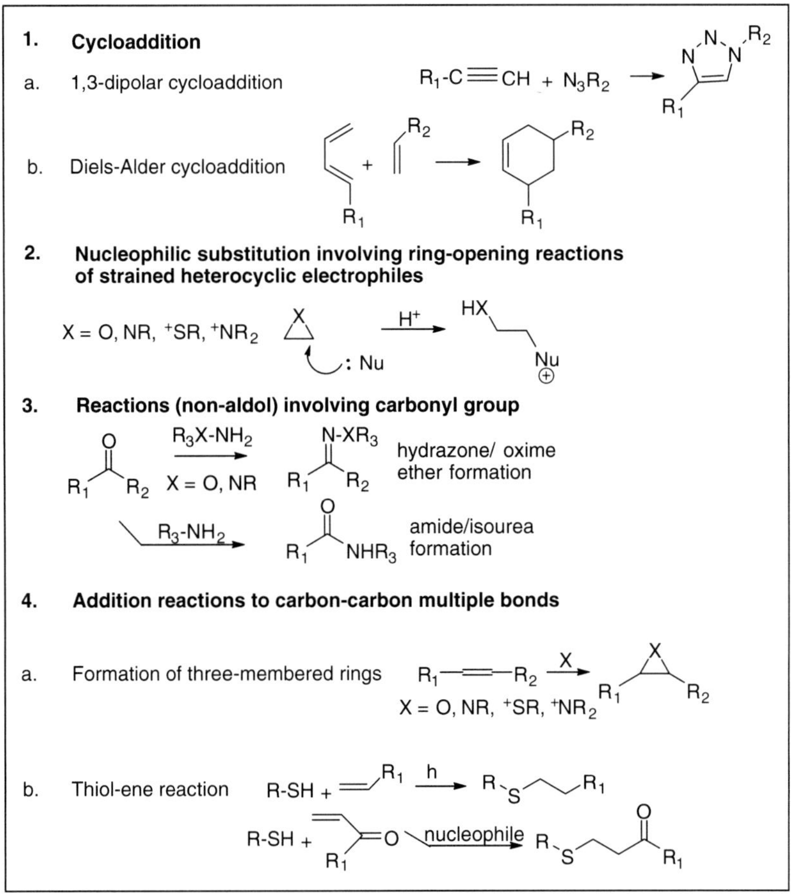

SCHEME 1.2 Click reactions according to Carvalho, Field, et al. [4].

click reaction product with the "open square." Figure 1.4 shows clearly that both functionalities involved in a click process should not be attached to a sacrificial unit. In such a case, the molecule of a sacrificial unit will react with another molecule of a sacrificial unit to form a polymer (Scheme 1.4a). Even if the same functionalities are attached to the sacrificial units, certain amounts of undesired symmetrical products will be formed (Scheme 1.4b).

Employing two different click reactions solves the problem (Scheme 1.5).

What chemical functionalities can act as sacrificial functionalities? The answer depends mainly on the structure of molecular units that are to be coupled. If one (or more) of the molecular units to be connected is a (poly)saccharide, such functionalities as acetals and esters should be rather avoided since polysaccharides usually have an abundance of such groups.

Let us take a look at a few applicable examples of click chemistry reactions. It has been already more than 10 years since the concept was introduced [8]. Since then,

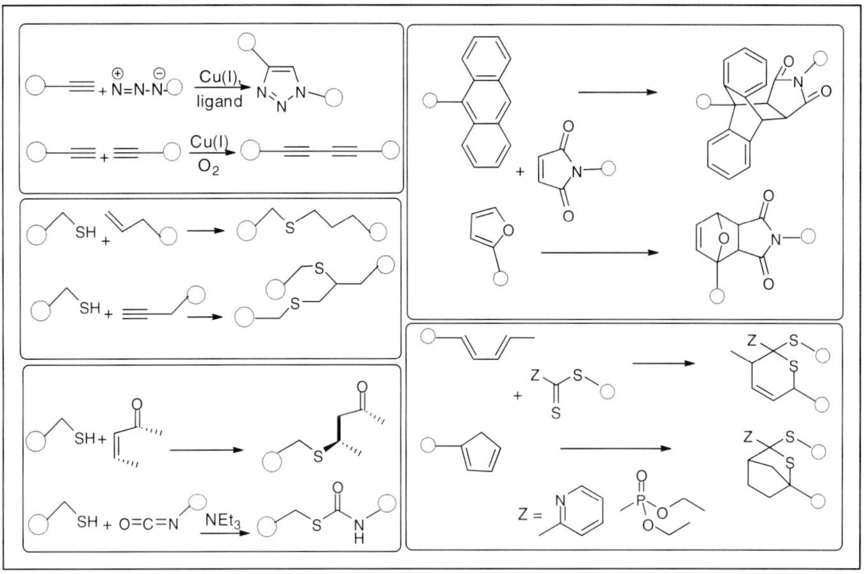

SCHEME 1.3 Click reactions in polymer chemistry according to Becer, Schubert, et al. [7].

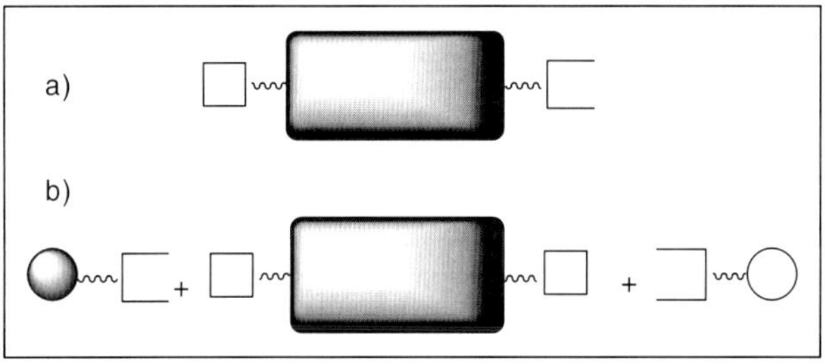

SCHEME 1.4 The use of the same click process on both sides of the sacrificial unit: (a) different functionalities attached to the sacrificial unit; (b) same functionalities on both sides of the sacrificial unit.

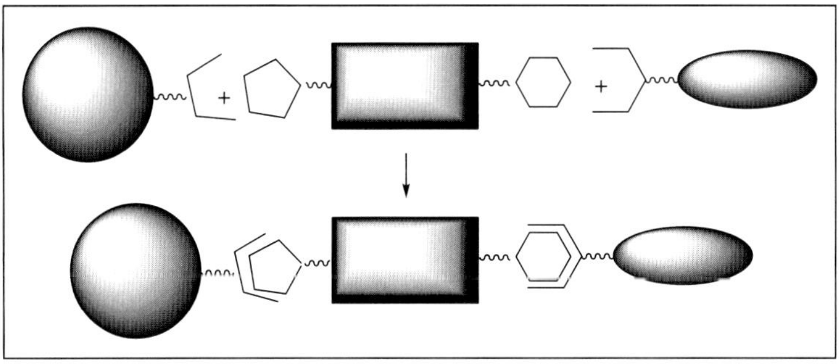

SCHEME 1.5 Use of two different click reactions.

thousands of papers employing click reactions have been published. Thus, there is a plethora of data describing the connection of large and small, natural and non-natural molecules equipped with a variety of functional groups. For many click reactions, their scope is well known and it is relatively easy to choose the one that will be effective at the given circumstance. The following carbohydrate examples are meant to serve as an illustration only.

1.2 COUPLING USING HUISGEN DIPOLAR CYCLOADDITION

Let us begin with [3 + 1] dipolar cycloaddition of azides to acetylenes. These reactions are usually very easy to perform, particularly since the use of copper-containing catalysts was introduced [9,10]. Since copper salts are not always applicable, Bertozzi et al. [11] and Boons et al. [12] developed methodologies employing no copper catalyst, but offering most of the advantages of CuAAC reactions. Both methodologies use a substituted cyclooctyne.

As we already mentioned, the Cu(I)-catalyzed [3 + 1] dipolar cycloaddition of azides to acetylenes (CuAAC) is, by far, the most popular click process. A few years ago, Marmuse, Nepogodiev, and Field [13] synthesized starch fragments analogs. The results of 1,3-dipolar cycloaddition of dipropargylated maltosides and azidoglucosides are shown in Figure 1.1. All reactions were carried out using $(Ph_3P)_3{\cdot}CuBr$

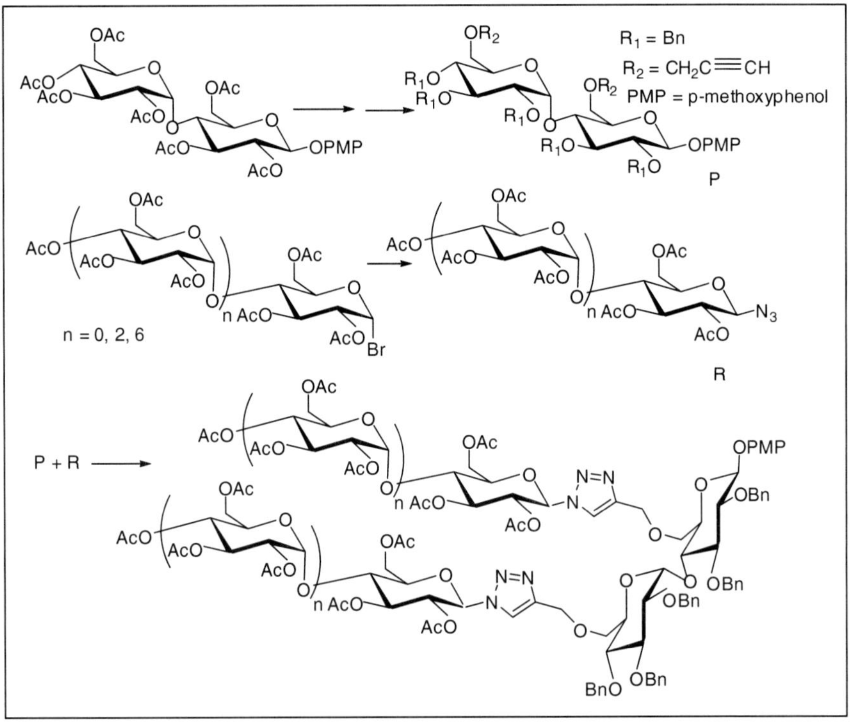

FIGURE 1.1 Example of rapid assembly of starch fragment analogs using CuAAC "click chemistry."

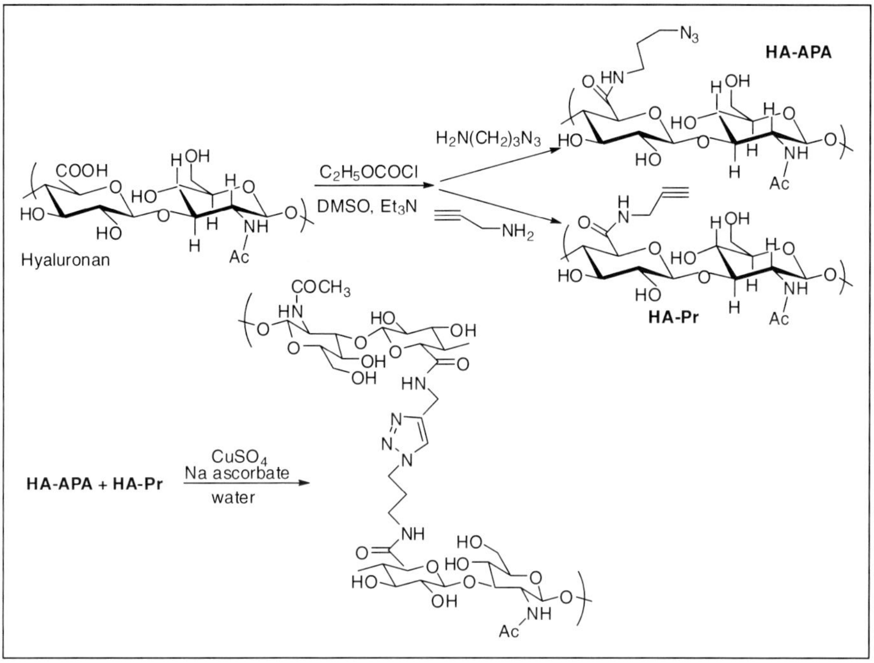

FIGURE 1.2 Use of CuAAC reaction as cross-linking to form HA-based hydrogels.

as a catalyst in the presence of DIPEA as a base for a relatively long reaction time (12 hours) at room temperature. The yields of cycloaddition reactions varied between 65% and 27%, decreasing with increasing length of the azidooligosaccharide chain.

Huerta-Angeles et al. [14] also synthesized truly large molecules by taking advantage of the Cu(I)-assisted AAC. Actually, they used the formation of the cyclic triazole as the cross-linking reaction to form hyaluronan (HA)-based hydrogels with well-defined 3D-molecular architecture for potential application in tissue engineering. Figure 1.2 shows the relevant chemistry.

Maillard and coworkers [15] synthesized a series of porphyrins each containing three glycosyl units using microwave activation. The products were designed as photodynamic therapy (PDT) agents. They are linked by a triazole group to chromophore in the aim to target tumor cells overexpressing lectin-type membrane receptors. Figure 1.3 shows the synthesis of the constructs. Importantly, zinc(II) cations in porphyrins are sufficiently stable under Cu-assisted azide acetylene cycloaddition click reaction conditions to avoid replacement by a copper(II) ions. The products turned out to be less active than analogs containing no triazole rings. The yields were very reasonable.

Chapleur and coworkers [16] synthesized more than a dozen neoglycopeptides via a direct functionalization of cysteine. The employed monosaccharides include acetylated and unprotected glucose, mannose, galactose, and 6-deoxy-6-fluoroglucose. Interestingly, the reaction of the fluoro compound equipped with the azido group and

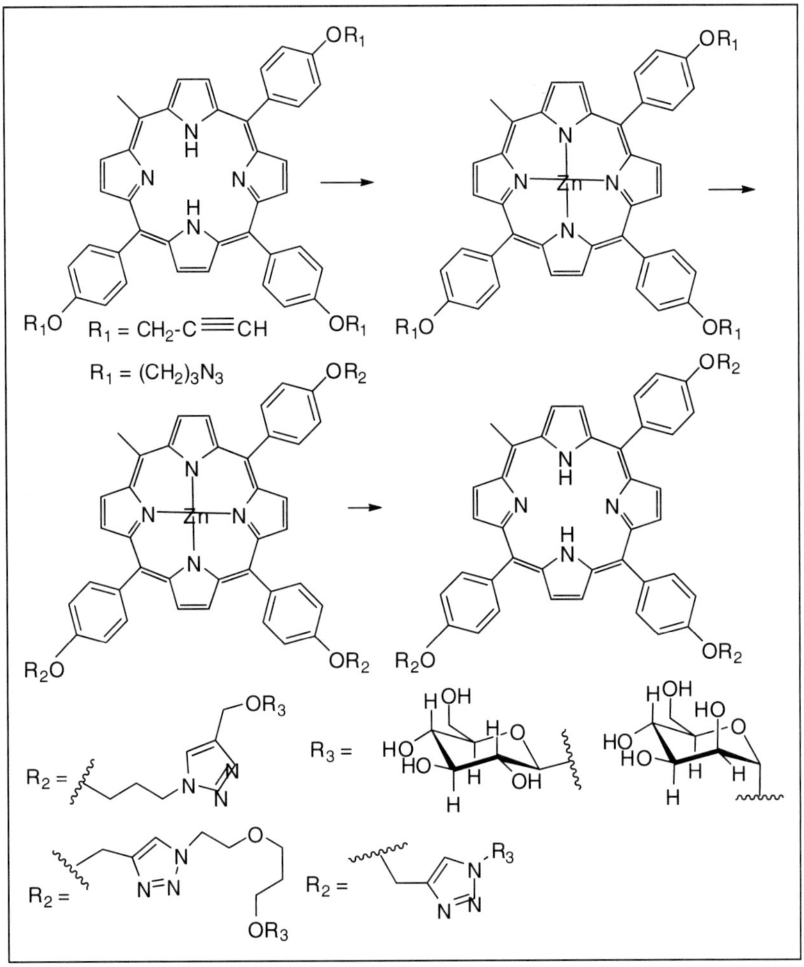

FIGURE 1.3 Porphyrin-based constructs synthesized by Maillard et al.

the tetrapeptide, Arg-Gly-Asp-Cys (RGDC) substituted with the propargyl group gave the highest yield. The synthetic strategy is shown in Figure 1.4.

1.3 COUPLING USING THIOL-ENE COUPLING CLICK CHEMISTRY

In recent years, another reaction, thiol-ene coupling (TEC), joined the exclusive club of click chemistry processes [17–19]. It is worth noting that the TEC reaction has been known for more than 100 years [20]. The addition of thiols to simple alkenes (and alkynes) is almost always free radical and can be initiated either thermally or photolytically.

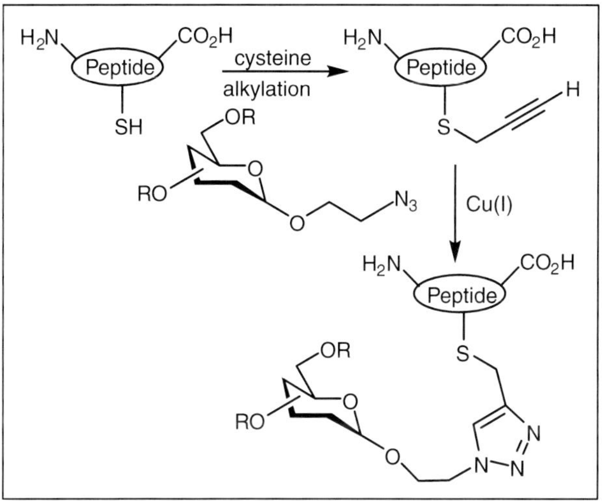

FIGURE 1.4 Chapleur et al.'s synthetic strategy for neoglycopeptides construction.

Fiore, Marra, and Dondoni [21] took advantage of photoinduced coupling of anomeric sugar thiols with sugar alkenes to synthesize 1,6-linked *S*-disaccharides in good to excellent yields (76–92%) and high diastereoselectivities (up to 99%). Figure 1.5 shows a typical example of the synthesized disaccharide.

Stenzel and coworkers [22] synthesized a block copolymer containing di(ethylene glycol) methyl ether methacrylate (DEGMA) and 2-hydroxyethyl methacrylate (HEMA) by reversible addition-fragmentation chain transfer (RAFT) polymerization, which was subsequently modified with glucothiose. Glucothiose was introduced via UV-promoted TEC process. The resulting glycosylated block copolymer led to the formation of thermoresponsive micelles, a potential candidate for targeted drug delivery. Figure 1.6 illustrates the methodology.

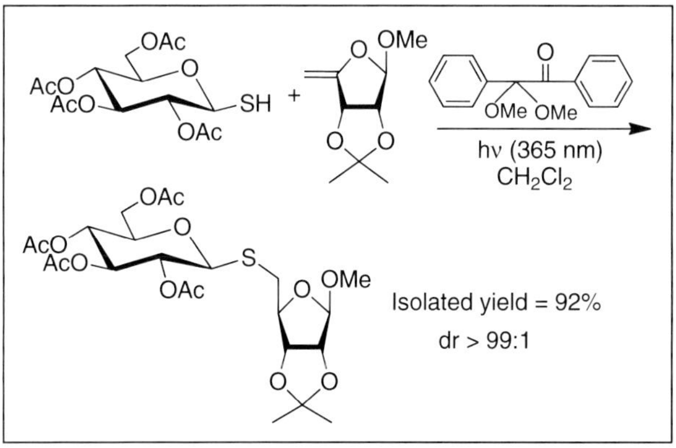

FIGURE 1.5 Example of Dondoni et al.'s [21] photoinduced TEC as a click ligation tool for thiodisaccharide synthesis.

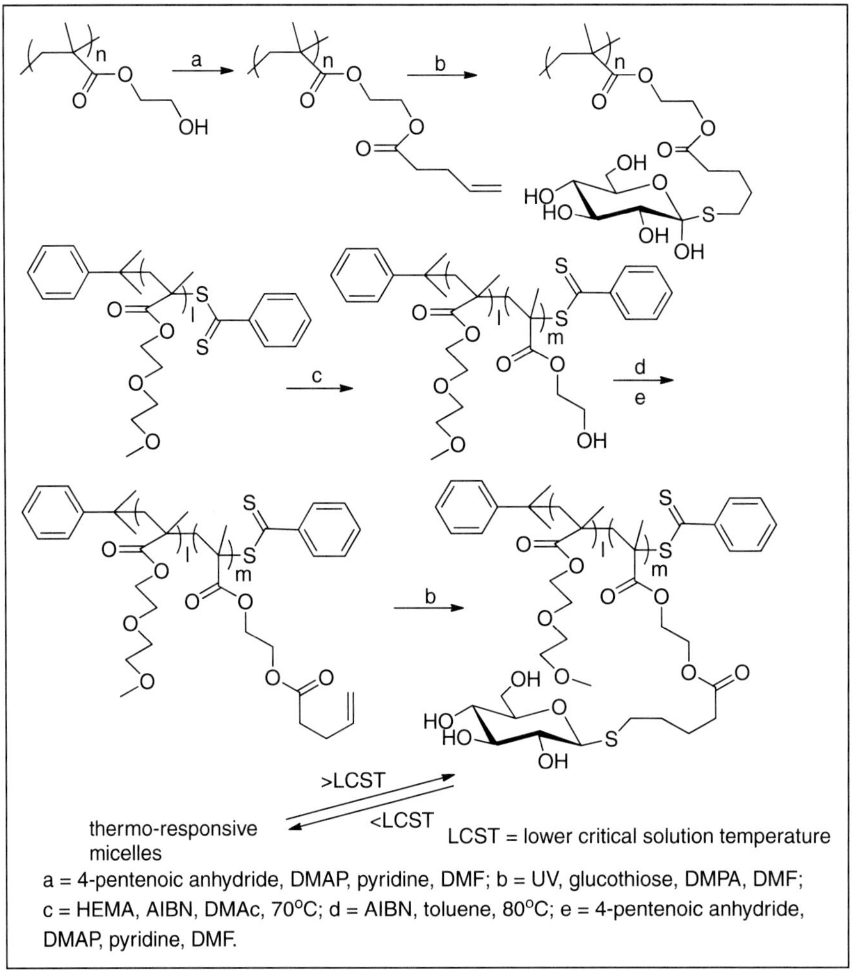

FIGURE 1.6 Stenzel et al.'s [22] synthetic strategies for the preparation of glucose functionalized (co)polymers.

Most Michael additions of thiols to double bonds conjugated to carbonyl and similar groups are reversible. Nevertheless, some of the Michael additions of thiols have been shown to offer stable and easy-to-isolate products. For example, Chandrasekaran et al. [23] synthesized several thioglycosides and other thiosugar analogs using benzyltriethylammonium tetrathiomolybdate [$(BnNEt_3)_2MoS_4$] or ammonium tetrathiomolybdate as a sulfur-transfer reagent. The reagent reacts with sugar halides to give sugar disulfides, which then undergo reductive cleavage *in situ* to provide the corresponding thiolates, followed by Michael addition to give the corresponding thioglycosides or other monosaccharides containing sulfur. Alternatively, the authors also made sugar enones to react with disulfides to form thiosugars, often with

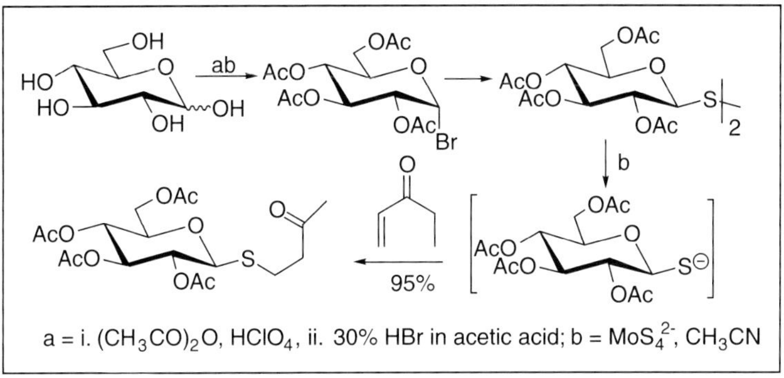

FIGURE 1.7 Thio-glucoside forming tandem sulfur transfer/reduction/Michael addition assisted by tetrathiomolybdate [23].

excellent diastereoselectivity (but sometimes the ratio of diastereoisomers was 1:1). The process is simple, offers good to excellent yields, and is performed in one pot. While the authors did not consider the process to belong to click chemistry, it definitely shows most features of the click chemistry. Figures 1.7 and 1.8 illustrate the methodology.

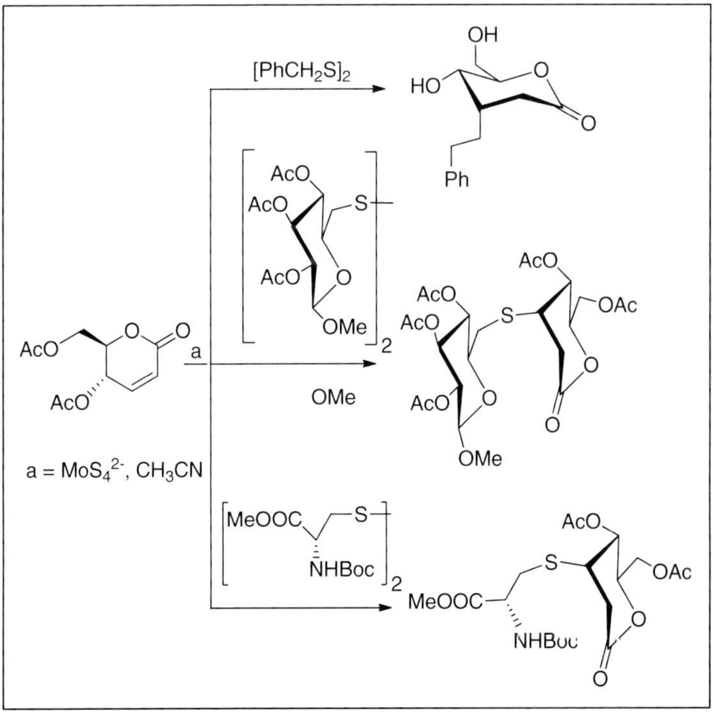

FIGURE 1.8 Synthesis of 3-deoxy-3-thiosugar derivatives via Michael addition according to Reference 23.

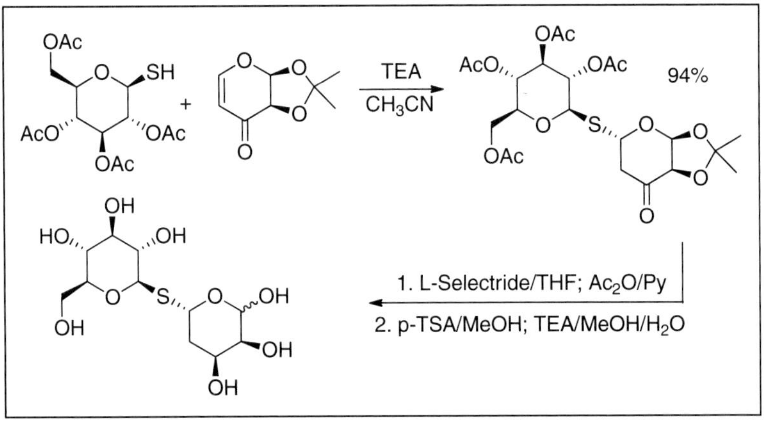

FIGURE 1.9 Witczak et al.'s [24] synthesis of thiodisaccharide using Michael addition of monosaccharide with a thiol functionality.

Witczak et al. [24] synthesized an interesting thiodisaccharide presented in Figure 1.9. The synthesis takes advantage of a click chemistry reaction, the (Michael) addition of a thiol to the enone system.

1.4 PROCESSES WHERE BOTH CuAAC AND TEC CLICK CHEMISTRIES ARE USED

Perrier et al. [25] synthesized highly branched and hyperbranched glycopolymers via RAFT polymerization and click chemistry which includes Cu(I)-catalyzed Huisgen 1,3-cycloaddition of azides and alkynes (CuAAC), thiol-ene addition, and thiol-yne addition. The authors explain that in some circumstances, the drawback of the RAFT approach is the poor availability and compatibility of specific functional monomers. They circumvent the problem by introducing functionalities via post-polymerization modification employing click chemistry. The TEC reactions were performed in the presence of HCl using glucothione sodium salt as the carbohydrate component and 2,2-dimethoxy-2-phenylacetophenone (DMPA) as a photoinitiator, under UV at room temperature. 2-azidoethyl-β-D-galactopyranoside served as a carbohydrate component of CuAAC reactions. Figure 1.10 shows the relevant products.

Haddleton and coworkers [26] also synthesized polymers capable of forming click products deriving from both CuAAC and thiol-ene reactions. They took advantage of the catalytic chain transfer polymerization (CCTP) to form alkyne-functional macromonomers which were subsequently functionalized with sugar azides (CuAAC) and thiols. Interestingly, the addition of thiol was a Michael-type reaction and benzyl mercaptan in the presence of dimethylphenylphosphine (DMPP) was employed as a thiol. Figure 1.11 shows this convenient synthesis of end-functionalized glycopolymers and the utilized carbohydrate azides.

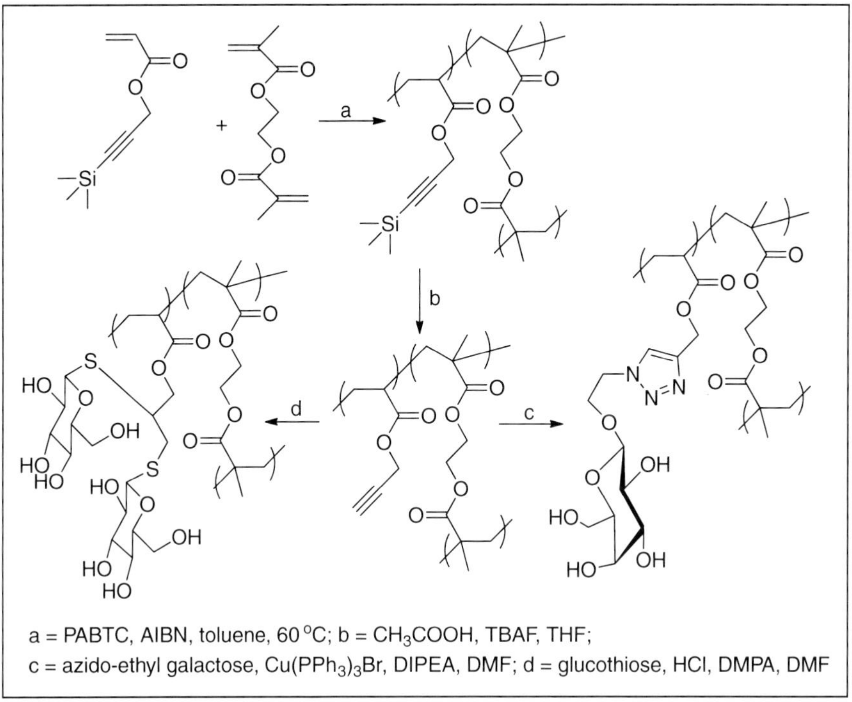

FIGURE 1.10 Synthetic strategy for the preparation of highly branched glycopolymers devised by Perrier et al. [25].

Marra, Dondoni, et al. [27] also use both CuAAC and TEC chemistries when synthesizing calix[4]arene-based *S*-glycoclusters. They describe the dual clustering at the upper and lower rim of a calix[4]arene with two different sugars (galactose and glucose) via sequential copper(I)-catalyzed azide–alkyne cycloaddition and photoinduced TEC. Figure 1.12 illustrates the approach.

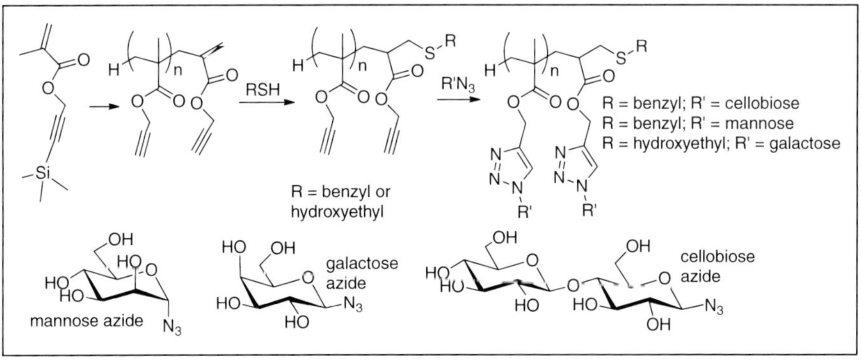

FIGURE 1.11 Haddleton and coworkers' [26] synthetic approach to end-functionalized glycopolymers and structures of utilized carbohydrate azides.

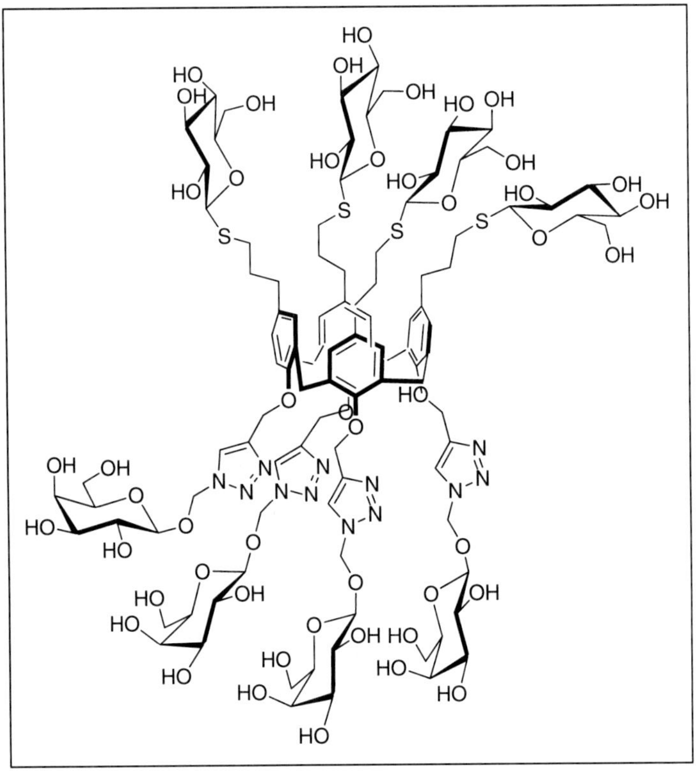

FIGURE 1.12 Structure of one of the dual clusters at the upper and lower rim of a calix [4] arene with two different sugars (galactose and glucose) synthesized by Dondoni et al. [27].

1.5 DECOUPLING

It should be very strongly emphasized that the number of literature examples of decoupling large molecules is very limited. At this point, we must rely on examples of decoupling in which small molecules were split to smaller fragments. We hope that soon there will be many examples of decoupling reactions applied to large molecules consisting of two or, at least, one macromolecule. It will help researchers to understand the scope and limitations of various reactions and will minimize guessing (or extrapolating) whether a specific coupling or decoupling reaction is applicable in a specific situation.

The following examples show some of the decoupling which includes carbohydrates sacrificial functionalities that can be found in the literature.

1.5.1 Enzymatic Cleavage

Let us start with decoupling processes that are triggered by enzymes. It is well known that enzymes are highly sensitive to the structure of reactants. Thus, one has to be

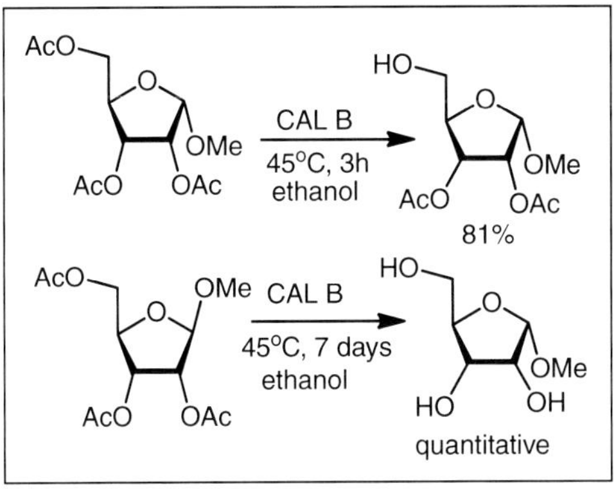

FIGURE 1.13 Highly regioselective and exhaustive deacetylation using lipases described by Iglesias and coworkers [28].

exceptionally cautious when trying to expand the use of specific enzymes to other categories of compounds. It applies particularly to reactions performed on small molecules. It may not translate well to much larger (macro) molecules even when the specific part of the molecule is the same. We can only hope that with time our understanding of specific enzymatic processes will improve and there will be the abundance of literature data enabling successful application of enzymatic decoupling to specific situations. The following examples derive from small molecule chemistry, but hopefully will find application in decoupling processes as well.

Iglesias and coworkers [28] studied alcoholysis of methyl 2,3,5-tri-*O*-acetyl-α-D-ribofuranoside and methyl 2,3,5-tri-*O*-acetyl-β-D-ribofuranoside using the *Candida antarctica lipase* B (CAL B). The reaction of the α-anomer lead to the isolation of the monodeacetylated product. The reaction of the β-anomer lead to a variety of partially deacetylated products but eventually provided the fully deacetylated product in a quantitative yield (Fig. 1.13).

Italian and Spanish researchers [29] performed selective hydrolysis of peracetylated β-monosaccharides using immobilized lipases from different sources [*Thermomyces lanuginosa* (TLL), *Aspergillus niger* (ANL), and *Candida antarctica* B (CAL B)]. Figure 1.14 shows some of the results. The shown yields represent isolated yields. The results illustrate a crucial role played by the procedure applied to the immobilization of enzymes. The paper is an example of a larger series of papers devoted to studying regioselectivity of various enzymatic hydrolyses [30] and references quoted therein. The acquired knowledge was employed to a very elegant synthesis of linear oligosaccharides by a simple monoprotective, chemoenzymatic approach.

Baba and Yoshioka [31] prepared 1-β-*O*-acyl glucuronides of three nonsteroidal, anti-inflammatory drugs, diclofenac (DF), mefenamic acid (MF), and (*S*)-naproxen (NP) (their structures are shown in Fig. 1.15). Next, after some screening, they employed selected commercial enzymes—lipase AS Amano LAS: (from ANL), and porcine liver esterase (PLE). LAS hydrolyses the *O*-acetyl groups with high

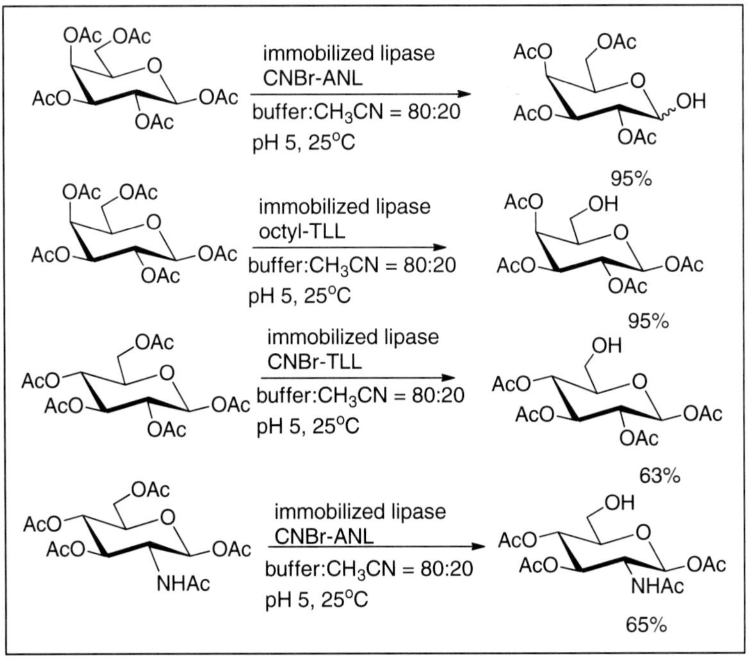

FIGURE 1.14 Specific and regioselective hydrolysis of different 1,2,3,4,6-penta-*O*-acetyl-D-glycopyranoses using immobilized lipases [29].

chemoselectivity to provide the methyl ester equipped with three OH groups and PLE hydrolyses the methyl ester groups with high chemoselectivity to provide glucuronic acid containing three acetyl groups. Any combination of two esters leads to high yields (90% or more) of glucuronic acids with unchanged substituent at the aglycone position (Fig. 1.15).

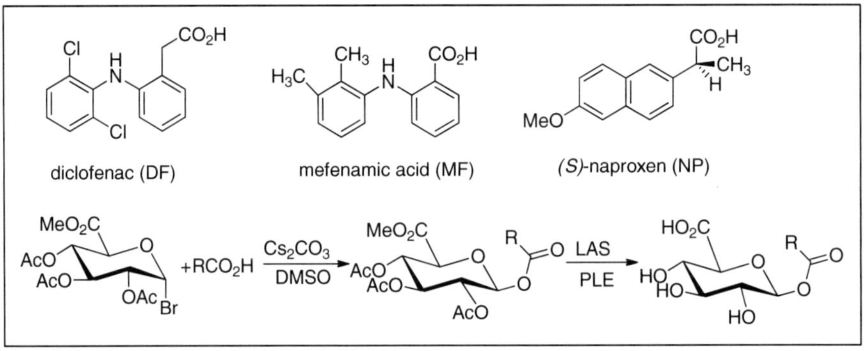

FIGURE 1.15 Chemo-enzymatic synthesis of 1-β-*O*-acyl glucuronides described by Baba and Yoshioka [31].

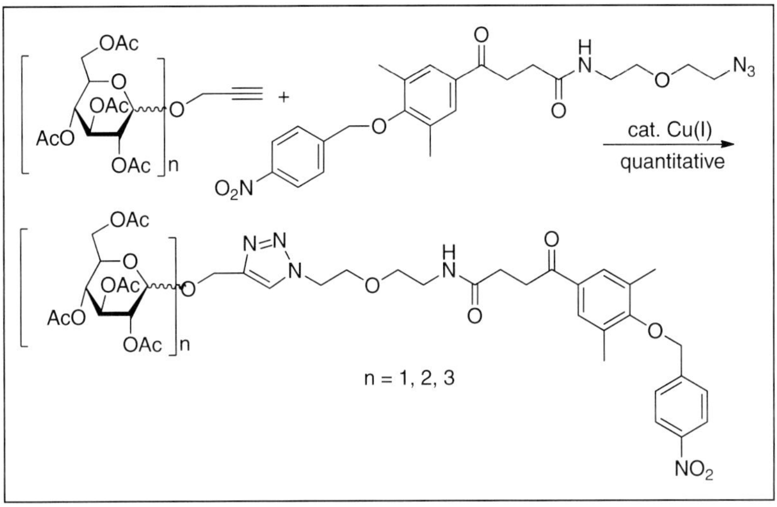

FIGURE 1.16 Maki and Ishida's [32] synthesis of compounds containing carbohydrate and photocleavable parts.

1.5.2 Photocleavage

Maki and Ishida [32] designed and synthesized photocleavable molecules for laser desorption ionization-mass spectrometry (LDI-MS). The authors envisaged that a photocleavable molecule, which affords an MS-detectable ion upon irradiation without matrix assistance, would simplify the ionization mechanism and be a reliable and selective labeling device for LDI-MS. As expected, the connection takes advantage of the click chemistry. The final step of the synthetic procedure is shown in Figure 1.16. The cleavage takes place under the laser pulse at 337 nm.

Ju and coworkers [33] designed and synthesized a 3′-modified photocleavable fluorescent nucleotide, 3′-*O*-allyl-dUTP-PC-Bodipy-FL-510 (PC-Bodipy, photocleavable 4,4-difluoro-4-bora-3α,4α-diaza-s-indacene), as a reversible terminator for DNA sequencing by synthesis (SBS). Figure 1.17 shows the design and synthesis of the product.

1.5.3 Chemical Cleavage

Manabe, Ueki, and Ito [34] developed a very convenient method of deprotecting propargyl ethers using the samarium–amine–water system. The method requires little time (usually 15 minutes), reactions take place at room temperature, offer good yields, and chemoselectivity (Fig. 1.18). The method was applied to the solid state synthesis of oligosaccharides (Fig. 1.19).

At this point, it is impossible to say if the same decoupling can be applied to a propargyl group in which one of the hydrogen atoms (in either of the two possible positions) was replaced with a large (macromolecular) substituent.

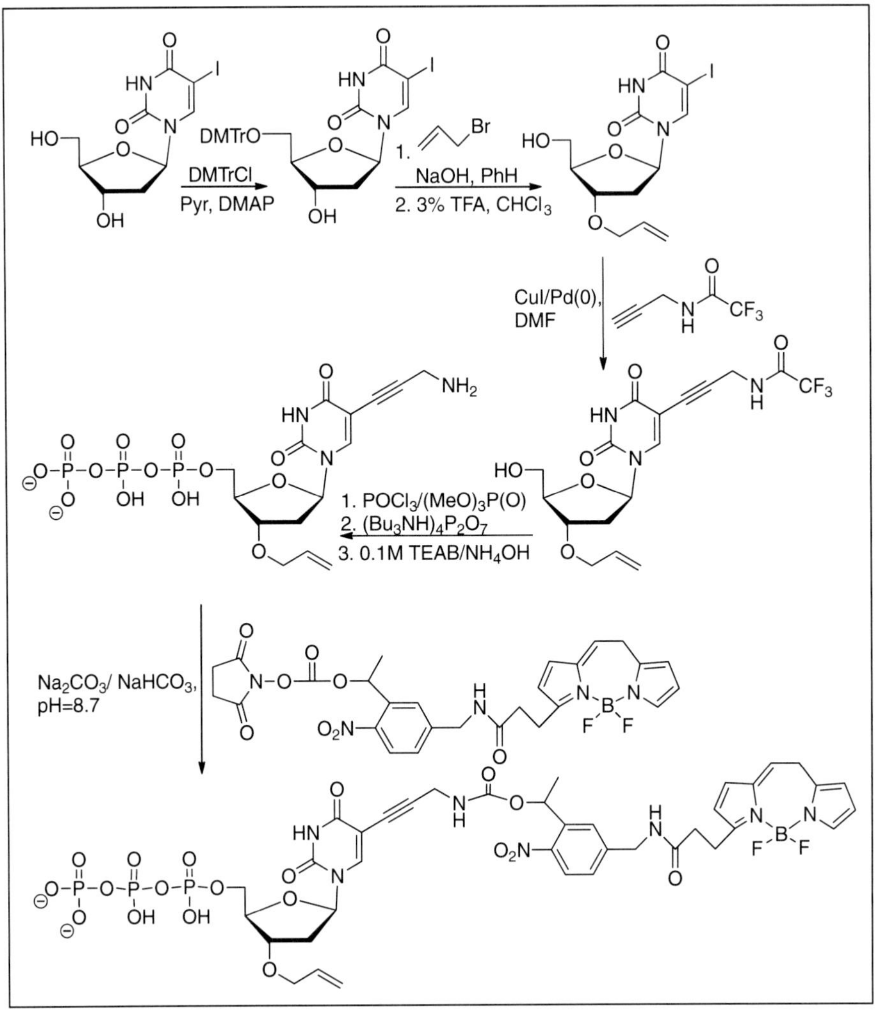

FIGURE 1.17 Ju and coworkers' [33] design and synthesis of a 3′-*O*-allyl photocleavable fluorescent nucleotide as a reversible terminator for DNA sequencing by synthesis.

Chen and coworkers [35] developed a novel method of selective deacylation using dioxomolybdenum dichloride as a catalyst. The conditions are mild and the yields are very high. Figure 1.20 shows more spectacular examples of removal of acyl group in the presence of other protecting functionalities.

Padrón, Bermejo, and collaborators [36] used as simple reagent as *p*-toluenesulfonic acid (or 10-camphorsulfonic; CSA) to selectively cleave acetates in the presence of benzoates and *p*-bromobenzoates. The reactions are performed in DCM/methanol at 40°C and yields are high. The examples shown in Table 1.1 illustrate the methodology.

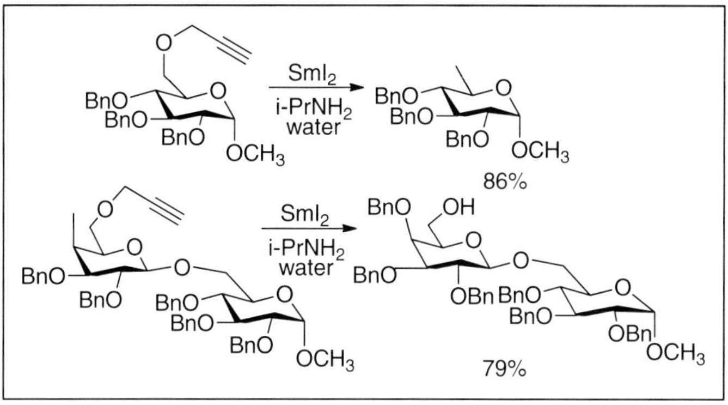

FIGURE 1.18 Examples of selective deprotection of propargyl ethers using samarium–amine–water system [34].

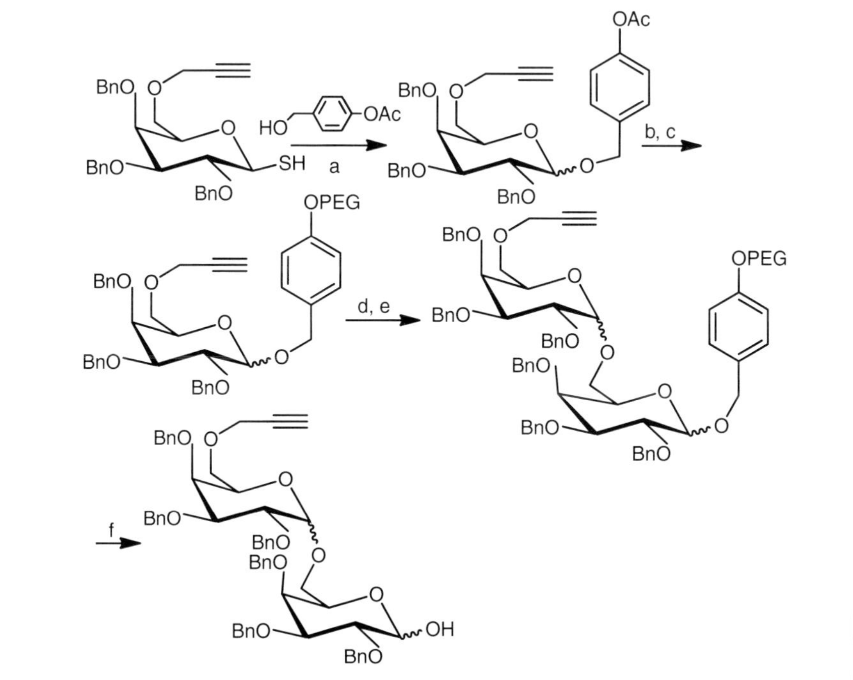

Reagents and conditions: a = NIS, TfOH, CH_2Cl_2, 40°C, 30 min, quant.; b = NaOMe, MeOH, rt, 2 h, 93%; c = MPEG-Br, Cs_2CO_3, CH_3CN, 14 h, 80°C, 92%; d = SmI_2, i-$PrNH_2$, H_2O, THF, rt, 15 min, 63%; e = NIS, TfOH, CH_2Cl_2, 40°C, 8 h, 94%; f = TFA:CH_2Cl_2 1:4, rt, 1 h, 64%.

FIGURE 1.19 Polymer-supported disaccharide synthesis according to Manabe, Ueki, and Ito [34].

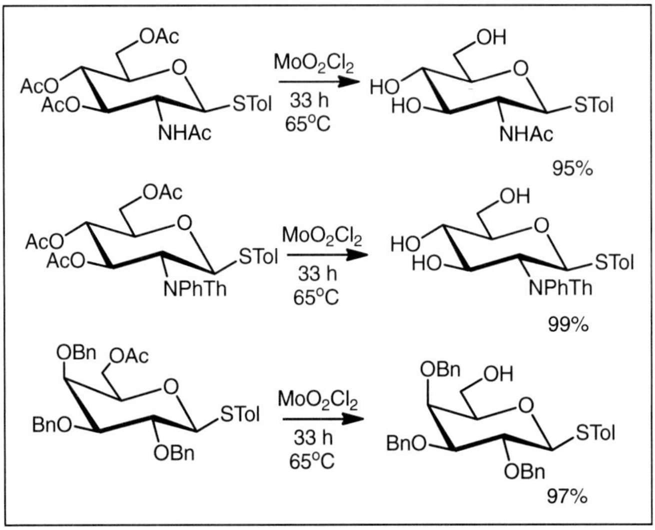

FIGURE 1.20 Effective deacetylation developed by Chen et al.

TABLE 1.1 Selective Deacetylation Described by Padrón, Bermejo et al.

Entry	Substrates	Conditions	Products[a]	Yield[b]
1	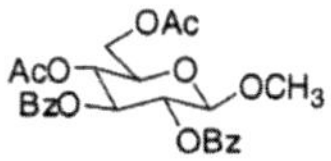	**p-TsOH (2 equiv)** **40°C, 7 h** **CSA (2 equiv)** **rt, 72 h**	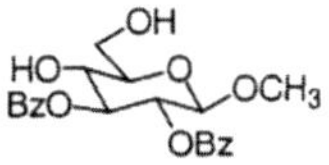	91% 88%
2	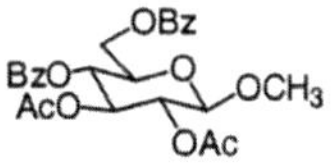	**p-TsOH (2 equiv)** **rt, 24 h**	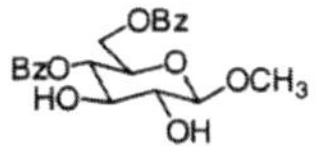	65%
3	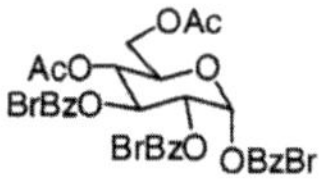	**p-TsOH (2 equiv)** **rt, 24 h**	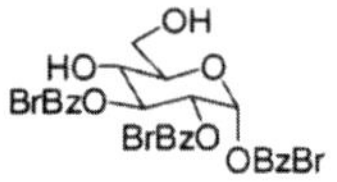	90%
4	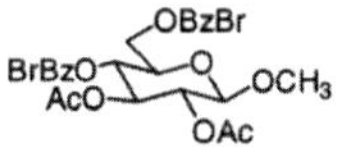	**p-TsOH (2 equiv)** **40°C, 24 h**	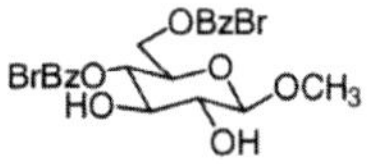	70%
5	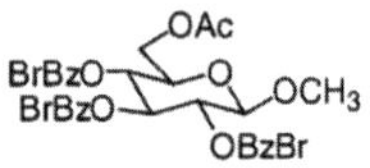	**p-TsOH (1 equiv)** **40°C, 24 h**	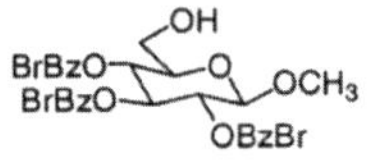	88%
6	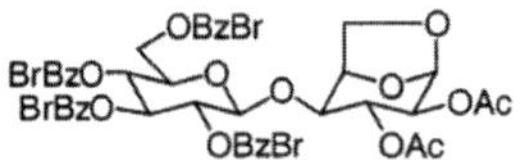	**p-TsOH (2 equiv)** **40°C, 24 h**	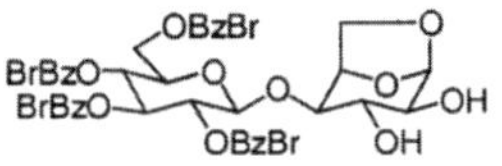	70%

[a]All products were characterized by ^{1}H NMR and ^{13}C NMR.

[b]Isolated yields.

Source: Reproduced with permission from Reference 36.

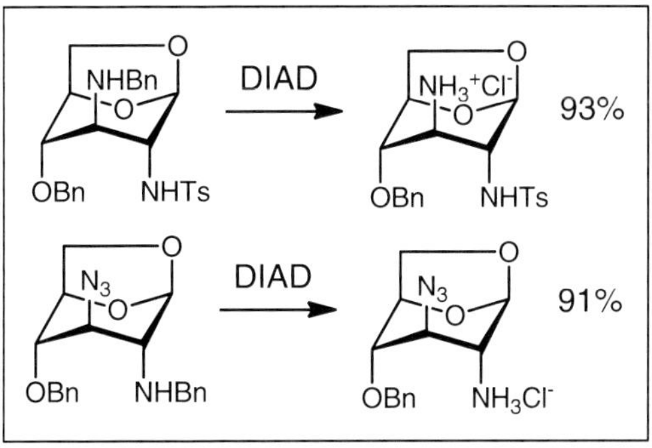

FIGURE 1.21 Selective N-debenzylation of benzylamines by DIAD according to Kroutil et al. [37].

Kroutil and collaborators [37] developed conditions enabling chemoselective cleavage of benzylamines in the presence of such functionalities as azide, *O*-benzyl, and *N*-tosyl using diisopropyl azidodicarboxylate (DIAD). Figure 1.21 shows some examples of this high-yielding methodology.

Spencer et al. [38] compared the rate of hydrogenation of benzyl ethers containing different substituents in the aromatic ring. The differences are significant. For example, the reaction of the compound with the benzyl group containing trifluoromethyl group in the para position is about five times slower than the reaction of a compound with unsubstituted benzyl group. Introduction of *t*-butyl group into the para position of the benzyl ether increases the rate of hydrogenation about 25 times. The researchers found that the naphthylmethyl (NAP) group is exceptionally selective. Figure 1.22 shows the carbohydrate examples.

It is worth adding that the NAP group can be introduced either using NAP bromide [38, 39] or by forming the cyclic naphthyl acetal followed by hydrogenation

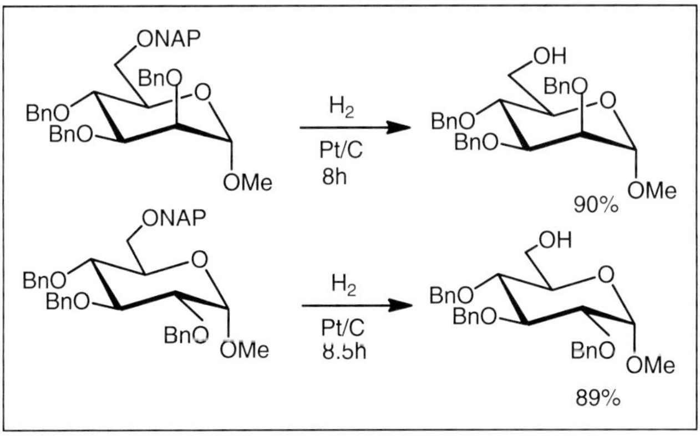

FIGURE 1.22 Selective removal of the NAP group in the presence of benzyl ethers according to Spencer et al. [38].

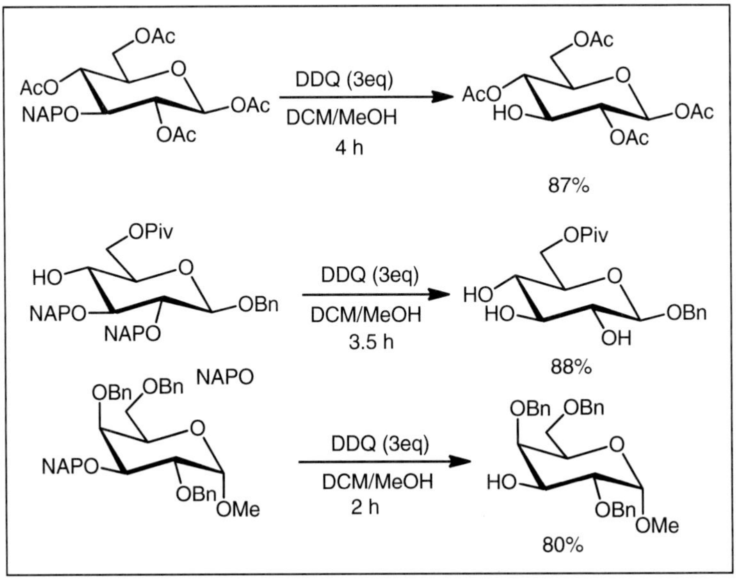

FIGURE 1.23 Selective removal of NAP groups in the presence of benzyl ethers using DDQ according to Matta et al. [39].

[40]. Interestingly, the NAP group can be cleaved also at room temperature using entirely different conditions than Spencer and coworkers. Matta et al. [39] developed a method based on DDQ. Figure 1.23 shows the conditions and yields. Recently, Boons et al. [41] took advantage of this approach when synthesizing an unusual phosphoglycopeptide derived from α-dystroglycan. Yields were equally high.

Kovensky and coworkers [42] demonstrated quantitative yields of the allyl group removal from the anomeric position of allyl galactopyranosiduronic acid derivatives. The authors employ a two-step procedure consisting of DABCO and $(Ph_3P)_3RhCl$ followed by mercuric-assisted cleavage. Their work is an excellent example of difficulties one often encounters when employing a specific reactant to a novel structure. It turned out that standard literature procedures lead to mixtures consisting of the starting material, the desired product, and a significant amount of the allyl group oxidation product. Most alternatives did not offer any improvement. Eventually, the researchers developed a very successful method as shown in Figure 1.24.

Recently, Finnish and Hungarian scientists [43] have shown the effectiveness of utilizing flow chemistry to deprotection of benzyl/benzylidene protected carbohydrates. While the required conditions (that were not optimized) include a relatively high pressure of hydrogen (40 bar) and elevated temperature (80°C), the method offers excellent yields and does not affect such protecting groups as acetates and silyl ethers. Figure 1.25 shows the results. The presented yields are isolated yields.

Das and collaborators [44] developed a method enabling an effective and mild removal of a trityl group in the presence of various protecting groups. The method employs silica-supported sodium hydrogen sulfate. Figure 1.26 shows the results.

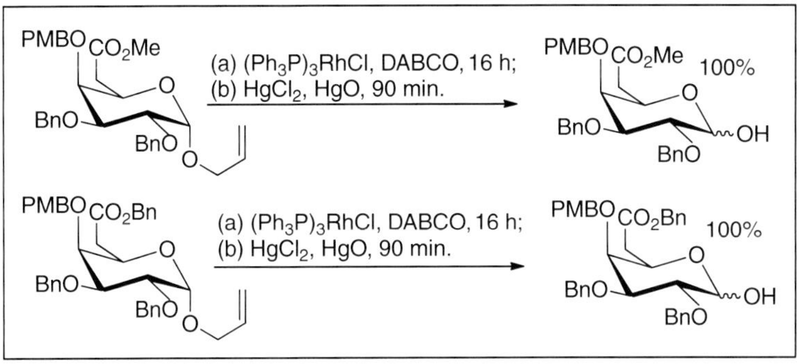

FIGURE 1.24 Barbier et al.'s [42] two-step method to remove the allyl group from allyl galactopyranosiduronic acid derivatives using DABCO and $(Ph_3P)_3RhCl$ followed by mercuric-assisted cleavage.

Sharma et al. [45] describe a novel, zirconium chloride-based methodology enabling a highly selective deprotection of *t*-butyldimethylsilyl (TBS) ethers in the presence of *t*-butyldiphenylsilyl (TPS) ethers. Figure 1.27 shows the only carbohydrate example from the paper. Other examples are equally impressive. While the difference between the TPS and TBS ethers seems to be very minute, one can accomplish a spectacular selectivity in deprotection of these species. However, without

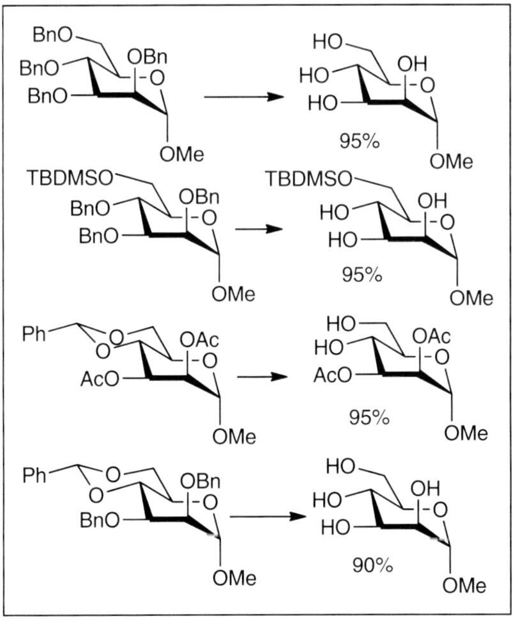

FIGURE 1.25 Application of flow chemistry to deprotection of benzyl-/benzylidene-protected carbohydrates [43].

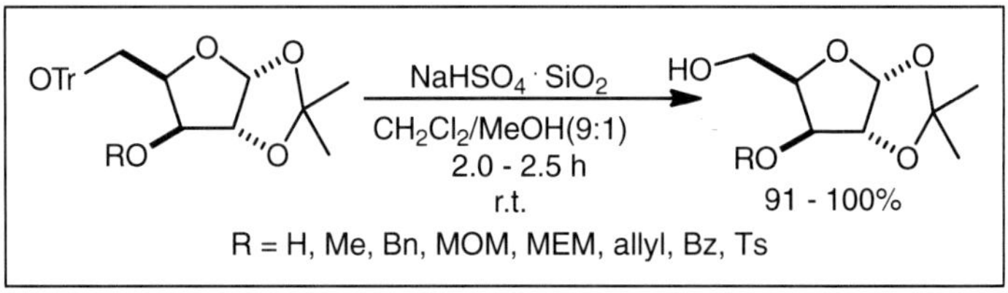

FIGURE 1.26 Chemoselective deprotection of trityl ethers using silica-supported sodium hydrogen sulfate developed by Das et al. [44].

appropriate experiments, it is impossible to say whether similar selectivity is achievable when, for example, one of the hydrogen atoms in a *t*-butyl group is replaced with a large macromolecular unit.

Another accomplishment in deprotection of silyl ethers has been developed by Chen, Le, Lin, and coworkers [46] who deprotect primary silyl ethers in the presence of the secondary ones and other protecting groups such as benzyl ethers. The methodology requires that the silyl ethers are treated with a catalytic amount of CBr_4 in methanol under photochemical reaction conditions. Selected results are shown in Figure 1.28.

Pale and coworkers [47] developed a new type of a protecting group-bis (4-methoxy-phenyl) methyl group. The relevant ethers are formed (in acetonitrile) and deprotected (in ethanol) in the presence of Cu(II) bromide. The yields are high. Figure 1.29 shows the carbohydrate deprotection (decoupling) example. It is worth noting that the size of the protecting group suggests that it may be equally effective when large substituent is attached to one of the methoxyphenyl groups.

Sridhar and Chandrasekaran [48] developed a novel protecting group for amines and alcohols, propargyloxycarbonyl (Poc) group. It can be easily removed under neutral conditions using tetrathiomolybdate MoS_4^{2-} in acetonitrile at room temperature.

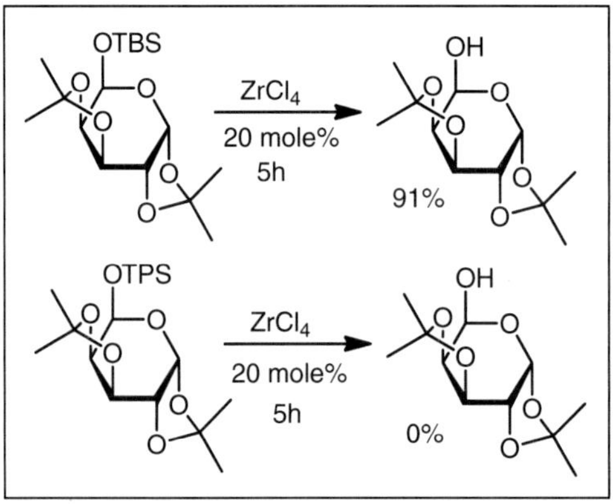

FIGURE 1.27 Sharma et al.'s deprotection of *t*-butyldimethylsilyl ethers in the presence of *t*-butyldiphenylsilyl ethers.

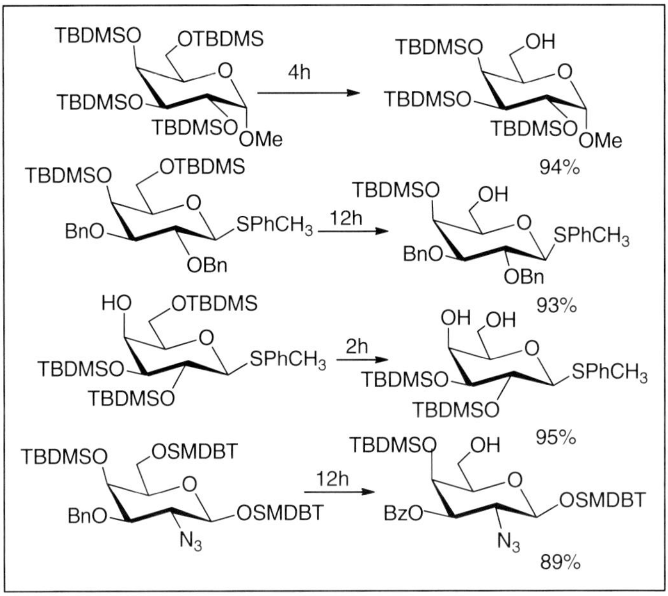

FIGURE 1.28 Chen, Lee, Lin, and coworkers' selective deprotection of secondary silyl ethers.

Such groups as benzylidene acetals, benzyl ethers, acetyl and levulinoyl esters, and allyl and benzyl carbonates are left untouched. Importantly, the new protective group (Poc) is compatible with acidic, basic, and also glycosylation conditions. While all the examples relate to the gluco configuration, there are no reasons to believe that other configurations would not give similar results. Figure 1.30 shows a few examples of Poc deprotection.

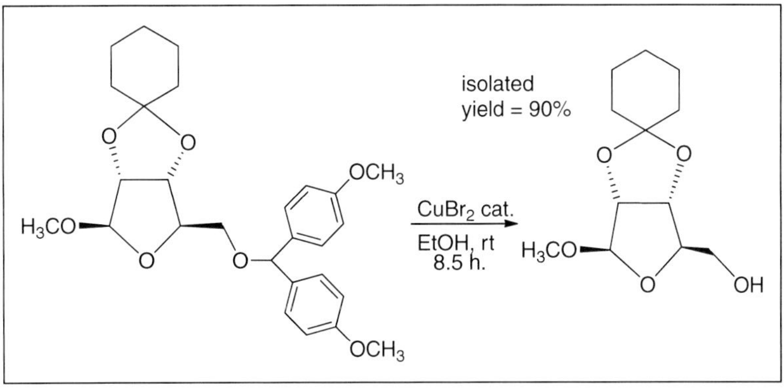

FIGURE 1.29 Deprotection of a new protecting group-bis (4-methoxyphenyl) methyl group developed by Pale and coworkers [47].

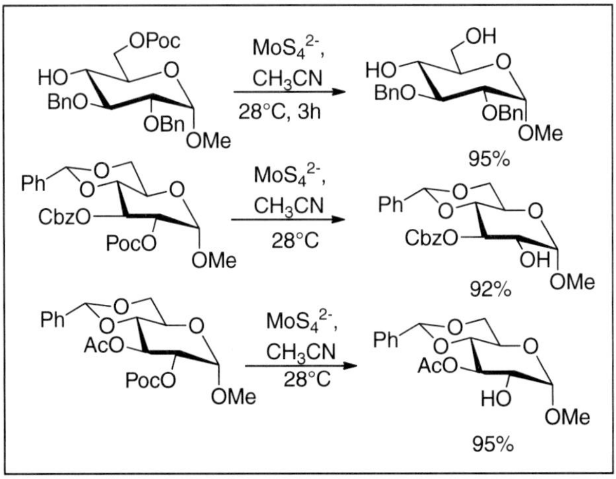

FIGURE 1.30 High-yielding removal of propargyloxycarbonyl (Poc) group developed by Sridhar and Chandrasekaran [48].

1.6 CONCLUSION

We hope that the present review convincingly shows click chemistry to be an exceptionally useful methodology not only applicable to coupling larger molecular units but also when such coupling must be later somehow reversed. The proposed strategy of using sacrificial units is simple and should become the approach of choice when the units must be later decoupled. The review shows that there are several excellent click reactions and some of them such as azide alkyne addition or thiol addition to unsaturated systems have been applied very successfully to the synthesis of various large molecules including macromolecular units containing carbohydrates. Nevertheless, the field of decoupling is much less often visited, and thus, there is an acute need for novel methods of decoupling. In addition, we hope that future availability of a variety of sacrificial units will encourage chemists to take advantage of them. It will enable much better assessment of what the limitations of specific decoupling methods are. The area of research that should strongly benefit from such knowledge is the sequencing of complex, natural polysaccharides.

REFERENCES

1. Bielski, R.; Witczak, Z. J. *Chem. Rev.* **2012**. DOI: org/10.1021/cr200338q.
2. Brantley, J. N.; Wiggins, K. M.; Bielawski, C. W. *Science* **2011**, *333*, 1606–1609.
3. Barner-Kowollik, C.; Du Prez, F. E.; Espeel, P.; Hawker, C. J.; Junkers, T.; Schlaad, H.; Van Camp, W. *Angew. Chem. Int. Ed.* **2010**, *50*, 60–62.
4. Aragão-Leoneti, V.; Campo, V. L.; Gomes, A. S.; Field, R. A.; Carvalho, I. *Tetrahedron* **2010**, *66*, 9475–9792.

5. Wilkinson, B. L.; Bornaghi, L.; Houston, T. A.; Poulsen, S.-A. Click Chemistry in Carbohydrate Based Drug Development and Glycobiology. In *Drug Design Research Perspective*; Kaplan, S. P., Ed.; Nova Science Publishers: New York, **2007**; pp 57–102.
6. Elchinger, P-H.; Faugeras, P-A.; Boënsl B.; Brouillette, F.; Montplaisir, D.; Zerrouki, R.; Lucas, R. *Polymers* **2011**, *3*, 1607–1651.
7. Kempe, K.; Krieg, A.; Becer, C. R.; Schubert, U. S. *Chem. Soc. Rev.* **2012**, *41*, 176–191.
8. Kolb, H. C.; Finn, M. G.; Sharpless, K. B. *Angew. Chem. Int. Ed.* **2001**, *40*, 2004–2021.
9. Torne, C. W.; Christensen, C.; Meldal, M. *J. Org. Chem.* **2002**, *67*, 3057–3064.
10. Rostovtsev, V. V.; Green, L. G.; Fokin, V. V.; Sharpless, K. B. *Angew. Chem., Int. Ed.* **2002**, *41*, 2596–2599.
11. Baskin, J. M.; Prescher, J. A.; Laughlin, S. T.; Agard, N. J.; Chang, P. V.; Miller, I. A.; Lo, A.; Codelli, J. A.; Bertozzi, C. R. *Proc. Natl. Acad. Sci. USA.* **2007**, *104*, 16793–16797.
12. Ning, X. H.; Guo, J.; Wolfert, M. A.; Boons, G. J. *Angew. Chem., Int. Ed.* **2008**, *47*, 2253–2255.
13. Marmuse, L.; Nepogodiev, S. A.; Field, R. A. *Org. Biomol. Chem.* **2005**, *3*, 2225–2227.
14. Huerta-Angeles, G.; Šmejkalová, D.; Chláková, D.; Ehlová, T.; Buffa, R.; Velebný, V. *Carbohydrate Polymers* **2011**, *84*, 1293–1300.
15. Garcia, G.; Naud-Martin, D.; Carrez, D.; Croisy, A.; Maillard, P. *Tetrahedron* **2011**, *67*, 4924–4932.
16. Vala, C.; Chrétien, F.; Balentova, E.; Lamandé-Langle, S.; Chapleur, Y. *Tetrahedron Lett.* **2011**, *52*, 17–20.
17. Dondoni, A. *Angew. Chem. Int. Ed.* **2008**, *47*, 8995–8997.
18. Hoyle, C. E.; Bowman, C. N. *Angew. Chem. Int. Ed.* **2010**, *49*, 1540–1573.
19. Dondoni, A.; Marra, A. *Chem. Soc. Rev.* **2012**, *41*, 573–586.
20. Posner, T. *Chem. Ber.* **1905**, *38*, 646–657.
21. Fiore, M.; Marra, A.; Dondoni, A. *J. Org. Chem.* **2009**, *74*, 4422–4425.
22. Chen, G.; Amajjahe, S.; Stenzel, M. H. *Chem. Commun.* **2009**, *14*, 1198–1200.
23. Sridhar, P. R.; Prabhu, K. R.; Chandrasekaran, S. *Eur. J. Org. Chem.* **2004**, 4809–4815.
24. Witczak, Z. J.; Lorchak, D.; Nguyen, N. *Carbohydr. Res.* **2007**, *342*, 1929–1933.
25. Semsarilar, M.; Ladmiral, V.; Perrier, S. *Macromolecules* **2010**, *43*, 1438–1443.
26. Nurmi, L.; Lindqvist, J.; Randev, R.; Syrett, J.; Haddleton, D. M. *Chem. Commun.* **2009**, 2727–2729.
27. Fiore, M.; Chambery, A.; Marra, A.; Dondoni, A. *Org. Biomol. Chem.* **2009**, *7*, 3910–3913.
28. Iñigo, S.; Porro, M. T.; Montserrat, J. M.; Iglesias, L. E.; Iribarren, A. M. *J. Mol. Catal. B: Enzymatic.* **2005**, *35*, 70–73.
29. Palomo, J. M.; Filice, M.; Fernandez-Lafuente, R.; Terreni, M.; Guisan, J. M. *Adv. Synth. Catal.* **2007**, *349*, 1969–1976.
30. Filice, M.; Palomo, J. M.; Bonomi, P.; Bavaro, T.; Fernandez-Lafuente, R.; Guisan, J. M.; Terreni, M. *Tetrahedron* **2008**, *64*, 9286–9292.
31. Baba, A.; Yoshioka, T. *Org. Biomol. Chem.* **2006**, *4*, 3303–3310.
32. Maki, T.; Ishida, K. *J. Org. Chem.* **2007**, *72*, 6427–6433.
33. Ruparel, H.; Bi, L.; Li, Z.; Bai, X.; Kim, D. H.; Turro, N. J.; Ju, J. *Proc. Natl. Acad. Sci. USA.* **2005**, *102*, 5932–5937.
34. Manabe, S.; Ueki, A.; Ito, Y. *Tetrahedron Lett.* **2008**, *49*, 5159–5161.

35. Liu, C-Y.; Chen, H-L.; Ko, C-M.; Chen, C-T. *Tetrahedron* **2011**, *67*, 872–876.
36. González, A. G.; Brouard, I.; León, F.; Padrón, J. I.; Bermejo, J. *Tetrahedron Lett.* **2001**, *42*, 3187–3188.
37. Kroutil, J.; Trnka, T.; Černý, M. *Synthesis.* **2004**, 446–450.
38. Gaunt, M. J.; Yu, J.; Spencer, J. B. *J. Org. Chem.* **1998**, *63*, 4172–4173.
39. Xia, J.; Abbas, S. A.; Locke, R. D.; Piskorz, C. F.; Alderfer, J. L.; Matta, K. L. *Tetrahedron Lett.* **2000**, *41*, 169–173.
40. Lipták, A.; Borbás, A.; Jánossy, L.; Szilágyi, L. *Tetrahedron Lett.* **2000**, *41*, 4949–4953.
41. Mo, K-F.; Fang, T.; Stalnaker, S. H.; Kirby, P. S.; Liu, M.; Wells, L.; Pierce, M.; Live, D. H.; Boons, G-J. *J. Am. Chem. Soc.* **2011**, *133*, 14418–14430.
42. Barbier, M.; Grand, E.; Kovensky, J. *Carbohydr. Res.* **2007**, *342*, 2635–2640.
43. Ekholm, F. S.; Mándity, I. M.; Fülöp, F.; Leino, R. *Tetrahedron Lett.* **2011**, *52*, 1839–1841.
44. Das, B.; Mahender, G.; Kumar, V. S.; Chowdhury, N. *Tetrahedron Lett.* **2004**, *45*, 6709–6711.
45. Sharma, G. V. M.; Srinivas, B.; Krishna, P. R. *Lett. Org. Chem.* **2005**, *2*, 57–60.
46. Chen, M-Y.; Lu, K-Ch.; Lee, A. S-Y.; Lin, Ch-Ch. *Tetrahedron Lett.* **2002**, *43*, 2777–2800.
47. Mezaache, R.; Dembelé, Y. A.; Bikard, Y.; Weibel, J-M.; Blanc, A.; Pale, P. *Tetrahedron Lett.* **2009**, *50*, 7322–7326.
48. Sridhar, P. R.; Chandrasekaran, S. *Org. Lett.* **2002**, *4*, 4731–4733.

PART II

THIO-CLICK CHEMISTRY OF CARBOHYDRATES

2

THIO-CLICK CHEMISTRY IN GLYCOSCIENCE: OVERVIEW AND PERSPECTIVES

ZBIGNIEW J. WITCZAK

2.1 INTRODUCTION

The new concept by Sharpless, Finn, and Kolb for conducting organic reactions capable of producing wide libraries of fully functionalized synthetic molecules has been classified as click chemistry. The traditional azide–alkyne click chemistry approaches with multiple types of clickable functional groups were the leading examples of the early strategy.

Recently, thiol-ene chemistry became a well-accepted part of the click reaction toolbox and is rapidly developing in many strategic areas of thio-conjugation. In carbohydrate chemistry the most important substrates are glycosyl thiols. Few years ago, they became available in both anomeric forms, which created new opportunities for the expansion of the existing techniques for thio-conjugation to form products deriving from both stereo isomers.

2.2 THIO-BASED CLICK REACTIONS

The addition of a thiol to the double bond is known as the thiol-ene reaction. The scope of the thiol-ene approach is quite impressive with respect to the endless applicability of reactive thiols and almost all molecules with a double bond (ene) moiety. The reactions take place under various experimental conditions, such as, base or acid

Click Chemistry in Glycoscience: New Developments and Strategies, First Edition.
Edited by Zbigniew J. Witczak and Roman Bielski.

catalysis and radical induction (thermal or photochemical). Most of these approaches are applicable to polymer chemistry [1,2].

In carbohydrate chemistry, a reactive sugar thiol can be added to a double bond of unsaturated sugar quite easily, always with the formation of thio-adduct in high yields.

Since several review articles have dealt with many aspects of thio-functionalization [3,4] *via* click reaction in polymer chemistry and there is no extensive compilation in synthetic carbohydrate chemistry, the aim of this chapter is to review only synthetic methods involving genuine thio-click approaches either *via* thiol-ene coupling (TEC) or thiol Michael addition enone (TMAE) to the conjugated systems of carbohydrate enones.

The application of TEC and thiol-yne coupling (TYC) approaches to glycoconjugation in carbohydrate chemistry was recently reviewed by Dondoni and Marra [5,6]. Other pertinent articles by Davis [7] deal exclusively with the design and formation of disulfide linkers and its incorporation into specific biological systems.

In carbohydrate chemistry, functional group manipulations are highly dependent on the availability of the position for all thio-click reactions and are designated to form either internal or external thio-linking functional group. The external thio-functionalized group or linker could be further manipulated to build macromolecular assemblies with specific moieties containing blocked or exposed thiol groups suitable for additional chemical ligation or glycoconjugation. These structural thio-conjugation strategies *via* click chemistry are the most promising among elaborate new developments in the multidisciplinary chemistry/biology methodologies. The thio-click chemical reactions toolbox is depicted in Scheme 2.1.

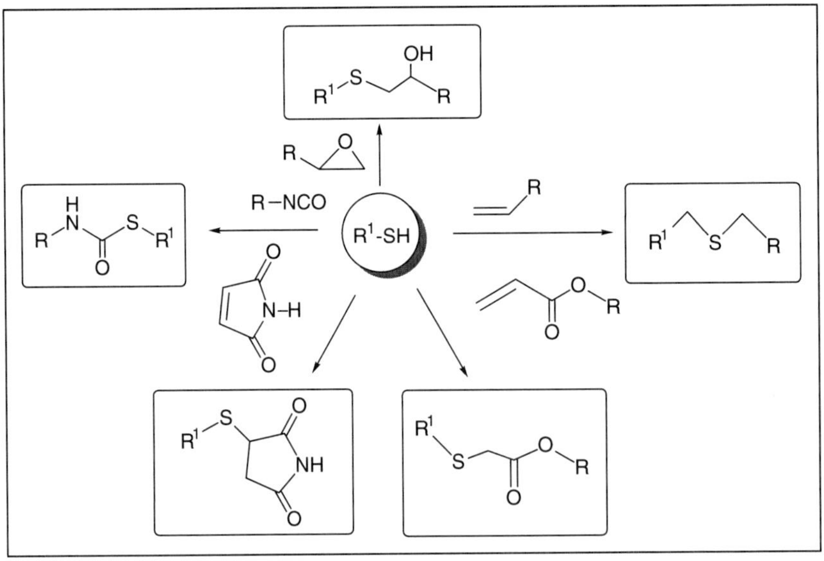

SCHEME 2.1 Thio-click chemical reaction toolbox.

2.3 *S*-THIODISACCHARIDE SYNTHESIS

S-thiodisaccharides are simple functionalized sugars bearing sulfur bridge in one of the strategically important positions, C-1, C-4 (head–tail) or C-1, C-6 (head–tail). Other less strategically important structural connections *via* sulfur bridge include C-1, C-1 (head–head) and C-1, C-2 or C-1, C-3 and are summarized in Scheme 2.2.

Several methods for the synthesis of *S*-thiodisaccharides have been published. Most of them involve SN_2-type substitution of thiolate anions on glycosyl halides, or 1-thio-donors on acceptors bearing a good leaving group. None of these approaches are considered thio-click reactions.

The alternative formation of thio-bridge *via* TMEA addition to the conjugate system can clearly fit the definition of thio-click approach, despite their development before the official birth of click chemistry. Our group was the first to develop this attractive strategy [8–13] of the synthesis of *S*-thiodisaccharides, followed by others [14–17]. These strategies were compiled and reviewed [18–21]. Many other strategies for the synthesis of thiodisaccharides were published [22–24].

Chandrasekaran and coworkers [23] developed a new methodology for the *S*-bridge formation of many thioglycosides by employing benzyltriethylammonium tetrathiomolybdate. The proposed mechanism postulates the tetrathiomolybdate reaction with sugar halides to proceed with the formation of sugar disulfides, which then undergo reductive cleavage *in situ* to form the corresponding thiolates, followed by Michael addition to produce the corresponding thioglycosides. This particular methodology always exhibits excellent diastereoselectivity.

One of the important reactions mediated by tetrathiomolybdate is an intramolecular disulfide formation/reduction/Michael addition sequence (one-pot reaction) producing 1,5- and 1,6-thioenones as depicted in Scheme 2.3. This type of intermolecular thio-bridge formation can be clearly classified as thio-click reaction. The example offered leads to the formation of the internal bridge (intramolecular click reaction) but the same approach has been successful in coupling of two molecules as well (see Chapter 1).

Ćerny and coworkers [24] reported coupling α-glucosyl thiols with the unsaturated tosylhexenose in the presence of anhydrous potassium carbonate in acetone. This

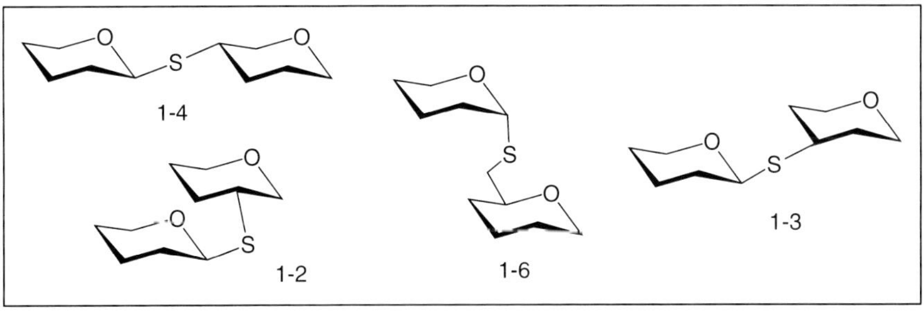

SCHEME 2.2 Strategical connections of thiodisaccharides.

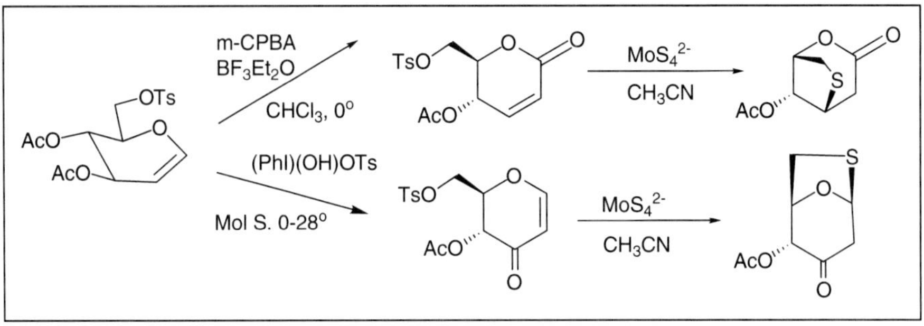

SCHEME 2.3 An intramolecular disulfide formation/reduction Michael addition sequence mediated by tetrathiomolybdate.

highly efficient methodology gives access to a variety of unsaturated thiodisaccharides. It is interesting to note that the anhydrous potassium carbonate in dry acetone proved to be the best base catalyst, whereas lithium carbonate in dry acetone or DBU in oxolane were not effective. The tosyloxy group was readily replaced by thiol nucleophiles at room temperature to form 4-thiolevoglucosenones. The reaction proceeds *via* stereoselective allylic rearrangement and the steric interactions of 1,6-anhydro bridge may hinder the access of the thiol nucleophile toward C-2 and C-4 from the *endo* side. On the other hand, substitution at C-2 with retention of configuration is disfavored by repulsive interaction between the thiol and the O-5-ring oxygen. The combination of the thiol addition followed by the allylic rearrangement is a novel, elegant example of the click chemistry. The reaction is presented in Scheme 2.4.

The newer thio-click chemistry strategies *via* TEC approach applied to the synthesis of thiodisaccharides were originally published by the Dondoni group [25] and reviewed in Chapter 3. Some other examples of thio-click formation of

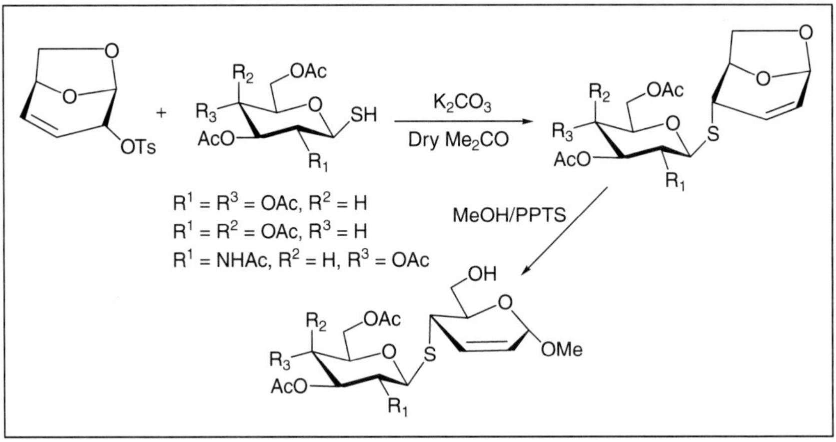

SCHEME 2.4 Coupling of tosyl hexenose with β-glycosyl thiols.

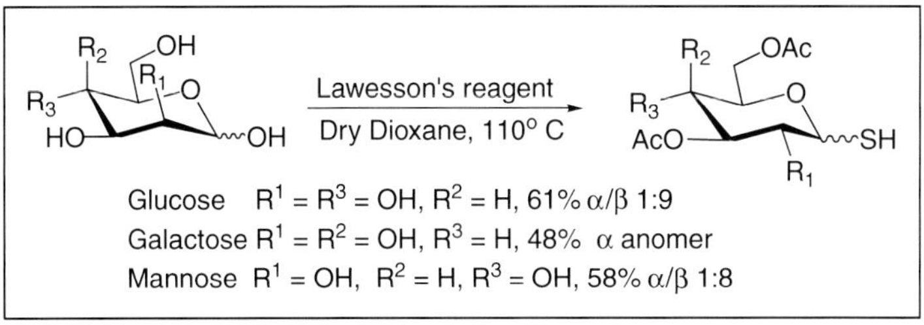

SCHEME 2.5 Davis synthesis of glycosyl thiols from unprotected sugars with Lawesson's reagent [26].

S-thiodisaccharides *via* TEC and TMEA are also discussed in Chapter 1 and for that reason are not included in this chapter.

2.4 SYNTHESIS OF GLYCOSYL THIOLS

As mentioned before, sugar thiols are essential substrates for thio-click reactions in carbohydrate chemistry. The reactivity of glycosyl thiols is sufficient enough for practical application in many TEC or TYC coupling. These key building blocks for construction of many thiodisaccharides, thiooligosaccharides and *S*-glycoconjugates are also quite stable and their anomeric configuration can be effectively preserved during the thioglycosylation reactions. Notably, the anomeric thiols of 1-thiopyranoses anomerize much more slowly than the corresponding oxygen analogs, particularly under basic conditions or when heated. This particular character can be exploited for the synthesis of thioglycosides including thiodisaccharides and oligothiosaccharides. The high nucleophilicity of the thiol group under basic conditions enables base-promoted *S*-glycosylation. This great practical utility and potential is mainly attributable to glycosyl thiols with the thiol group in the equatorial position. The glycosyl thiols with the opposite anomeric configuration (with the thiol group in the axial position) were essentially unknown until recently [25–33].

Davis and coworkers [26] (Scheme 2.5) reported an important strategy of direct formation of glycosyl thiols using Lawesson's reagent in the reaction with protected or unprotected anomeric lactols/reducing sugars. However, in this procedure configurationally impure glycosyl thiols as anomeric mixture were often obtained. Therefore, the development of a direct and highly stereoselective methodology for the synthesis of α-glucosyl thiols seems to be of utmost importance.

The Zhou group [27, 28] (Scheme 2.6) developed an excellent approach to α-glucosyl thiols starting from 1,6-anhydro sugar templates with different protecting group patterns and wide structural diversity. The series of α-glucosyl thiols were synthesized and most importantly no trace of β-anomers was produced in all the reactions.

This ring-opening protocol is a significant milestone development because of the overall yield of the final products, and its exclusive α-anomeric purity. The choice of an excellent sulfur nucleophile, bis(trimethylsilyl) sulfide (BTMSS) in combination

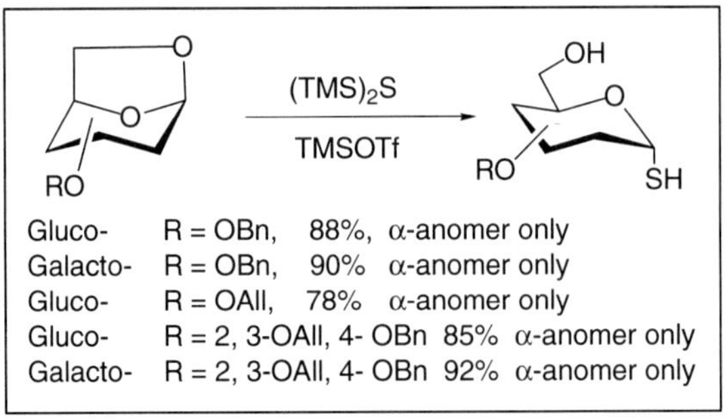

SCHEME 2.6 Zhu synthesis of α-glycosyl thiols from 1,6-anhydro sugars [27, 28].

with catalytic amount of trimethylsilyl triflate at elevated temperature proved to be essential for the opening of the 1,6-anhydro ring with an excellent stereoselectivity. The methodology is based on the previous application of BTMSS in the acid-catalyzed ring opening of *O*-benzyl protected 1,2-anhydro-glucopyranose (Brigl's anhydride benzyl analog) originally developed by the Schmidt group [29].

Bundle and coworkers [30] reported a simple strategy of nucleophilic displacement of the anomeric α-bromide of peracetylated D-mannopyranose with thioacetate prepared *in situ* by mixing thioacetic acid and potassium *tert*-butoxide. This protocol leads exclusively to the formation of the strategically very important 1-thio-β−mannopyranoside as depicted in Scheme 2.7.

Knap and coworkers [31, 32] (Scheme 2.8) reported an efficient strategy for the synthesis of α-GlcNAc and α-GalNAc thiols. The procedure is based on the treatment of 2-acetamido-2-deoxy-1,3,4,6-tetra-*O*-acetyl-β-D-galactopyranose or glucopyranose with Lawesson's reagent in toluene at 80°C with the formation of thiazoline derivatives, followed by hydrolysis under acidic conditions (TFA in wet MeOH). The hydrolysis occurred exclusively at the iminium carbon and produced α-GlcNAc and α-GalNAc thiols in quantitative yields.

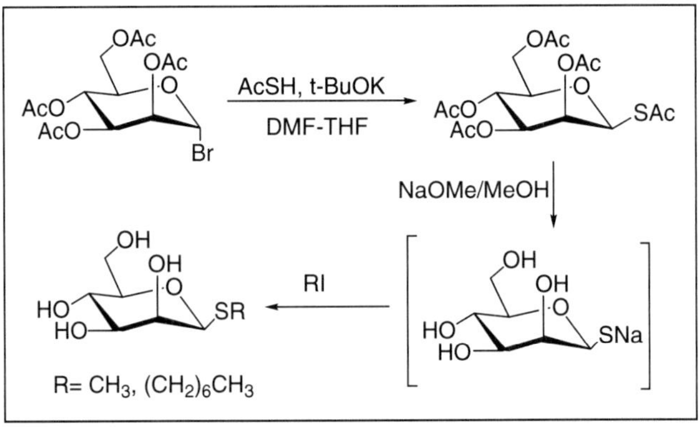

SCHEME 2.7 Bundle synthesis of 1-thio-β-mannopyranose [30].

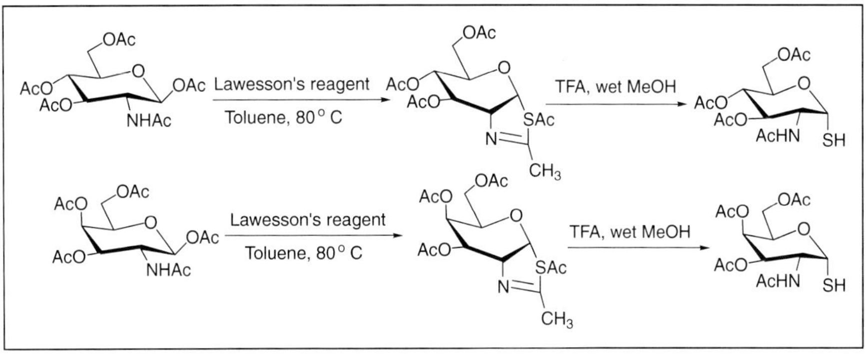

SCHEME 2.8 Knapp synthesis of α-GlcNAc and α-GalNAc [31, 32].

Ibatulin and coworkers [33, 34] (Scheme 2.9) developed a novel strategy for the synthesis of thio-xylooligosaccharides using a triethylamine–promoted coupling of isothiouronium bromides with axial sugar triflates. This simple methodology of utilizing strongly nucleophilic masking group of isothiouronium bromides for thiols proceeds with good yield and under relatively mild conditions and short reaction times. The high yields of the final product combined with the availability of the reagents are clearly very advantageous in comparison to other known methods for 1,2-trans thiooligosaccharide synthesis. For this particular reason, this methodology can also be considered a thio-click reaction.

The references to selected glycosyl thiols are listed in Table 2.1.

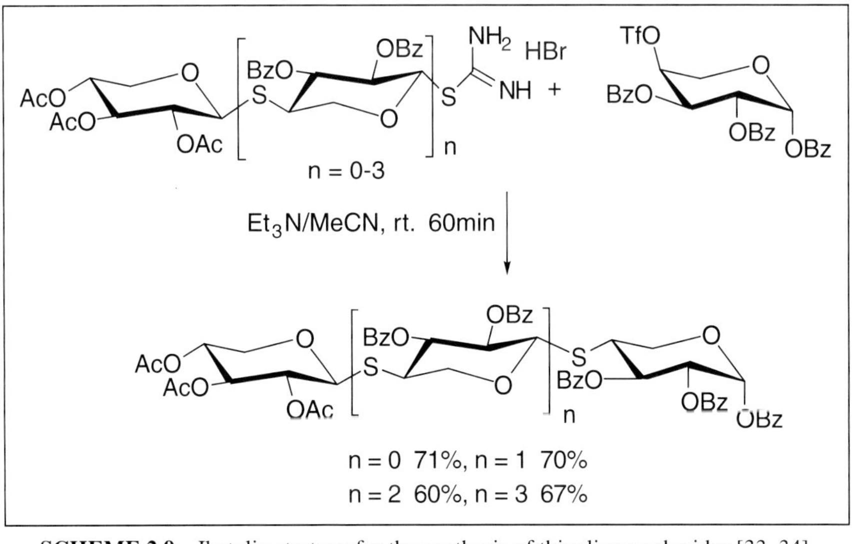

SCHEME 2.9 Ibatulin strategy for the synthesis of thiooligosaccharides [33, 34].

TABLE 2.1 Glycosyl Thiols

Glycosyl Thiols	Reference
1-thio-2-acetamido-2-deoxy-β-D- glucopyranose	26
1-thio-2-acetamido-2-deoxy-β-D- glucopyranose	26
2,3,4,6-tetra-*O*-acetyl-1-*S*-acetyl-1-thio-β-D-mannopyranose	30
2,3,4,6-tetra-*O*-acetyl-1-*S*-acetyl-1-thio-α-D-glucopyranose	26
2,3,4,6-tetra-*O*-acetyl-1-*S*-acetyl-1-thio-α-D-galactopyranose	26
2,3,4,6-tetra-*O*-acetyl-1-*S*-acetyl-1-thio-α-D-galactopyranose	26
1-thio-2,3,4,-tri-*O*-benzyl-6-hydroxy-α-D-glucopyranose	27
1-thio-2,3,4,-tri-*O*-benzyl-6-hydroxy-α-D-galactopyranose	27
1-thio-2,3,4,-tri-*O*-allyl -6-hydroxy-α-D-glucopyranose	27
1-thio-2,3,-di-*O*-allyl -4-*O*-benzyl- 6-hydroxy-α-D-glucopyranose	27
1-thio-2,3,-di-*O*-allyl -4-*O*-benzyl- 6-hydroxy-α-D-galactopyranose	27
1-thio-3,4-di-*O*-benzyl-2,6-di-hydroxy-α-D-glucopyranose	27
1-thio-2,4-di-*O*-benzyl-3,6-di-hydroxy-α-D-glucopyranose	27
1-thio-2,3-di-*O*-benzyl-4,6-di-hydroxy-α-D-glucopyranose	27
1-thio-3-*O*-benzyl-2,4,6-tri-hydroxy-α-D-glucopyranose	27
1-thio-2-*O*-benzyl-3,4,6-tri-hydroxy-α-D-glucopyranose	27
2-acetamido-2-deoxy-3,4,6-tri-*O*-acetyl-1-*S*-acetyl-1-thio-α-D-galactopyranose	32
2-acetamido-2-deoxy-3,4,6-tri-*O*-acetyl-1-*S*-acetyl-1-thio-α-D-glucopyranose	31

2.5 SYNTHESIS OF S-LINKED GLYCOSPHINGOLIPIDS

Traditional thio-click methods have been widely used in the synthesis of important *S*-glycosyl amino acids. Among many methods, [35–37] the most useful and practical is coupling glycosyl thiols with fully protected iodides under phase-transfer catalysis conditions.

Zhu et al. [35, 36] reported the first synthesis of thioglycoside analog of α-GalCer by using a nonconventional protocol depicted in Scheme 2.10. The synthesis is based on a newly developed strategy for the preparation of α-glycosyl thiols reported by Schmidt and coworkers [29]. The α-galactosyl thiol precursor was prepared from 1,6-anhydrosugar substrate by ring opening with commercial BTMSS in the presence of trimethylsilyl trifluoro-methanesulfonate (TMSOTf) as a catalyst (as described earlier in the previous paragraph). In the second step, the α-galactosyl thiol is coupled with the electrophilic sphingoid iodide under phase-transfer catalysis [36] to produce an α-linked product in very good yield. The selection of tetrabutylammonium hydrogen sulfate (TBAHS) in ethyl acetate and an aqueous solution of sodium hydrogen carbonate at pH 8.5 for the phase-transfer catalyst conditions proved to be extremely efficient. This new approach constitutes an excellent methodology to the brand new family of thio sugars with α-configuration at the anomeric center. The target α-*S*-Gal-Cer was produced through 14 steps sequence in 3% overall yield.

Murphy and coworkers [37, 38] reported synthesis of thioglycoside-based α-GalCer analog. A chelation-induced anomerization strategy developed earlier by the Murphy group [39] was applied elegantly to this synthesis as is depicted in

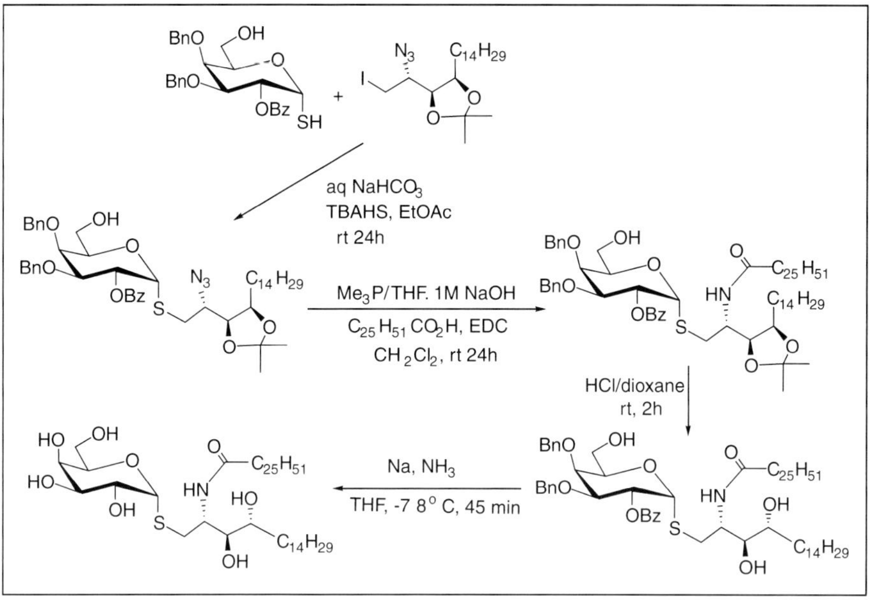

SCHEME 2.10 Zhu strategy for the synthesis of α-thioglycoside analog of KRN7000.

Scheme 2.11. The target uronic-based glycolipids were constructed with high stereoselectivity.

Other methods [40–42] for introduction of sulfur heteroatom into multivalent structures including glycoclusters were recently published and some of them are reviewed by the authors of Chapters 6 and 8.

2.6 CONCLUSION AND PERSPECTIVES

The variety of new methods for thio-functionalization of small carbohydrate molecules and more complex biomolecules via thio-click approach presented in this review constitute a very attractive strategy for the synthesis of multiple targets of glycoconjugates, including glycoclusters on various scaffolds, glycopeptides, and glycoproteins.

Moreover, these powerful thio-click functionalization strategies will likely dominate and compete with all other synthetic approaches available for molecular ligation or glycoconjugation.

The recent availability of axial α-glycosyl thiols also opened the enormous opportunity for the fully stereoselective approach of thio-click functionalization for *S*-glycosylation and protein labeling and ligation.

Furthermore, interdisciplinary attempts to utilize the high potential of thio-click chemistry, either TEC, TYC, or TMEA in many areas of glycoscience/glycobiology will be forthcoming. These methodologies will create a serious challenge for better

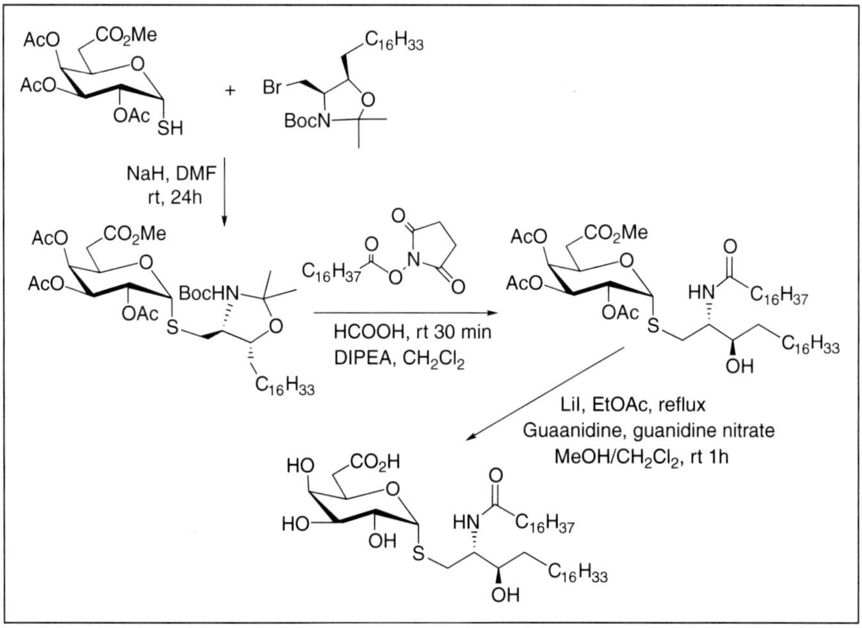

SCHEME 2.11 Murphy strategy for the synthesis of α-S-glycosphingolipids based on uronic acids.

applications in complex and multiple constructions of molecular units for biological ligation and specifically designated multiple-step glycoconjugation.

REFERENCES

1. Hoyle, C. A.; Lowe, A. B.; Bowman, C. N. *Chem. Soc. Rev.* **2010**, *39*, 1355.
2. Hoyle, C. A.; Bowman, C. N. *Angew. Chem. Int. Ed.* **2010**, *49*, 1540.
3. Lowe, A. B.; Harvison, M. A. *Aust. J. Chem.* **2010**, *63*, 1251.
4. Slavin, S.; Burns, J.; Hadletton, D. M.; Becer, C. R. *Eur. Polymer J.* **2011**, *47*, 435.
5. Dondoni, A.; Marra, A. *Chem. Soc. Rev.* **2012**, *41*, 573.
6. Dondoni, A. *Angew. Chem. Int. Ed.* **2008**, *47*, 8995.
7. Bernardes, G. J. L.; Grayson, E. J.; Thompson, S.; Chalker, J. M.; Errey, J. C.; El oualid, F.; Claridge, T. D. W.; Davis, B. G. *Angew. Chem. Int. Ed.* **2008**, *47*, 2244.
8. Witczak, Z. J.; Sun, J.; Mielguj, R. *Bioorg. Med. Chem. Lett.* **1995**, *5*, 2169.
9. Witczak, Z. J.; Chabra, R.; Chen, H.; Xie, X-Q. *Carbohydr. Res.* **1997**, *301*, 167.
10. Witczak, Z. J.; Chen, H.; Kaplon, P. *Tetrahedron Asymmetry* **2000**, *11*, 519.
11. Witczak, Z. J.; Chabra, R.; Boryczewski, D. *J. Carb. Chem.* **2000**, *19*, 543.
12. Witczak, Z. J.; Kaplon, P.; Dey, P. M. *Carbohydr. Res.* **2003**, *338*, 1.
13. Witczak, Z. J.; Lorchak, Nguyen, N. *Carbohydr. Res.* **2007**, *342*, 1929.

14. Becker, B.; Thimm, J.; Thiem J. *J. Carbohydr. Chem.* **1996**, *15*, 1179.
15. Uhrig, M. L.; Varela, O. *Aust. J. Chem.* **2002**, *55*, 155.
16. Uhrig, M. L.; Manzano, V. E.; Varela, O. *Eur. J. Org. Chem.* **2006**, 162.
17. Uhrig, M. L.; Varela, O. *Carbohydr. Res.* **2002**, *337*, 2069.
18. Witczak, Z. J. *Curr. Med. Chem.* **1999**, *6*, 165.
19. Robina, I.; Vogel, P.; Witczak, Z. J. *Curr. Org. Chem.* **2001**, *6*, 159.
20. Witczak, Z. J.; Culhane, J. M. *Appl. Microbiol. & Biotech.* **2005**, *69*, 237–244.
21. Pachamuthu, K.; Schmidt, R. R. *Chem. Rev.* **2006**, *106*, 160.
22. Ellis, D.; Norman, S. E.; Osborn, H. M. I. *Tetrahedron* **2008**, *64*, 2832.
23. Sridhar, P. R.; Prabhu, K. R.; Chandrasekaran, S. *Eur J. Org Chem.* **2004**, 4809.
24. Bartos, P.; Cerny, M.; Tislerova, I.; Trnka T. *Collect. Czech. Chem. Commun.* **2000**, *65*, 1737.
25. Fiorre, M.; Marra A.; Dondoni, A. *J. Org. Chem.* **2009**, *74*, 4422.
26. (a) Bernardes, G. J. L.; Gamblin, D. P.; Davis, B. G. *Angew. Chem. Int. Ed.* **2006**, *45*, 4007. (b) Gamblin, D. P.; Garnier, P.; van Kastern, S.; Oldham, N. J. Firbanks, A. J.; Davis, B. G. Angew. *Chem. Int. Ed.* 2004, *43*, 823.
27. Zhu, X.; Dere, R. T.; Jiang, J.; Zhang, L.; Wang, X. *J. Org. Chem.* **2011**, *76*, 10187.
28. Dere, R. T.; Wang, Y.; Zhu, Z. *Org. Biomol. Chem.* **2008**, *6*, 2061.
29. Dere, R. T.; Kumar, A.; Kumar, V.; Zhu, X.; Schmidt, R. R. *J. Org. Chem.* **2011**, *76*, 7539.
30. Yu, H. N.; Ling, C. C.; Bundle, D. R. *J. Chem. Soc., Perkinn. Trans.* **2001**, *1*, 832.
31. Knapp, S.; Myers, D. S. *J. Org. Chem.* **2001**, *66*, 3636.
32. Knapp, S.; Myers, D. S. *J. Org.Chem.* **2002**, *67*, 2995.
33. Ibatulin, F. M.; Shabalin, K. A.; Janis, J. V.; Shavva, A. G. *Tetrahedron Lett.* **2003**, *44*, 7691.
34. Ibatulin, F. M.; Shabalin, K. A.; Janis, J. V.; Selivanov, S. I. *Tetrahedron Lett.* **2001**, *42*, 4565.
35. Dere, R. T.; Zhu, X. *Org. Lett.* **2008**, *10*, 4641.
36. Zhu, X.; Schmidt, R. R. *Chem. Eur. J.* **2004**, *10*, 875.
37. Pilgrim, W.; Murphy, P. V. *Org. Lett.* **2009**, *11*, 939.
38. O'Reilly, C.; Murphy, P. V. *Org. Lett.* **2011**, *13*, 5168.
39. O'Brien, C.; Polakova, M.; Pitt, N.; Tosin, M.; Murphy, P. V. *Chem. Eur. J.* **2007**, *13*, 902.
40. Liakatos, A.; Kiefel, M. J.; von Itzstein, M. *Org. Lett.* **2003**, *5*, 4365.
41. Cagnoni A. J.; Varela, O.; Gouin, S. G.; Kovensky, J.; Uhrig, M. L. *J. Org. Chem.* **2011**, *76*, 3064–3077.
42. Floyd, N.; Vijayakrishnan, B.; Koeppe, J. R.; Davis, B. *Angew. Chem. Int. Ed.* **2009**, *48*, 7798.

3

FREE-RADICAL THIOL-ENE AND THIOL-YNE COUPLINGS AS CLICK PROCESSES FOR GLYCOCONJUGATION

ALESSANDRO DONDONI AND ALBERTO MARRA

3.1 INTRODUCTION

Since the formulation of the click chemistry concept by Sharpless and coworkers [1] in 2001 and the discovery of the copper-catalyzed azide–alkyne cycloaddition (CuAAC) as the quintessential click reaction [2] in 2002, there has been an explosive growth in the number of applications of this reaction in the most disparate fields of synthetic chemistry, including material and life sciences. No better start of the new millennium could have been foreseen by synthetic chemists. In fact, the advent of the click chemistry concept was very likely determined by the need to put some order in the plethora of new synthetic methods that had been discovered in the second half of last century. The set of criteria that a process must meet in the context of click chemistry was defined by Sharpless as follows:

> *"The reaction must be modular, wide in scope, give very high yields, generate only inoffensive byproducts that can be removed by nonchromatographic methods, and be stereospecific (but not necessarily enantioselective). The required process characteristics include simple reaction conditions (ideally, the process should be insensitive to oxygen and water), readily available starting materials and reagents, the use of no solvent or a solvent that is benign (such as water) or easily removed, and simple product isolation. Purification—if required—must be by nonchromatographic methods,*

Click Chemistry in Glycoscience: New Developments and Strategies, First Edition.
Edited by Zbigniew J. Witczak and Roman Bielski.

such as crystallization or distillation, and the product must be stable under physiological conditions. Click processes proceed rapidly to completion and also tend to be highly selective for a single product: we think of these reactions as being spring-loaded for a single trajectory."

A summary of these criteria is presented in Figure 3.1.

In contrast to some belief, click chemistry is not limited to CuAAC or its copper-free version—referred to as strain-promoted azide–alkyne cycloaddition (SPAAC) [3]—because other spring-loaded reactions own the essential "click" attributes. Of special importance are click reactions occurring in the absence of metal catalysts [4] because this allows their extensive use for bioconjugation without damaging the delicate molecules that are involved. Among these metal-free reactions are the sister reactions consisting of the additions of thiols to inactivated alkenes (thiol-ene coupling, TEC) [5] and alkynes (thiol-yne coupling, TYC) [5b, 6]. Under photochemical conditions both reactions occur by a free-radical chain reaction mechanism while they are promoted by traces of a photoinitiator (Scheme 3.1). Thus, TEC allows the introduction of a single thiol fragment across the terminal carbon–carbon double bond in an anti-Markovnikov fashion (monohydrothiolation) to give a thioether while TYC enables the addition of two thiol residues across the carbon–carbon triple bond and leads to a dithioether derivative (bis-hydrothiolation). Both photoaddition reactions can proceed at near-visible wavelengths (λ = 365–405 nm) and therefore can be conducted in sealed glass vials using a handheld UV lamp. In general the reactions occur with fast kinetics with reaction times of the order of minutes. No special care is needed to exclude air and moisture while pure water or mixtures of water and polar solvents (Dimethylformamide [DMF], methanol) can be employed as reaction media. In general only a slight excess of thiol is used with respect to the number of alkene or alkyne groups to be hydrothiolated and therefore the only byproduct that is formed is the corresponding disulfide arising from the radical homocoupling of thiol residues. Significantly, both reactions lead to the formation of robust carbon–sulfur bonds that in addition of being flexible and sterically nondemanding, are stable under a wide range of chemical environments such as basic and acid media. Thus, while the click status of TEC and TYC can be anticipated on the light of the above features, it

High yields
Stable compounds
Modular
Wide in scope
Chemoselective
Single reaction trajectory
Oxygen and water compatible
No solvent or a benign solvent
Easy purification
Stereospecific
Readily available reagents

FIGURE 3.1 Sharpless criteria for a click process.

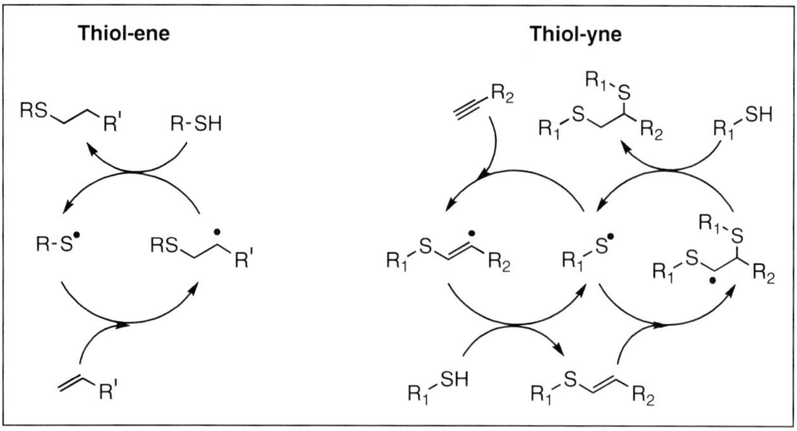

SCHEME 3.1 Mechanisms of photoinduced thiol-ene and thiol-yne coupling.

appeared to be quite important to validate this prerogative by using these reactions in a field dealing with complex, multifunctionalized and delicate molecules such as carbohydrates. Therefore, this chapter illustrates the results obtained in our own laboratory on the use of both TEC and TYC in the construction of various *S*-linked glycoconjugates. These include oligosaccharides, glycodendrimers and glycoclusters, glycosyl aminoacids and glycopeptides. The glycoconjugation of a protein is described as well. The aim of our work was to extend the scope of these reactions in organic and bioorganic chemistry as their use in these fields is quite limited [7] in comparison to the numerous applications in polymer and material sciences [8, 9].

3.2 *S*-LINKED OLIGOSACCHARIDES

Oligosaccharides and glycoconjugates deeply influence many fundamental biological processes in living organisms [10]. They mediate a variety of events, including inflammation, immunological response, fertilization, cancer metastasis, viral, and bacterial infections [11]. Hence, usable quantities of natural carbohydrates with a well-defined structure and composition are needed to study those phenomena at molecular level. There is, however, substantial difficulty to obtain complex natural *O*-linked oligosaccharides in a pure and homogeneous form from natural sources because of the presence of mixtures of glycosylated species (glycoforms). Moreover these compounds display an intrinsic instability toward chemical and enzymatic degradation due to the labile exocyclic carbon–oxygen bond. Therefore, synthetic efforts have been directed toward the supply of carbon- and sulfur-linked isosteres, that is, compounds bearing a methylene group or a sulfur atom in place of the oxygen atom of the glycosidic bond that is present in natural products (Fig. 3.2). Thus, simple *C*-disaccharides [12] and *S*-disaccharides [13] have been mainly prepared because these products can be used as probes of recognition specificity and may provide important insight into the mechanism of glycoside elaboration by carbohydrate-processing enzymes. In turn, these glycomimetics may become effective inhibitors of

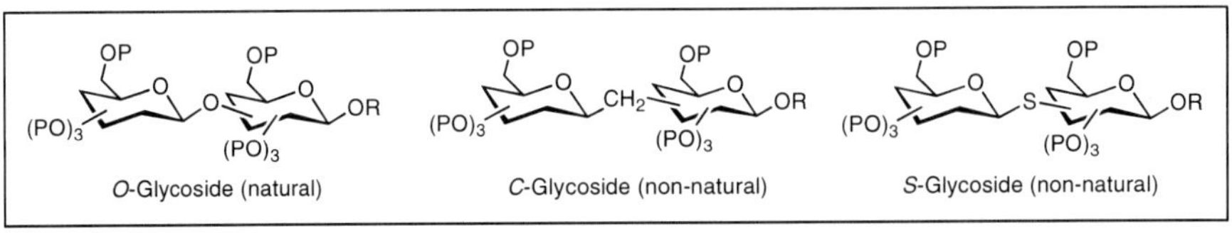

FIGURE 3.2 Natural (*O*-linked) and synthetic (*C*- and *S*-linked) disaccharides.

those enzymes and therefore evolve into lead compounds of pharmaceutical relevance. It has to be noted, however, that for biological studies, *S*-oligosaccharides may be more useful than *C*-analogs because the former products represent the smallest step away from natural *O*-glycosides in the backbone space [14].

While several chemical methods have been established for the synthesis of *S*-disaccharides [13], thiol-ene coupling between two sugar fragments was reported for the first time by our group in 2009 [15]. Reaction conditions were optimized by studying the model reaction between the readily available peracetylated 1-thio-β-D-glucose **1** and the diacetonide-protected galactose-derived alkene **2** (Scheme 3.2). The reaction between these reagents occurred readily at room temperature by irradiation at λ_{max} 365 nm in the presence of 2,2-dimethoxy-2-phenylacetophenone (DPAP) as the initiator. Significantly, the conversion of the alkene **2** as established by ^{1}H NMR analysis was higher than 97% and the yield of isolated *S*-disaccharide **3** was 91%. The same reaction was carried out by unfocused sunlight irradiation (8 h) to give the product **3** in 77% isolated yield.

Also the photoinduced couplings of sugar thiol **1** with hex-5-enopyranoside **4** and pent-4-enofuranoside **5** were examined in detail because these reactions afforded *S*-disaccharides **6** and **7**, respectively, that is, exact *S*-isosteres of 1,6- and 1,5-linked natural *O*-disaccharides (Scheme 3.3). Noteworthy, both reactions proceeded with high diastereoselectivity (de >99%) to give essentially a single product and therefore they represented striking examples of asymmetric thiol-ene coupling. Finally, the substrate scope of TEC in *S*-disaccharide synthesis was investigated by using other sugar thiols and sugar alkenes so that a collection of *S*-linked disaccharides with various elements of diversity (substituent, configuration) was prepared. All products **6–14** were obtained in good isolated yields, thus confirming the fidelity and efficiency of TEC while substantial structural changes of the reagents were introduced (Fig. 3.3).

Synthesis of thiodisaccharide **6**. The reaction was carried out in a glass vial (diameter: 1 cm; wall thickness: 0.65 mm), closed with a natural rubber septum, located 2.5 cm away from the UVA lamp (λ_{max} 365 nm). To a solution of thiol **1** (44 mg, 0.12 mmol) and sugar alkene **4** (30 mg, 0.10 mmol) in anhydrous CH_2Cl_2

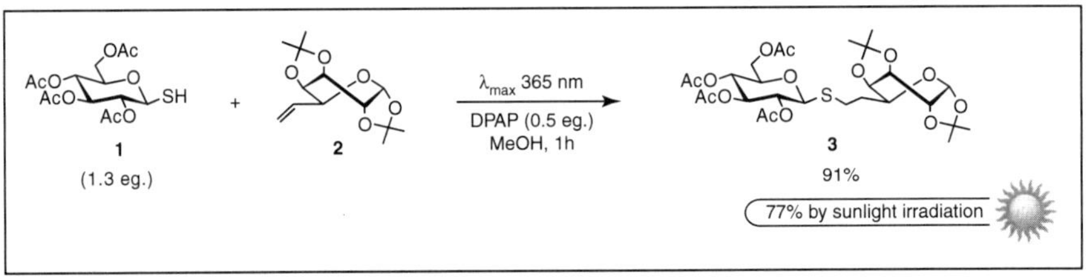

SCHEME 3.2 Synthesis of an *S*-disaccharide via thiol-ene coupling (TEC).

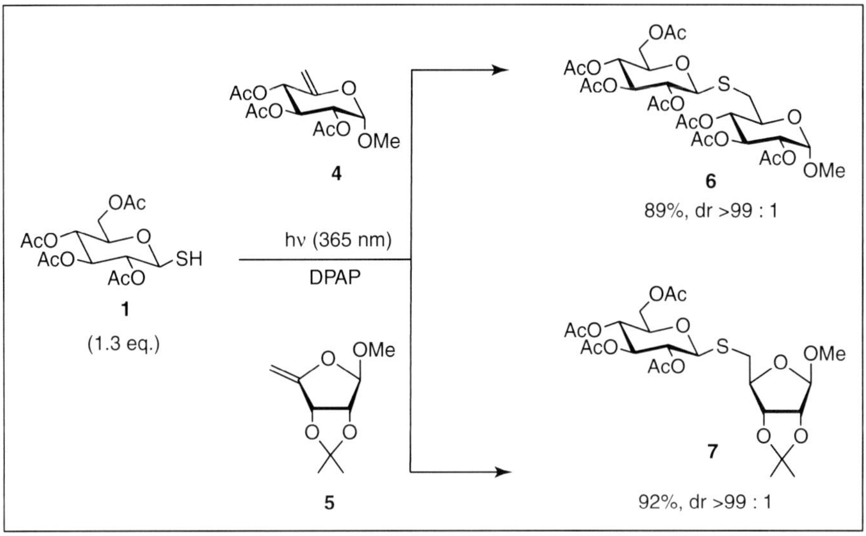

SCHEME 3.3 Synthesis of exact *S*-disaccharide isosteres via TEC.

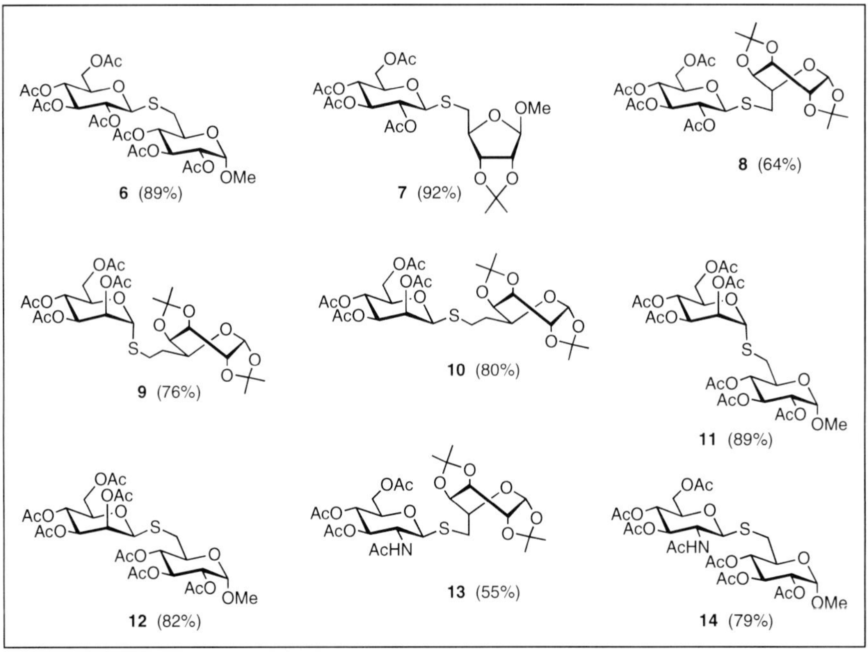

FIGURE 3.3 Collection of *S*-disaccharides prepared via TEC.

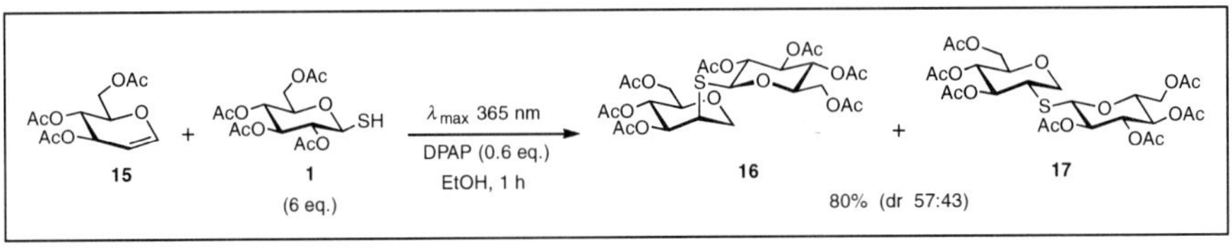

SCHEME 3.4 TEC between a sugar thiol and a glycal.

(2 mL) was added DPAP (3.1 mg, 0.012 mmol). The solution was irradiated at room temperature for 15 minutes under magnetic stirring, then concentrated. The pale yellow residue was eluted from a column of silica gel with 1:1 cyclohexane-AcOEt to give **6** (59 mg, 89%) as a white solid; $[\alpha]_D = +113.0$ (*c* 1.3, $CHCl_3$).

Another TEC-based approach to *S*-linked disaccharides entailed the coupling of glycosyl thiols with glycals [16]. TEC is currently carried out on terminal alkenes whereas in this case the challenge laid in performing the reaction on an endocyclic carbon–carbon double bond. Only the free-radical addition of thioacetic acid to D-glucal was previously reported in 1970 [17]. This reaction was scarcely stereoselective as it afforded a mixture of two diastereomeric thioacetates, that is, one with the AcS group axial and the other with the AcS group equatorial, in 70:30 ratio. Thus, it was not surprising that the photoinduced reaction (λ_{max} 365 nm) between D-glucal triacetate **15** and peracetylated glucosyl thiol **1** (6 eq.) in the presence of DPAP produced the *S*-disaccharide diastereomers **16** and **17** in 57:43 ratio and 80% isolated overall yield (Scheme 3.4). Hence, the reaction was essentially lacking in diastereoselectivity while it appeared to be totally regioselective with the thiyl radical attacking exclusively the C2 of glucal.

It was suggested that the reaction proceeded through the formation of two anomeric glycosyl radical (AGR) intermediates stabilized by the adjacent oxygen atom, namely AGR_1 **19** in which the RS group (R = sugar) is axial and AGR_2 **20** in which the same group is equatorial (Scheme 3.5). In the present case these intermediates should

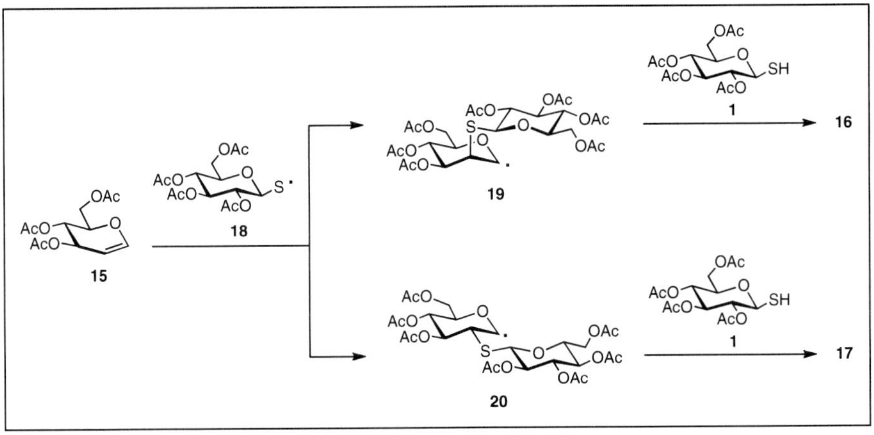

SCHEME 3.5 Stereochemistry of the addition of thiyl radical **18** to D-glucal **15**.

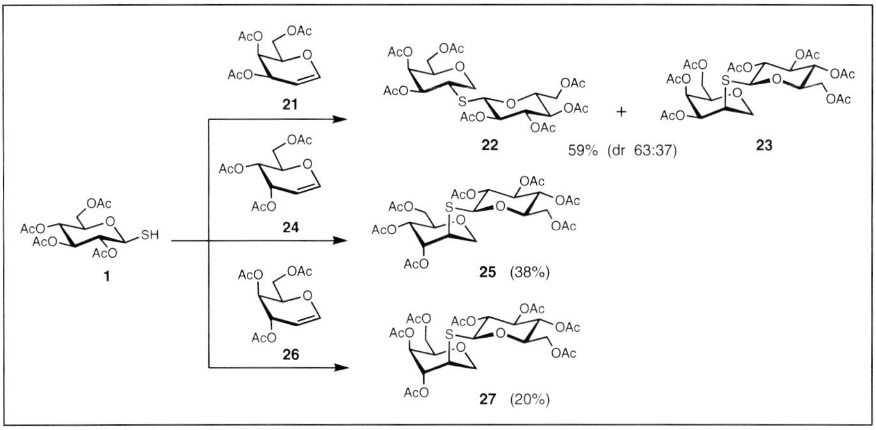

SCHEME 3.6 Photoinduced addition of glucosyl thiol **1** to three different glycals.

have comparable energies and consequently they lead to the products **16** and **17** in essentially equal amounts.

Some degree of stereoselectivity was instead observed in the addition of glucosyl thiol **1** to D-galactal triacetate **21** (Scheme 3.6). This reaction carried out under the above conditions afforded the diastereomeric *S*-disaccharides **22** and **23** in 63:37 ratio and 59% isolated overall yield. Thus, the preference for the formation of the derivative **22**, in which the RS group occupies the equatorial position, was ascribed to the shielding of the upper face of **21** by the OAc group at C4. Steric effects turned out to play a key role on the stereochemistry as well as the kinetic of the free-radical reaction when the OAc group at C3 of the glycal was axially configured. The reactions of **1** with the D-allal **24** and D-gulal **26** were totally stereoselective as they afforded exclusively diastereomers **25** and **27** although in fair (38%) and low (20%) yield, respectively (Scheme 3.6). In both cases a substantial amount of unreacted glycal (60–70%) was recovered. These findings indicated that in both cases there was a substantial decrease of reactivity and total inhibition to the equatorial attack of the thiyl radical at C2 due to the steric hindrance by the axial OAc group at C3; consequently, a trans-diaxial addition took place with the exclusive formation of a single AGR intermediate. Evidently, this stereochemical course was in contrast with the presence of the axial OAc group at C4 in D-gulal **27** and this explained the dramatic negative effect on reactivity.

Synthesis of thiodisaccharides **16** *and* **17**. To a solution of glucosyl thiol **1** (480 mg, 1.32 mmol) and D-glucal **15** (60 mg, 0.22 mmol) in CH_2Cl_2 (25 μL) and ethanol (500 μL) was added DPAP (34 mg, 0.13 mmol). The solution was irradiated (λ_{max} 365 nm) at room temperature for 1 hour under magnetic stirring, then concentrated. A solution of the residue in pyridine (2 mL) and Ac_2O (2 mL) was kept at room temperature for 2 h, then concentrated. The residue was eluted from a column of silica gel with 1:1 cyclohexane-AcOEt to give a mixture (195 mg) of **16** and **17** in a 57:43 ratio and 80% yield (^{1}H NMR analysis) contaminated by ca. 30% of the disulfide derived from thiol **1**. Analytical samples of **16,** $[\alpha]_D = -72.0$ (*c* 1.5, $CHCl_3$), and **17**,

$[\alpha]_D = -18.8$ (c 0.4, $CHCl_3$), were obtained by column chromatography purification (1.5:1 AcOEt-cyclohexane) of the above mixture.

In conclusion, it appeared that photoinduced TEC of sugar thiols with glycals was totally regioselective as it furnished only sulfides arising from the attack of the thiyl radical at C2 of the glycal. On the other hand, the reaction showed variable stereoselectivity and it afforded axially and equatorially substituted diastereomers in different ratios depending on the glycal stereochemistry at C3 and C4. Nevertheless, the synthetic value of this route to a new family of *S*-disaccharides appeared to have been sufficiently established.

3.3 CALIX[4]ARENE-BASED GLYCOCLUSTERS

It is firmly established that a myriad of key physiological and pathological events occur through interactions between carbohydrate and sugar-binding proteins called lectins [18]. The first step of influenza virus infection, for example, involves the viral hemagglutinin-mediated adhesion of the virion to sialic acid residues that are present on the surface of endothelial cells. A similar approach is exploited by bacteria and parasites. Such interactions, however, are usually weak, with K_D values of the order of mM for monosaccharides. However, when a set of multivalent saccharides is clustered together with the right structure and disposition, the interaction becomes strong and specific: multiple weak interactions result in strong adhesion. Notably, this augmented association is more than what would be expected on the basis of the increased carbohydrate local concentration: this phenomenon has been referred to as the glycoside cluster effect [19]. It is quite obvious that in order to intervene efficiently on these interactions, multivalent external agents have to be employed. In this context, one can envisage the use of glycoclusters firmly anchored to a conformationally rigidified platform such as a calix[4]arene, as suitable systems for protein recognition and consequently protein activity inhibition. This is especially important in those cases in which protein–carbohydrate interactions are detrimental such as viral and bacterial infections, inflammations, and cancer metastasis.

Following previous work from our laboratory in which calix[4]arene-based *O*- and *C*-glycoclusters were synthesized by different ligation strategies [20], the preparation of calix[4]arene-based *S*-glycoclusters via TEC was carried out. In all cases examined, the main problem to be addressed was to find suitable conditions so that four concomitant TECs occurred efficiently on the same calix[4]arene scaffold. This would avoid the formation of partially glycosylated products. Thus, the photoinduced reaction of tetra-allyl calix[4]arene **28** was carried out by using a significant excess of sugar thiol **1**, that is, 3 eq. per allyl group (Scheme 3.7) [21]. Under these conditions glycosylation of calix[4]arene **28** was complete to give the tetravalent calix[4]arene-based *S*-glycocluster **29** in 78% isolated yield. From this yield it can be calculated that the average yield of hydrothiolation of each allyl group of **28** was 95%, thus confirming the high efficiency of TEC. Finally, it was observed that the photoinduced reaction did not induce any change to the calixarene structure as the NMR spectrum of the product **29** showed that the macrocycle was in its original cone conformation.

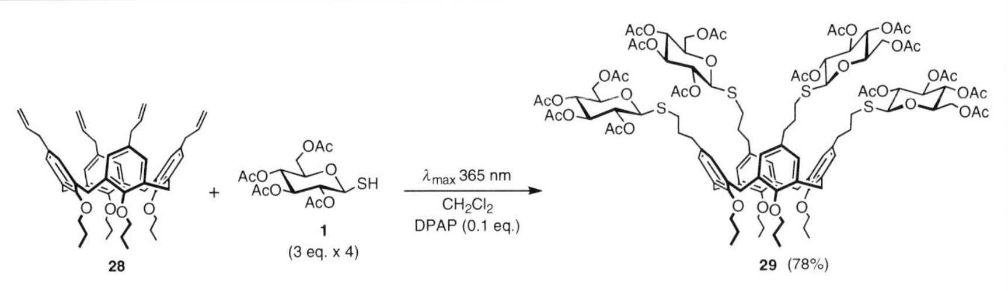

SCHEME 3.7 Glycoclustering on a calixarene platform via TEC.

Given the high efficiency of TEC in the multiple hydrothiolation of the tetra-allylated calix[4]arene **28**, we afforded the more challenging endeavor of installing tetravalent *S*-glucoside clusters at both the calix[4]arene upper and lower rims using the octa-allyl derivative **30** as a substrate reacting with the model sugar thiol **1** (Scheme 3.8) [21]. Also this reaction was carried out under standard conditions, that is, irradiation at λ_{max} 365 nm in the presence of the radical initiator DPAP and using 3 eq. of **1** per allyl group of **30**. The total hydrothiolation of **30** was demonstrated by the NMR analysis of the reaction mixture showing the absence of the signals in the range 5.8–6.4 ppm corresponding to the alkene protons. Accordingly the octavalent *S*-glucoside cluster **31** was isolated in 67% yield corresponding to a 95% average yield for each TEC.

Synthesis of the octavalent S*-glucoside cluster* **31**. To a solution of sugar thiol **1** (436 mg, 1.20 mmol) and calixarene **30** (37 mg, 0.05 mmol) in anhydrous CH_2Cl_2 (1 mL) was added DPAP (31 mg, 0.12 mmol). The solution was irradiated (λ_{max} 365 nm) at room temperature for 1 hour under magnetic stirring, then concentrated. The residue was treated with Ac_2O (1 mL) and pyridine (1 mL) at room temperature for 2 hours and then concentrated. The new residue was eluted from a column of silica gel with AcOEt-cyclohexane (from 3:1 to 1:0) to give **31** (122 mg, 67%) as a white foam; $[\alpha]_D = -44.7$ (*c* 0.7, $CHCl_3$).

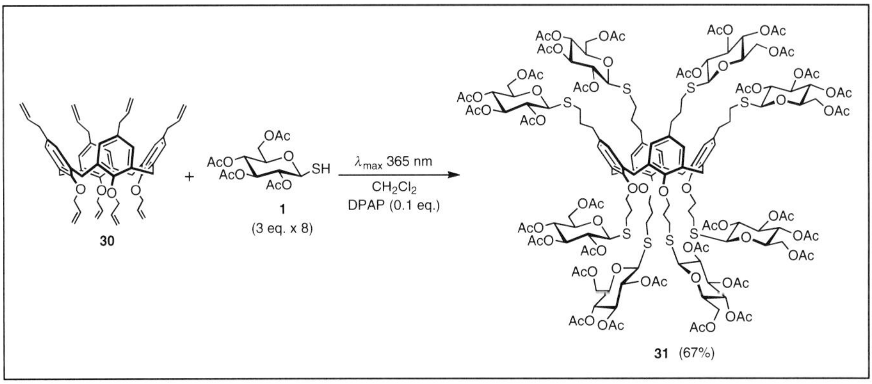

SCHEME 3.8 Glycoclustering at the upper and lower rims of calixarene platform via TEC.

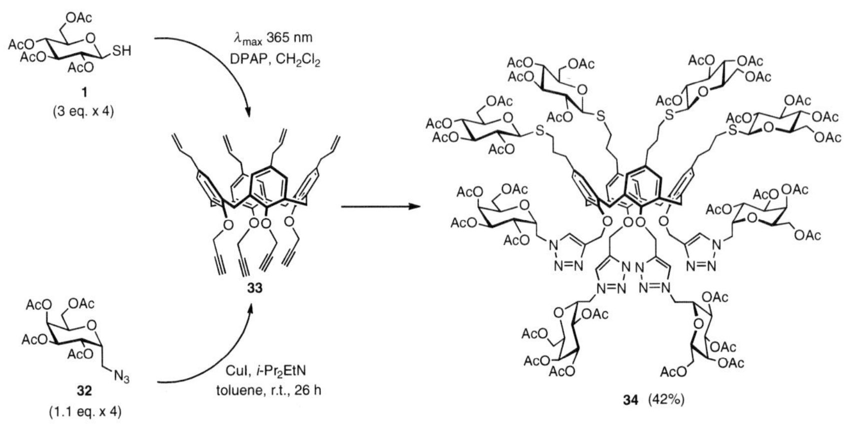

SCHEME 3.9 Upper and lower rim calixarene glycoclustering by sequential CuAAC and TEC.

As a final experiment in this topic, the dual clustering around the calix[4]arene scaffold by two different glycoside fragments was examined. Thus, it was intended to preparing a calix[4]arene-based glycocluster that could recognize two different lectins at the same time. To this aim, we considered the use of sequential CuAAC and TEC on an alkene and alkyne functionalized calix[4]arene derivative. The orthogonality of these two click processes had not been assessed in previous studies. According to this plan, the CuI-catalyzed reaction between galactosylmethyl azide **32** and tetra-*O*-propargyl functionalized calix[4]arene **33** was first carried out (Scheme 3.9) [21]. As expected, the reaction was totally chemoselective as it afforded an intermediate (not shown) featuring the triazole-linked *C*-glycocluster at the lower rim while it contained the unaltered set of *C*-allyl groups at the upper rim. Then, the photoinduced TEC between this intermediate and the β-D-glucosyl thiol **1** was carried out to give product **34**. This compound featured two different tetravalent glycoclusters attached to the lower and upper rims of the same calix[4]arene scaffold, one being constituted of galactose residues and the other of glucose residues. As galactose and glucose display an affinity for different lectins [22], **34** can play a double role in molecular recognition processes. Before closing this section, it is worth noting that CuAAC and TEC are fully orthogonal in the sequential approach employed while a reversal of the reaction sequence, that is, the occurrence of TEC first, is not possible because the potential consumption of the alkynyl groups by the concomitant TYC.

Synthesis of glycocluster **34**. A mixture of calixarene **33** (44 mg, 0.06 mmol), sugar azide **32** (102 mg, 0.26 mmol), freshly distilled *N,N*-diisopropylethylamine (210 μL, 1.20 mmol), CuI (11.4 mg, 0.06 mmol), and anhydrous toluene (1.20 mL) was sonicated in an ultrasound cleaning bath for 1 minute, then magnetically stirred in the dark at room temperature for 26 hours, diluted with AcOEt (ca. 20 mL), filtered through a pad of Celite to remove most of the copper salts, and concentrated. A solution of the resulting yellow syrup in pyridine (2 mL) and acetic anhydride (2 mL) was kept at room temperature for 3 hours, then concentrated. The residue was eluted from a column of silica gel with 2:1 AcOEt-cyclohexane, then AcOEt, to give the triazole-linked tetravalent *C*-glycocluster (77 mg, 56%) as syrup. To a solution of the latter glycocluster (35 mg, 0.015 mmol) and sugar thiol **1** (67 mg, 0.18 mmol) in anhydrous CH_2Cl_2 (0.5 mL) was added DPAP (4.5 mg, 0.018 mmol). The solution was irradiated (λ_{max} 365 nm) at room temperature for 1 hour under magnetic stirring, then concentrated. The residue was treated with Ac_2O (1 mL) and pyridine (1 mL) at room temperature for 2 hours and then concentrated. The residue was eluted from a column of silica gel with AcOEt, then 2:1 AcOEt-acetone, to give **34** (43 mg, 76%) as a syrup; $[\alpha]_D = +20.1$ (*c* 0.7, $CHCl_3$).

3.4 SILSESQUIOXANE-BASED GLYCOCLUSTERS

Cube-octameric silsesquioxanes (COSS, $R_8Si_8O_{12}$), more commonly referred to as polyhedral oligosilsesquioxanes (POSS) [23], are organic–inorganic hybrid materials that are attracting increasing interest due to their presentation of eight equivalent R groups in a globular structure. In this context, in 2004, Lee and coworkers described the photoinduced reaction of the commercially available octavinyl POSS **35** with *O*-mannosyl and *O*-lactosyl γ-thiobutyramides **36a,b** to give the corresponding octavalent glycoconjugates **37a,b** (Scheme 3.10) [24].

The rather harsh photochemical conditions (λ_{max} 254 nm) as well as the long reaction time (48 hours) employed by Lee stimulated our group to develop a TEC-based approach to silsesquioxane-based glycoclusters under mild conditions. It was also considered to use suitable sugar thiols to give more stable products in respect to the *O*-glycoside derivatives prepared by Lee. Initial coupling experiments between **35** and excess of a model sugar thiol 1-thio-β-D-glucose (up to 4 eq. per ene of **35**) under our standard photochemical conditions (λ_{max} 365 nm in the presence of DPAP as radical initiator) were unsatisfactory as only partially glycosylated POSS derivatives were obtained. This finding together with the observation that Lee employed thiols in which the sulfhydryl group was attached to the sugar residue through a linker,

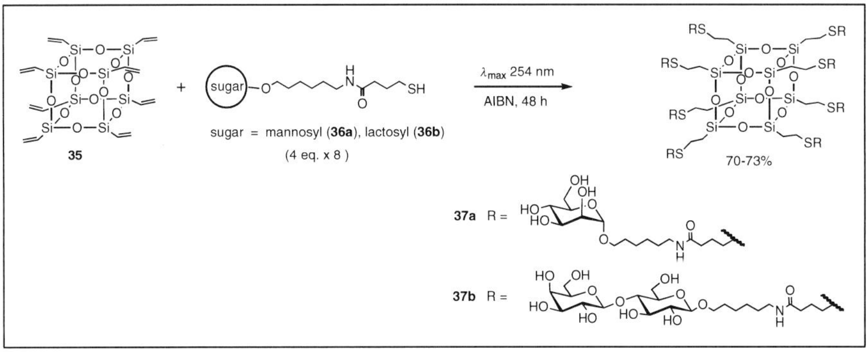

SCHEME 3.10 POSS-based glycoconjugate prepared by Lee via TEC.

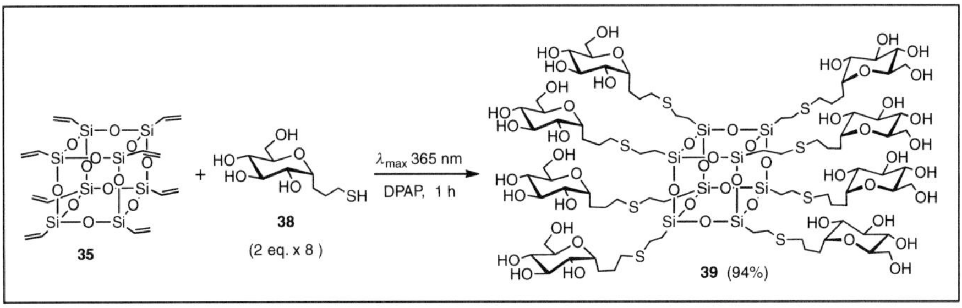

SCHEME 3.11 POSS-based *C*-glycoconjugate prepared by TEC.

suggested that the reaction between **1** and **35** was suffering of a substantial steric congestion. Therefore, it was set out to circumvent this obstacle by introducing a suitable tether holding the sulfhydryl group of the carbohydrate. This turned out to be a winning move. Indeed, the photoinduced reaction of octavinyl POSS **35** with propylthio α-D-*C*-glucopyranoside **38** in the presence of DPAP as initiator was complete in 1 hour to give the octavalent glycocluster **39** in 94% isolated yield (Scheme 3.11) [25]. Significantly, all the sugar residues of this glycocluster featured a *C*-glycosidic bond. Therefore, it can be foreseen that **39** is stable under a wide range of chemical environments.

Synthesis of C-*glycoconjugate* **39**. A solution of octavinyl POSS **35** (10 mg, 15.8 μmol), thiol **38** (60 mg, 0.25 mmol), and DPAP (6.5 mg, 25.3 μmol) in DMF (300 μL) and tetrahydrofuran (THF) (100 μL) was irradiated (λ_{max} 365) at room temperature for 1 hour under magnetic stirring and then concentrated. The residue was eluted from a column of Sephadex LH-20 with 3:1 MeOH-H_2O to give **39** (38 mg, 94%) as a syrup; $[\alpha]_D = +54.4$ (*c* 1.5, H_2O).

In a more challenging effort, the coupling of octavinyl POSS **35** with the alkyl-tethered sialic acid thiol **40** was carried out (Scheme 3.12) [25] as no examples of sialylation of POSS had been previously reported. There was special interest to perform the sialoconjugation in view of the potential viral inhibitory activity of the resulting POSS-based globular sialocluster as observed in the case of sialoclusters anchored to calixarene scaffolds [26]. Key in this experiment was the choice of a suitable solvent capable to dissolve all the reagents, including the hydrophobic silsesquioxane **35** and hydrophilic carbohydrate **40**. Thus, the photoinduced reaction was conducted in a 1:1 DMF–THF mixture to give the fully sialylated POSS conjugate **41**. Nevertheless, the yield of isolated **41** was rather modest (35%) due to the difficulty of handling such amphiphilic products. It has to be noted that this compound could not be fully characterized by MALDI-TOF spectrometry due to the failure to obtain the relevant spectrum under a variety of conditions.

In a second instance, a complementary approach was carried out in which suitable arms were introduced in the POSS platform. To this aim, the eight vinyl groups of POSS **35** were elongated by introducing a PEGylated linker bearing a terminal allyl group to give the PEG-linked octaene silsesquioxane **43** (Scheme 3.13) [25].

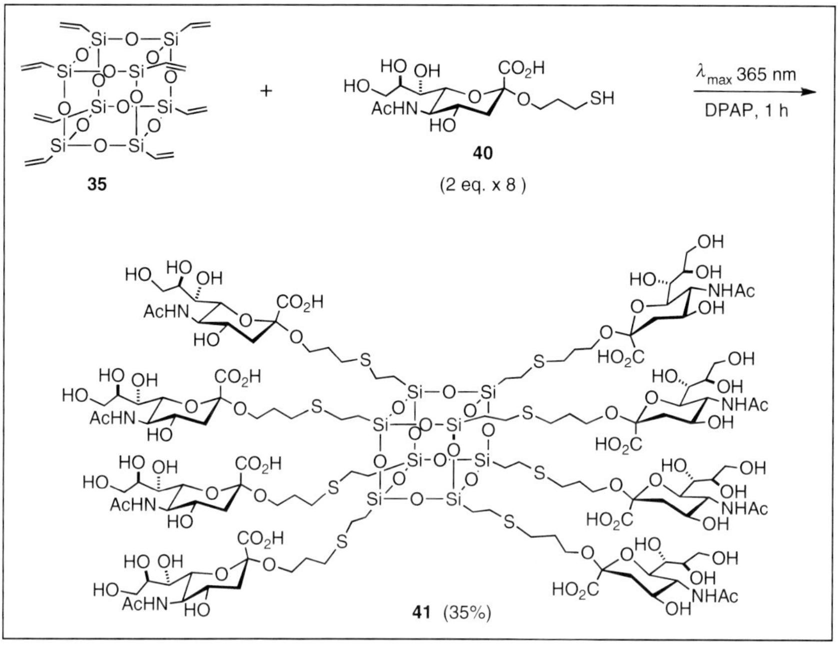

SCHEME 3.12 Synthesis of the POSS-based octavalent sialocluster **41** via TEC.

Synthesis of the PEG-linked octaene silsesquioxane **43**. A solution of octavinyl POSS **35** (80 mg, 126.4 μmol), 2-[2-(2-hydroxyethoxy)ethoxy]-1-ethanethiol (336 mg, 2.02 mmol), and DPAP (16 mg, 63.2 μmol) in DMF (1.4 mL) and THF (0.7 mL) was irradiated (λ_{max} 365) at room temperature for 45 minutes under magnetic stirring and then concentrated. The residue was eluted from a column of Sephadex

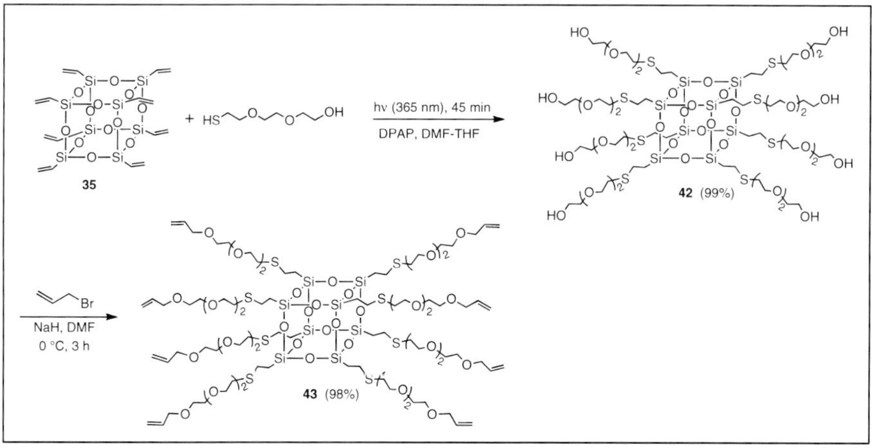

SCHEME 3.13 PEGylation of octavinyl POSS **35**.

LH-20 with 1:1 MeOH–H_2O to give **42** (245 mg, 99%) as a syrup. To a stirred, cooled (0°C) solution of **42** (49 mg, 25.0 μmol) in anhydrous DMF (2 mL) was added sodium hydride (NaH) (16 mg, 0.40 mmol, of a 60% dispersion in oil) and then allyl bromide (35 μL, 0.4 mmol). The mixture was stirred at 0°C for 3 hours, then diluted with 1 M phosphate buffer at pH 7 (0.5 mL), warmed to room temperature, diluted with H_2O (15 mL), and extracted with AcOEt (3 × 30 mL). The combined organic phases were dried (Na_2SO_4) and concentrated. The residue was eluted from a column of Sephadex LH-20 with MeOH to give **43** (56 mg, 98%) as a syrup.

Then, this new functionalized POSS was reacted with the model sugar thiol 1-thio-β-D-glucose **44,** that is the thiol bearing the sulfhydryl group directly linked to the anomeric carbon and which failed to give a complete hydrothiolation of **35**. Under the standard photoinduced reactions, that is, irradiation at λ_{max} 365 nm at room temperature and using a significant excess of **44**, the hydrothiolation of the POSS derivative **43** was complete to give the silsesquioxane-based octavalent glycocluster **45** (Scheme 3.14).

The same successful TEC on **43** was carried out with lactosyl and sialyl thiols to give the corresponding POSS-based glycoclusters **46a** and **46b** in good isolated yields (Fig. 3.4).

In conclusion, TEC turned out to be a valuable tool for performing eight concomitant hydrothiolation reactions by sugar thiols on two different types of octavalent alkene functionalized POSS. Conditions have been established to achieving the complete hydrothiolation of the POSS derivatives thus allowing for a single, totally glycosylated product to be isolated. The mild and neutral reaction

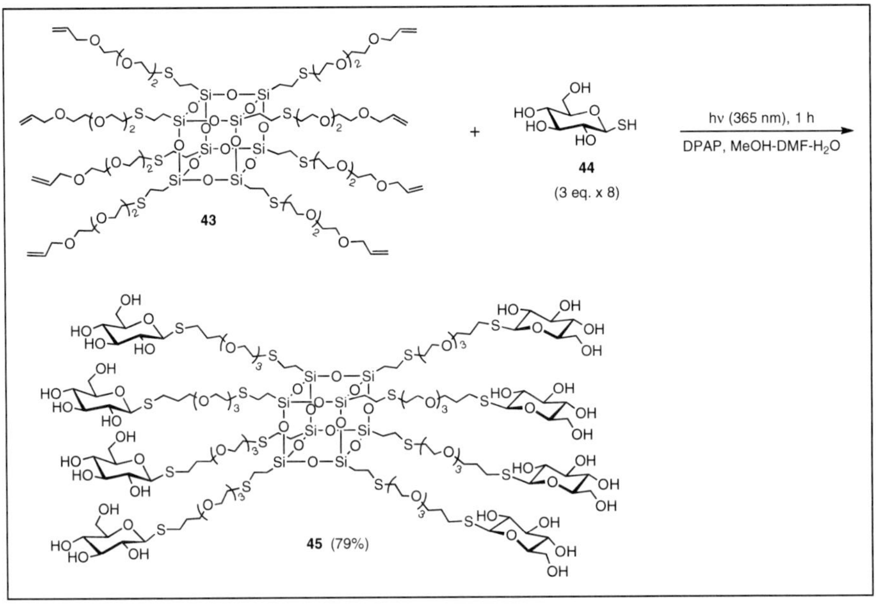

SCHEME 3.14 Glycoclustering on the PEGylated POSS **43** via TEC.

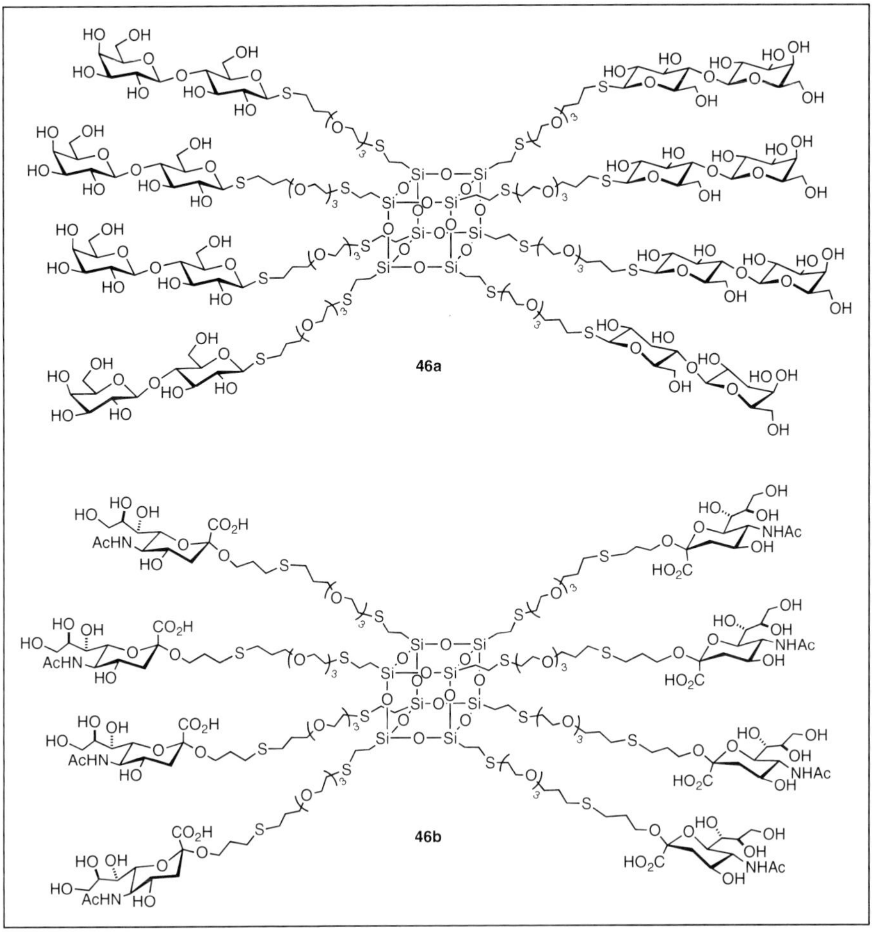

FIGURE 3.4 Other POSS-based glycoclusters derived from PEGylated POSS **43**.

conditions under which all TECs have been carried out leave unaltered the structure of the POSS moiety as evidenced by ^{29}Si NMR spectroscopy showing a sharp peak at -66 ± 1 ppm.

3.5 GLYCODENDRIMERS

Glycodendrimers are promising candidates for a number of applications, including their use in enantioselective catalysis [27], as glycoclusters for multivalent interaction with proteins such as lectins [28], and as hyperbranched glycomimetics of bioactive carbohydrates found as a part of the glycocalyx surrounding cells [29]. While Lindhorst and coworkers reported on glycodendron construction by a linear approach involving sequential reactions, one of which was in fact a free-radical thiol-ene

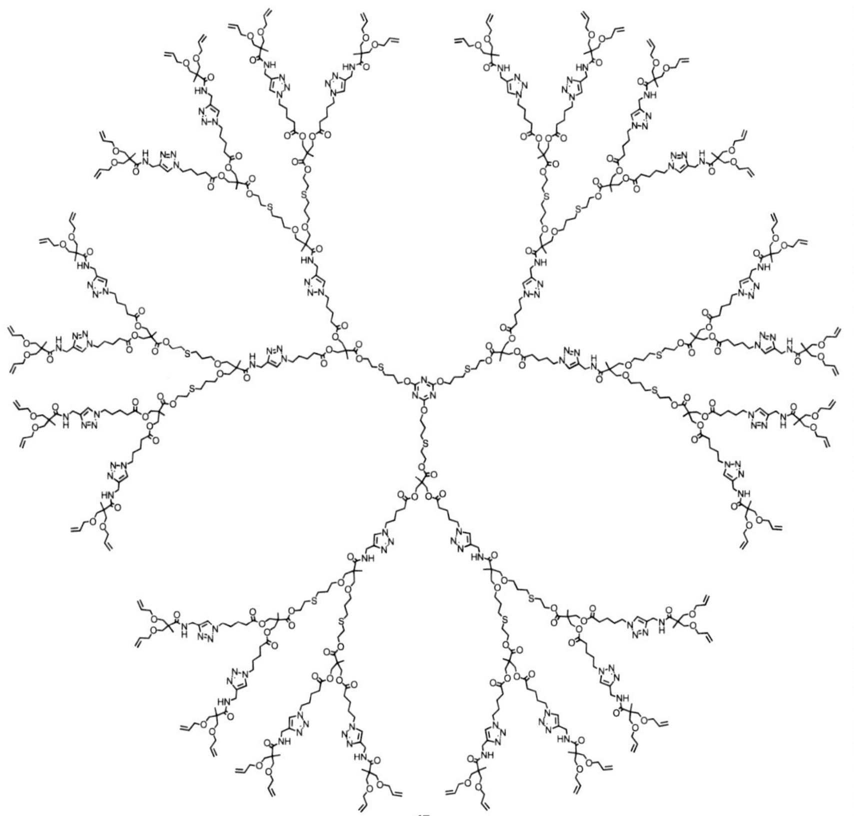

47

FIGURE 3.5 Alkene functional dendrimer [G4]ene$_{48}$ **47** prepared in the Hawker laboratory.

coupling [29, 30], we set out to develop an approach wherein concomitant TECs between sugar thiols and the alkene groups of a preformed dendrimer would take place in a single step [31]. In particular, as we planned to use the fourth generation alkene functional dendrimer [G4]ene$_{48}$ **47** prepared in the Hawker laboratory (Fig. 3.5) [32], the challenge laid in performing 48 regioselective and high–yielding TECs on the same substrate. It was hoped that each reaction was nearly quantitative because, for example, even a yield of 95% would give the multifunctionalized product in only $(0.95)^{48} = 8.5\%$ yield. Therefore, conditions were sought to avoid the formation of undersubstituted byproducts that usually prove to be chromatographically similar to the target compound. The first successful reaction was carried out using dendrimer **47** and excess of 1-thio-β-D-glucose **44** (4 eq./ene) in DMF as the solvent and under conditions of high substrate concentration (Scheme 3.15) [31]. After irradiation of the mixture at λ_{max} 365 nm for 1 h, ^{1}H NMR analysis revealed that all 48 ene groups of **47** were consumed. Accordingly, the glycodendrimer **48a** was obtained in excellent isolated yield (94%). The robustness of TEC also in this program was

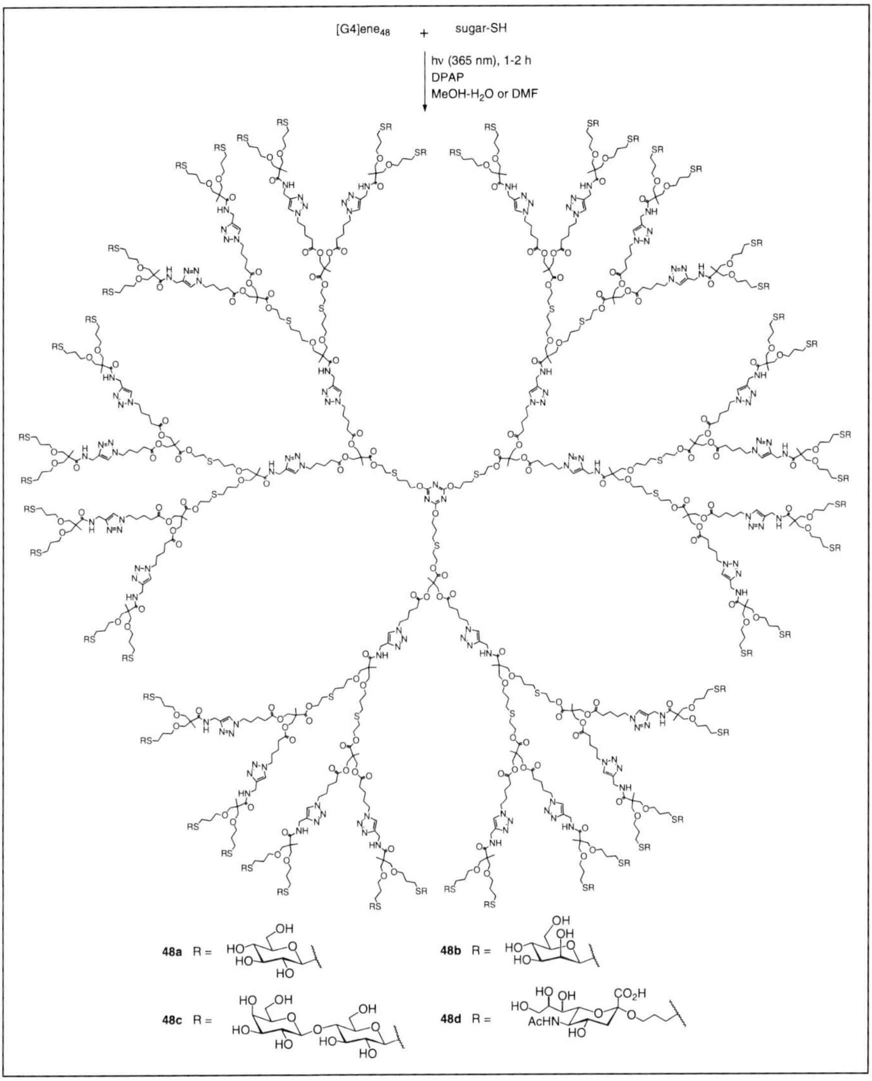

SCHEME 3.15 Glycodendrimers prepared via TEC between sugar thiols and alkene functional dendrimer [G4]ene$_{48}$ **47**.

validated by coupling **47** with three additional glycosyl thiols, namely 1-thio-β-D-mannose, 1-thio-β-D-lactose and an alkyl-tethered sialyl thiol. Also these reactions afforded the corresponding glycodendrimers **48b d** in nearly quantitative yields. In conclusion, this work represented one of the most striking examples of the power of TEC for glycoconjugation and paved the route for the attachment of other biomolecules to **47**. Indeed, the introduction of peptides in **47** was reported in the same paper [31].

Synthesis of glycodendrimer **48a**. A solution of dendrimer **47** (10 mg, 0.69 μmol), glucosyl thiol **44** (26 mg, 0.13 mmol), and DPAP (3.4 mg, 13.3 μmol) in DMF (70 μL) was irradiated (λ_{max} 365 nm) at room temperature for 1 hour under magnetic stirring and then concentrated. The residue was eluted from a column of Sephadex LH-20 (1 × 30 cm) with 1:1 H_2O–MeOH to give pure **48a** (15.5 mg, 94%) as a white foam; $[\alpha]_D = -6.2$ (*c* 1, H_2O).

3.6 AMINO ACID, PEPTIDE, AND PROTEIN GLYCOCONJUGATES

As more than 50% of proteins in human cells are glycosylated [33], it is now amply documented that carbohydrate residues profoundly affect protein folding, immunogenicity, and stability toward proteases as well as control biological properties and activities [34]. The latter functions operate mainly in cellular processes that are at the basis of vital biological events some of which are detrimental such as inflammation, cancer metastasis, viral and bacterial infections, while some others are beneficial such as immune response and fertilization. Studies to clarify the mechanism of such events and eventually control their occurrence are at the forefront of modern glycobiology. Unfortunately, the isolation of glycopeptides and glycoproteins with well-defined oligosaccharide arrays from natural sources is very difficult with current techniques due to the presence of various glycoforms and the labile *O*- and *N*-glycosidic bonds through which the carbohydrate residues are linked to the peptide/protein chain. As a consequence, synthetic non-natural glycopeptides and glycoproteins with a well-defined carbohydrate structure and composition remain in high demand [35]. These homogeneous materials may serve to improve our understanding of the role of the natural counterparts in biological events and can be interesting candidates for drug developments.

We entered in this widely investigated field by examining in a first instance the use of TEC in the preparation of *S*-glycosyl amino acids, that is, compounds in which the sugar and amino acid moieties are connected through a carbon–sulfur bond. These amino acids can be used for the assembly of synthetic *S*-linked glycopeptides [36], that is, products that are stable toward *O*- and *N*-glycosidases, the enzymes that promote the hydrolytic cleavage of *O*- and *N*-glycosidic bonds of natural compounds. Glycopeptides with these non-native linkages are suitable as probes for biochemical studies and leads in drug discovery such as, for example, the preparation of synthetic vaccines. While the classical method for the synthesis of *S*-glycosyl amino acids relies on the coupling of an activated sugar derivative with a protected cysteine residue [13], we envisaged the use of TEC as a modular approach that could allow the assembly of compounds with a tether of variable length holding the glycosyl residue and the glycinyl group [37]. This strategy is exemplified in Scheme 3.16 showing the photoinduced coupling of 2-acetamido-glucosyl thiol **49** with *N*-protected vinyl and allyl glycinate **50** and **51**. Both reactions proceeded readily upon irradiation at λ_{max} 365 nm in the presence of DPAP as a radical initiator and afforded the corresponding *S*-glycosyl amino acids **52** and **53** respectively without any loss of stereocenter

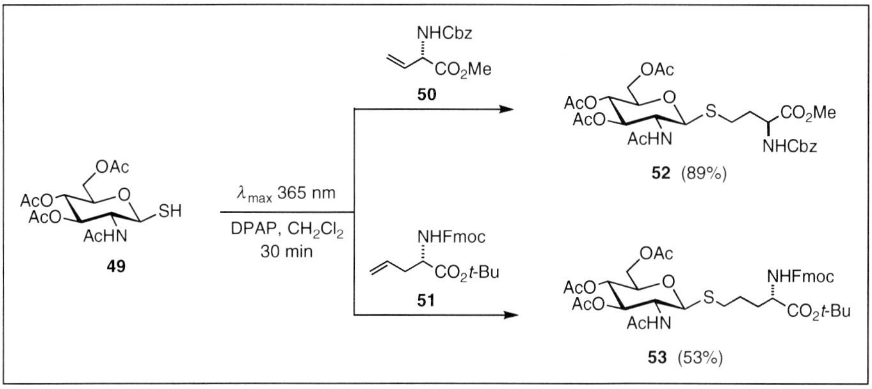

SCHEME 3.16 TEC-based synthesis of *S*-glycosyl amino acids.

integrity [37]. The scope of this approach was further demonstrated by the successful coupling of alkenyl glycines **50** and **51** with peracetylated glucosyl and lactosyl thiols.

Synthesis of S-glycosyl amino acids **52** *and* **53**. DPAP (0.01 mmol) was added to a solution of alkenyl glycines **50** or **51** (0.10 mmol) and glycosyl thiol **49** (0.12 mmol) in CH_2Cl_2 (1 mL). The solution was irradiated (λ_{max} 365 nm) at room temperature for 30 minutes under magnetic stirring and then concentrated. The residue was eluted from a column of silica gel (AcOEt-cyclohexane) to give the *S*-glycosyl amino acid **52** (89%), $[\alpha]_D = -12.2$ (*c* 0.9, $CHCl_3$), or **53** (53%), $[\alpha]_D = -47.5$ (*c* 1.2, $CHCl_3$).

Interestingly, one of the glycosyl amino acids thus obtained was used in the assembly of an *S*-linked glycopeptide via solution phase chemistry (Scheme 3.17). Succinctly, the *N*-protected glucosyl amino acid **54** was coupled with phenylalanine ethyl ester under basic conditions and in the presence of a condensing agent to give the *N*-Fmoc-protected dipeptide **55**. This was liberated of the *N*-Fmoc protective group by basic treatment and the free amine was condensed with *N*-Boc alanine under basic conditions and in the presence of a condensing agent. The pure final tripeptide **56** featuring a pendant glucoside residue linked through a robust thioether linkage was isolated by chromatography in 75% yield. Significantly, this compound turned out to be a single diastereomer by NMR analysis, thus indicating that all stereocenters of the reactants employed remained unaltered throughout the whole synthetic sequence.

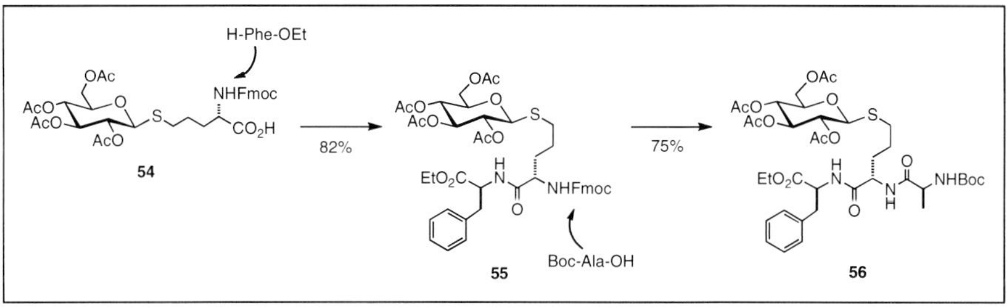

SCHEME 3.17 Assembly of a non-natural *S*-glycopeptide.

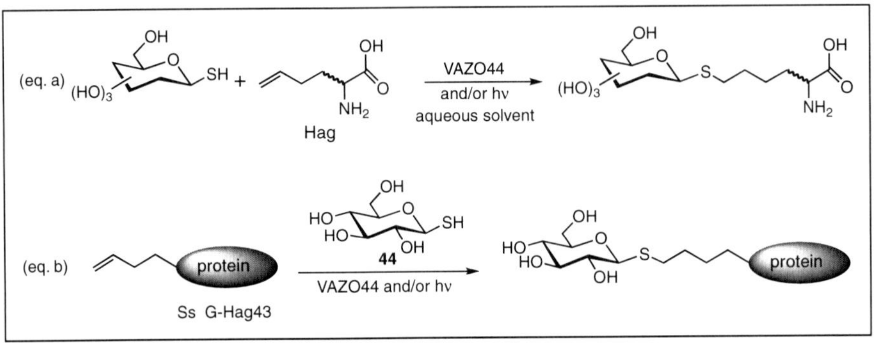

SCHEME 3.18 Davis glycosylation of Hag (eq. a) and Hag-containing protein (eq. b)

It is worth mentioning at this point that while our work described above was in progress, Davis and coworkers reported the synthesis of *S*-glycosyl amino acids via photoinduced addition of glycosyl thiols to homoallylglycine (Hag) (Scheme 3.18, equation a) [38]. This work was propaedeutic to the development of a convergent approach to *S*-glycosylated proteins via TEC. In fact, Hag was site-specifically introduced into a number of proteins by gene sequence design and the resulting alkenyl proteins were allowed to react with glycosyl thiols to give the target *S*-glycosylated proteins in almost quantitative yields (Scheme 3.18, equation b). Notably, as a demonstration of the power of TEC, the glycoconjugation reaction to a virus-like nanoparticle, obtained by self-assembly of the Qβ-(Hag16) protein, occurred with complete consumption of all 180 ene functional groups, thus establishing a record for a multiple hydrothiolation of the same substrate via TEC.

Given this knowledge, we were motivated to develop our own TEC and TYC-based strategies for post-synthetic glycosylation of peptides and proteins. To this aim we set out to follow two approaches both starting from cysteine-containing peptides and proteins (Scheme 3.19). One approach entailed the introduction of a chemical handle bearing an alkene tag followed by TEC between this group and a sugar thiol

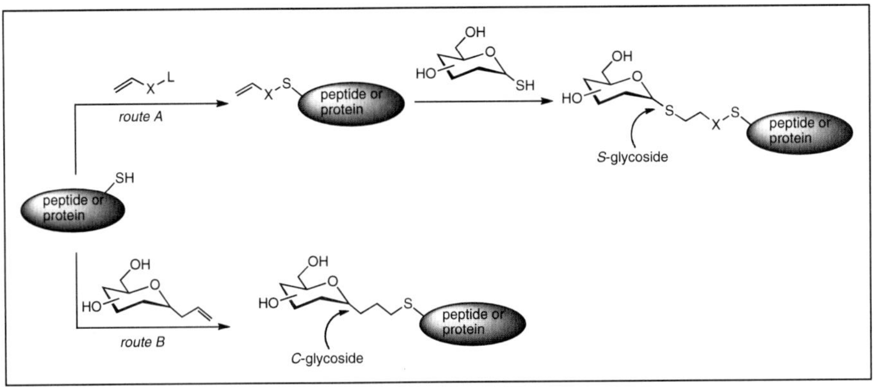

SCHEME 3.19 Routes toward post-synthetic peptide and protein glycosylation via TEC.

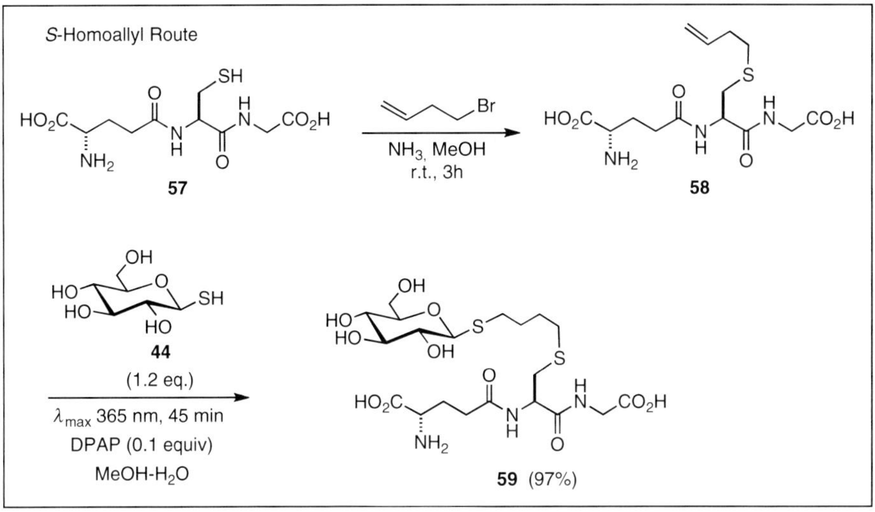

SCHEME 3.20 Post-translational approach to glutathione-based *S*-glycopeptide.

to give an *S*-linked glycoconjugate (route A). The other approach, instead, consisted of a single step involving TEC between the unmodified peptide or protein and an allyl *C*-glycoside to give in this case a *C*-linked glycoconjugate (route B).

In the first instance, we explored the viability of route A using the tripeptide glutathione Glu-Cys-Gly **57** as a cysteine-containing substrate (Scheme 3.20) [37]. To this end, we introduced a homoallyl chain in **57** by reaction with 4-bromo-1-butene in the presence of NH_3. Notably, this reaction proceeded readily and specifically to give exclusively *S*-butenyl glutathione **58** in almost quantitative yield due to the superior nucleophilicity of the cysteine sulfhydryl group with respect to the terminal amino group of glutamic acid. Then, the crude alkene-tagged **58** was submitted to the photoinduced reaction (λ_{max} 365 nm) with glycosyl thiol **44** in the presence of DPAP to give the *S*-glycoconjugate **59** in excellent isolated yield (97%). It has to be noted that *S*-butenylation of glutathione **57** was a key factor for the success of this approach. This functionalization was selected after some experimentation on cysteine as a model substrate. In fact, the *S*-allylation of cysteine by allyl bromide occurred readily and quantitatively but the photoinduced reaction of the *S*-allyl cysteine thus formed with the thiol **44** gave the expected thioether in very low yield together with several byproducts.

Synthesis of S-*glucosyl glutathione* **59**. NH_4OH (1 mL of a 28% solution in H_2O) and butenyl bromide (16 μL, 0.14 mmol) were added to a cooled (0°C), stirred solution of glutathione **57** (40 mg, 0.13 mmol) in MeOH (1 mL). The solution was stirred at 0°C for 1 hour, then warmed to room temperature, stirred for an additional 1 hour, and concentrated. A stirred solution of the crude **58**, glucosyl thiol **44** (30 mg, 0.15 mmol), and DPAP (4 mg, 0.015 mmol) in 1:2 $MeOH-H_2O$ (1.6 mL) was irradiated (λ_{max} 365 nm) at room temperature for 45 minutes and then concentrated. The residue was eluted from a column of Sephadex LH-20 with

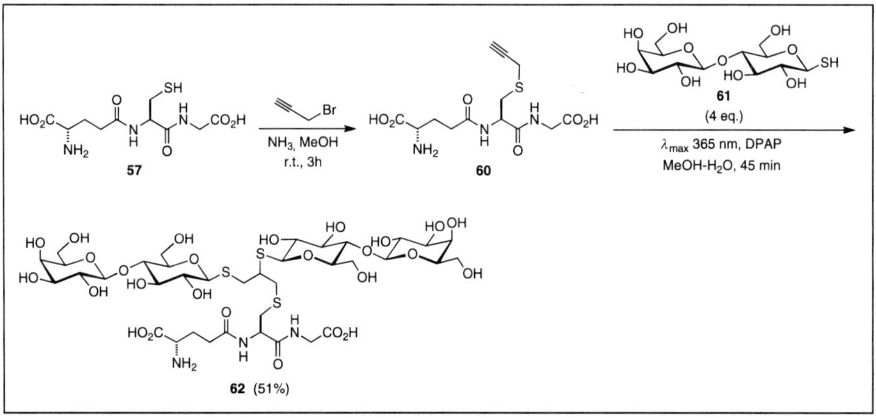

SCHEME 3.21 Glutathione dual glycosylation via thiol-yne coupling.

1:2 MeOH-H_2O to give **59** (70 mg, 97%) as an amorphous solid; $[\alpha]_D = +13.4$ (*c* 0.5, H_2O).

The same strategy was followed in the intent of introducing two sugar fragments on the same peptide via free-radical thiol-yne chemistry (TYC) according to the mechanism presented in Scheme 3.1. We thought that peptide double glycosylation could affect much more substantially than single glycosylation the peptide structure and biological activity. Also, in this case the whole process consisted of two operations; that is, peptide modification by introduction of a chemical handle bearing an alkynyl tag and then the execution of the photoinduced TYC with a sugar thiol. Suitable conditions for performing these two operations in one-pot sequence were established by using glutathione **57** as a substrate (Scheme 3.21) [39]. Succinctly, *S*-alkynylation of **57** was effectively carried out using propargyl bromide and NH_3. Then a mixture of the crude *S*-propargyl glutathione **60** and an excess of lactosyl thiol **61** (4 eq.) was irradiated (λ_{max} 365 nm) in the presence of catalytic DPAP. Also in this case the photoreaction was conducted in a glass vial at room temperature without caution to exclude air and moisture. Indeed, the double hydrothiolation took place with great efficiency with conversion >95% into the corresponding *S*-lactosyl peptide **62**. NMR analysis revealed that this product was actually a mixture of two diastereomers in ca. 1:1 ratio but no efforts were made for their separation. The same procedure was successfully followed for *S*-glycosylation of other cysteine-containing peptides including the synthetic tetrapeptide Arg-Gly-Asp-Cys and octapeptides Asp-Leu-Tyr-Cys-Tyr-Glu-Gln-Leu and Phe-Gly-Trp-Cys-Tyr-Lys-Leu-Val. The structures of the *S*-glycopeptides **63–66** thus obtained are shown in Figure 3.6. It is worth noting in this respect that the latter higher peptides contained basic lysine residues and, therefore, there was the possibility of nonselective propargylation. This event did not occur due to the superior nucleophilicity of the cysteine sulfhydryl group with respect to the basic nitrogen atoms of other amino acid residues.

Synthesis of glycopeptide **62**. NH_4OH (0.35 mL of a 28% solution in H_2O) and propargyl bromide (4.3 μL, 0.06 mmol) were added to a cooled (0°C), stirred solution

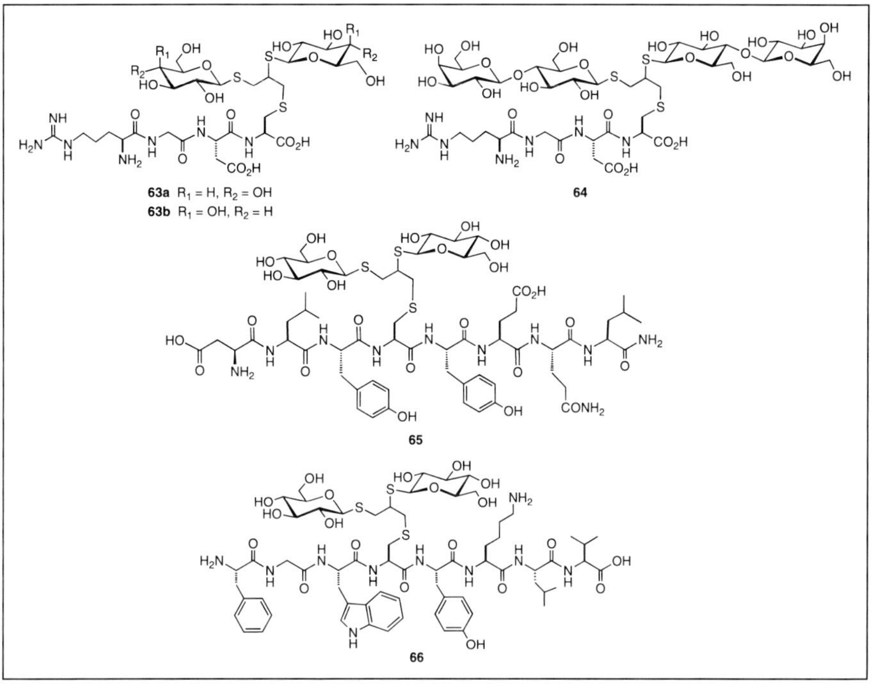

FIGURE 3.6 Dually glycosylated peptides prepared via TYC.

of glutathione **57** (17 mg, 0.055 mmol) in MeOH (0.35 mL). The solution was stirred at 0°C for 1 hour, then warmed to room temperature, stirred for an additional 2 hours, and then concentrated. A stirred solution of the crude product, lactosyl thiol **61** (79 mg, 0.22 mmol), and DPAP (5.5 mg, 0.022 mmol) in 2:1 MeOH-H_2O (1.6 mL) was irradiated (λ_{max} 365 nm) at room temperature for 45 minutes and then concentrated. The residue was eluted from a column of Sephadex LH-20 with MeOH to give **62** (mixture of diastereomers, 30 mg, 51%) as an amorphous solid.

As early experiments on bis-glycosylation of cysteine via TYC demonstrated that the vinyl thioether intermediate that formed in the early stage of the reaction (see mechanism in Scheme 3.1) showed some stability, the sequential hydrothiolation of alkynyl peptides by two different thiyl radicals generated from different thiols was examined. Thus, a solution of *S*-propargyl glutathione **60** in MeOH was treated with glucosyl thiol **44** (1.1 eq.) and sensitizer DPAP and the mixture was irradiated at λ_{max} 365 nm for 15 minutes (Scheme 3.22) [39]. The ^{1}H NMR spectrum showed the total consumption of the alkynyl derivative **60** (absence of the signal at 2.26 ppm corresponding to ≡CH) and the formation of the vinyl thioether intermediate as a mixture of *E* and *Z* isomers (presence of vinyl proton signals in the range 5.6–6.3 ppm). Then, a solution of biotin thiol **67** (2 eq.) in MeOH was added and the mixture was again irradiated at λ_{max} 365 nm for 45 minutes. The work-up of the reaction mixture and chromatography allowed the glycosylated and biotinylated

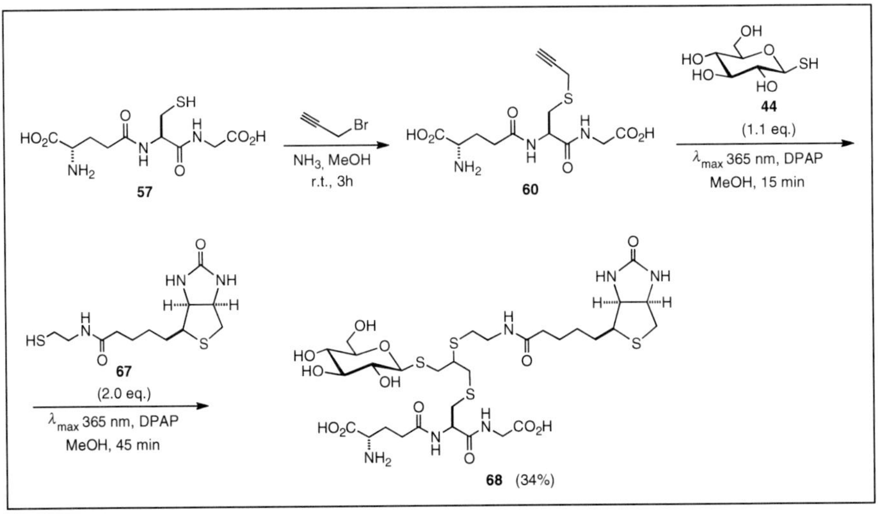

SCHEME 3.22 Glycosylation and biotinylation of *S*-propargyl glutathione via TYC.

dithioether **68** to be isolated in 34% yield. Thus, glycosylation and biotin labeling of glutathione **57** via TYC was demonstrated. This was an important result in view of the planned dual conjugation of proteins exposing a cysteine residue (see below).

Parallel to the development of route A, the viability of route B was also investigated (see Scheme 3.19). This route was quite attractive because it involved unmodified cysteine-containing peptides or proteins. After some experimentation with cysteine and glutathione **57**, TEC between the synthetic nonapeptide Thr-Ala-Leu-Asn-Cys-Asn-Asp-Ser-Leu **69** and a large excess of allyl *C*-galactoside **70** (30 eq.) was carried out (Scheme 3.23) [40]. As an innovative condition, this photoinduced reaction employed a mixture of DMSO and phosphate buffer (pH 7.4) as the solvent, that is, a nondenaturating aqueous reaction medium that was compatible with the planned protein glycosylation. Under these conditions, the chemoselective coupling between the cysteine sulfhydryl group of **69** and the allyl group of **70** took place quite efficiently to give the target *C*-glycopeptide **71**.

Next, glycosylation of a cysteine-containing protein was considered as a continuation of this program. To this end, the 66 kDa globular bovine serum albumin (BSA, **72**) was selected since this was known to contain a single free cysteine residue at position 34 (Scheme 3.24) [40]. Thus, the coupling of BSA **72** with allyl α-D-*C*-galactoside **70** was carried out under the optimized conditions established for glycosylation of nonapeptide **69.** The only change was that the reaction was complete only after 5 minutes. However, with great surprise, the MALDI-TOF analysis of the *C*-glycosylated protein **73** thus obtained showed that not one, as expected, but three galactoside residues had been introduced. Tryptic digestion and spectrometric analysis revealed that BSA **72** had reacted by using not only the sulfhydryl-free cysteine 34 but also the two sulfhydryl groups formed by cleavage of the 75–91 disulfide bridge. It was suggested that this cleavage was due to an electron transfer from

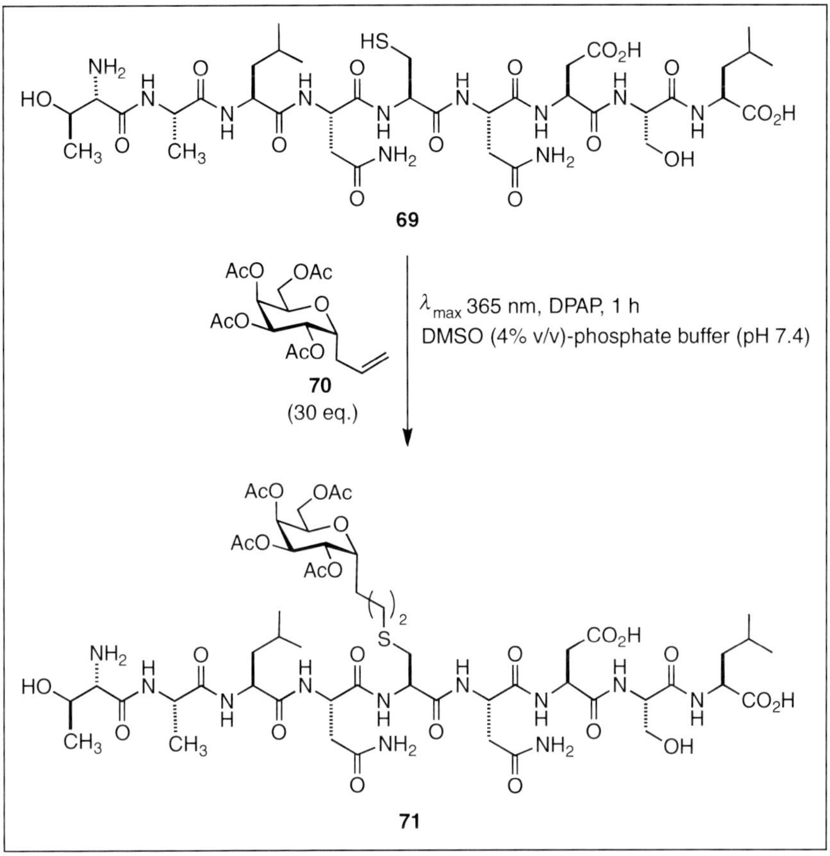

SCHEME 3.23 Glycosylation of cysteine-containing nonapeptide via TEC.

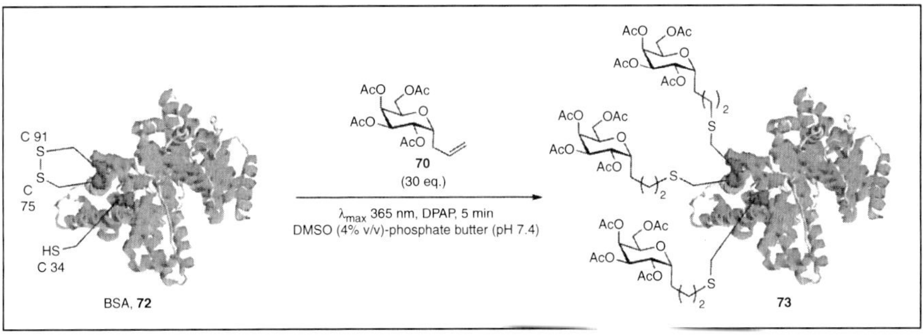

SCHEME 3.24 Multiple glycosylation of bovine serum albumin (BSA) via TEC.

photoexcited tryptophan residues present in BSA. Indeed, prolonged irradiation of the reaction mixture up to 2 hours induced the introduction of even seven glycoside residues. Nevertheless, this result was considered from a positive standpoint as it provided the first example of site-selective protein modification (hyperglycosylation) induced by a photocleavage reaction. A great advantage of this approach over other existing methods for protein glycoconjugation is that it can be applied to proteins in their *unmodified* native form without the prior introduction of any chemical tag. In conclusion, the Davis approach (see Scheme 3.18) and our own work presented above demonstrated the great potential of TEC for protein modification. The Davis method afforded protein glycoconjugates in which the sugar residues were linked to the protein by an *S*-glycosidic bond while our method furnished compounds with a *C*-glycosidic linkage. Both these flexible yet robust linkages should not perturb the topical characteristics of the protein while they ensure resistance toward enzymatic degradation.

Synthesis of the BSA glycoconjugate **73**. A solution of allyl *C*-galactoside **70** (18 mg, 47.3 μmol) and DPAP (2 mg, 7.8 μmol) in DMSO (0.4 mL) was slowly added to a solution of BSA **72** (104 mg, 1.58 μmol) in 20 mM phosphate buffer (pH 7.4, 10 mL). The mixture was irradiated (λ_{max} 365 nm) at room temperature for 5 minutes under magnetic stirring, then purified by using the high-performance Zeba Desalt Spin Column (0.5 mL) to give **73**. The buffer was exchanged to 6 M urea in 100 mM ammonium bicarbonate, pH 8.0.

In a second instance, the modification of BSA **72** via TYC was considered as another important goal. As it was established that TYC proceeded by a kinetic that allowed the dual hydrothiolation with two different thiols to be performed (see Scheme 3.22), we set out to modify native BSA **72** by introducing a sugar and a fluorescein fragment via TYC. The presence of fluorescein as a label would permit easy tracking of the glycosylated protein in biological assays. Thus, a one-pot procedure for the glycosylation and fluorolabeling of BSA **72** was established (Scheme 3.25) [41]. The first step comprised photoinduced coupling (λ_{max} 365 nm) of **72** with a large excess (33 eq.) of the alkynyl sugar **74** in the presence of catalytic DPAP. The reaction was carried out in DMSO-H_2O at pH 7.4 with an irradiation time of 5 minutes. These conditions were selected with the aim of achieving the total transformation of **72** into the postulated intermediate **75** featuring three vinyl thioether groups between three sugar residues and the protein. This formulation was advanced on the basis of the reasonable hypothesis that initial and fast coupling could take place between three units of alkynyl sugar **74** and the BSA sulfhydryl groups at position 34, 75, and 91, the latter two arising from the photocleavage of the cystine disulfide bridge. Thus, the crude reaction mixture was liberated of the excess of the alkynyl sugar **74** and then treated with a very large excess (160 eq.) of the fluorescein thiol **76**. To this new mixture, DPAP and fresh solvent (DMSO-H_2O) were added and the resulting solution was irradiated for 10 minutes. The product of this final reaction was isolated and dried and was assigned the structure **77** on the basis of MALDI-TOF analysis. Notably, as it was demonstrated that BSA reacted readily under photochemical conditions with alkynes using its cysteine/cystine residues, it was suspected that this protein could react analogously with strained cycloalkynes, that is, reagents that are used in the

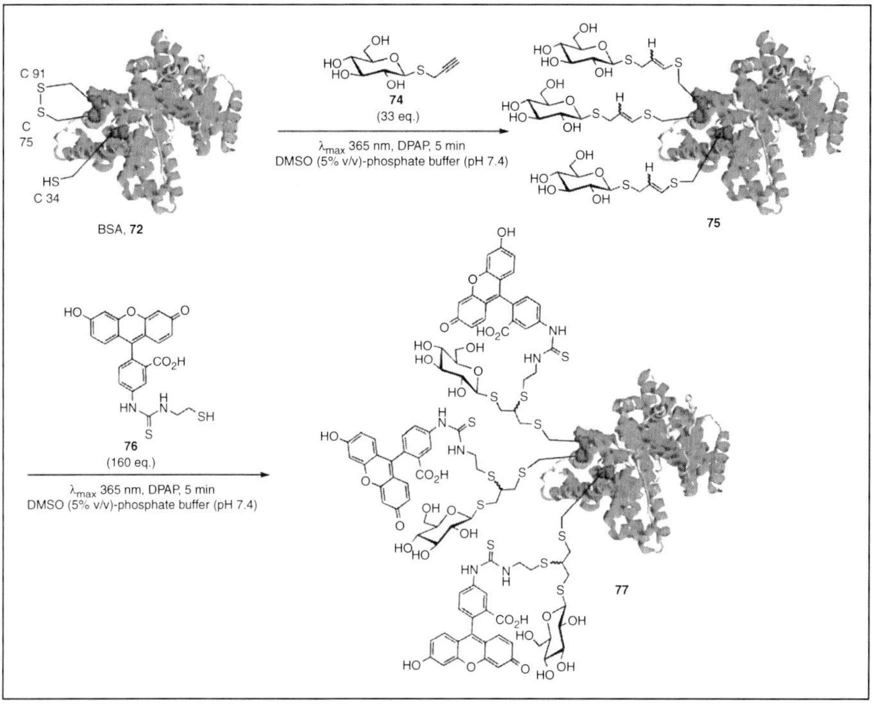

SCHEME 3.25 Glycosylation and fluorescein labelling of bovine serum albumin (BSA) via sequential TYC and TEC.

so-called copper-free SPAAC approach developed by Bertozzi, Boons [3] and their coworkers. Actually, this turned out to be the case. In fact, it was demonstrated that a dibenzocyclooctyne derivative reacted with BSA in the *presence* and also in the *absence* of light. It was inferred that this side reaction might constitute a limitation on the use of cycloalkynes as partners of SPAAC in *in vivo* experiments due to the presence of cysteine-containing proteins.

Synthesis of glycoprotein **77**. A solution of **74** (3.5 mg, 15.0 μmol) and DPAP (0.3 mg, 1.3 μmol) in DMSO (150 μL) was slowly added to a solution of commercial BSA **72** (30 mg, 0.46 μmol) in 20 mM phosphate buffer at pH 7.4 (3 mL). The mixture was irradiated (λ_{max} 365 nm) at room temperature for 5 minutes under magnetic stirring, then filtered twice through Ultrafree-MC microcentrifuge filter (nominal MW limit of 5000 Da) for 30 minutes at 5000 rpm. A solution of **76** (34 mg, 73 μmol) and DPAP (1.9 mg, 7.3 μmol) in DMSO (150 μL) was added to a solution of the protein glycoconjugate **75** in 20 mM phosphate buffer at pH 7.4 (3 mL). The mixture was irradiated again (λ_{max} 365 nm) at room temperature for 10 minutes under magnetic stirring, then filtered twice through Ultrafree-MC microcentrifuge filter (nominal MW limit of 5000 Da) for 30 minutes at 5000 rpm to give **77**.

On closing this section, it may be concluded that two methodologies, one based on TEC and the other on TYC, were established for the modification of BSA **72** at three different sites. Thus, multiple glycosylation and fluorolabeling of **72** have been carried out. Noteworthy, both methodologies employ the unmodified protein in its native form. Therefore, it can be deduced that a route has been paved toward the dual conjugation of proteins bearing one or more cysteine residues. Although free cysteine is relatively rare in protein sequences, in principle, modern molecular biology methods allow the introduction of this amino acid anywhere in a protein [42]. Alternatively, sulfhydryl groups can be introduced in natural proteins by a totally chemical method. This consists of the nucleophilic ring opening of the readily available *N*-acetylhomocysteine thiolactone by the ε-NH_2 groups of abundant lysine residues [43].

3.7 CONCLUSION

The above results taken from research carried out in our laboratory illustrate the efficiency of TEC and TYC in the manipulation of complex and delicate substrates such as carbohydrates, peptides, and proteins. The click status of TEC has been amply validated by the successful applications in a number of different fields. The reaction represents a metal-free click process that allows the assembly of complex substrates to be carried in an efficient and simple manner. In most of the cases examined the products were formed with nearly quantitative yields, while their isolation required chromatographic techniques. Unfortunately, this fails to meet one of the "click" criteria. Significantly, also the less exploited TYC is emerging as a valuable metal-free click ligation tool. For some aspects TYC is complementary to TEC due to its potential for the introduction of two different fragments on the same substrate. Moreover, given the numerous methods for terminal alkyne synthesis that have been made available for the execution of CuAAC, TYC can be considered a complementary click method for which a wide range of applications can be foreseen (Scheme 3.26).

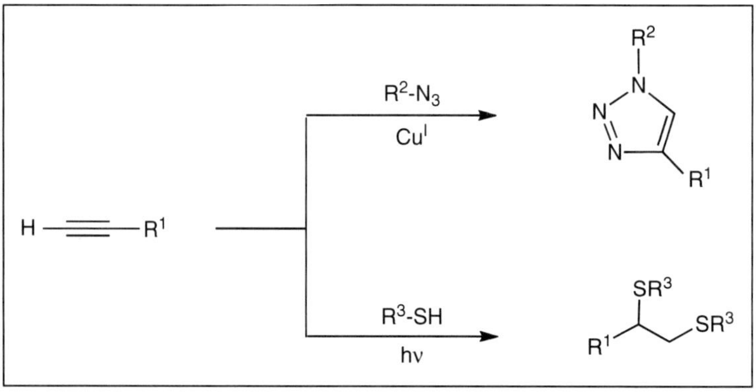

SCHEME 3.26 Click ligation tools with alkynes: CuAAC and TYC.

Note added in proof. A new polyhedral oligomeric silsesquioxane (POSS) derivative featuring eight PEGylated chains functionalized with terminal propargyl groups (POSS-yne) was synthesized and its photoinduced free-radical coupling with 2-acetamido-glucosyl thiol was carried out. This approach enabled the introduction of two sugar residues across each alkynyl group to give a globular hexadecavalent glycocluster installed on the cubic platform (see below). The binding properties of this glycocluster toward tetrameric wheat germ agglutinin (WGA) were determined by Enzyme-Linked Lectin Assay (ELLA). Notably, a measured IC_{50} at 2 nM was obtained from which the value of the relative potency per number of sugar unit (rp/n) was calculated to be close to 10^6. These results revealed the occurrence of a dramatic multivalent effect [44].

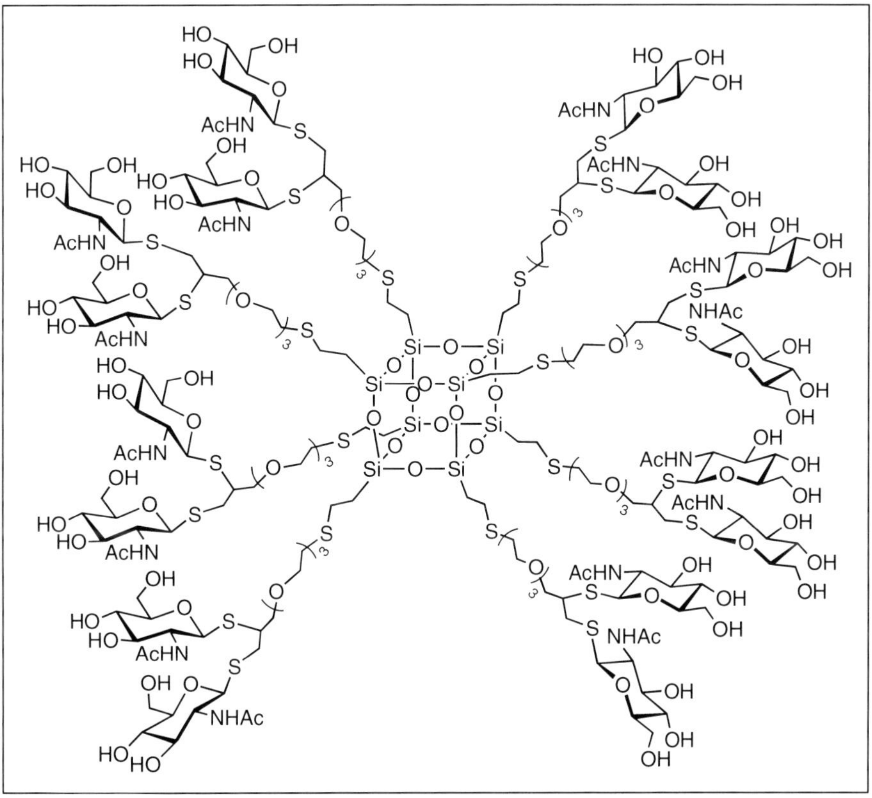

REFERENCES

1. Kolb, H. C.; Finn, M. G.; Sharpless, K. B. *Angew. Chem. Int. Ed.* **2001**, *40*, 2004–2021.
2. (a) Tornøe, C. W.; Christensen, C.; Meldal, M. *J. Org. Chem.* **2002**, *67*, 3057–3062. (b) Rostovtsev, V. V.; Green, L. G.; Fokin, V. V.; Sharpless, K. B. *Angew. Chem. Int. Ed.* **2002**, *41*, 2596–2599.

3. (a) Agard, N. J.; Prescher, J. A.; Bertozzi, C. R. *J. Am. Chem. Soc.* **2004**, *126*, 15046–15047. (b) Baskin, J. M.; Bertozzi, C. R. *Aldrichimica. Acta* **2010**, *43*, 15–23. (c) Ning, X. H.; Guo, J.; Wolfert, M. A.; Boons, G. J. *Angew. Chem. Int. Ed.* **2008**, *47*, 2253–2255. (d) Boons, G.J. *Carbohydrate Chemistry: Chemical and Biological Approaches*; Lindhorst, T. K., Reuter, A. P, Eds.; RSC: London, 2010; Vol. 36, pp 152–167.
4. Becer, C. R.; Hoogenboom, R.; Schubert, U. S. *Angew. Chem. Int. Ed.* **2009**, *48*, 4900–4908.
5. (a) Posner, T. *Chem. Ber.* **1905**, *38*, 646–657. (b) Griesbaum, K. *Angew. Chem. Int. Ed.* **1970**, *9*, 273–287.
6. Bader, H.; Cross, L. C.; Heilbron, I.; Jones, E. R. H. *J. Chem. Soc.* **1949**, 619–623.
7. (a) Dondoni, A. *Angew. Chem. Int. Ed.* **2008**, *47*, 8995–8997. (b) Dondoni, A.; Marra, A. *Chem. Soc. Rev.* **2012**, *41*, 573–586.
8. (a) Hoyle, C. E.; Bowman, C. N. *Angew. Chem., Int. Ed.* **2010**, *49*, 1540–1573. (b) Lowe, A. B. *Polym. Chem.* **2010**, *1*, 17–36.
9. (a) Hoogenboom, R. *Angew. Chem., Int. Ed.* **2010**, *49*, 3415–3417. (b) Lowe, A. B.; Hoyle, C. E.; Bowman, C. N. *J. Mater. Chem.* **2010**, *20*, 4745–4750.
10. (a) Varki, A. *Glycobiology.* **1993**, *3*, 97–130. (b) Lee, Y. C.; Lee R. T., *Acc. Chem. Res.* **1995**, *28*, 321–327. (c) Dwek, R. A. *Chem. Rev.* **1996**, *96*, 683–720. (d) Bertozzi, C. R.; Kiessling, L. L. *Science* **2001**, *291*, 2357–2364. (e) Rudd, P. M.; Elliott, T.; Cresswell, P.; Wilson, I. A.; Dwek, R. A. *Science* **2001**, *291*, 2370–2376.
11. (a) Witczak, Z. J. *Curr. Med. Chem.* **1995**, *1*, 392–405. (b) Sears, P.; Wong, C.-H. *Chem. Commun.* **1998**, 1161–1170. (c) Lis, H.; Sharon, N. *Chem. Rev.* **1998**, *98*, 637–674.
12. (a) Postema, M. H. D. *C-Glycoside Synthesis*, CRC Press: Boca Raton, **1995**; pp 379. (b) Liu, L.; McKee, M.; Postema, M. H. D. *Curr. Org. Chem.* **2001**, *5*, 1133–1167.
13. Pachamuthu, K.; Schmidt, R. R. *Chem. Rev.* **2006**, *106*, 160–187.
14. Bundle, D. R.; Rich, J. R.; Jacques, S.; Yu, H. N.; Nitz, M.; Ling, C.C. *Angew. Chem. Int. Ed.* **2009**, *48*, 7798–7802.
15. Fiore, M.; Marra, A.; Dondoni, A. *J. Org. Chem.* **2009**, *74*, 4422–4425.
16. Staderini, S.; Chambery, A.; Marra, A.; Dondoni, A. *Tetrahedron Lett.* **2012**, *53*, 702–704.
17. Igarashi, K.; Honma, T. *J. Org. Chem.* **1970**, *35*, 606–610.
18. (a) Marth, J.; Lowe, B.; Varki, A. *Essentials of Glycobiology*; Varki, A., Cummings, R. D., Esko, J. D., Freeze, H. H., Hart, G. W., Marth, J., Eds.; Cold Spring Harbor Laboratory Press: New York, **1999**. pp 101–113 and 253–266. (b) Sears, P.; Wong, C-H. *Cell. Mol. Life Sci.* **1988**, *54*, 223–252. (c) Gabius, H. J.; André, S.; Kaltner, H.; Siebert, H. C. *Biochim. Biophys. Acta* **2002**, *1572*, 165–177.
19. (a) Lee, Y. C.; Lee, R. T. *Acc. Chem. Res.* **1995**, *28*, 321–327. (b) Quesenberry, M. S.; Lee, R. T.; Lee, Y. C. *Biochemistry.* **1997**, *36*, 2724–2732. (c) Lundquist, J. J.; Toone, E. J. *Chem. Rev.* **2002**, *102*, 555–578. (d) Chabre, Y. M.; Roy, R. *Adv. Carbohydr. Chem. Biochem.* **2010**, *63*, 165–393.
20. Dondoni, A.; Marra, A. *Chem. Rev.* **2010**, *110*, 4949–4977.
21. Fiore, M.; Chambery, A.; Marra, A.; Dondoni, A. *Org. Biomol. Chem.* **2009**, *7*, 3910–3913.
22. Ambrosi, M.; Cameron, N. R.; Davis, B. G. *Org. Biomol. Chem.* **2005**, *3*, 1593–1608.
23. (a) Baney, R. H.; Itoh, M.; Sakakibara, A.; Suzuki, T. *Chem. Rev.* **1995**, *95*, 1409–1430. (b) Cordes, D. B.; Lickiss, P. D.; Rataboul, F. *Chem. Rev.* **2010**, *110*, 2081–2173.

24. Gao, Y.; Eguchi, A.; Kakehi, K.; Lee, Y. C. *Org. Lett.* **2004**, *6*, 3457–3460.

25. Lo Conte, M.; Staderini, S.; Chambery, A.; Berthet, N.; Dumy, P.; Renaudet, O.; Marra, A.; Dondoni, A. *Org. Biomol. Chem.* **2012**, *10*, 3269–3277.

26. Marra, A.; Moni, L.; Pazzi, D.; Corallini, A.; Bridi, D.; Dondoni, A. *Org. Biomol. Chem.* **2008**, *6*, 1396–1409.

27. (a) Schmitzer, A.; Perez, E.; Rico-Lattes, I.; Lattes, A. *Tetrahedron Lett.* **1999**, *40*, 2947–2950. (b) Schmitzer, A.; Perez, E.; Rico-Lattes, I.; Lattes, A. *Tetrahedron Asymmetry* **2003**, *14*, 3719–3730.

28. Lis, H.; Sharon, N. *Chem. Rev.* **1998**, *98*, 637–674.

29. Heidecke, C. D.; Lindhorst, T. K. *Chem. Eur. J.* **2007**, *13*, 9056–9067.

30. Kieburg, C.; Dubber, M.; Lindhorst, T. K. *Synlett.* **1997**, 1447–1449.

31. Lo Conte, M.; Robb, M. J.; Hed, Y.; Marra, A.; Malkoch, M.; Hawker, C. J.; Dondoni, A. *J. Polym. Sci., Part A: Polym. Chem.* **2011**, *49*, 4468–4475.

32. Antoni, P.; Robb, M. J.; Campos, L. Montañez, M.; Hult, A.; Malmström, E.; Malkoch, M.; Hawker, C. J. *Macromolecules* **2010**, *43*, 6625–6631.

33. Apweiler, R.; Hermjakob, H.; Sharon, N.; *Biochim. Biophys. Acta* **1999**, *1473*, 4–8.

34. (a) Seitz, O. *ChemBioChem.* **2000**, *1*, 214–246. (b) Wong C. H. *J. Org. Chem.* **2005**, *70*, 4219–4225.

35. Pratt, M. R.; Bertozzi, C. R. *Chem. Soc. Rev.* **2005**, *34*, 58–68.

36. Marcaurelle, L. A.; Bertozzi, C. R. *Chem. Eur. J.* **1999**, *5*, 1384–1390.

37. Fiore, M.; Lo Conte, M.; Pacifico, S.; Marra, A.; Dondoni, A. *Tetrahedron Lett.* **2011**, *52*, 444–447.

38. Floyd, N.; Vijayakrishnan, B.; Koeppe, J. R.; Davis, B. G. *Angew. Chem., Int. Ed.* **2009**, *48*, 7798–7802.

39. Lo Conte, M.; Pacifico, S.; Chambery, A.; Marra, A.; Dondoni, A. *J. Org. Chem.* **2010**, *75*, 4644–4647.

40. Dondoni, A.; Massi, A.; Nanni, P.; Roda, A. *Chem. Eur. J.* **2009**, *15*, 11444–11449.

41. Lo Conte, M.; Staderini, S.; Marra, A.; Sanchez-Navarro, M.; Davis, B. G.; Dondoni, A. *Chem. Commun.* **2011**, *47*, 11086–11088.

42. (a) Bontempo, D.; Heredia, K. L.; Fish, B. A.; Maynard, H. D. *J. Am. Chem. Soc.* **2004**, *126*, 15372–15373. (b) Heredia, K. L.; Maynard, H. D. *Org. Biomol. Chem.* **2007**, *5*, 45–53. (c) Chalker, J. M.; Bernardes, G. J. L.; Lin, Y. A.; Davis, B. G. *Chem. Asian J.* **2009**, *4*, 630–640. (d) Li, M.; De, P.; Li, H.; Sumerling, B. S. *Polym. Chem.* **2010**, *1*, 854–859.

43. Benesch, R.; Benesch, R. E. *Proc. Natl. Acad. Sci. USA* **1958**, *44*, 848–853.

44. Marra, A.; Staderini, S.; Berthet, N.; Dumy, P.; Renaudet, O.; Dondoni, A. *Eur. J. Org. Chem.*, DOI: 10.1002/ejoc.201201453.

PART III

CARBOHYDRATE CLICK CHEMISTRY FOR NOVEL SYNTHETIC TARGETS

4

THE DEVELOPMENT AND APPLICATION OF CLICKABLE LIPID ANALOGS FOR ELUCIDATING AND HARNESSING LIPID FUNCTIONS

MICHAEL D. BEST

4.1 INTRODUCTION TO THE ADVANTAGES OF APPLYING CLICK CHEMISTRY FOR STUDYING LIPIDS

In many respects, lipids represent overachieving constituents within the context of complex biological systems. From what initially appeared to be relatively simple structures, it is now apparent that lipids govern numerous critical biological processes. For instance, lipids directly regulate the function and localization of a litany of proteins, and in doing so, control vital physiological pathways via roles including protein–membrane-binding events [1,2] and the covalent lipidation of protein targets [3]. In addition, lipids control the transport of materials via vesicle budding and fusion [4]. Furthermore, advancements resulting from studies of lipid biological activities have fueled interest in pursuing applications that exploit these properties. For example, lipids have been heavily pursued for a wide range of endeavors such as the delivery of drugs, genes, and imaging agents [5–8], as well as the development of model systems that can be used to mimic the native membrane environments of lipids [9–12]. As such, the interest in understanding and harnessing the power of lipids has grown considerably over the years. However, these efforts have been met with a number of technical challenges that result from the nature of lipids, such as

Click Chemistry in Glycoscience: New Developments and Strategies, First Edition.
Edited by Zbigniew J. Witczak and Roman Bielski.

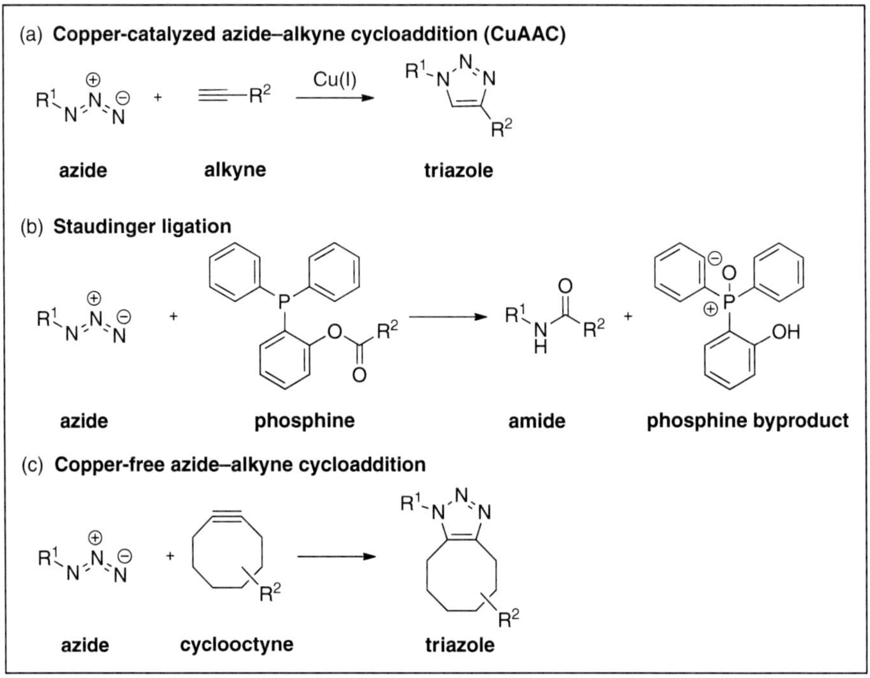

FIGURE 4.1 Bioorthogonal reactions that have been developed for the selective derivatization of tagged biomolecules.

the massive diversity of molecules that arise from subtle variations in structure, and the complexities associated with the environments of cellular membranes.

In order to overcome these obstacles for understanding and mimicking lipid function, the concept of click chemistry has proven to be an invaluable tool [13]. Click chemistry has been the subject of numerous recent reviews [14–17], including those focused on specific areas that this chemistry has impacted, such as bioorthogonal labeling studies [18–20], medicinal chemistry [17, 21, 22], supramolecular chemistry [23], and materials science [24, 25]. Therefore, we will only briefly introduce the basic principles of this topic herein. In 2002, Sharpless and coworkers outlined the strategy of click chemistry, which involved the development of reactions that would proceed rapidly in high yields under ambient conditions to a single desired product through the reaction of readily available functional groups, and in doing so tolerate the presence of a range of functional groups on the starting materials [26]. The prototypical click reaction was the copper-catalyzed azide–alkyne cycloaddition (CuAAC, Fig. 4.1a)—a modification to the original 1,3-dipolarcycloaddition reported by Huisgen [27]—that was reported concurrently by both Sharpless and coworkers [28], as well as Meldal and coworkers [29].

One of the key advantages of CuAAC is that the azide and alkyne reactive partners employed for this reaction are orthogonal to biological systems, since these functional groups show minimal reactivity with or presence within biological systems. Thus, this chemistry allows for the selective labeling of tagged biomolecules within complex biological samples, a powerful approach referred to as bioorthogonal labeling [18–20]. Other reactions are also effective for labeling within complex samples, and indeed the Staudinger ligation reaction, reported by Bertozzi and coworkers prior to the initial reports of CuAAC, accomplished this goal through the selective reaction of azides with functionalized phosphines to ultimately produce an amide linkage through the trapping of an aza-ylide intermediate (Fig. 4.1b) [30–36]. While CuAAC exhibits many beneficial properties, the original conditions of this reaction led to problems, particularly during bioorthogonal labeling, including the toxicity of the copper catalyst within living systems as well as side products resulting from the oxidation of biomolecules [37]. However, these issues have been minimized through the development of copper-chelating ligands that sequester the copper catalyst [38–43]. In addition, Bertozzi and coworkers also developed the copper-free click chemistry variation, which exploited a cyclooctyne moiety for a strain-promoted cycloaddition with azide that circumvented the need for copper catalyst (Fig. 4.1c) [44–53].

Due to these beneficial properties, the CuAAC, Staudinger ligation, and copper-free click reactions have each been applied to undertake a wide array of studies and applications to elucidate and exploit the biological properties of lipids. Herein, we will provide an overview of these studies, including approaches to the derivatization of intact membranes, generation of functionalized lipid structures, *in situ* labeling and imaging of lipid probes within complex biological systems, and activity-based labeling and characterization of protein–lipid-binding interactions and covalent lipidation. These approaches show strong promise for the efficient elucidation and harnessing of lipid properties.

4.2 APPLICATION OF CLICK CHEMISTRY FOR THE FUNCTIONALIZATION OF MEMBRANE SURFACES

The ability to functionalize intact membrane assemblies in a controlled manner has generated significant interest due to the wide range of applications that can be achieved using these systems. This includes the applicability of membranes to the delivery of drugs, genes, and imaging agents [5–8] since assemblies such as liposomes represent biologically inert containers that can be used to carry encapsulated cargo to desired locations, which can also be followed by content release. Functionalized membranes are also used as model systems for the study of lipids in a biologically relevant context [9–12]. Click chemistry has recently been explored as a bioorthogonal solution to circumvent the challenges associated with the selective labeling of membranes within the context of complex biological systems (Fig. 4.2).

In 2006, Schuber and coworkers implemented CuAAC to introduce mannose moieties onto the surface of preformulated liposomes [54]. This was performed through the design and synthesis of **1**, containing an alkyne linked to a glycerolipid backbone via a tetraethylene glycol spacer (Fig. 4.3a). This lipid was incorporated

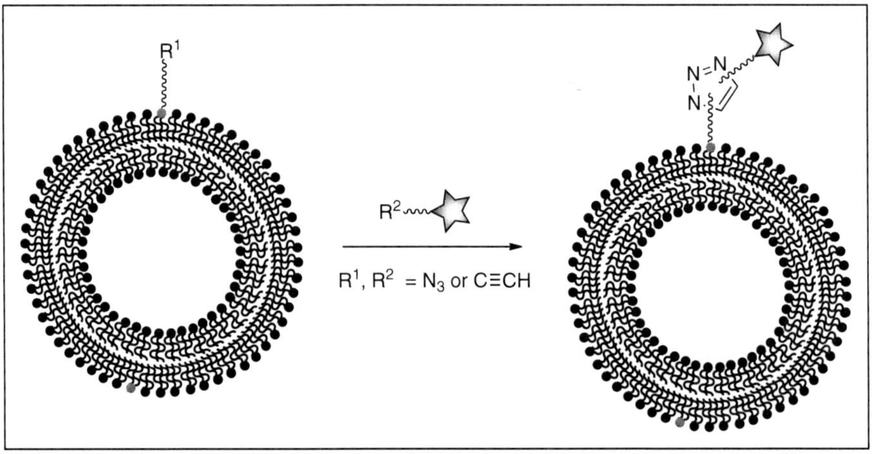

FIGURE 4.2 Cartoon for the derivatization of intact liposomes using bioorthogonal chemistry.

into liposomes containing phosphatidylcholine (PC), phosphatidylglycerol (PG), and cholesterol. Next, CuAAC was explored for functionalizing membrane surfaces through reaction with synthetic α-thiomannose-azide conjugate **2**. In these studies, the authors found that high concentrations of the traditional copper sulfate and sodium ascorbate CuAAC components (2.3 mM and 51.5 mM, respectively) were necessary to achieve even modest yields for conjugation (25%). As a result, copper chelation was exploited to as a strategy for improvement, and the use of the water-soluble ligand bathophenanthroline disulfonate led to around 80% derivatization within 1 hour, and almost complete reaction after 24 hours. Importantly, the authors used dynamic light scattering (DLS) and fluorescence-based dye leakage assays to assess the integrity of membranes during conjugation. In these studies, changes in liposome size (~50% increases in diameter) were observed using traditional CuAAC conditions, indicating membrane disruption. However, this effect was counteracted through ligand addition and by controlling the order of reactant addition such that no vesicle leakage or size changes were observed. Finally, the resulting mannose-decorated liposomes were found to successfully bind target proteins through the agglutination of the lectin Concanavalin A.

Also in 2006, Kros and coworkers reported the use of a Förster resonance energy transfer (FRET) system to characterize click-chemistry-based membrane functionalization [55]. In this study, phosphatidylethanolamine (PE) analog **3**, bearing an alkyne, was incorporated into liposomes composed of PC and also lissamine-rhodamine-tagged PE as a FRET acceptor tag (Fig. 4.3b). Upon reaction with lysine derivative **4** bearing both an azide as well as an NBD (FRET donor) label, using half an equivalent of copper bromide catalyst, FRET activation was observed through enhancement of the rhodamine (acceptor) emission signal relative to controls lacking copper. Fluorescence-based analysis showed that the reaction was complete after approximately 4 hours. DLS experiments were also used to demonstrate that the sizes of the

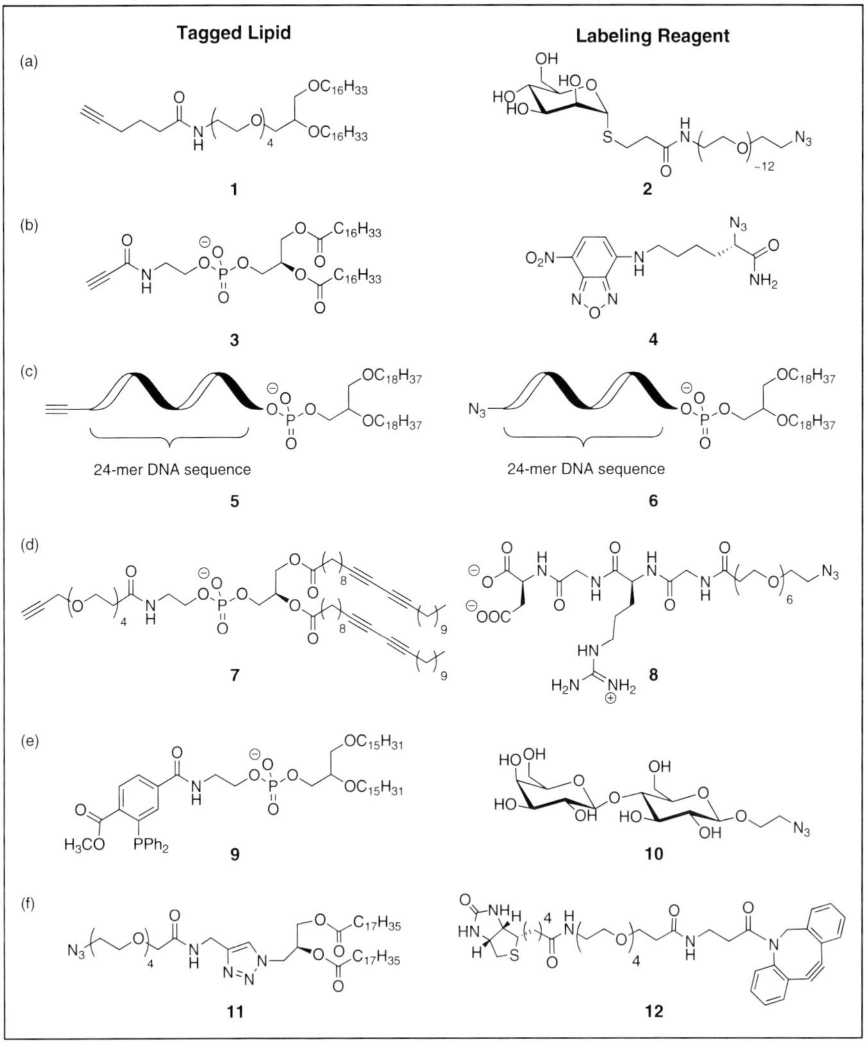

FIGURE 4.3 Examples of tagged lipids and reagents used for bioorthogonal labeling of intact membranes.

liposomes did not change during the reaction, and HPLC showed that 57% of alkyne **4** reacted, suggesting that conjugation only occurred on the outer membrane surface. In addition, by using dialysis to remove unreacted N_3-Lys(NBD), derivatization could be readily observed by the naked eye through a colorimetric assay.

Boxer and coworkers have applied click chemistry membrane functionalization to study DNA-mediated-membrane-binding interactions and fusion [56]. In this study, intact liposomes were immobilized onto supported lipid bilayers (SLBs) using a sense/antisense DNA-templated CuAAC reaction, which was implemented as an orthogonal approach to liposome anchoring that enhanced control over DNA-binding interactions. To do so, liposomes containing both a DNA–lipid conjugate consisting of a 24-mer attached to PE, as well as a separate azide-tagged PE derivative, were affixed onto SLB containing a PE analog bearing the complementary 24-mer sequence that terminated in an alkyne moiety (depicted generically as **5**, Fig. 4.3c). The liposomes also contained a second combination of sense/antisense DNA–lipid conjugates that were employed to enforce liposome–liposome-binding interactions and fusion. Labeled liposomes and SLBs were incubated in the presence of high salt concentrations to induce DNA sense/antisense liposome–SLB-binding interactions along with copper sulfate, sodium ascorbate, and the copper-ligand TTMA to enforce covalent attachment via CuAAC. When the salt concentration was lowered to release DNA–DNA contacts, liposomes remained attached to the surface via the triazole linkage. Following this, the salt concentration was again increased to activate inter-liposome binding via the DNA strands presented on the membrane surface. Alternate approaches were used to vary the liposome–SLB connection point, both by switching the presence of the azide and alkyne between the SLB and liposome and also through the use of DNA–lipid conjugates containing terminal azides (depicted generically as **6**) and alkynes for conjugation. Importantly, the authors noted that templating the CuAAC reaction during liposome deposition was necessary to achieve appreciable yields of conjugation. Using fluorescence microscopy via fluorophore-tagged liposomes, this approach allowed for direct observation of liposome diffusion and liposome–liposome binding in this system, in which the liposomes were limited to movement in a parallel plane by SLB attachment.

While the initial reports of CuAAC-based derivatization of intact membranes demonstrated that this approach can be highly effective, there were challenges that needed to be overcome. In particular, low conjugation yields and membrane decomposition were circumvented primarily through the use of copper-chelating ligands. The potential pitfalls, in combination with challenges associated with the application of CuAAC for in vivo studies, including copper toxicity and delivery, have led researchers to pursue alternate strategies for bioorthogonal labeling of intact membranes. In these reports, alterations to both the membrane composition and the conjugation reaction have been pursued.

In one such study, Cai and coworkers reported the development of clickable polymerized liposomes for in vivo delivery applications in order to overcome issues associated with liposome stability and lipid phase segregation. In this case, polymerized liposomes were accessed through irradiation of alkyne-tagged PE analog **7** bearing polydiacetylene moieties within the acyl chains (Fig. 4.3d). Despite the enhanced stability of polymerized liposomes, a copper-chelating ligand was still necessary for CuAAC to enhance yields and avoid decomposition of the unsaturated lipids that result from polymerization. For the copper-chelating ligand, TBTA was not effective, which was attributed to poor water solubility, and bathophenanthroline disulfonate was avoided due to concerns over sensitivity of its copper complex to air oxidation. As a result, the authors implemented a tris-(TEG-triazole)-based ligand that was found to yield a 12-fold enhancement in the rate of CuAAC. The conditions for CuAAC were then optimized since high concentrations of copper and ascorbic acid were found to result in the decomposition of polymerized liposomes as judged by dye leakage studies. Optimized conditions, obtained through conjugation with fluorogenic azidocoumarin, included copper sulfate (0.3 mM), ascorbic acid (6 mM), and ligand, in which conjugation was complete within 1 hour. The yield of this reaction was calculated to be 23%, although this could be an underestimation due to fluorescence quenching, and is also limited since only lipids on the outer leaflet of the bilayer should react. With these conditions, CuAAC was used to introduce both an RGD peptide as an integrin-targeting ligand, through reaction with **8**, as well as a fluorescent label onto the polymerized liposomes, and fluorescence imaging demonstrated successful delivery to cell membranes.

In an effort to pursue alternate bioorthogonal conjugation strategies, Sun and coworkers evaluated the Staudinger ligation for derivatization of intact liposomes, which is beneficial as it does not require additional catalyst for conjugation [57]. To do so, PE analog **9** was devised in which the traditional Staudinger reagent, a triphenylphosphine bearing a proximal ester for covalent trapping during reaction with azide, was appended (Fig. 4.3e). The authors noted that the synthesis of **9** was complicated by phosphine oxidation when this compound was treated with aqueous methanol, but that this problem could be counteracted through reduction or avoided altogether when using only organic solvents. Next, the Staudinger ligation was evaluated for conjugation of azide-tagged lactose analog **10** onto PC/cholesterol liposomes containing **9** (5 mol%). During the reaction, DLS and dye leakage assays were performed, which both demonstrated that membrane integrity was maintained. In addition, the introduction of lactose onto liposomal surfaces was quantified using the phenol–sulfuric acid test, which indicated that 80% of **9** on the outer leaflet of the membrane (40% total) were modified. Finally, a lectin-binding assay using β-galactose-binding lectin with detection of aggregation via DLS was used to showcase the accessibility of lactose units on the surface.

Our group has also been interested in developing bioorthogonal methods for labeling intact membranes that circumvent the copper catalyst required for CuAAC, for which we recently reported an approach involving copper-free click chemistry in collaboration with Popik and coworkers [58]. While a variety of strained cyclooctynes have been shown to be effective for click chemistry in the absence of copper catalyst, we exploited the aza-dibenzocyclooctyne (ADIBO) system for this purpose due to the robust nature, desirable kinetic properties, and ease of synthetic preparation of these reagents [49, 50]. To assess conjugation, we chose to implement an assay involving liposome immobilization onto streptavidin-coated microplates. This was performed by incorporating azido-TEG-glycerolipid **11** along with a fluorescent NBD-PE conjugate into liposomes, followed by incubation with ADIBO-TEG-biotin reagent **12** (Fig. 4.3f). Following washes, the resulting microplate-immobilized biotinylated liposomes could be detected in high-throughput fashion using a fluorescence plate reader. In these studies, a dose-dependent relationship was observed between the amount of added ADIBO-TEG-biotin and the resulting fluorescence signal, indicating successful liposome derivatization. Controls, including the replacement of the ADIBO reagent with an acyclic alkyne, the use of liposomes lacking **11** and preincubation of the microplate with biotin, all resulted in no signal, indicating that both **11** and **12** are necessary for derivatization and ruling out nonspecific deposition.

While this section has focused primarily on the labeling of intact membranes, it should also be noted that membrane functionalization has also benefited in cases where click chemistry has been used to facilitate the synthesis of lipid analogs that are subsequently incorporated into liposomes. For example, prior to our report involving copper-free click chemistry, we exploited a simple azido-glycerolipid as a modular lipid scaffold for the convenient synthesis of functionalized lipids [59]. This strategy was applied to the development of a microplate-based assay for the detection of protein–membrane-binding interactions. In doing so, both biotinylated and fluorescence-tagged lipids were produced that were exploited to immobilize whole liposomes onto streptavidin-coated microplates, and for control studies to ensure that liposome deposition resulted from the added biotin and not from nonspecific adhesion. This assay then proved to be effective for sensitive detection of protein–membrane binding, as demonstrated for the binding of protein kinase C(PKC)α to microplate-immobilized liposomes containing diacylglycerol (DAG), PS, and PC.

From these studies, a variety of strategies have emerged that are effective for derivatizing membranes in bioorthogonal fashion. In addition, pitfalls involving the decomposition of membranes under the conditions for functionalization have been overcome through multiple routes including the use of copper-chelating ligands for CuAAC, the stabilization of the membrane through polymerization and the use of the Staudinger ligation, and copper-free click reaction for modification. The resulting strategies are primed to impact the application of membrane surfaces for delivery and biological studies. For example, click chemistry has recently been applied to trigger the fusion of liposomes containing complementary reactive tags, which has implications for both liposome delivery and release [60].

4.3 CLICK CHEMISTRY AND THE GENERATION OF FUNCTIONALIZED LIPID ANALOGS

In addition to the advantages for bioorthogonal labeling, since click chemistry by definition involves reactions that proceed in high yield under ambient conditions and tolerate a range of functional groups, it has also been beneficial as a synthetic tool for accessing a number of lipid analogs that are useful for a broad range of applications. This becomes invaluable for challenging synthetic targets, exemplified by an early report by Moroder and coworkers involving the synthesis of lipopeptides [61]. In this report, azido-PE **13** was conveniently used to access lipopeptides such as **14** through reaction with alkynyl peptides (Fig. 4.4). Subsequently, native chemical ligation was used to grow the peptide chain via a terminal cysteine residue on **14** to produce **15**, which was performed in micellar solutions of 2% octyl-β-glucopyranoside. The authors next used this approach to synthesize a lipidated peptide fragment from human prion protein bearing a fluorescent dansyl label. Fluorescence microscopy was then exploited to show fast uptake and intracellular distribution of the resulting lipopeptide analog in HeLa cells.

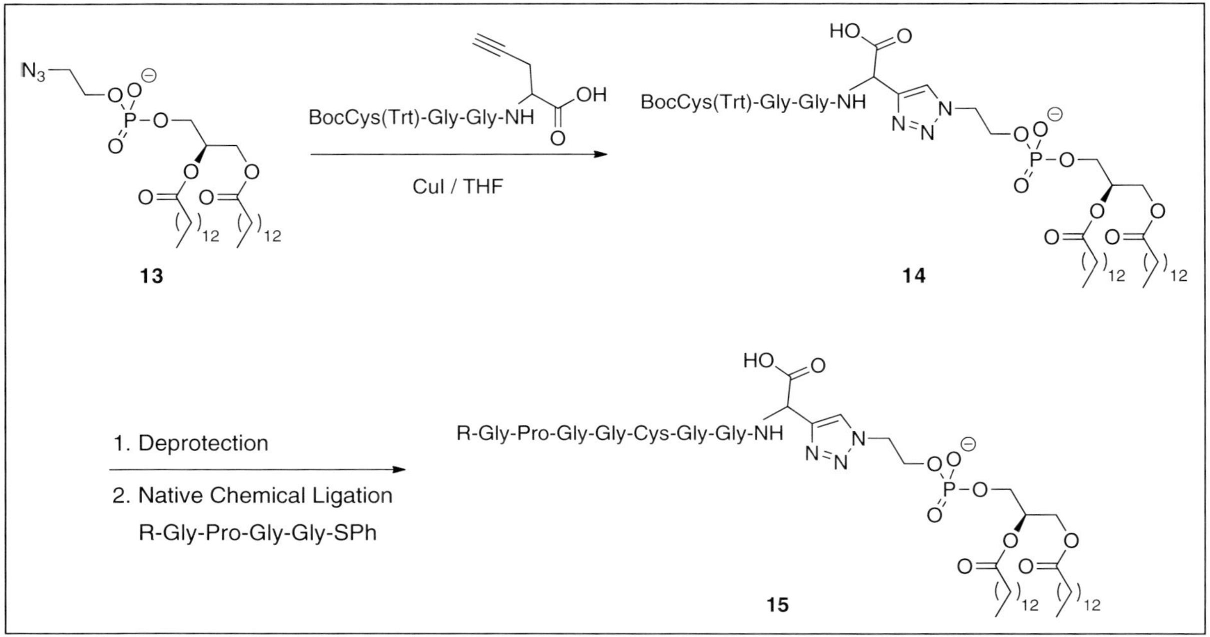

FIGURE 4.4 Production of lipopeptide analogs using CuAAC to introduce peptides onto the lipid scaffold, followed by native chemical ligation to elongate the peptide chain.

Our group has been interested in using click chemistry to facilitate the production of functionalized lipid probes of use for characterizing protein-membrane-binding interactions (Fig. 4.5). Our initial report in this area focused on the generation of DAG probes due to the important roles this lipid has in activating targets such as PKC isoforms, which are critical to cell growth. We specifically targeted functionalization at the lipid headgroup, such that the added reporter group would be proximal to the protein-binding domain. Towards this end, we designed and synthesized azido-DAG **16** as a modular scaffold that was derivatized via CuAAC in the final step of synthesis through reactions with alkynes of type **17** to produce probes **18a–e**, bearing either a fluorescence or photoaffinity tag [62]. The resulting probes were evaluated for PKCα binding through incorporation into liposomes for a surface plasmon resonance (SPR) assay, during which all probes bound with similar affinity to naturally DAG, demonstrating the validity of these compounds as probes. Following this initial report, we extended this approach to develop probes corresponding to another important signaling lipid, phosphatidic acid (PA) [63]. To do so, reporter groups were introduced onto protected azido-PA derivative **19** in order to produce probes **20a–d,f–g**. In this case, to validate protein-binding properties, we performed SPR assays using an individual C2 domain of PKCα, since the full-length protein contains multiple lipid-binding domains. The results demonstrated that while affinities dropped off somewhat for probes **20a–d,f–g** compared to synthetic PA, these probes nevertheless bound PKCα-C2 with high affinity, indicating that this strategy for probe development was generally effective.

In addition to the prior headgroup-derivatized lipid analogs, click chemistry has also been beneficial for functionalization of lipid acyl chains. Smith and coworkers initially exploited CuAAC for the synthesis of PC-derived bolaamphiphiles, which consist of two lipid headgroups connected by long hydrophobic chains that span the traditional membrane bilayer [64]. In the synthesis, CuAAC was used to efficiently produce a range of bolaamphiphiles of type **21**; in this case, through the reaction of PC derivative **22** containing an alkyne at the terminus of the sn-2 acyl chain with multiple bis-azide linkers (Fig. 4.6). It should also be mentioned that **22** was conveniently produced in a chemoenzymatic manner from PC by using phospholipase A2 to remove the sn-2 acyl chain, followed by coupling with an alkynyl-fatty acid. Following synthesis, it was observed that bolaamphiphiles of type **21** were not capable of forming stable vesicles. This was attributed to the high polarity of the triazoles, which likely reside primarily at the aqueous interface rather than the hydrophobic membrane core. In fact, upon further investigation, these compounds were found to promote dye leakage from PC/cholesterol liposomes, indicating that they destabilize membranes.

Solaiman and coworkers reported another study in which click chemistry was performed using synthetic lipids derivatized within the acyl chains to produce novel lipid analogs [65]. In this report, 17-azido stearic acid and 3-azido decanoic acid were initially synthesized as clickable fatty acids. In the latter case, an indirect route involving the opening of a β-lactam with azide was used to circumvent issues with the susceptibility of 3-substituted carboxylic acid intermediates to eliminations. With these azido-fatty acids in hand, CuAAC was then implemented to synthesize multiple dimer lipids through binary combinations of triacylglycerols and/or glycolipids,

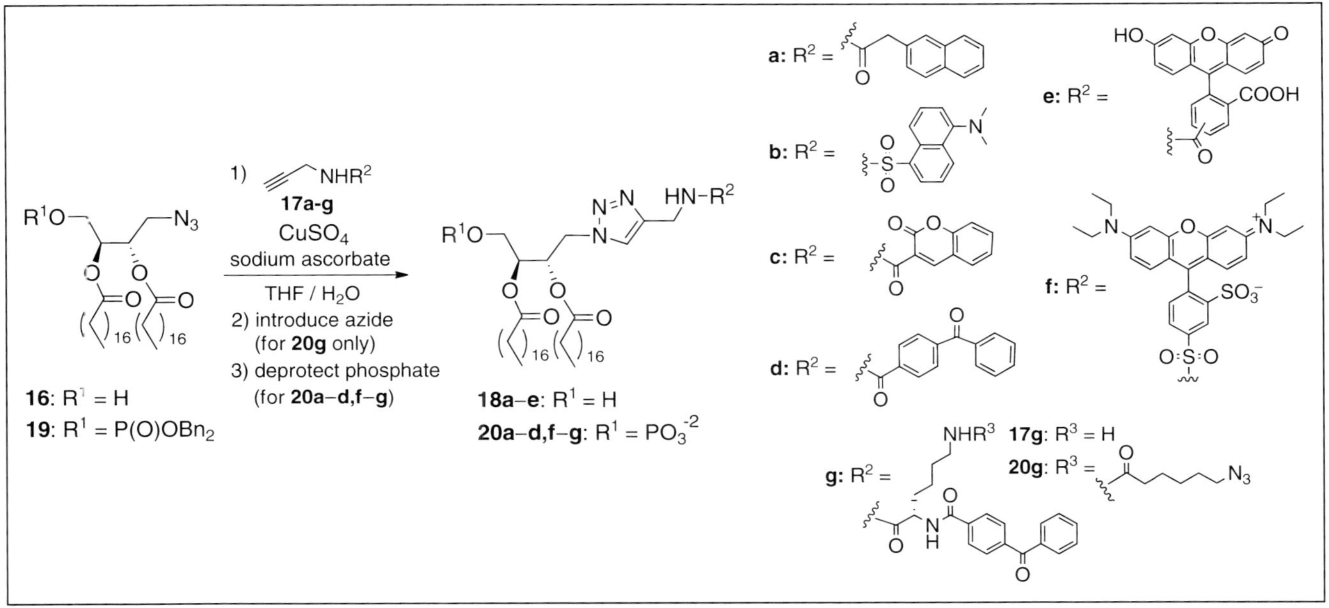

FIGURE 4.5 Functionalized DAG and PA probes accessed through CuAAC.

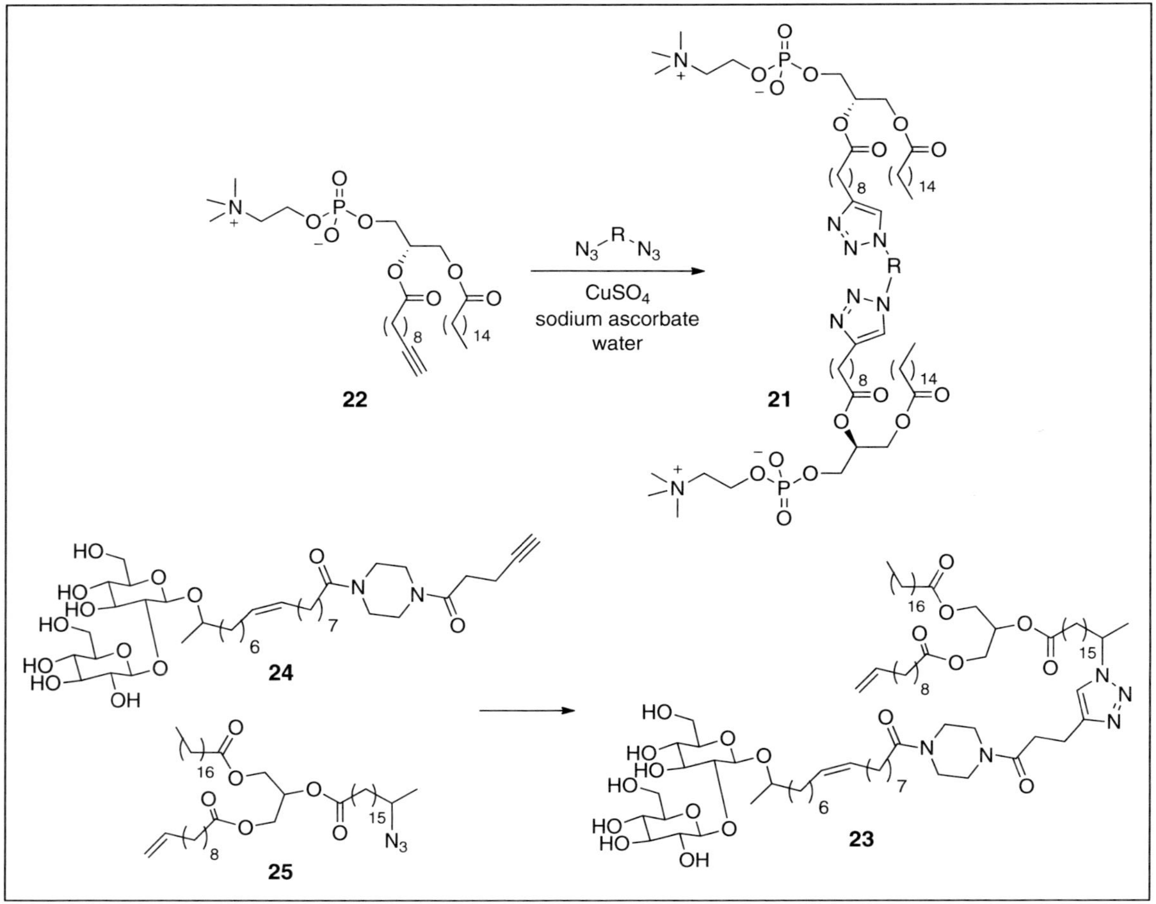

FIGURE 4.6 The use of CuAAC for the tethering of tagged lipid components.

which is exemplified by the production of **23** from alkynyl-glycolipid **24** and azido-triacylglycerol analog **25**.

While advances in the synthesis of traditional carbohydrate and lipid motifs have allowed the generation of highly complex glycolipids, click chemistry has furthered the cause of producing functionalized analogs. A prime example pertains to the work of Guo and coworkers in the development of **26**, a protected alkyne-tagged analog of monophosphoryl Lipid A (Fig. 4.7) [66]. This glycolipid is of significant interest due to the pivotal role this compound plays in endotoxicity through the activation of the toll-like receptor 4. As a result, synthetic Lipid A conjugates have been pursued to develop new immunostimulants to combat sepsis, although the synthesis of such analogs is hindered by the complex structure of this glycolipid. Thus, the production of clickable derivatives **26** shows promise for enhancing the rapid production of analogs with useful properties. The convenient modification of this analog was demonstrated through coupling with GM3-alkyne **27** to access conjugate **28**.

As a result of the diversity of lipid structures that have been generated using click chemistry, it is clear that this is a generally effective strategy for the pursuit of complex lipids for a wide range of applications. In addition, a few important concepts have arisen from this work, including the location for tag placement and the properties of the resulting triazoles. It has been shown that click chemistry can be effectively used to derivatize lipids at both the headgroup and within the acyl chains. Here, the choice of tag location will often be dependent on the particular structures and applications that are targeted. However, the properties of the subsequent triazoles, and in particular, their polar nature, play into this as well by affecting the physicochemical properties of the resulting lipid analogs. These previous lessons should be considered when developing the optimal synthetic targets in future studies.

4.4 DIRECT *IN SITU* LABELING AND IMAGING OF SYNTHETICALLY TAGGED LIPID ANALOGS

Another area of study that has benefited from click-chemistry-based derivatization involves *in situ* labeling and imaging of lipids within the complex environments of biological samples, which exploits the bioorthogonal nature of this reaction (Fig. 4.8a). Lipids are dynamic biomolecules that continually undergo modification via metabolic pathways and movement throughout cellular membranes. The localization of each particular lipid, which is regulated in a spatial and temporal manner through processes such as enzymatic modification and diffusion, controls important biological activities, including fusion and signaling. Thus, the elucidation of the details of lipid localization is necessary to fully understand subsequent biological properties. However, due to the relatively simple structures and specific properties of lipids, it is difficult to track these molecules in a native environment without perturbing the system. Recent advancements using click chemistry have shown promise for overcoming these challenges.

In a recent report, Schultz and Neef used synthetic alkyne-tagged PA analogs to image the dynamics of this lipid in both fixed and living cells (Fig. 4.8b) [67].

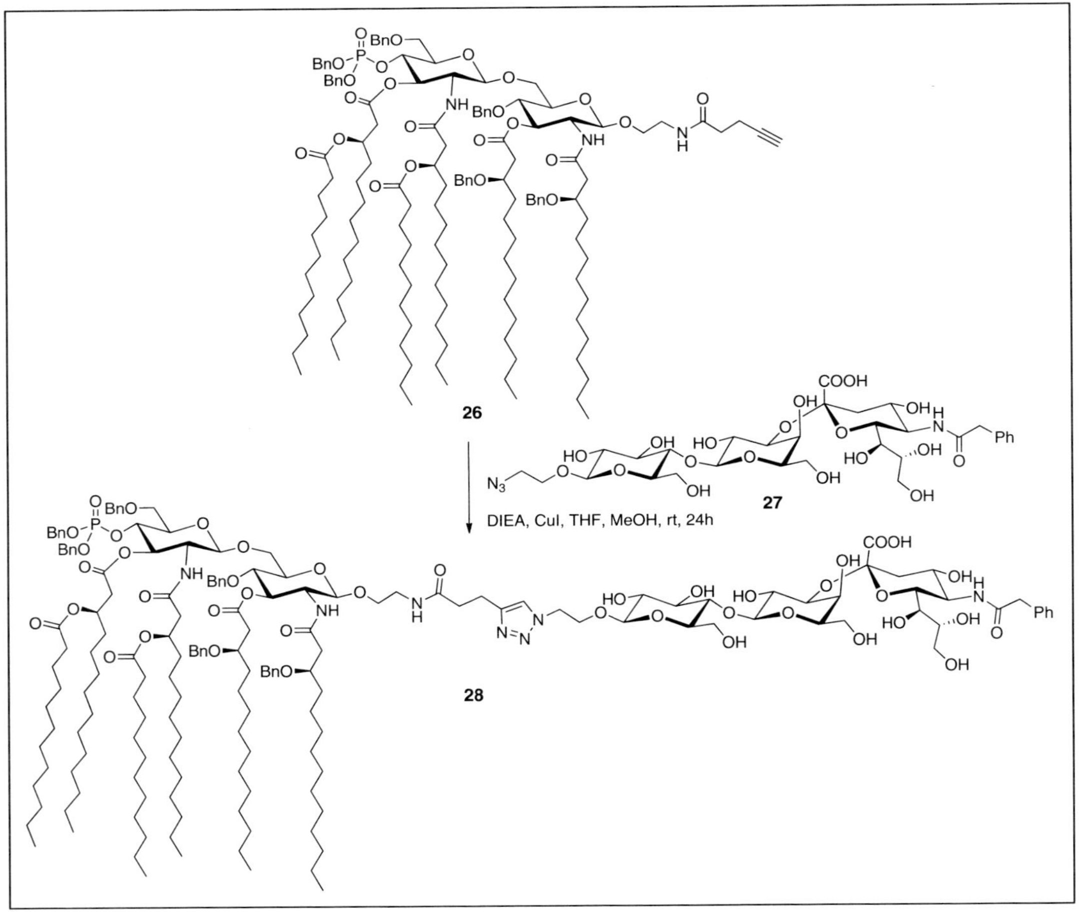

FIGURE 4.7 Synthesis and functionalization of an alkynyl-Lipid A analog.

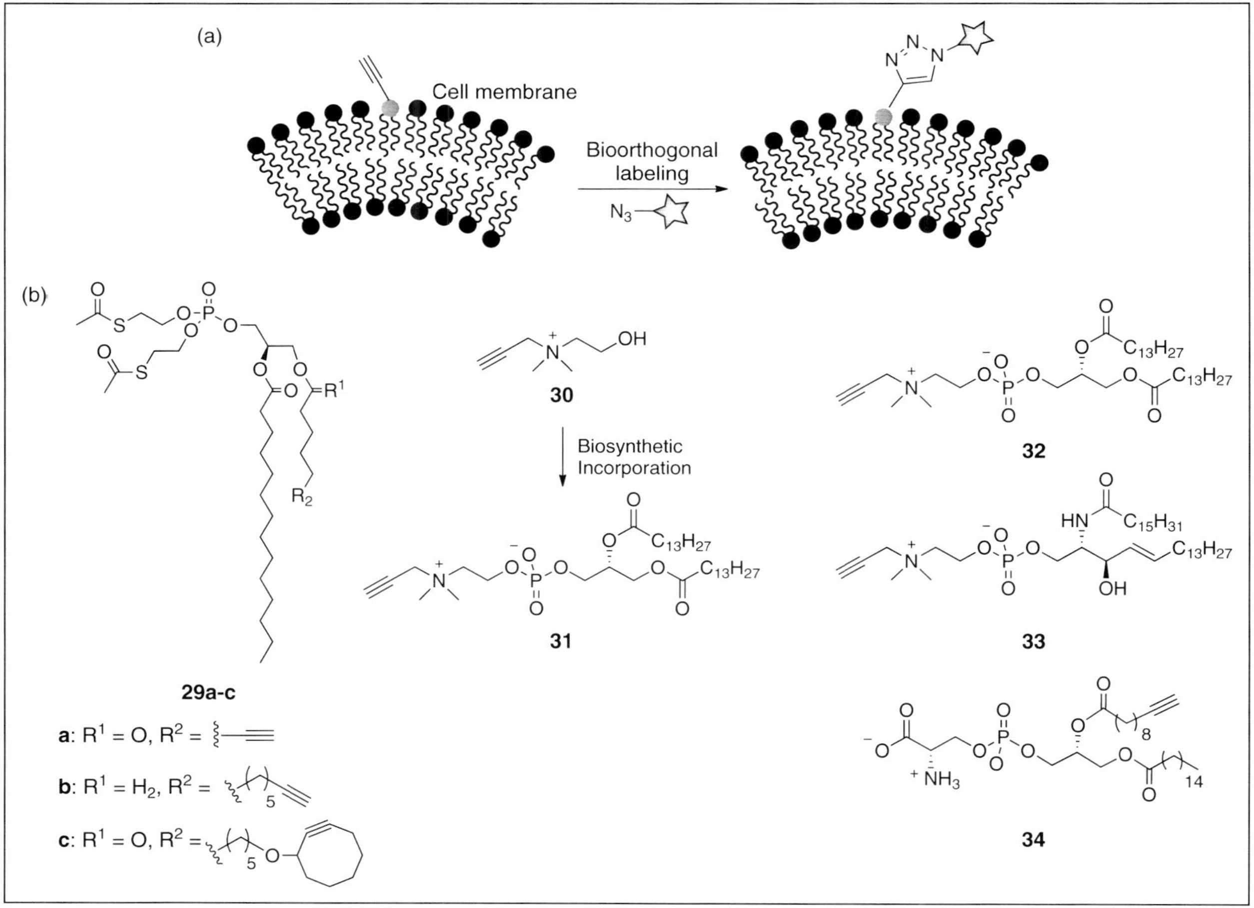

FIGURE 4.8 Strategies for *in situ* labeling of tagged lipid analogs. (a) Cartoon depicting bioorthogonal labeling of lipids in complex environment. (b) Structures that have been developed for *in situ* labeling studies. (See color plate section.)

Three different PA analogs were used in this study, which included either a propargyl moiety for CuAAC (**29a,b**) or a cyclooctyne for copper-free click chemistry (**29c**). In addition, the alkynyl acyl chain was attached either via the natural ester chain (**29a,c**) or using an ether linkage to circumvent enzymatic hydrolysis (**29b**). Finally, S-acetylthioethyl-(SATE)-protected phosphate moieties were used to facilitate transfer across membranes into cells, after which these groups are removed by hydrolysis to reveal the traditional phosphate headgroup. When using alkyne-tagged lipids **29a** and **29b**, cells were fixed prior to CuAAC due to the toxicity of copper. Subsequent labeling using a fluorogenic coumarin reagent resulted in essentially uniform labeling for both compounds across membranes within the cells. Notably, the authors also used CuAAC to introduce fluorophores onto PA derivatives prior to cell treatment, and found that this altered the imaging results, with fluorescence intensity varying among intracellular membranes. These results show that the added fluorophore impacts the distribution of the lipid, which would invalidate results from studies using initially fluorescence-tagged lipids. In addition, fluorescence recovery after photobleaching (FRAP) studies showed that these lipids were not mobile after cell fixing. Next, to achieve labeling in living cells, cyclooctyne-PA analog **29c** was implemented, which yielded no toxicity after several hours.

Salic and coworkers took an alternate approach to in vivo labeling of phospholipids by instead implementing an alkyne-tagged precursor that is converted to the target lipid via metabolic pathways [68]. In doing so, propargyl–choline (**30**) was developed for conversion into alkyne-tagged PC lipid analogs of type **31** when introduced into cells. Upon incubation with **30** followed by fluorescence tagging using CuAAC, cell membranes were found to be labeled, which was affected by the addition of detergent and abrogated through the use of phospholipase C, providing evidence that the labeled species were in fact PC lipid analogs. The resulting labeled PC derivatives were localized in different cellular membranes, including the plasma membrane, Golgi, mitochondria, and ER. This distribution was further confirmed through immunoelectron microscopy. Finally, mass spectrometry was then used to probe lipid structures, indicating that **30** did not affect the distribution of non-PC lipids, and was introduced into all classes of PC within the cell, although incorporation into sphingomyelin (SM) was slower than for PC.

In an additional study targeting PC and also SM, Cairo and coworkers reported the design, synthesis, and derivatization of analogs of these lipids (**32** and **33**, respectively) bearing an alkyne appended to the lipid headgroup [69]. In order to minimize structural perturbation, the alkyne was introduced in the form of a propargyl group in place of one methyl on the choline moiety, leading to presentation at the lipid headgroup. Following synthesis, it was demonstrated that benzoxadiazole fluorophores could be conjugated to these compounds in high yields using CuAAC. Finally, to assess the ability of alkynyl-SM analog **33** and its benzoxadiazole conjugate to mimic natural SM, these compounds were evaluated as substrates for sphingomyelinase using a TLC-based assay. These structures exhibited 50% and 70% enzymatic conversion, respectively, relative to controls, indicating that they act as substrates, although they are less active than natural SM.

In 2008, Smith and coworkers reported chemoenzymatic syntheses of PS analogs bearing alkyne or azide acyl chains, such as **34**, towards applications involving imaging and delivery [70]. This was pursued to exploit transport systems that pass PS across cell membranes in order to deliver the PS analogs and any attached cargo into cells. In this case, the bioorthogonal tags were incorporated within lipid acyl chains to avoid the potential of affecting the recognition of the lipid headgroup by transport machinery. The synthetic strategy entailed incorporation of the tagged acyl chain onto commercial lysophosphocholine (LPC), followed by Phospholipase D hydrolysis to PA analogs, and finally introduction of the serine headgroup. While initial conjugation using CuAAC resulted in low yields ($<5\%$), ultrasonication was found to substantially increase efficacy to 45–90%.

While the design and application of clickable lipids for in vivo applications such as imaging represents a recent advancement, a number of probe strategies corresponding to different lipids have now been reported. Furthermore, initial labeling studies have shown promise for the ability to selectively label these compounds in a native environment, and by doing so, advancing the understanding of the dynamics of important lipid-based biological processes.

4.5 ACTIVITY-BASED CHARACTERIZATION OF LIPID-BINDING AND LIPID-MODIFYING PROTEINS

One of the prominent roles of biologically active lipids pertains to their involvement in noncovalent protein–lipid-binding interactions, which can occur in different settings. For example, certain lipids act as ligands that trigger the transient docking of soluble peripheral proteins onto the membrane surface [1,2]. These binding events commonly regulate both protein function and localization, and have been found to be aberrant in a number of prominent diseases. In other situations, integral membrane proteins are known to bind to particular lipids that assist in embedding the protein within the membrane in particular conformations. In addition, due to the important roles of lipids, enzymes involved in lipid biosynthesis control important pathways. In recent years, there has been significant interest in the development of probe-based strategies that allow for efficient identification and characterization of proteins that bind and modify lipids. Such studies generally utilize the approach of activity-based protein profiling, in which probes are employed for the collective labeling, purification, and analysis of target proteins out of complex samples such as cell extracts, live cells, and organisms (Fig. 4.9a) [71]. The development of lipid activity probes typically involves the modification of the ligand structure in order to enforce covalent labeling of target proteins, and the introduction of a secondary handle that is exploited to label and manipulate tagged proteins. Since protein–lipid-binding interactions are noncovalent in nature, photoaffinity tags [72] are commonly implemented to achieve probe–protein conjugation via photo-cross-linking. In addition, clickable functional groups have proven invaluable as latent tags for post-derivatization of labeled proteins by taking advantage of the bioorthogonal nature of this reaction. Using these strategies, multiple click-chemistry-based lipid activity probes have been reported of late.

In one study of this type, de Kroon and coworkers utilized functionalized PC probes of type **35a–b** to characterize proteins that bind to this lipid (Fig. 4.9b) [73, 74]. These compounds each included one of a series of photoaffinity moieties

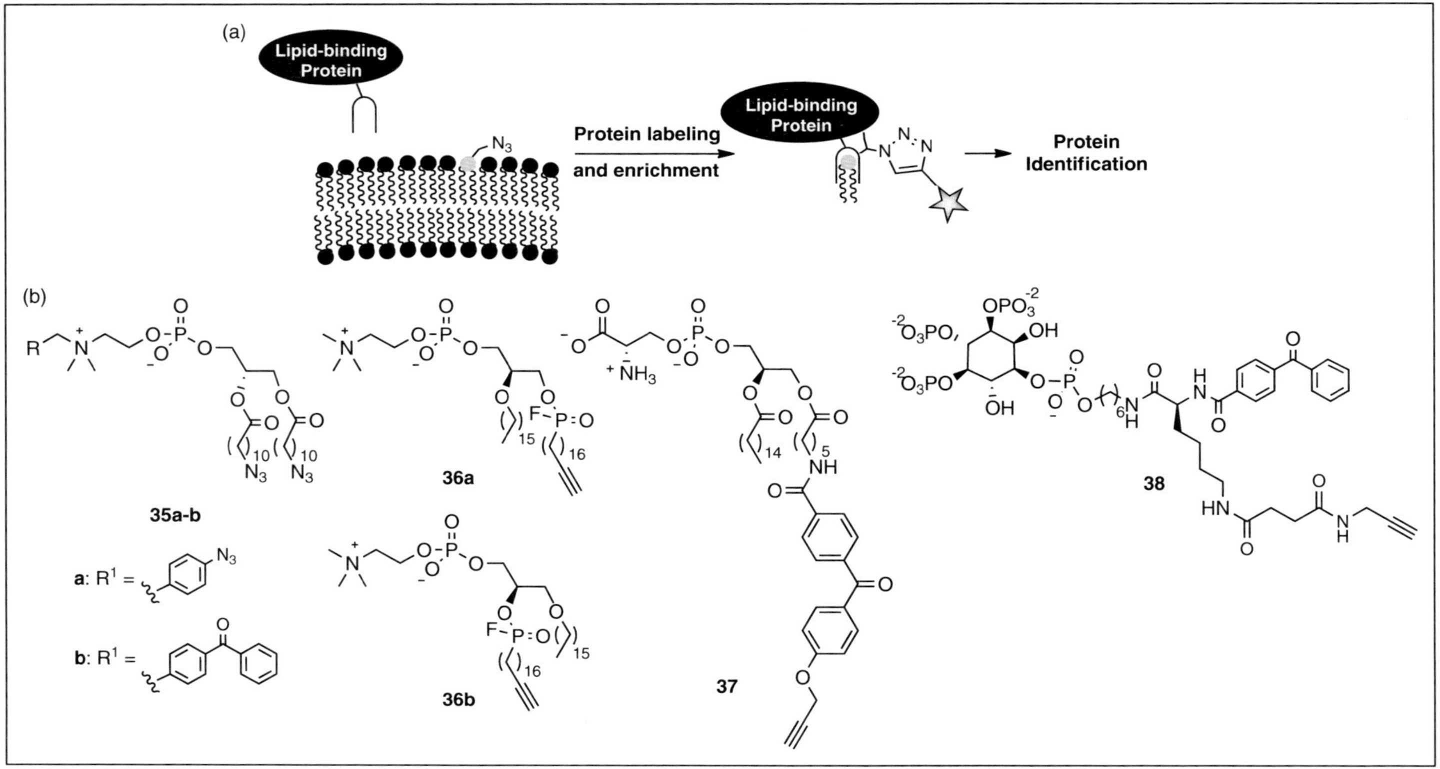

FIGURE 4.9 The development of activity-based lipid probes. (a) Cartoon depicting the application of lipid activity probes for the labeling and identification of target proteins. (b) Examples of lipid activity probes that have been reported. (See color plate section.)

(including phenyl azide (**35a**) or benzophenone (**35b**)) added in place of a methyl on the choline headgroup, as well as azide secondary tags at the terminus of the lipid acyl chains. A probe bearing a diazirine photo-cross-linking was also studied, but was abandoned when it did not yield significant labeling in preliminary studies. Initially, these compounds were exploited for model membrane studies using the protein apocytochrome c, in which protein labeling was conveniently characterized since lipid-cross-linking enhanced carbonate wash resistance. With probes **35a** and **35b** both being successful for cross-linking this purified protein target, studies were next extended to the labeling of proteins in inner mitochondrial membranes, in which reaction with a fluorescent TAMRA-alkyne was used to visualize labeled proteins. Interestingly, both probes were found to label different, but partially overlapping, sets of proteins, suggesting that the choice of photoaffinity tag affected the proteins that were labeled in each case. Benzophenone probe **35b** was found to yield enhanced background labeling in the absence of irradiation, and a cross-linking profile that was more similar to Coomassie staining, suggesting that this probe may result in more nonspecific protein labeling. Next, probe **35a** was passed on for studies aimed at discovering cross-linked proteins, which were performed through CuAAC to introduce biotin onto labeled proteins, followed by affinity purification and LC-MS/MS identification. This yielded 27 proteins that exhibited a $\geq$twofold enhancement in signal when using probe **35a**, including some proteins that were previously known to bind to PC and other novel candidate-binding partners.

In an effort to efficiently characterize enzymes that modify lipids, Cravatt and Tully recently reported studies employing PC activity probes **36a–b** to study phospholipases [75]. In this case, a fluorophosphonate functional group, which was known to act as an electrophilic group for the covalent labeling of enzymes, and particularly serine hydrolases, was incorporated in place of the traditional ester linkage of the acyl chain at either the *sn*-1 (**36a**) or *sn*-2 (**36b**) position. In addition, an alkyne was incorporated at the terminus of the modified acyl chain as a latent clickable secondary tag. Initial labeling studies were performed with purified enzymes DDHD1, which had a known specificity as a phospholipase A_1 that selectively cleaves the *sn*-1 acyl chain, and DDHD2, a phospholipase with unknown specificity. In these studies, DDHD1 was labeled by probe **36a,** but not by **36b**, which is consistent with the substrate requirements of this enzyme and indicates the specificity of this probe for enzymatic targets. On the other hand, DDHD2 was labeled similarly by both probes, suggesting that this enzyme may possess both phospholipase A_1 and A_2 activities. Next, **36a** and **36b** were implemented for the labeling of proteins in mouse-tissue proteomes, where each exhibited clear selectivity in the labeling of different sets of proteins. Finally, these probes were exploited to characterize the poorly understood enzyme calcium-dependent transacylase from rat cortical membrane proteomes, which was found to be inhibited to a greater extent by **36a** (IC_{50} = 3 nM) compared to **36b** (IC_{50} = 165 nM). These studies clearly indicate that the activity–probe strategy is effective for the characterization of protein targets from complex samples.

Other phospholipids have also been targeted for the development of activity probes. Bong and Bandyopadhyay designed and synthesized multiple PS probes, exemplified by **37**, to analyze proteins that bind to this lipid [76]. This is of interest due to the transfer of PS to the outer leaflet of the bilayer during apoptosis, which triggers cell surface recognition events that result in cell death. In the design for probes of type **37**, both a benzophenone photoaffinity and an alkyne secondary tag were included within the *sn*-1 acyl chain. These probes were studied for the labeling of prothrombin-1, which was assessed through CuAAC followed by in-gel fluorescence detection. Here, it was found that a series of PS probes, including those with inverted stereochemistry, all labeled prothrombin-1 in a concentration-dependent manner, suggesting that the stereochemistry does not affect recognition.

Our group has been interested in developing activity probes corresponding to lipids such as PA (**20g**, Fig. 4.5) and the phosphatidylinositol polyphosphates (PIP_ns) [77]. We recently reported the application of activity probe **38** corresponding to the headgroup of phosphatidylinositol-(3,4,5)-trisphosphate (PI(3,4,5)P3, Figure 4.9b) [78]. The PIPns were targeted since these lipids bind and activate a litany of protein targets, and these interactions are commonly aberrant in various diseases [79–83]. We first sought to demonstrate the efficacy of this compound for labeling a known protein target based on a specific binding interaction. In these studies, probe **38** was effective for the labeling of the PH domain of Akt (Akt-PH), as judged in-gel fluorescence scanning. Control studies were next undertaken to show that this labeling event results from a specific binding interaction. First, heat denaturation of the protein target was observed to abrogate in-gel labeling due to the loss of protein structure. Additionally, co-incubation of Akt-PH with activity probe **38** along with an unlabeled headgroup analog as a competing ligand also led to blockage of the labeling event. Following validation of specific labeling, probe **38** was exploited for the characterization of target proteins in cancer cell extracts, followed by enrichment of probe-labeled proteins and mass spectrometry-based proteomics. This enabled the identification of 265 putative protein targets that were labeled by this probe, including at least 44 known PI(3,4,5)P3-binding proteins in addition to numerous novel candidate targets. From these early examples, it is clear that lipid activity probes provide an efficient means for the collective characterization of protein targets from complex samples.

4.6 PROBING COVALENT PROTEIN LIPIDATION

An area in which clickable lipid probes have been exploited for extensive biological investigations involves the elucidation of covalent protein lipidation [3]. A large number of eukaryotic proteins undergo the covalent addition of lipid units, which plays a key role in regulating both the function and subcellular localization of the modified protein, generally by promoting membrane anchoring or cycling between different membrane compartments. A common modification is the addition of saturated fatty acid chains, including the cotranslational *N*-myristoylation of glycine, producing an amide linkage, and the post-translational *S*-palmitoylation of cysteine, resulting in a thioester attachment. The latter case differs among examples of covalent lipidation as palmitoylation is reversible, allowing for dynamic regulation of lipidation. Prenyl groups are also added to a number of proteins through the post-translational addition

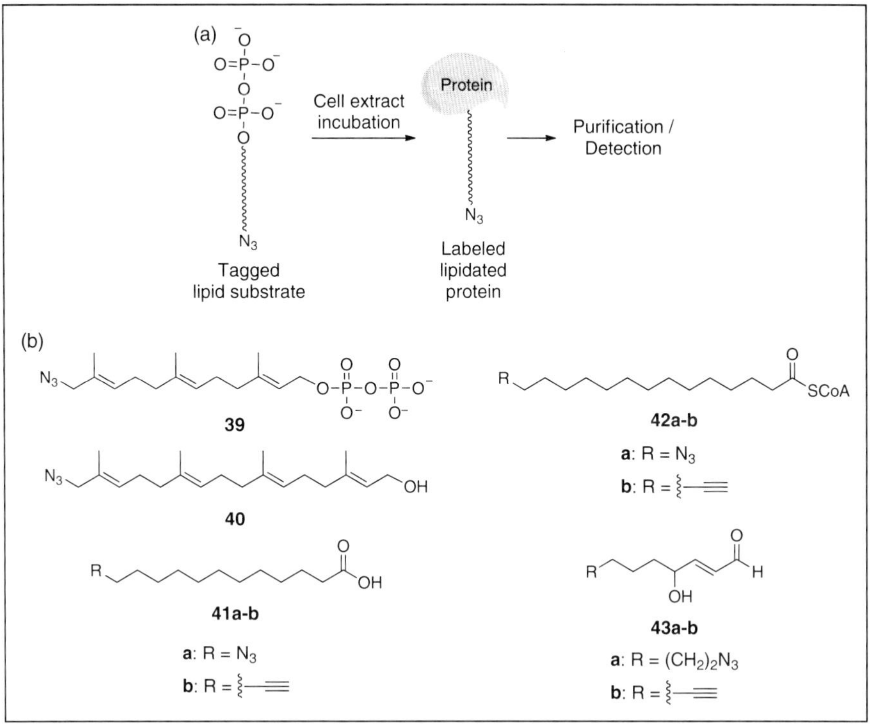

FIGURE 4.10 Investigation of covalent protein lipidation using bioorthogonal labeling. (a) Cartoon depicting the covalent labeling of target using azide-tagged lipid substrate analogs. (b) Examples of structures used to infiltrate covalent protein modification using tagged lipid substrate derivatives. (See color plate section.)

of *S*-farnesyl or *S*-geranylgeranyl moieties onto cysteine. The importance of covalent protein lipidation is represented by the role it plays in regulating proteins in the Ras superfamily, which undergo a switching mechanism controlled by the identities of added lipid chains [84]. When Ras is only farnesylated, it is localized at the endoplasmic reticulum (ER) membrane. Upon the introduction of one or more palmitoyl groups, cycling to the inner leaflet of the plasma membrane is triggered, where Ras becomes activated for its pivotal role in promoting cell growth. Due to the significance of protein lipidation, numerous recent studies have exploited the incorporation of lipid moieties bearing azide or alkyne onto proteins to study covalent lipid addition (Fig. 4.10a). Due to the large volume of work in this area, and the fact that these efforts have been the subject of recent reviews [85–87], we will only briefly summarize this work herein.

In an early example of using bioorthogonal reporters to characterize protein lipidation, Zhao and coworkers reported studies employing azido-farnesyl pyrophosphate (**39**) as a substrate analog for farnesyltransferases [88]. Initially, **39** was shown to be an effective substrate in vitro using the RC peptide as a representative sequence that undergoes farnesylation, as detected using mass spectrometry (Fig. 4.10b). Following this, experiments were performed in vivo by following the farnesylation of Ras and Hdj-2. In this case, lovastatin was used as an inhibitor of farnesyl pyrophosphate biosynthesis to block natural farnesylation. As a result, lovastatin addition resulted in a shift of the gel bands associated with these proteins, which was reversed by the introduction of **39**, indicating successful lipidation with this azido-tagged analog. Subsequently, global analysis of protein farnesylation was pursued using a bioorthogonal labeling strategy. For these studies, whole cell lysates were subjected to **38**, followed by biotinylation of the resulting labeled proteins using the Staudinger ligation, which was analyzed by in-gel labeling and mass spectrometry. By using the latter technique for sequencing, 21 proteins were identified, 17 of which contained the CAAX consensus sequence for farnesylation, and in addition, three endogenously biotinylated proteins and Annexin A2.

Probes bearing clickable tags have now been applied for extensive studies involving global analysis of protein prenylation [89–96] and fatty acylation [97–102] using probes exemplified by those that introduce azido-geranylgeranyl (**40**), azido- and alkynyl-myristate (**41a–b**), and azido- and alkynyl-palmitate (**42a–b**) groups onto target proteins. In addition, studies have focused on particular systems to elucidate aspects of lipidation, including the myristoylation of proteins associated with apoptosis [103], palmitoylation of mitochondrial proteins [104], palmitoylation in T-cell activation [105], palmitoylation of histone H3 variants [106], the dependence of IFITM3 antiviral activity on S-palmitoylation [107], palmitoylation of neuronal proteins [108], and dynamic imaging of palmitoylation [109]. Furthermore, similar approaches have proven effective for characterizing protein cholesterylation [110] as well as bacterial lipoproteins [111]. Otherwise, covalent lipidation has been adaptable in order to label a broader range of proteins [112, 113]. For example, Ting and coworkers identified a minimalist 22-amino acid sequence of the enzyme lipoic acid ligase that can be added to the N- or C-terminus of various proteins to introduce an azidoacyl chain [114]. Finally, the fact that covalent lipidation introduces a bioorthogonal tag onto target proteins at a specific location has been exploited for covalent modification of proteins, such as through immobilization onto surfaces in a geometrically defined fashion [115–118].

While prior studies focused on the covalent modification of proteins via physiological pathways, there has also been interest in using similar approaches to characterize aberrant covalent modification events. Marnett and coworkers used a bioorthogonal labeling strategy to characterize covalent adducts of 4-hydroxynonenal (HNE), an electrophilic species that results from oxidative stress [119, 120]. Towards this end, synthetic azido- and alkynyl-HNE analogs **43a–b** were prepared and studied for the covalent labeling of proteins, which was initially analyzed using three short peptide sequences with different known adduction sites (His, Lys, or Cys). While most of the probe reactions provided the same results as natural HNE, the exception was the labeling of Cys with azido-HNE **43a**, which yielded 50% less adduct. Probes **43a–b** were also shown to have similar toxicity profiles compared to natural HNE. Next, the conditions for post-labeling via CuAAC were optimized, and sodium ascorbate was reported to be a significantly better reducing agent for promoting triazole formation when compared to tris-(2-carboxyethyl)phosphine (TCEP). Subsequently, these compounds were exploited to characterize covalent protein adducts from RKO cell lines using biotinylation via either CuAAC or the Staudinger ligation, followed by in-gel and mass spectrometric analysis. In these studies, the authors noted problems eluting biotin-conjugated proteins from beads when using the Staudinger ligation, and thus derivatization via CuAAC was found to be more effective. Also, alkynyl-HNE **43b** was observed to yield less nonspecific labeling compared to azido-HNE **43a**. Proteomic experiments resulted in the labeling of a number of proteins involved in stress responses, such as heat shock proteins.

4.7 CONCLUSION

From the examples described in this chapter, it is clear that, following the initial formulation of the philosophy of click chemistry, this strategy has quickly led to significant impacts regarding the study and application of lipid structures. In particular, the bioorthogonal nature of these recently discovered reactions has allowed researchers to infiltrate the highly complex environments of cellular membranes for efforts aimed at further understanding and harnessing the activities of target lipids. From the direct labeling and imaging of specific lipid and membrane structures in living samples to the ability to fish out proteins that are modified through interactions with lipids, studies facilitated by click chemistry show great promise for a paradigm shift in the methods by which these enigmatic molecules are analyzed. However, many of the studies discussed herein represent the initial wave of reports that showcase the efficacy of click-chemistry-based studies. In the future, the success of these approaches will be judged by the extent to which they are successfully implemented to answer important questions pertaining to the activity and properties of lipids, as well as the applications that they enable for exploiting these properties. Nevertheless, the early reports point to a bright future for click-chemistry-based lipid studies.

REFERENCES

1. Cho, W. H.; Stahelin, R. V. *Ann. Rev. Biophys. Biomol. Struct.* **2005**, *34*, 119.
2. Lemmon, M. A. *Nat. Rev. Mol. Cell. Biol.* **2008**, *9*, 99.
3. Resh, M. D. *Nat. Chem. Biol.* **2006**, *2*, 584.
4. Jahn, R.; Lang, T.; Sudhof, T. C. *Cell.* **2003**, *112*, 519.
5. Brochu, H.; Polidori, A.; Pucci, B.; Vermette, P. *Curr. Drug. Deliv.* **2004**, *1*, 299.
6. Ferrara, K. W.; Borden, M. A.; Zhang, H. *Acc. Chem. Res.* **2009**, *42*, 881.
7. Mintzer, M. A.; Simanek, E. E. *Chem. Rev.* **2009**, *109*, 259.
8. Vermette, P.; Meagher, L.; Gagnon, E.; Griesser, H. J.; Doillon, C. J. *J. Cont. Rel.* **2002**, *80*, 179.
9. Cho, W. W.; Bittova, L.; Stahelin, R. V. *Anal. Biochem.* **2001**, *296*, 153.

10. Narayan, K.; Lemmon, M. A. *Methods* **2006**, *39*, 122.
11. Chan, Y. H. M.; Boxer, S. G. *Curr. Opin. Chem. Biol.* **2007**, *11*, 581.
12. Jelinek, R.; Silbert, L. *Mol. Biosyst.* **2009**, *5*, 811.
13. Best, M. D.; Rowland, M. M.; Bostic, H. E. *Acc. Chem. Res.* **2011**, *44*, 686.
14. Wu, P.; Fokin, V. V. *Aldrichim. Acta* **2007**, *40*, 7.
15. Debets, M. F.; van der Doelen, C. W. J.; Rutjes, F.; van Delft, F. L. *ChemBioChem.* **2010**, *11*, 1168.
16. Mamidyala, S. K.; Finn, M. G. *Chem. Soc. Rev.* **2010**, *39*, 1252.
17. Moses, J. E.; Moorhouse, A. D. *Chem. Soc. Rev.* **2007**, *36*, 1249.
18. Sletten, E. M.; Bertozzi, C. R. *Angew. Chem., Int. Edit.* **2009**, *48*, 6974.
19. Best, M. D. *Biochemistry* **2009**, *48*, 6571.
20. Salisbury, C. M.; Cravatt, B. F. *QSAR. Comb. Sci.* **2007**, *26*, 1229.
21. Kolb, H. C.; Sharpless, K. B. *Drug. Disc. Tod.* **2003**, *8*, 1128.
22. Tron, G. C.; Pirali, T.; Billington, R. A.; Canonico, P. L.; Sorba, G.; Genazzani, A. A. *Med. Res. Rev.* **2008**, *28*, 278.
23. Miljanic, O. S.; Dichtel, W. R.; Aprahamian, I.; Rohde, R. D.; Agnew, H. D.; Heath, J. R.; Stoddart, J. F. *QSAR. Comb. Sci.* **2007**, *26*, 1165.
24. Hawker, C. J.; Fokin, V. V.; Finn, M. G.; Sharpless, K. B. *Aust. J. Chem.* **2007**, *60*, 381.
25. Lutz, J. F. *Angew. Chem., Int. Ed.* **2007**, *46*, 1018.
26. Kolb, H. C.; Finn, M. G.; Sharpless, K. B. *Angew. Chem., Int. Ed.* **2001**, *40*, 2004.
27. Huisgen, R. *1,3-Dipolar Cycloaddition Chemistry*; Padwa, A., Ed.; Wiley: New York, **1984**; Vol. *1*, p 1.
28. Rostovtsev, V. V.; Green, L. G.; Fokin, V. V.; Sharpless, K. B. *Angew. Chem., Int. Ed.* **2002**, *41*, 2596.
29. Torne, C. W.; Christensen, C.; Meldal, M. *J. Org. Chem.* **2002**, *67*, 3057.
30. Saxon, E.; Bertozzi, C. R. *Science* **2000**, *287*, 2007.
31. Kohn, M.; Breinbauer, R. *Angew. Chem., Int. Ed.* **2004**, *43*, 3106.
32. Staudinger, H.; Meyer, J. *Helv. Chim. Acta.* **1919**, *2*, 635.
33. Nilsson, B. L.; Kiessling, L. L.; Raines, R. T. *Org. Lett.* **2000**, *2*, 1939.
34. Nilsson, B. L.; Kiessling, L. L.; Raines, R. T. *Org. Lett.* **2001**, *3*, 9.
35. Saxon, E.; Armstrong, J. I.; Bertozzi, C. R. *Org. Lett.* **2000**, *2*, 2141.
36. Soellner, M. B.; Nilsson, B. L.; Raines, R. T. *J. Org. Chem.* **2002**, *67*, 4993.
37. Hong, V.; Presolski, S. I.; Ma, C.; Finn, M. G. *Angew. Chem., Int. Edit.* **2009**, *48*, 9879.
38. Lewis, W. G.; Magallon, F. G.; Fokin, V. V.; Finn, M. G. *J. Am. Chem. Soc.* **2004**, *126*, 9152.
39. Chan, T. R.; Hilgraf, R.; Sharpless, K. B.; Fokin, V. V. *Org. Lett.* **2004**, *6*, 2853.
40. Sen Gupta, S.; Kuzelka, J.; Singh, P.; Lewis, W. G.; Manchester, M.; Finn, M. G. *Bioconjugate Chem.* **2005**, *16*, 1572.
41. Rodionov, V. O.; Presolski, S. I.; Diaz, D. D.; Fokin, V. V.; Finn, M. G. *J. Am. Chem. Soc.* **2007**, *129*, 12705.
42. Rodionov, V. O.; Presolski, S. I.; Gardinier, S.; Lim, Y. H.; Finn, M. G. *J. Am. Chem. Soc.* **2007**, *129*, 12696.

43. del Amo, D. S.; Wang, W.; Jiang, H.; Besanceney, C.; Yan, A. C.; Levy, M.; Liu, Y.; Marlow, F. L.; Wu, P. *J. Am. Chem. Soc.* **2010**, *132*, 16893.
44. Soares, E. V.; Hebbelinck, K.; Soares, H. *Can. J. Microbiol.* **2003**, *49*, 336.
45. Agard, N. J.; Prescher, J. A.; Bertozzi, C. R. *J. Am. Chem. Soc.* **2004**, *126*, 15046.
46. Codelli, J. A.; Baskin, J. M.; Agard, N. J.; Berozzi, C. R. *J. Am. Chem. Soc.* **2008**, *130*, 11486.
47. Sletten, E. M.; Bertozzi, C. R. *Org. Lett.* **2008**, *10*, 3097.
48. Ning, X. H.; Guo, J.; Wolfert, M. A.; Boons, G. J. *Angew. Chem., Int. Edit.* **2008**, *47*, 2253.
49. Kuzmin, A.; Poloukhtine, A.; Wolfert, M. A.; Popik, V. V. *Bioconjugate Chem.* **2010**, *21*, 2076.
50. Debets, M. F.; van Berkel, S. S.; Schoffelen, S.; Rutjes, F.; van Hest, J. C. M.; van Delft, F. L. *Chem. Commun.* **2010**, *46*, 97.
51. Baskin, J. M.; Prescher, J. A.; Laughlin, S. T.; Agard, N. J.; Chang, P. V.; Miller, I. A.; Lo, A.; Codelli, J. A.; Bertozzit, C. R. *Proc. Natl. Acad. Sci. U. S. A.* **2007**, *104*, 16793.
52. Laughlin, S. T.; Baskin, J. M.; Amacher, S. L.; Bertozzi, C. R. *Science* **2008**, *320*, 664.
53. Laughlin, S. T.; Bertozzi, C. R. *Proc. Natl. Acad. Sci. USA* **2009**, *106*, 12.
54. Hassane, F. S.; Frisch, B.; Schuber, F. *Bioconjugate Chem.* **2006**, *17*, 849.
55. Cavalli, S.; Tipton, A. R.; Overhand, M.; Kros, A. *Chem. Commun.* **2006**, *30*, 3193.
56. van Lengerich, B.; Rawle, R. J.; Boxer, S. G. *Langmuir* **2010**, *26*, 8666.
57. Zhang, H. L.; Ma, Y.; Sun, X. L. *Chem. Commun.* **2009**, *21*, 3032.
58. Bostic, H. E.; Smith, M. D.; Poloukhtine, A. A.; Popik, V. V.; Best, M. D. *Chem. Commun.* **2012**, *48*, 1431.
59. Losey, E. A.; Smith, M. D.; Meng, M.; Best, M. D. *Bioconjugate Chem.* **2009**, *20*, 376.
60. Loosli, F.; Doval, D. A.; Grassi, D.; Zaffalon, P.-L.; Favarger, F.; Zumbuehl, A. *Chem. Commun.* **2012**, *48*, 1604.
61. Musiol, H. J.; Dong, S. L.; Kaiser, M.; Bausinger, R.; Zumbusch, A.; Bertsch, U.; Moroder, L. *ChemBioChem.* **2005**, *6*, 625.
62. Smith, M. D.; Gong, D. H.; Sudhahar, C. G.; Reno, J. C.; Stahelin, R. V.; Best, M. D. *Bioconjugate Chem.* **2008**, *19*, 1855.
63. Smith, M. D.; Sudhahar, C. G.; Gong, D.; Stahelin, R. V.; Best, M. D. *Mol. Biosyst.* **2009**, *5*, 962.
64. O'Neil, E. J.; DiVittorio, K. M.; Smith, B. D. *Org. Lett.* **2007**, *9*, 199.
65. Zerkowski, J. A.; Nunez, A.; Strahan, G. D.; Solaiman, D. K. Y. *J. Am. Oil Chem. Soc.* **2009**, *86*, 1115.
66. Tang, S. C.; Wang, Q. L.; Guo, Z. W. *Chem. Eur. J.* **2010**, *16*, 1319.
67. Neef, A. B.; Schultz, C. *Angew. Chem., Int. Ed.* **2009**, *48*, 1498.
68. Jao, C. Y.; Roth, M.; Welti, R.; Salic, A. *Proc. Natl. Acad. Sci. USA* **2009**, *106*, 15332.
69. Sandbhor, M. S.; Key, J. A.; Strelkov, I. S.; Cairo, C. W. *J. Org. Chem.* **2009**, *74*, 8669.
70. Lampkins, A. J.; O'Neil, E. J.; Smith, B. D. *J. Org. Chem.* **2008**, *73*, 6053.
71. Cravatt, B. F.; Wright, A. T.; Kozarich, J. W. *Annu. Rev. Biochem.* **2008**, *77*, 383.
72. Dorman, G.; Prestwich, G. D. *Trends. Biotechnol.* **2000**, *18*, 64.

73. Gubbens, J.; Ruijter, E.; de Fays, L. E. V.; Damen, J. M. A.; de Kruijff, B.; Slijper, M.; Rijkers, D. T. S.; Liskamp, R. M. J.; de Kroon, A. *Chem. Biol.* **2009**, *16*, 3.
74. Gubbens, J.; de Kroon, A. *Mol. Biosyst.* **2010**, *6*, 1751.
75. Tully, S. E.; Cravatt, B. F. *J. Am. Chem. Soc.* **2010**, *132*, 3264.
76. Bandyopadhyay, S.; Bong, D. *Eur. J. Org. Chem.* **2011**, *4*, 751.
77. Gong, D.; Bostic, H. E.; Smith, M. D.; Best, M. D. *Eur. J. Org. Chem.* **2009**, *24*, 4170.
78. Rowland, M. M.; Bostic, H. E.; Gong, D.; Speers, A. E.; Lucas, N.; Cho, W.; Cravatt, B. F.; Best, M. D. *Biochemistry* **2011**, *50*, 11143.
79. Di Paolo, G.; De Camilli, P. *Nature* **2006**, *443*, 651.
80. Pendaries, C.; Tronchere, H.; Plantavid, M.; Payrastre, B. *FEBS. Lett.* **2003**, *546*, 25.
81. Vicinanza, M.; D'Angelo, G.; Di Campli, A.; De Matteis, M. A. *Cell. Mol. Life Sci.* **2008**, *65*, 2833.
82. Wymann, M. P.; Schneiter, R. *Nat. Rev. Mol. Cell. Biol.* **2008**, *9*, 162.
83. Best, M. D.; Zhang, H. L.; Prestwich, G. D. *Nat. Prod. Rep.* **2010**, *27*, 1403.
84. Wittinghofer, A.; Waldmann, H. *Angew. Chem., Int. Ed.* **2000**, *39*, 4193.
85. Hannoush, A. N.; Sun, J. *Nat. Chem. Biol.* **2010**, *6*, 498.
86. Hang, H. C.; Wilson, J. P.; Charron, G. *Acc. Chem. Res.* **2011**, *44*, 699.
87. Charron, G.; Wilson, J.; Hang, H. C. *Curr. Opin. Chem. Biol.* **2009**, *13*, 382.
88. Kho, Y.; Kim, S. C.; Jiang, C.; Barma, D.; Kwon, S. W.; Cheng, J. K.; Jaunbergs, J.; Weinbaum, C.; Tamanoi, F.; Falck, J.; Zhao, Y. M. *Proc. Natl. Acad. Sci. USA.* **2004**, *101*, 12479.
89. Charron, G.; Tsou, L. K.; Maguire, W.; Yount, J. S.; Hang, H. C. *Mol. Biosyst.* **2011**, *7*, 67.
90. Chan, L. N.; Hart, C.; Guo, L.; Nyberg, T.; Davies, B. S. J.; Fong, L. G.; Young, S. G.; Agnew, B. J.; Tamanoi, F. *Electrophoresis.* **2009**, *30*, 3598.
91. DeGraw, A. J.; Palsuledesai, C.; Ochocki, J. D.; Dozier, J. K.; Lenevich, S.; Rashidian, M.; Distefano, M. D. *Chem. Biol. Drug Des.* **2010**, *76*, 460.
92. Rose, M. W.; Xu, J. H.; Kale, T. A.; O'Doherty, G.; Barany, G.; Distefano, M. D. *Biopolymers* **2005**, *80*, 164.
93. Rose, M. W.; Rose, N. D.; Boggs, J.; Lenevich, S.; Xu, J.; Barany, G.; Distefano, M. D. *J. Peptide Res.* **2005**, *65*, 529.
94. Duckworth, B. P.; Zhang, Z. Y.; Hosokawa, A.; Distefano, M. D. *ChemBioChem.* **2007**, *8*, 98.
95. Wollack, J. W.; Silverman, J. M.; Petzold, C. J.; Mougous, J. D.; Distefano, M. D. *ChemBioChem.* **2009**, *10*, 2934.
96. Berry, A. F. H.; Heal, W. P.; Tarafder, A. K.; Tolmachova, T.; Baron, R. A.; Seabra, M. C.; Tate, E. W. *ChemBioChem.* **2010**, *11*, 771.
97. Hang, H. C.; Geutjes, E. J.; Grotenbreg, G.; Pollington, A. M.; Bijlmakers, M. J.; Ploegh, H. L. *J. Am. Chem. Soc.* **2007**, *129*, 2744.
98. Charron, G.; Zhang, M. Z. M.; Yount, J. S.; Wilson, J.; Raghavan, A. S.; Shamir, E.; Hang, H. C. *J. Am. Chem. Soc.* **2009**, *131*, 4967.
99. Yap, M. C.; Kostiuk, M. A.; Martin, D. D. O.; Perinpanayagam, M. A.; Hak, P. G.; Siddam, A.; Majjigapu, J. R.; Rajaiah, G.; Keller, B. O.; Prescher, J. A.; Wu, P.; Bertozzi, C. R.; Falck, J. R.; Berthiaume, L. G. *J. Lipid Res.* **2010**, *51*, 1566.

100. Hannoush, R. N.; Arenas-Ramirez, N. *ACS Chem. Biol.* **2009**, *4*, 581.

101. Martin, B. R.; Cravatt, B. F. *Nat. Methods* **2009**, *6*, 135.

102. Kostiuk, M. A.; Keller, B. O.; Berthiaume, L. G. *Methods in Enzymology; Mitochondrial Function, Partb Mitochondrial Protein Kinases, Protein Phosphatases and Mitochondrial Diseases*; Allison, W. S., Murphy, A. N., Eds.; Elsevier Academic Press Inc.: San Diego, **2009**; Vol. *457*, pp 149.

103. Martin, D. D. O.; Vilas, G. L.; Prescher, J. A.; Rajaiah, G.; Falck, J. R.; Bertozzi, C. R.; Berthiaume, L. G. *FASEB J.* **2008**, *22*, 797.

104. Kostiuk, M. A.; Corvi, M. M.; Keller, B. O.; Plummer, G.; Prescher, J. A.; Hangauer, M. J.; Bertozzi, C. R.; Rajaiah, G.; Falck, J. R.; Berthiaume, L. G. *FASEB J.* **2008**, *22*, 721.

105. Ladygina, N.; Martin, B. R.; Altman, A. *Advances in Immunology*; Alt, F. W., Austen, K. F., Honjo, T., Melchers, F., Uhr, J. W., Unanue, E. R., Eds.; Academic Press: New York, **2011**; Vol. 109, p 1.

106. Wilson, J. P.; Raghavan, A. S.; Yang, Y. Y.; Charron, G.; Hang, H. C. *Mol. Cell. Proteom.* **2011**, *10*.

107. Yount, J. S.; Moltedo, B.; Yang, Y. Y.; Charron, G.; Moran, T. M.; Lopez, C. B.; Hang, H. C. *Nat. Chem. Biol.* **2010**, *6*, 610.

108. Kang, R. J.; Wan, J. M.; Arstikaitis, P.; Takahashi, H.; Huang, K.; Bailey, A. O.; Thompson, J. X.; Roth, A. F.; Drisdel, R. C.; Mastro, R.; Green, W. N.; Yates, J. R.; Davis, N. G.; El-Husseini, A. *Nature* **2008**, *456*, 904.

109. Zhang, M. M.; Tsou, L. K.; Charron, G.; Raghavan, A. S.; Hang, H. C. *Proc. Natl. Acad. Sci. USA* **2010**, *107*, 8627.

110. Heal, W. P.; Jovanovic, B.; Bessin, S.; Wright, M. H.; Magee, A. I.; Tate, E. W. *Chem. Commun.* **2011**, *47*, 4081.

111. Rangan, K. J.; Yang, Y. Y.; Charron, G.; Hang, H. C. *J. Am. Chem. Soc.* **2010**, *132*, 10628.

112. Heal, W. P.; Wickramasinghe, S. R.; Leatherbarrow, R. J.; Tate, E. W. *Org. Biomol. Chem.* **2008**, *6*, 2308.

113. Heal, W. P.; Wickramasinghe, S. R.; Bowyer, P. W.; Holder, A. A.; Smith, D. F.; Leatherbarrow, R. J.; Tate, E. W. *Chem. Commun.* **2008**, *6*, 480.

114. Fernandez-Suarez, M.; Baruah, H.; Martinez-Hernandez, L.; Xie, K. T.; Baskin, J. M.; Bertozzi, C. R.; Ting, A. Y. *Nat. Biotechnol.* **2007**, *25*, 1483.

115. Gauchet, C.; Labadie, G. R.; Poulter, C. D. *J. Am. Chem. Soc.* **2006**, *128*, 9274.

116. Duckworth, B. P.; Xu, J. H.; Taton, T. A.; Guo, A.; Distefano, M. D. *Bioconjugate Chem.* **2006**, *17*, 967.

117. Nguyen, U. T. T.; Cramer, J.; Gomis, J.; Reents, R.; Gutierrez-Rodriguez, M.; Goody, R. S.; Alexandrov, K.; Waldmann, H. *ChemBioChem.* **2007**, *8*, 408.

118. Chen, Y.-X.; Triola, G.; Waldmann, H. *Acc. Chem. Res.* **2011**, *44*, 762.

119. Vila, A.; Tallman, K. A.; Jacobs, A. T.; Liebler, D. C.; Porter, N. A.; Marnett, L. J. *Chem. Res. Toxicol.* **2008**, *21*, 432.

120. Jacobs, A. T.; Marnett, L. J. *Acc. Chem. Res.* **2010**, *43*, 673.

5

CLICKING SUGARS ONTO SUGARS: OLIGOSACCHARIDE ANALOGS AND GLYCOCLUSTERS ON CARBOHYDRATE SCAFFOLDS

Maria Laura Uhrig and José Kovensky

5.1 INTRODUCTION

Many biologically relevant macromolecules are composed of carbohydrates. Oligosaccharides are commonly found as glycoconjugates having specific roles inside the cell, or exposed on the cell surface to the extracellular media, taking part in remarkable recognition features. The different arrangements of the residues, in terms of glycosylated positions, stereochemistry, and substitution, allow a huge diversity of structures. It is often said that they serve to "decorate" mainly proteins and lipids, as a way to signify the relevance of the structure diversity of carbohydrates in many pathological and physiological processes.

The recognition of the different oligosaccharide structures by specific receptors (i.e., proteins), trigger many biological events. For example, cell–cell recognition and adhesion processes, and also signal transduction pathways and their modulation, are direct consequences of the exposition of certain sugars linked in specific ways [1]. In the case of many proteins and lipids, their functionality depends exclusively on the presence or absence of the oligosaccharidic chain.

Among carbohydrate–protein interactions, those involving lectins modulate many biological events including inflammation [2], immune response, apoptosis, tumor metastasis or viral and bacterial infections [1c–f, 3, 4]. It is well known that lectins

Click Chemistry in Glycoscience: New Developments and Strategies, First Edition.
Edited by Zbigniew J. Witczak and Roman Bielski.

are carbohydrate-binding proteins, whose specificity depends on the size and shape of the recognition site. By interaction with their specific oligosaccharide ligands, intracellular lectins (for example, galectins) become functional, and thus, an optimal transmission of certain signals take place inside the cell [5]. Extracellular lectins are usually involved in adhesion of bacteria and other pathogens, as a first step in infection processes [6].

A better understanding of the role of sugar–protein interactions triggering such important biological effects requires carbohydrate-based molecules having defined structures. Moreover, the development of efficient synthetic methodologies to construct the desired molecules, as well as physiologically stable analogs, is of great interest. And when data about the structure–function relationship is available, analogs having specific geometries and properties can be designed.

The Huisgen 1,3-dipolar cycloaddition of azides and alkynes [7] to afford triazoles is probably the most powerful "click" reaction to connect two distinct fragments or residues of similar or different nature. Indeed, the reaction has shown a number of advantages over other methods. So, by using this reaction, carbohydrates have been linked to proteins, lipids, linkers, polymers and other sugars, giving rise to a heterogeneous family of new compounds with promising physicochemical and biological properties [8]. Interest for this reaction became clear after the discovery of the advantages of Cu(I) as catalyst reported independently by Sharpless [9] and Meldal [10] groups, the catalyzed reaction leads regioselectively under mild conditions to 1,4-disubstituted-1,2,3-triazoles in high yields. This constitutes a substantial improvement of the classical version of the thermal transformation, which affords mixtures of 1,4- and 1,5-disubstituted triazoles.

In this catalyzed version, known as the copper(I)-catalyzed azide–alkyne cycloaddition (CuAAC), the Cu(I) can be added as a salt (for example, CuCl), or generated *in situ* from $CuSO_4$ and ascorbic acid [11]. Although many groups still use Cu(I) salts and complexes as catalysts, the most popular methodology employed nowadays involves the $CuSO_4$/ascorbic acid combination, due to its simplicity and lower cost. Recent developments [12] showed that cyclooctynes react with azides in the absence of copper catalysts, and therefore, the cycloaddition reaction became compatible with living systems. However, its up-to-date limited use led us to consider it out of the scope of this review.

The CuAAC owes its usefulness in part to its high compatibility with functional groups (alcohols, carboxylic acids, amines) in different solvent systems including water. In the field of carbohydrate chemistry [13], the click reaction has been used for the synthesis of glycoconjugates [11, 14] and carbohydrate macrocycles [15] where a sugar possessing an azido function is grafted onto a saccharide [16], a peptide [17] or a polymeric chain [18]. Also, this methodology has recently been employed for the synthesis of glycosidase inhibitors [19].

An important fact to be considered is that the resulting triazole ring is stable to hydrolytic cleavage and virtually inert towards oxidation, reduction, and other metabolic transformations [8b, 20]. In addition, 1,2,3-triazoles are interesting pharmacophores in view of the topological and electronic similarities of triazoles and amide bonds [21], and they can improve the binding to biologically relevant molecules through hydrogen-bonding and/or dipole interactions [14, 22].

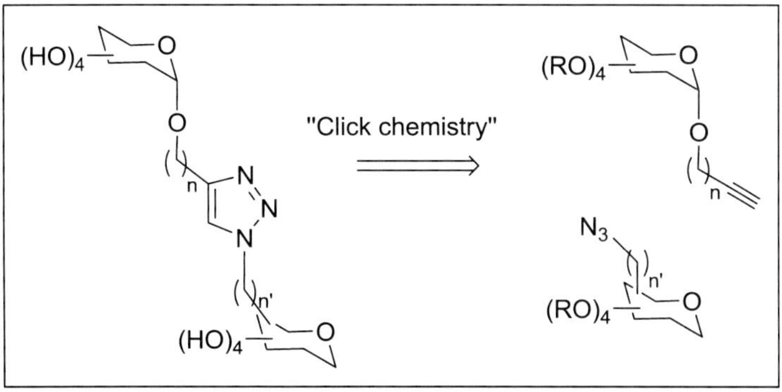

SCHEME 5.1 General retrosynthetic click reaction.

5.2 TRIAZOLE-LINKED OLIGOSACCHARIDE ANALOGS

In order to link a carbohydrate to another carbohydrate residue, through a covalent bond such as a 1,2,3-triazole ring, the modification of sugar backbones by functionalization with azide and alkynes groups, respectively, is required. This strategy leads to hybrid compounds, namely oligosaccharide mimics or analogs, with improved chemical properties (Scheme 5.1).

Santoyo-Gonzalez group investigated the reaction of propargyl, thiopropargyl, and aminopropargyl glycosides **1a–c** (Scheme 5.2) using the organic-soluble copper complexes $(Ph_3P)_3$.CuBr and $(EtO)_3$P.CuI. The S- and N-containing derivatives are valuable precursors of carbohydrate mimics with enhanced stability towards enzymatic hydrolysis, and they allowed the study of the influence of the glycosidic linkage in coupling, considering the electronic nature of each alkyne moiety [11].

It was observed that alkyne **1a**, which supports the more electronegative substituent, reacts well with $(Ph_3P)_3$.CuBr–N,N-diisopropylethylamine (DIPEA)

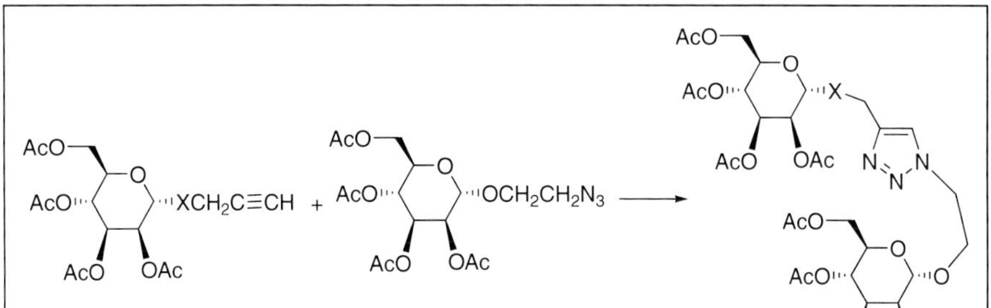

1a, R = O
1b, R – S
1c, R = N(Ac)

2a

3a, R = O
3b, R = S
3c, R = N(Ac)

SCHEME 5.2 Click reaction using organic-soluble copper complexes.

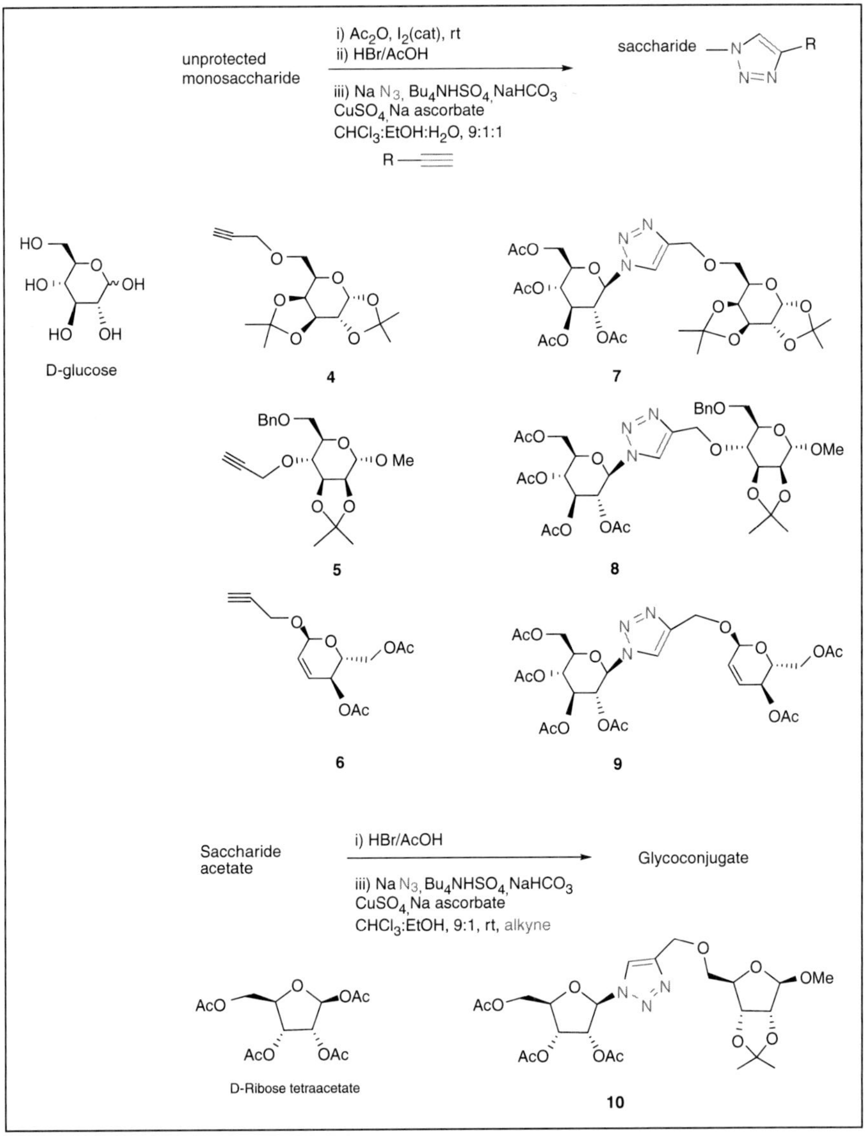

SCHEME 5.3 Cu(II)-ascorbate approach form unprotected or acetylated sugars.

(86% yield), but for **1b** and **1c,** the use of $(EtO)_3P.Cu(I)$–DIPEA led to higher yields (96% and 99%, respectively).

Interestingly, microwave irradiation was used here for the first time for CuAAC reactions, resulting in improved yields. Independently, Appukkutan et al. [23] reported that microwaves significantly accelerate the click reaction.

To overcome the instability of Cu(I), it was shown that it can be generated by *in situ* reduction of Cu(II) salts (e.g., copper sulfate) with sodium ascorbate [11, 24].

This methodology was applied in a one-pot approach from unprotected or acetylated monosaccharides reported by Wang (Scheme 5.3) [25]. Acetylation of D-ribose with acetic anhydride in the presence of a trace amount of iodine as catalyst was followed by brominolysis of the anomeric acetate. After removal of all volatiles and further conversion to the anomeric azide, an *in situ* CuAAC-promoted reaction afforded the triazolylglycoside **10** in 74% yield.

The functionalization of C-3 and C-5 of a sugar central unit with alkyne groups followed by the reaction of 1-azido mannofuranoside in the presence of a Cu(I) catalyst, generated 1,2,3-triazole-linked trisaccharide carbohybrid analog **11** (Scheme 5.4) [26].

Triazol-linked disaccharide analogs as **13** have also been prepared using water as solvent, by reaction of 1-azido-1-deoxy-2,3,4,6-tetra-*O*-acetyl-β-D-glucopyranose with the corresponding terminal alkynes in the presence of $CuSO_4$/ascorbic acid (Scheme 5.5) [27].

Wilkinson and coworkers studied the relative rate of cycloaddition by using a variety of alcohol co-solvents and reaction temperatures. It was found that the reaction proceeded with similar efficiency irrespective of the co-solvent used; however, mildly elevated temperatures (40°C) increased the speed of reaction significantly. They verified that, when using anomeric azides, the anomeric configuration is preserved during the reaction [28].

In our laboratory, we have developed a procedure to connect saccharide residues by click chemistry. This approach showed that different sugars bearing an azide on C-2, C-5 or C-6 (compounds **14–17**) can be used as building blocks in the synthesis of complex oligosaccharidic mimetics by reacting with alkynyl glycosides **18–21** (Scheme 5.6) [29]. The reaction tolerates the wide range of protecting groups

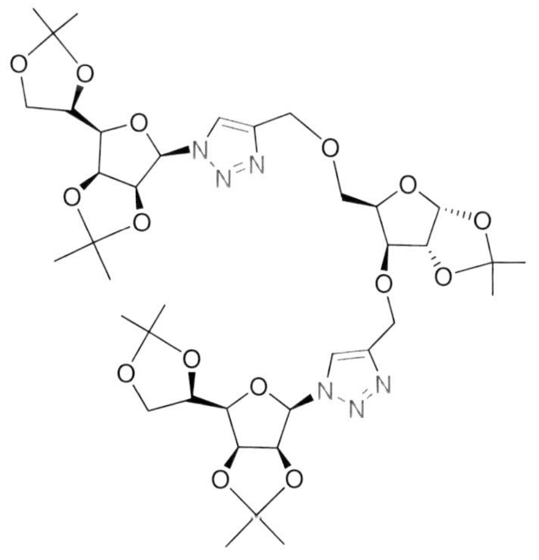

11

SCHEME 5.4 Trisaccharide carbohybrid analog.

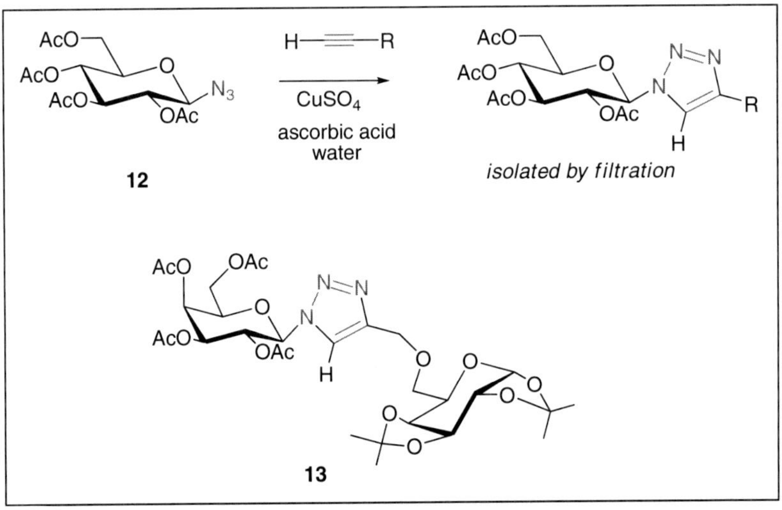

SCHEME 5.5 Click reaction in water.

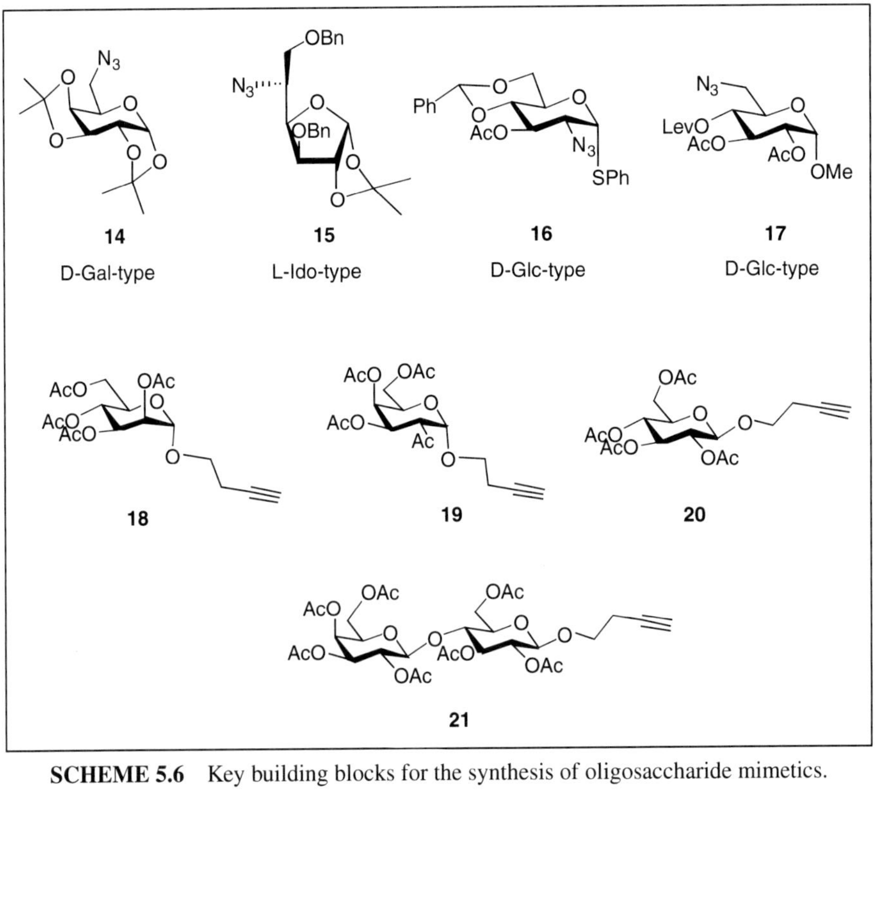

SCHEME 5.6 Key building blocks for the synthesis of oligosaccharide mimetics.

TABLE 5.1 Optimisation of the Click Reaction

Entry	Alkyne	DIPEA	Na Ascorbate	Catalyst	Solvent	Time	Yield
1	1 eq	3 eq	None	0.2 eq CuI	THF	2h	No reaction
2	1.2 eq	4 eq	None	0.2eq CuI	THF	4h	55%
3	1 eq	3 eq	None	2 eq CuI	CH_3CN	1h	49%
4	1 eq	None	0.4 eq	0.2eq $CuSO_4$	*sec*But	3h	No reaction
5	1 eq	None	0.4 eq	0.2eq $CuSO_4$	*sec*But/H_2O	1.1h	71%
6	1 eq	None	0.4 eq	0.2eq $CuSO_4$	DMF	2h	No reaction
7	1 eq	None	0.4 eq	0.2eq $CuSO_4$	DMF/H_2O	1.5h	88%

commonly used in carbohydrate chemistry. Moreover, the reaction was conducted at room temperature, using an equimolar ratio of reagents, and the corresponding products were obtained in high yields. Due to the high solubility of protected sugars in N,N-dimethylformamide (DMF), the addition of water to the solvent, up to 30%, greatly accelerated the cycloaddition reaction between **17** and **18** (Table 5.1). Furthermore, we changed the distance between the triazole-linked saccharide units by the introduction of an appropriate spacer (Scheme 5.7).

The reaction between glycosyl azides and alkynyl glycosides can be accelerated by the addition of Cu(II) sulfate pentahydrate and sodium ascorbate in 1:1 tert-Butanol/water, which results in the formation of triazole derivatives **33** and **35** in 78% and 85% yield, respectively (Scheme 5.8) [30].

Tripathi's group [31] synthesized 1,2,3-1*H*-triazolyl glycohybrids, such as **36** and **38**, by CuAAC coupling of glycosyl azides with 2,3-unsaturated alkynyl glycosides, obtained by Ferrier's rearrangement of glycals (Scheme 5.9). Some of these products (and others like **37** and **39**, containing coumarin residues) showed to be inhibitors of α-glucosidase, glycogen phosphorylase and glucose-6-phosphatase.

The reaction with internal alkynes, however, occurred only under thermal conditions (refluxing toluene) and hence, mixtures of the regioisomers were obtained (Scheme 5.10).

In an interesting approach to sustainable chemistry, the first study of a Cu(I)-catalyzed azide–alkyne click reaction in ionic liquids, was reported by Marra et al. [32]. The cycloaddition of a sugar azide with a sugar acetylene (Cu(I), DIPEA, 80°C) was achieved in high yields (up to 95%) using Ammoeng 110 and $[C_8dabco][N(CN)_2]$ among other ionic liquids (Scheme 5.11). The latter solvent was recycled in four subsequent reactions without loss of reaction efficiency. The authors investigated the role of DIPEA, used as an additive, whose presence seems to be crucial in the case of non-basic ionic liquids.

Nitrogen donors including bases such as DIPEA, and some solvents such as acetonitrile, have been reported to improve the reaction when Cu(I) is used as a catalyst by helping to prevent degradation of Cu(I) by disproportionation or oxidation [33].

A α-sialic acid azide derivative has been used as a substrate for the efficient preparation of 1,2,3-triazole derivatives of sialic acid. These products would act as

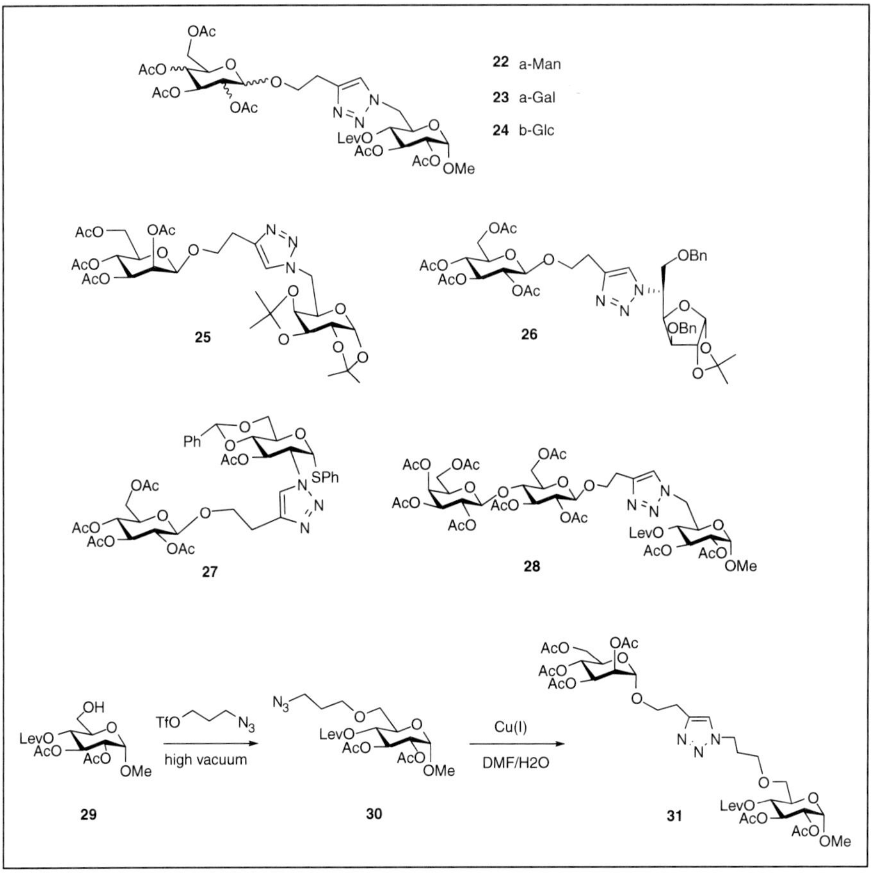

SCHEME 5.7 Clicking sugars with different spacer length.

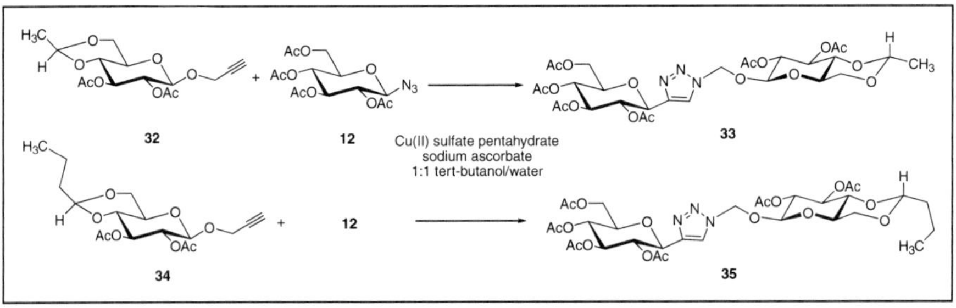

SCHEME 5.8 Click reaction in *tert*-butanol/water.

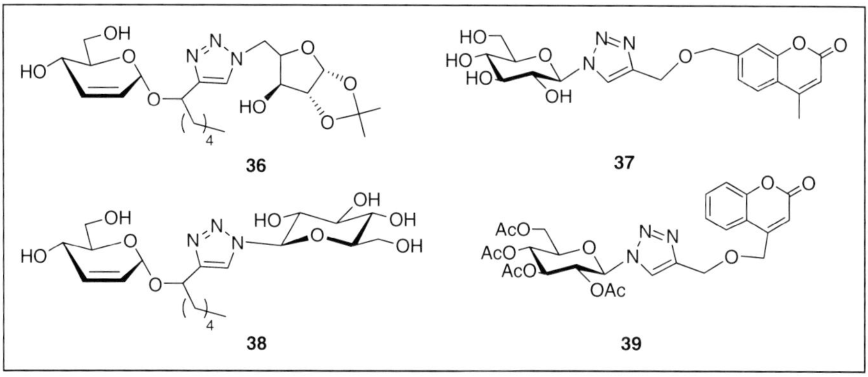

SCHEME 5.9 Triazolyl glycohybrids.

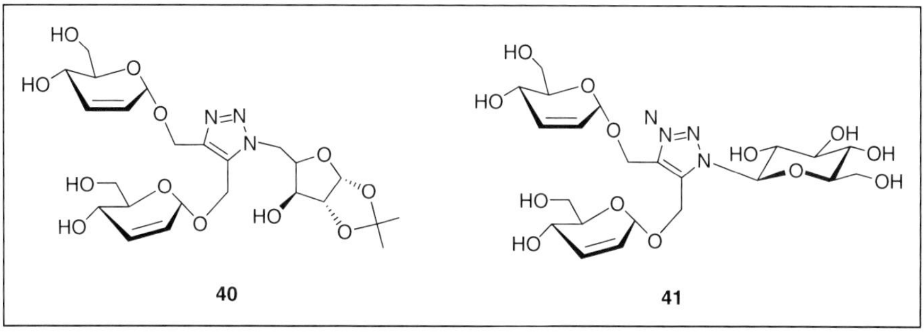

SCHEME 5.10 Thermal reaction with internal alkynes.

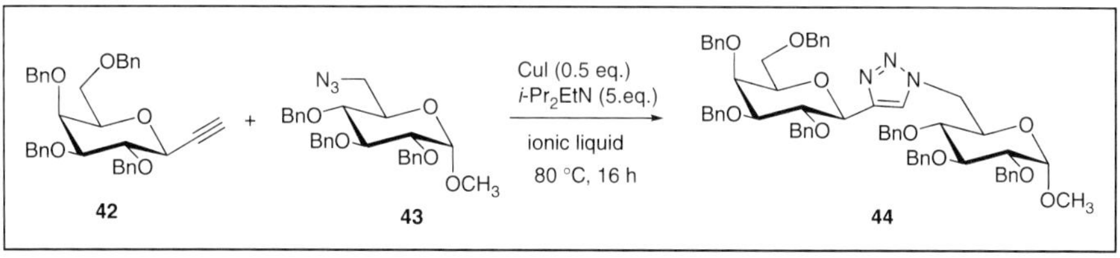

SCHEME 5.11 Click reaction in ionic liquids.

neuraminidase inhibitors to prevent the release of new virions [34]. Among them, the disaccharide mimic **45** (Scheme 5.12) showed high inhibition ($IC_{50} = 17$ mM), and the authors suggested that this analog would mimic the non-reducing end of the glycosidic chains of the glycoproteins present on the surface of epithelial cells.

A family of pseudo-glycoconjugates linked through a triazole ring produced by the Huisgen click reaction was prepared by reaction of glycosides having

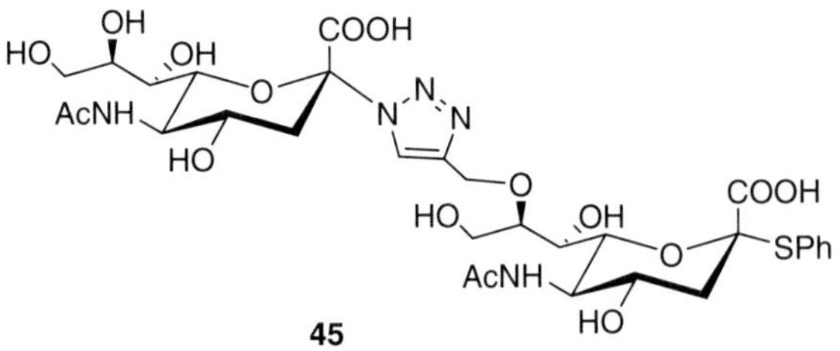

45

SCHEME 5.12 Sialic acid containing mimic.

alkynyl and azidoethyl groups as substituents in their anomeric chains. The reaction proceeded in the presence of Cu(I) and DIPEA, and 1,2,3-triazole bridged homo- and heterodimeric pseudo-oligosaccharides were obtained in 89–93% yields (Scheme 5.13). The methodology was validated using sugar precursors of *gluco-*, *lacto-*, and *xylo-*configuration [35].

Field and coworkers reported the application of click chemistry to the construction of amylopectin analogs composed of two linear maltooligosaccharide chains attached to a maltose (Scheme 5.14) or glucose (Scheme 5.15) unit [36]. The strategy involved

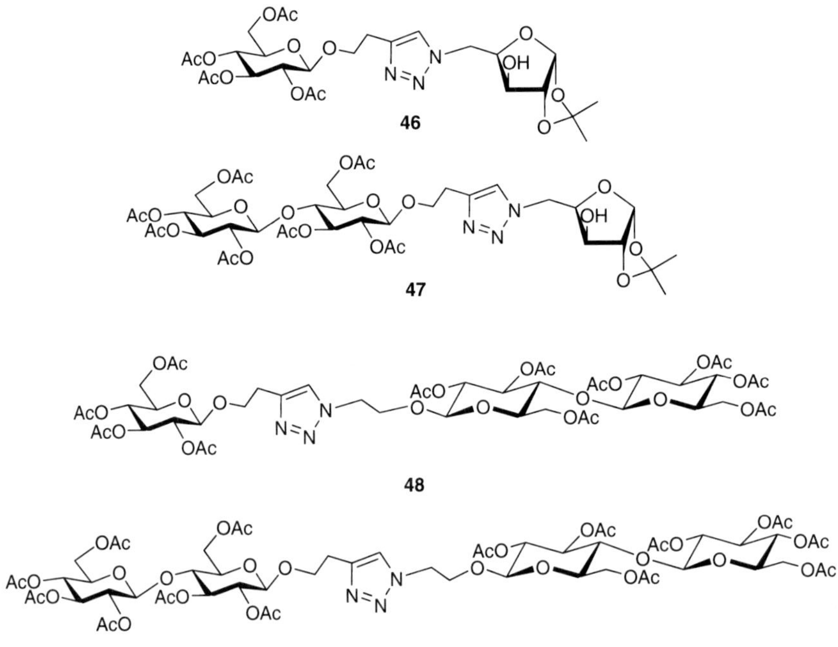

46

47

48

49

SCHEME 5.13 Pseudoglycoconjugates.

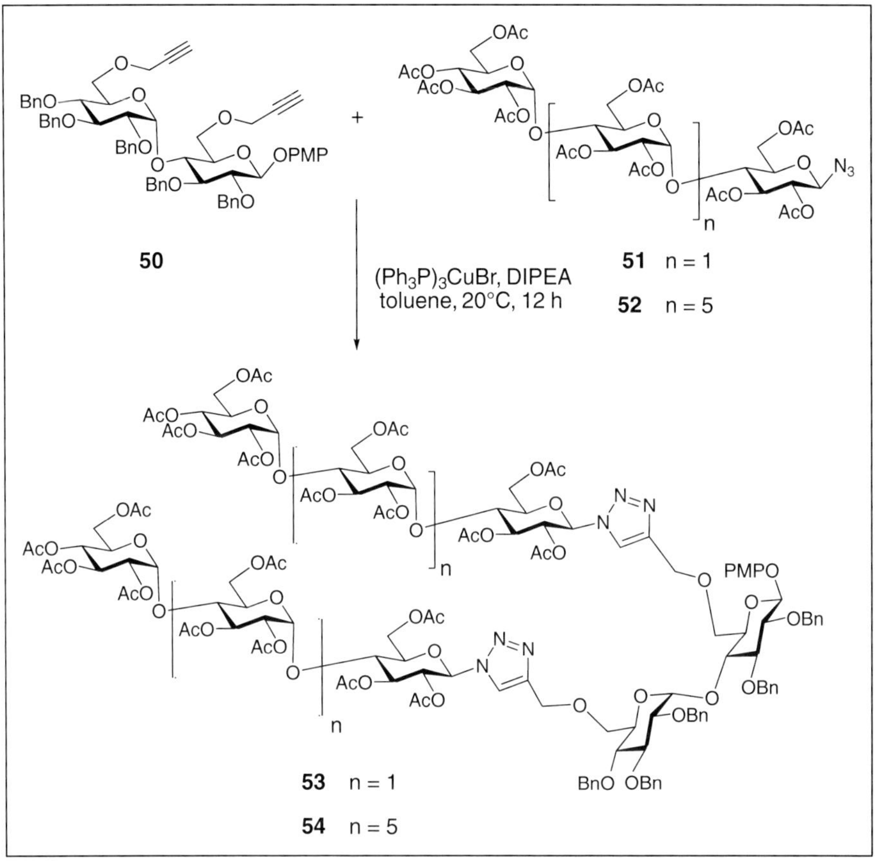

SCHEME 5.14 Synthesis of amylopectin analogs on maltose scaffold.

the synthesis of the regioselectively propargylated maltose derivatives **50** *via* the corresponding trityl precursors. On the other hand, peracetylated β-glycosyl azides of glucopyranose (**56**), maltotriose and maltoheptaose were prepared by reaction of the corresponding glycosyl bromides with Me_3SiN_3 in the presence of Bu_4NF. For the click reaction the authors used $(Ph_3P)_3$.CuBr as a catalyst in the presence of DIPEA as a base. Final compounds (**53**, **54**, **57**) were obtained in 27–65% yields, with yields decreasing when the length of the azidooligosaccharide chain increased.

1,2,3-Triazole glycosides designed as acarbose mimetics have been synthesized and tested as α-glucosidase inhibitors (Scheme 5.16) [19a]. The strategy involved the thermal cycloaddition of an internal asymmetric alkyne which led to a mixture of regioisomers (54% yield), which were isolated by chromatography (Scheme 5.17). These analogs have been included here because they belong to the family of sugar derivatives focused in this chapter, in spite of the fact that the cycloaddition reaction was not Cu(I) catalyzed.

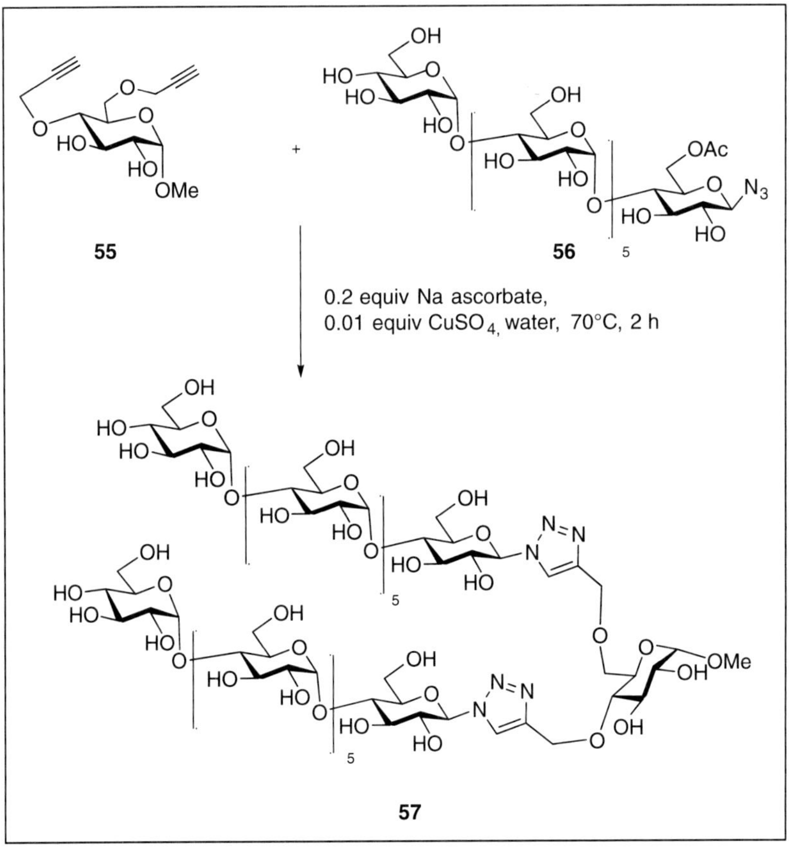

SCHEME 5.15 Synthesis of amylopectin analogs on glucose scaffold.

Triazole-linked mimetics of heparosan and chondroitin disaccharide and tetrasaccharides were synthesized by applying the CuAAC reaction to azido-glucuronic acid and propargylated N-acetyl glucosamine or N-acetyl galactosamine derivatives (Scheme 5.18). The transformations were achieved using Cu(I) in the presence of DIPEA in acetonitrile, leading, for example, to the chondroitin tetrasaccharide analog **67** in 77% yield [37].

Polysaccharides have also been used as starting materials to click sugar residues using the Huisgen reaction. Thus, regioselective bromination–azidation of curdlan, a β-D-(1→3)-glucan (**71**), afforded 6-azido-6-deoxycurdlan (**72**). Different molecules bearing a terminal alkyne group, such as β-lactoside, were clicked to this scaffold. Couplings were carried out at room temperature using $CuBr_2$ in the presence of ascorbic acid and in dimethyl sulfoxide (DMSO) as solvent. A base (triethylamine, propylamine, ammonia) was also added and yields were in the range of 63–78% (Scheme 5.19). The reaction was followed by ATR–FTIR spectroscopy, since azido groups show a clear and strong IR peak at around 2100 cm^{-1} [38].

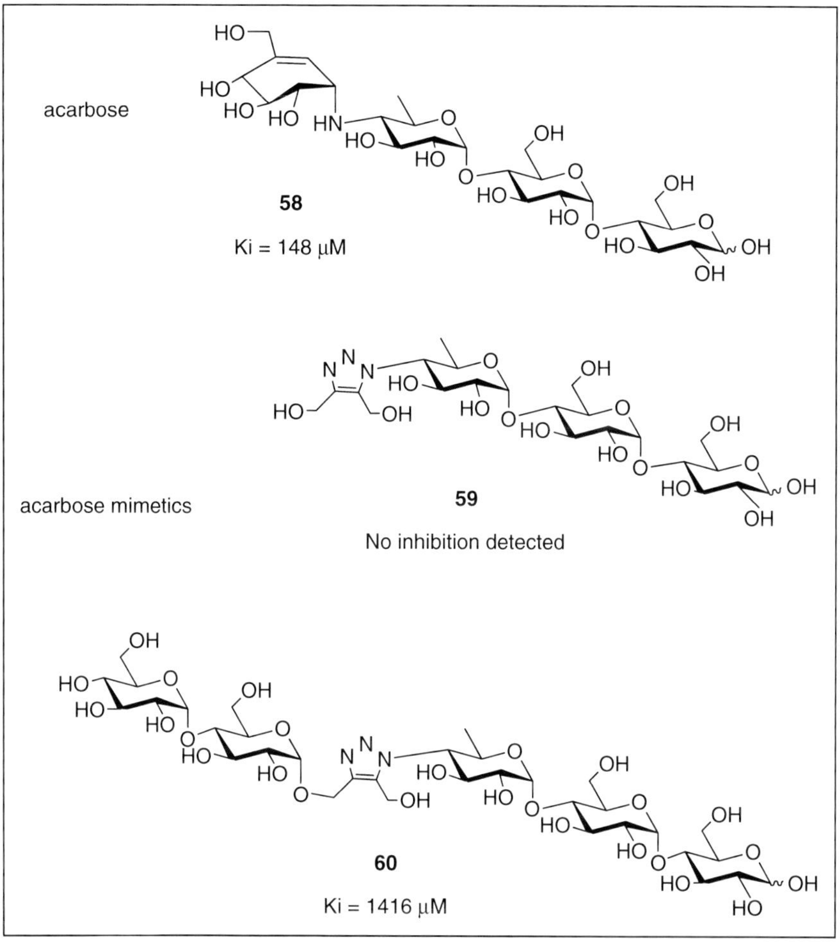

SCHEME 5.16 Acarbose mimetics.

C-glycosides have been also used as precursors in the synthesis of triazol-containing oligosaccharide analogs, resistant to enzymatic hydrolysis. Thus, the preparation of oligomannosides has been reported by Dondoni and coworkers [39], involving the iterative CuAAC reaction. Oligomers having alternating sugar residues and triazole rings have been obtained from ethynyl α-*C*-mannoside **75** and alkyl 6-azido-α-*C*-mannoside **76** (Scheme 5.20). The cycloaddition conditions involved the use of Cu(I) and DIPEA in toluene at room temperature. The strategy involved the transformation of the 6-hydroxy group of **75** into an azido group, by treatment with diphenylphosphoryl azide and 1,8-diazabicyclo[5.4.0]undec-7-ene (DBU). The azide was obtained in good yield and was the substrate in the following cycloaddition step. After deprotection, the free oligomannosides **77** were obtained in 33–45% overall yield from monosaccharide building blocks. The inhibition of mycobacterial

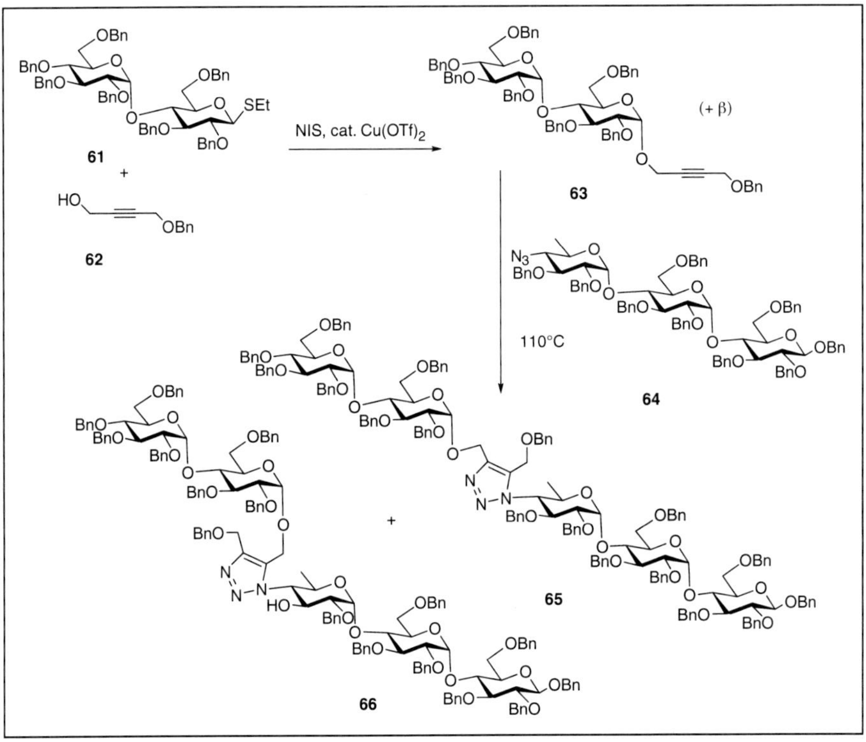

SCHEME 5.17 Synthesis of oligosaccharide analogs obtained by thermal cycloaddition.

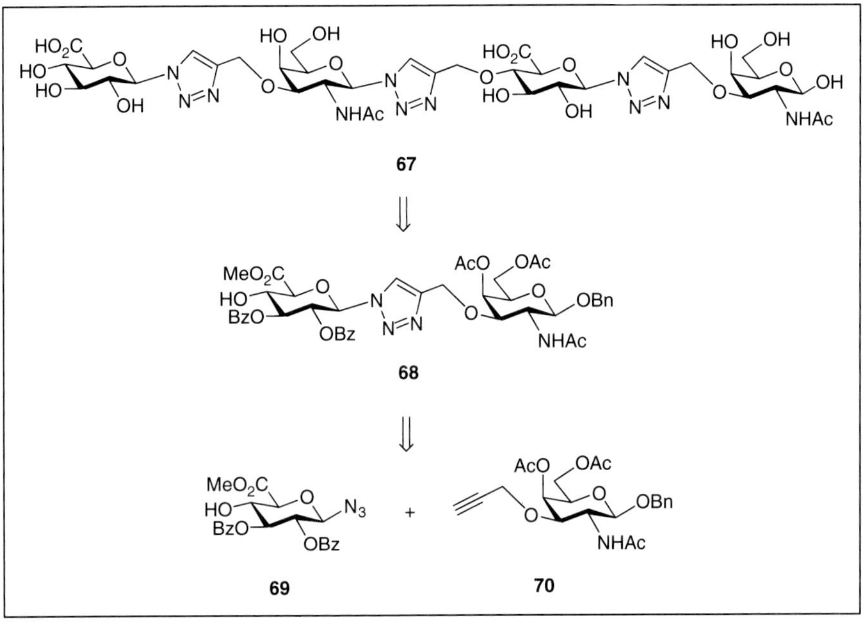

SCHEME 5.18 Retrosynthetic approach for the synthesis of GAGs analogs.

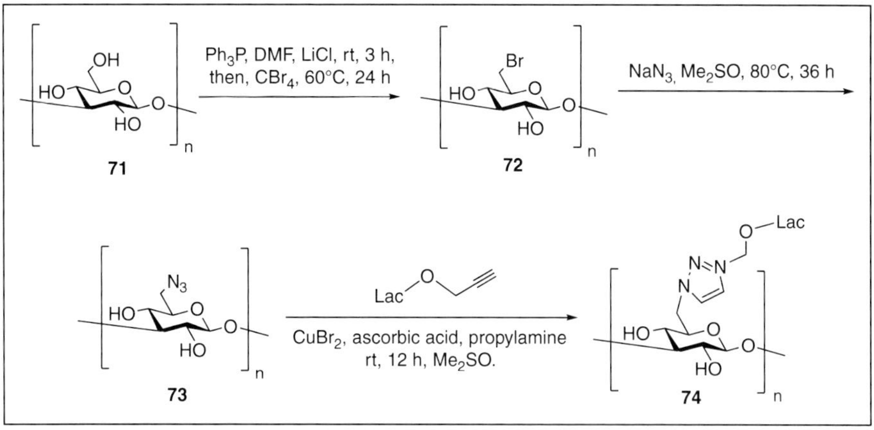

SCHEME 5.19 Synthesis of modified polysaccharides clicked with lactosides.

α-(1→6)-mannosyltransferases by the triazole-linked oligomannosides was determined, the most active compounds were the hexa- and octamannoside analogs.

5.3 CLICKED CYCLODEXTRINS

Glycocyclodextrins can be considered well-defined glyco-oligomers, having symmetric tridimensional structures. For these reasons, cyclodextrins have been selected as scaffolds for the synthesis of multivalent ligands by several groups of investigators [40]. Perglycosylated cyclodextrins (CDs), exposing a variety of sugar moieties such as β-D-glucose, β-D-galactose, α-D-mannose, N-acetyl-β-D-glucosamine and even sialyloligosaccharides, have been synthesized. A large variety of functional groups have been used in order to connect the sugar residues to the platform: among them, amide, urea, thiourea, and thioethers were constructed by photoaddition of thiols to

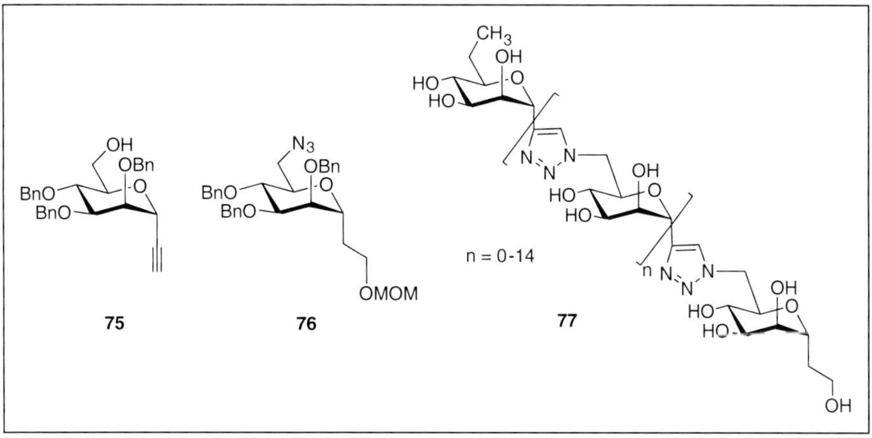

SCHEME 5.20 Triazol-linked C-mannosides.

allylic systems. Remarkably, there are not many examples of the use of the CuAAC methodology as key step in the construction of multivalent ligands.

Nevertheless, a family of heptavalent glycocyclodextrins (**86**, **87**) have been synthesized starting from the heptaazide per-*O*-acetylated β-cyclodextrin **83** and the 2-propargylglycosides of galactose or lactose, using $(EtO)_3P.Cu(I)$ (0.1 equiv. per azide group) in refluxing toluene (79–93% yield). Also, monoglycosylated CD (**81**, **82**) were obtained, which were also functionalized with a fluorescent dansyl residue (**88**) (Scheme 5.21). It has been shown that these neoglycoconjugates act as synthetic activators of cell adhesion processes, mimicking the lipopolysaccharide effects. Noticeably, multivalency did not significantly affect cell adhesion. In fact, the effects were similar for the monovalent and the heptavalent glycocyclodextrins [41].

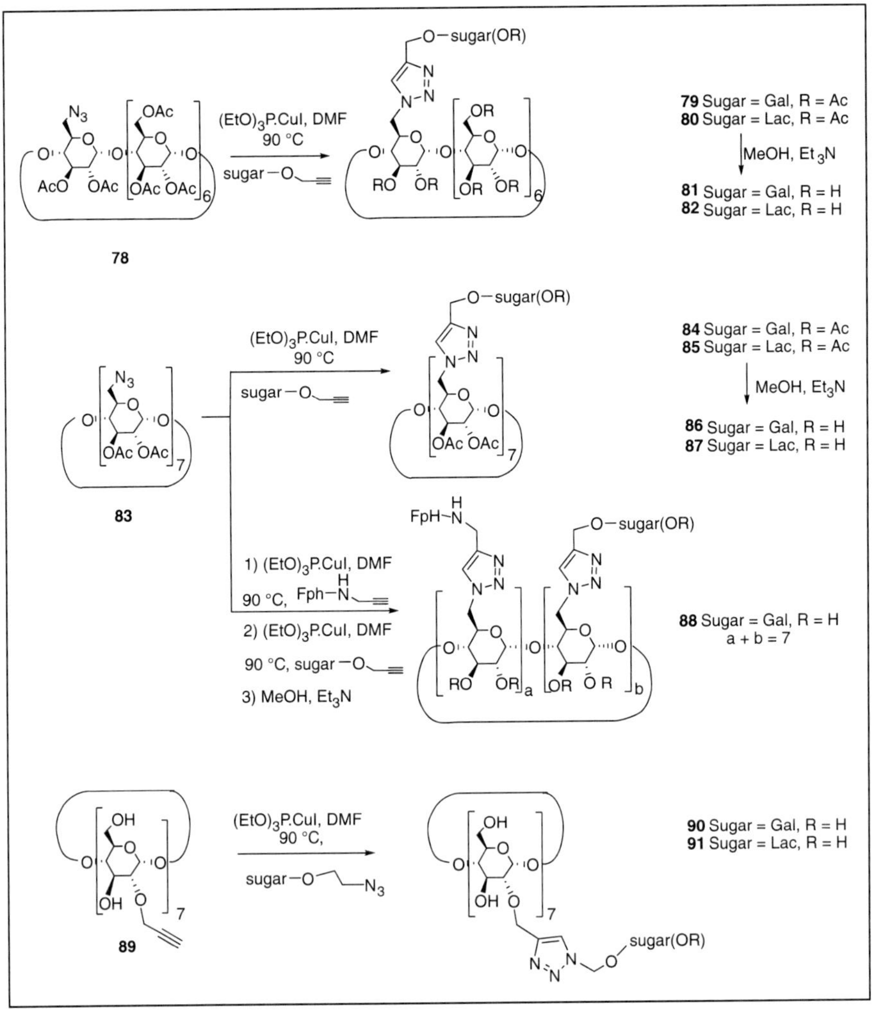

SCHEME 5.21 Synthesis of clicked cyclodextrins.

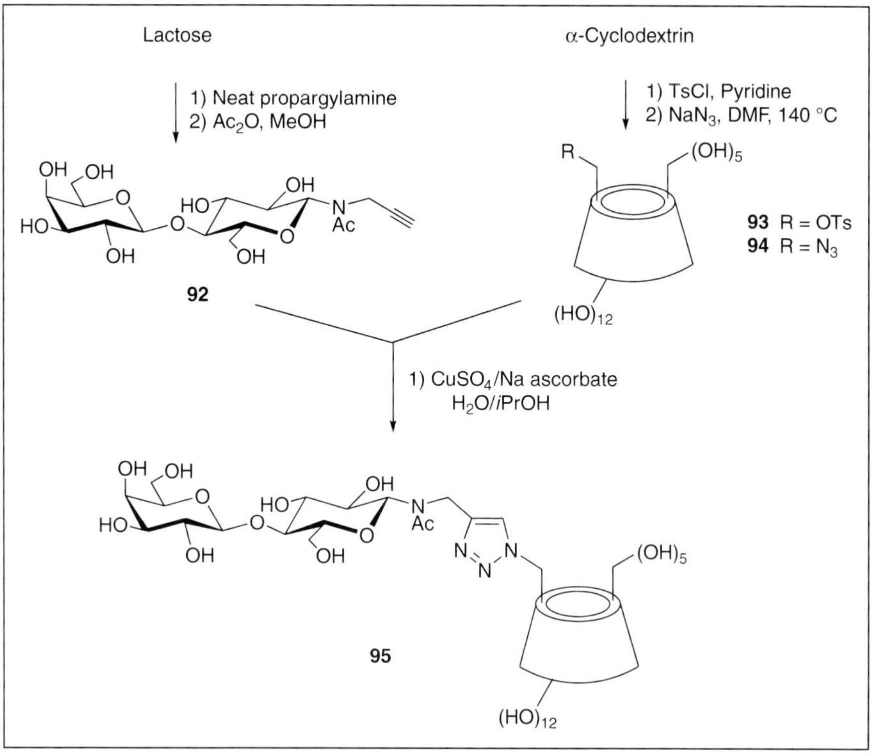

SCHEME 5.22 Synthesis of clicked lactosyl cyclodextrins.

Click-assembled glycorotaxanes have been synthesized from building blocks **95**, **96** and **97** as supramolecular systems based on host–guest inclusion complexes, constituted by a glycosylated-α-cyclodextrin conjugate and a decane guest (Scheme 5.22 and Scheme 5.23). Oligorotaxanes carrying up to five cyclodextrin units were isolated (67% yield for compound **100**) and characterized by MS and NMR experiments (1D five and diffusion-ordered spectroscopy [DOSY]). Their abilities to inhibit the binding of *Arachis hypogaea* agglutinin to asialofetuin was evaluated by enzyme-linked lectin assays (ELLA) [42].

5.4 TRIAZOLE-LINKED MACROCYCLES

An interesting application of the 1,3-dipolar cycloaddition is the synthesis of macrocycles, as mimetics of cyclodextrins and other cyclic oligomers capable to host different molecules. Earlier examples of cyclodextrin analogs obtained by thermal cycloaddition have been reported by Vasella and coworkers [43]. The desired intramolecular 1,3-dipolar cycloaddition took place at 110°C and gave the 1,4- and 1,5-substituted 1,2,3-triazoles **106** (24%) and **108** (31%), which were separated by

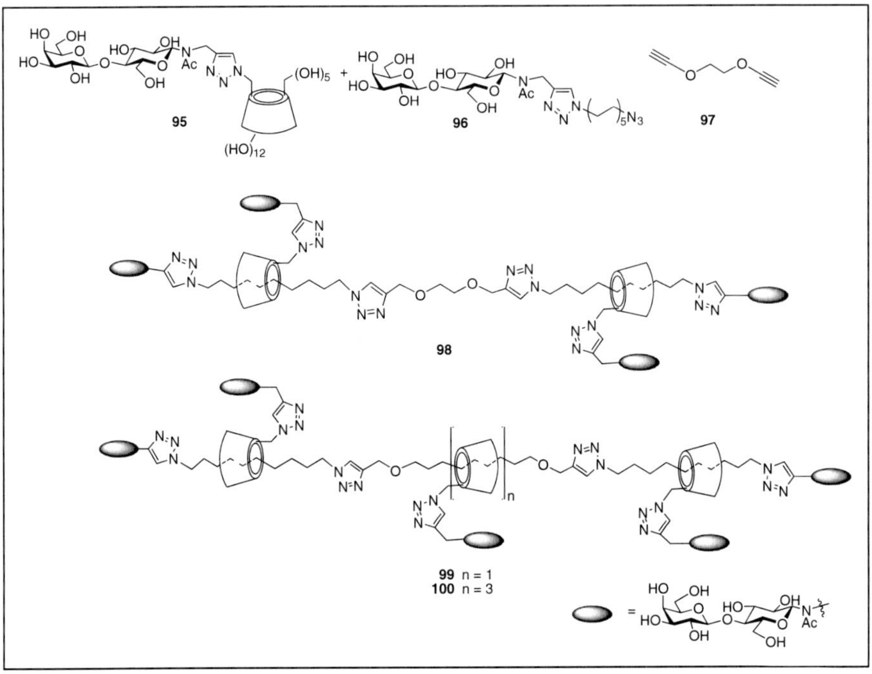

SCHEME 5.23 Synthesis of click-assembled glycorotaxanes.

chromatography. Deprotection with $FeCl_3$ in CH_2Cl_2 yielded the α-CD analogs **107** and **109** in 68% and 66% yield, respectively (Scheme 5.24 and Scheme 5.25).

Using CuAAC, Bodine et al. reported the synthesis of a cyclodextrin analog that exhibits association characteristics with the hydrophobic fluorophore 8-anilino-1-naphthalenesulfonate, similar to that observed for β-cyclodextrin [15b].

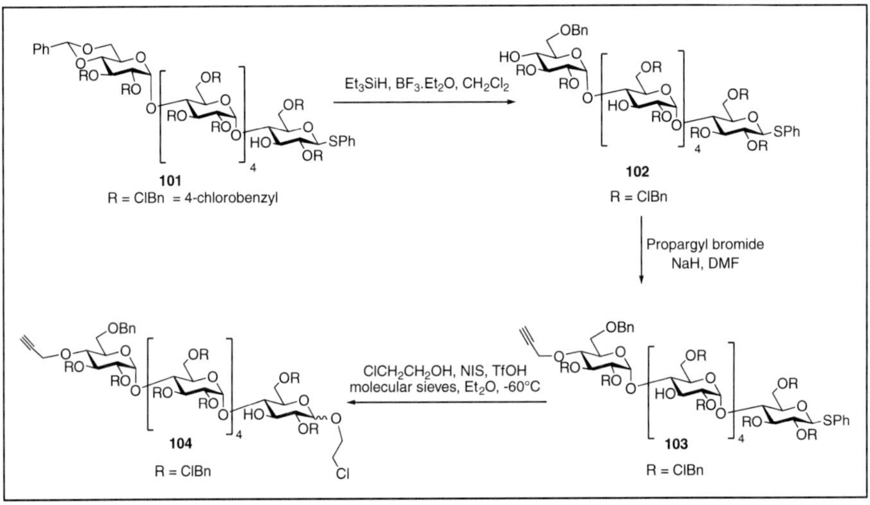

SCHEME 5.24 Synthesis of bifunctionalized oligosaccharides.

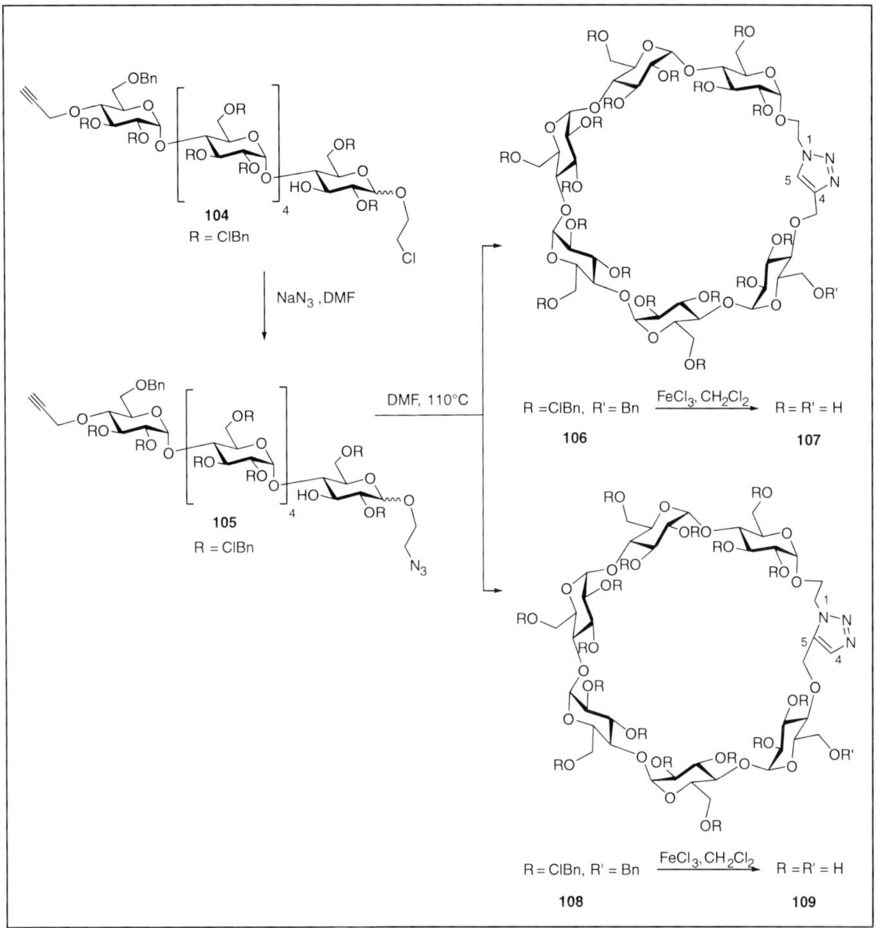

SCHEME 5.25 Preparation of triazol-linked macrocycles.

After surveying the efficiency of the reaction with a variety of Cu(I) sources ($CuSO_4$, sodium ascorbate; $CuSO_4$, Cu^0 powder; Cu(I).P(OEt)$_3$; Cu(I), additives (DBU, tris(triazolyl)amine), and solvents (DMF/H_2O, CH_3CN, toluene), a conveniently functionalized trisaccharide (**110**) was treated with Cu(I) and DBU in toluene at 50°C to give the concomitant cyclodimer **111** in 80% yield. The corresponding cyclotrimer was obtained in 15% yield (Scheme 5.26). In an extension of their work, the authors prepared smaller macrocycles from a mono- and disaccharide, respectively, able to act as molecular pores with small cavities (Scheme 5.27) [15a]

The copper-catalyzed dimerization of simple acetylated 2-propynyl and 3-butynyl 6-azido-6- deoxy-glycosides of the gluco, galacto and manno series has been reported [44]. It was observed that the copper-catalyzed dimerizations occurred in low to medium yields (14–54%) and showed to be dependent on the sugar moiety, the anomeric configuration, and the size of the formed ring (Scheme 5.28). The use of

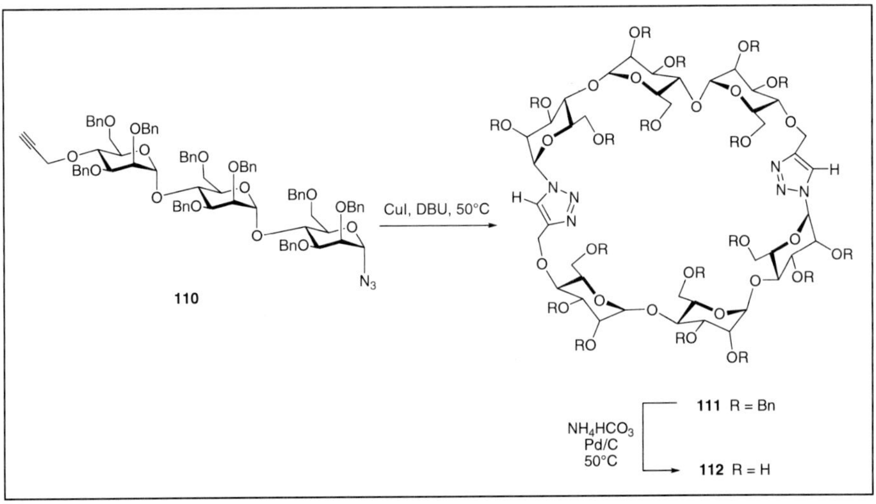

SCHEME 5.26 Synthesis of cyclodextrin analogs.

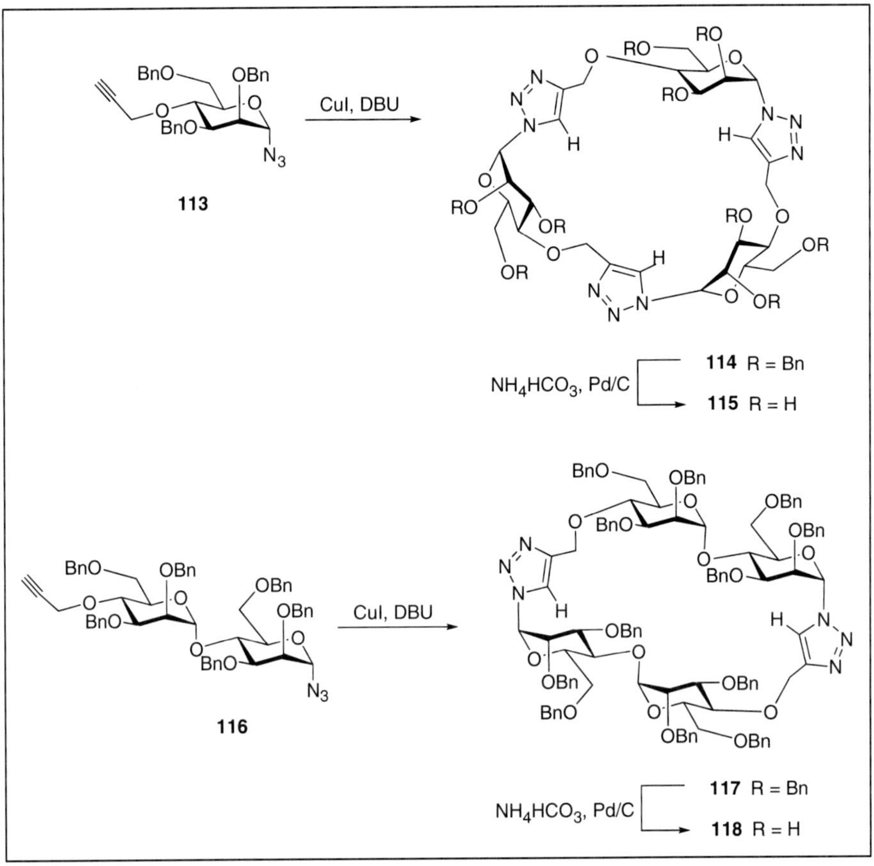

SCHEME 5.27 Obtention of small macrocycles.

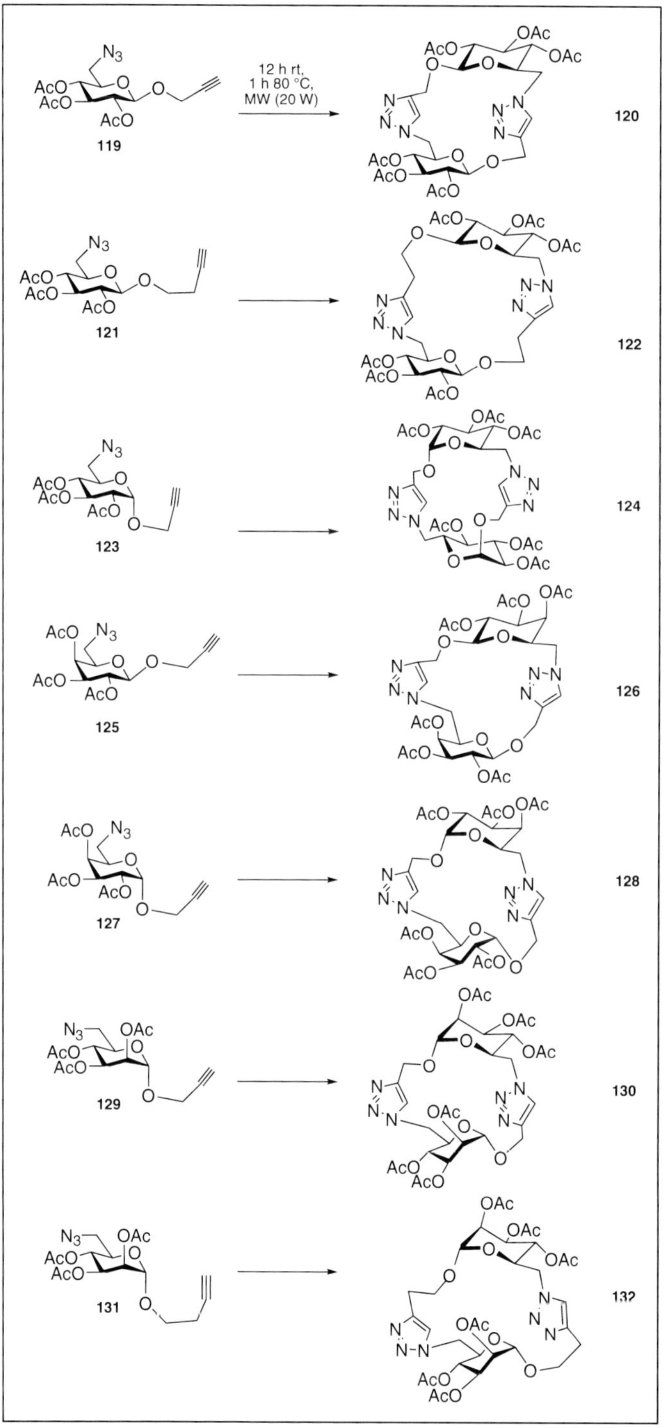

SCHEME 5.28 Microwave-assisted synthesis of macrocycles from bifunctionalized monosaccharides.

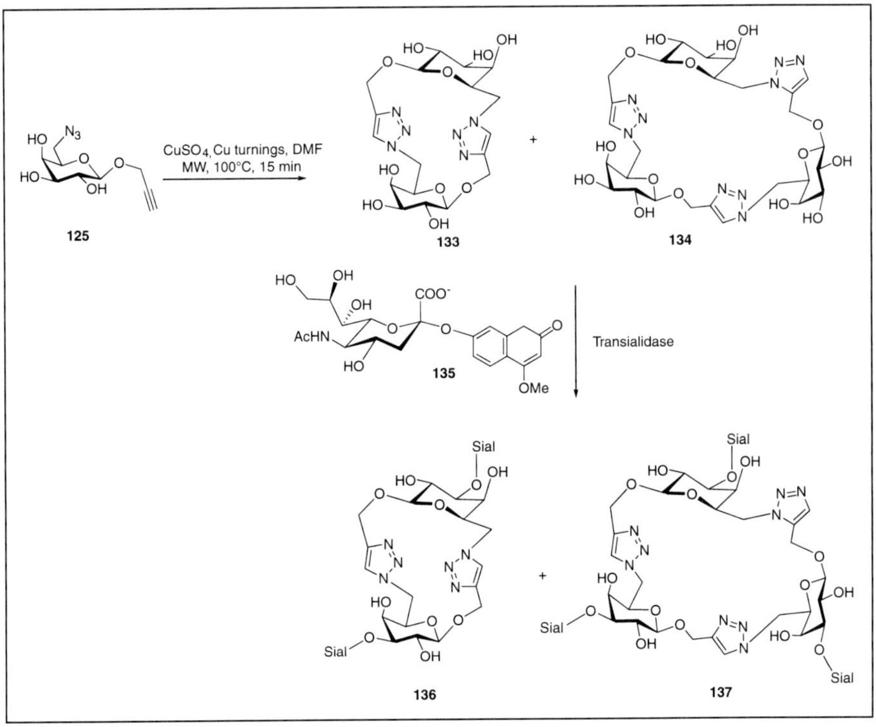

SCHEME 5.29 Cyclooligomerisation of unprotected 6-azidoalkyne derivatives.

microwave irradiation resulted in faster reactions; however, some decomposition of the starting material was also observed.

Cyclic pseudo-galactooligosaccharides were synthesized by cyclooligomerization of isomeric unprotected azido–alkyne derivatives of β-D-galactopyranose under Cu(I)-catalyzed ($CuSO_4$/Cu turnings in DMF) azide–alkyne 1,3-dipolar cycloaddition reaction conditions [45]. Dimers and trimers were the main isolated products in 20–35% yield. The cyclic compounds were treated with *Trypanosoma cruzi* trans-sialidase, leading to substitution at HO-3 of the galactose units (52–65% isolated yield). These results were explained by the outside exposition of secondary hydroxyl groups, as shown by molecular mechanics calculations (Schemes 5.29 and 5.30).

A chemoenzymatic approach was used in the synthesis of macrocyclic oligosaccharide analogs containing sialic acid units in the cycle [46]. Sialic acid derivatives functionalized with azide groups were enzymatically coupled to the non-reducing end of galactoside oligosaccharides containing propargyl groups. Starting from N-azidoacetylmannosamine, an azido derivative of cytidine monophosphate (CMP)-sialic acid was synthesized using a one-pot two-enzyme system. The resulting compound was clicked to propargyl-β-D-galactopyranoside, to give a CMP-activated disaccharide mimetic (CMP-Gal-triazole-Sia, **146**). In the presence of *Photobacterium damsela* α-(2→6)-sialyltransferase (Pd2,6ST), Gal-triazole-Sia was

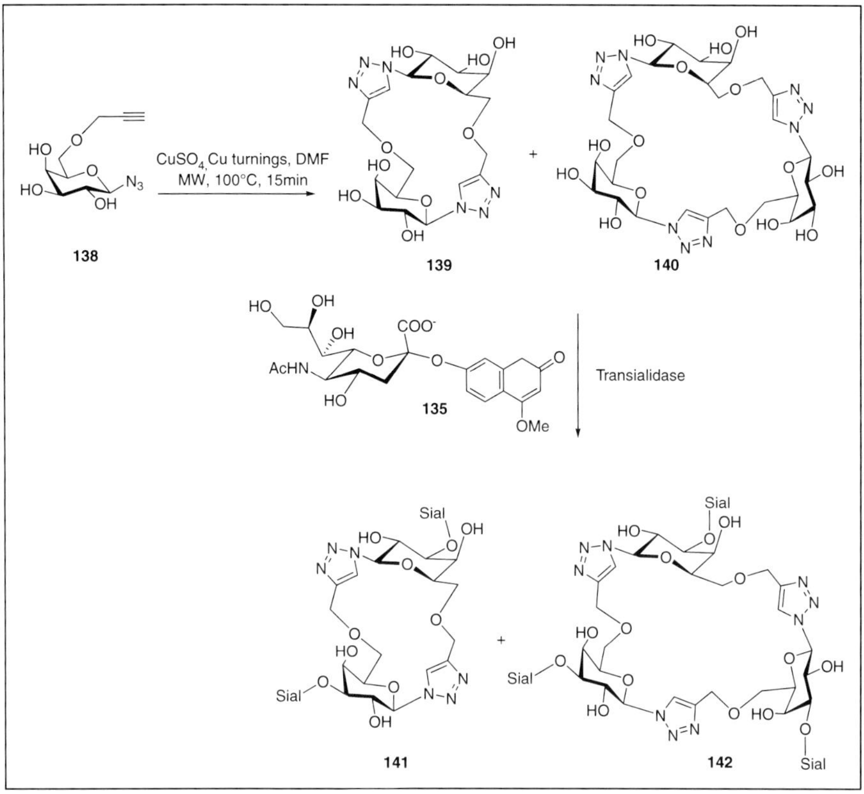

SCHEME 5.30 Cyclooligomerisation of unprotected 1-azidoalkyne derivatives.

transferred from its CMP-activated form to a propargyl-β-D-galactopyranoside to give trisaccharide **147**. Further sialylation of **147** resulted in the formation of a tetrasaccharide **148**. Macrocyclics of different sizes, containing two, three or four sialic acid residues were synthesized in moderate to excellent yields (30–91%) after intramolecular dipolar cycloaddition using Cu(I)–DIPEA in 1:1 CH_3CN–water (Scheme 5.31).

Recently, Dondoni's group has reported the synthesis of a new family of cyclodextrin-like macrocycles, containing two, four or six C-glucopyranoside residues. First, linear oligomers (**150**) have been obtained by sequential clicking of 6-azido ethinyl residues. Final intramolecular cyclization gave C_n-symmetric structures (74% yield for compound **151**) (Scheme 5.32) [47].

5.5 MULTIVALENT GLYCOCLUSTERS ON CARBOHYDRATE SCAFFOLDS

Carbohydrate–lectin interactions play a pivotal role in many biological events including inflammation [2], immune response, apoptosis, tumor metastasis, or viral and

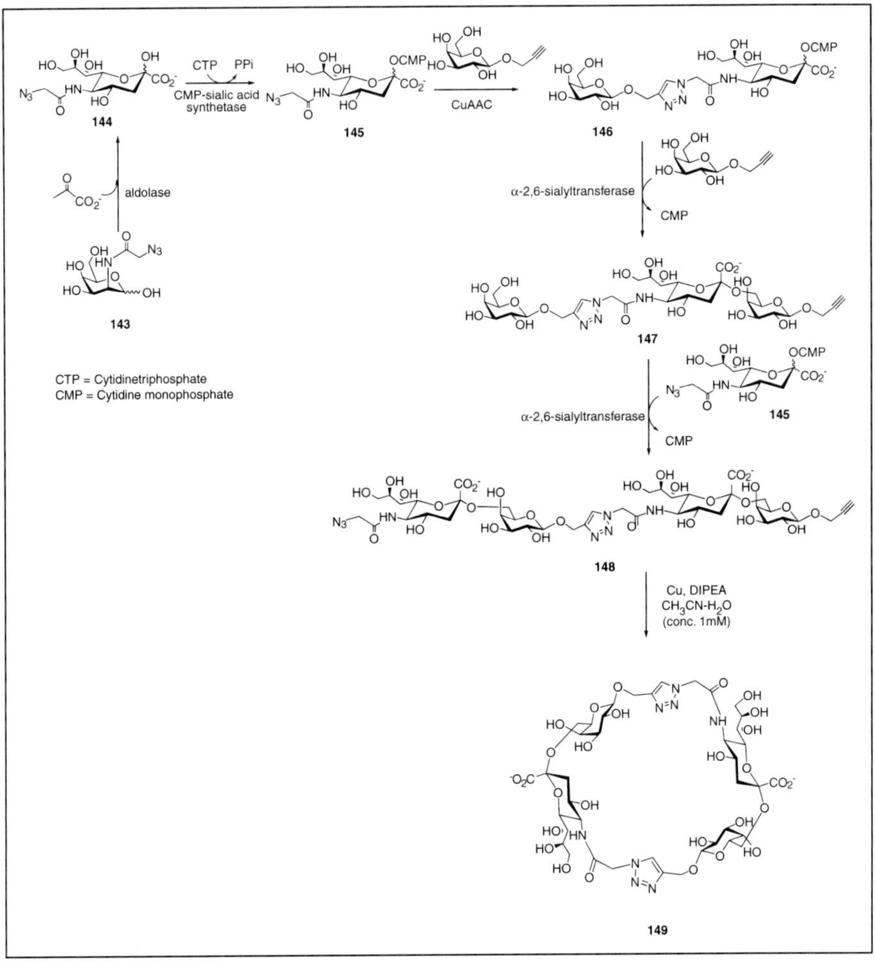

SCHEME 5.31 Chemoenzymatic approach for the synthesis of macrocyclic oligosaccharide analogs containing sialic acid units.

bacterial infections [1c–f, 3, 4]. The generally low affinity of monovalent ligands for lectins can be overcome by using polyfunctional scaffolds displaying structurally well-defined saccharide units. The affinity enhancements observed by using this strategy can be very impressive [48, 49]. A remarkable example is the decavalent ligand designed by Bundle and coworkers (STARFISH®), which inhibited Shiga-like toxins in enzyme-linked immunosorbent assays (ELISA) with a 10^6-fold enhancement over the monomeric species [50].

Depending both on the nature of the receptor and on the valency and geometry of the ligand, this gain in stability, the so-called multivalent or cluster effect [51], can be explained by the interplay of different mechanisms, including chelation binding

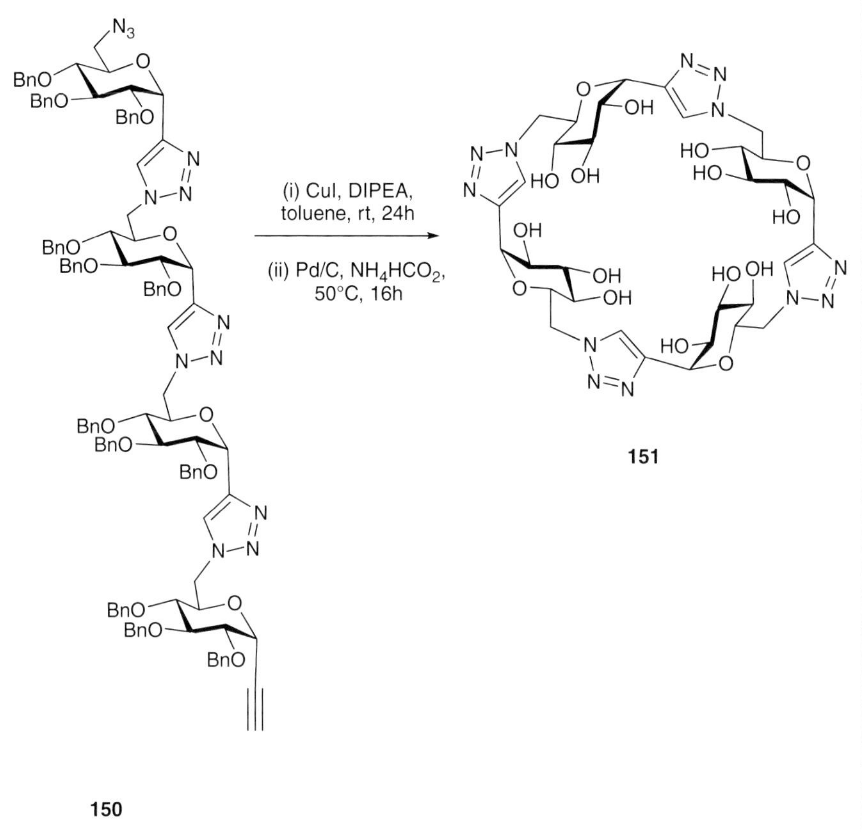

SCHEME 5.32 Synthesis of C-cyclodextrin-like analogs.

mode, clustering of receptors or statistical rebinding [52]. Hundreds of synthetic multivalent glycomimetics with diverse geometries, numbers of epitopes, and degrees of freedom have been synthesized. Such glycoconjugates have been classified as glycodendrimers, glycoclusters, glycopolymers and glycoproteins depending on the nature of their scaffold [40].

Compared to dendrimers or polymers, there are few reports on the use of carbohydrates as scaffolds in the synthesis of multivalent ligands [8, 53]. Some of the examples correspond to glycosylamide scaffolds and cyclomaltooligosaccharide (cyclodextrin)-centered glycoclusters, the latter already discussed in Section 5.3 [54, 55]. The use of unprotected non-reducing mono or disaccharides as templates to graft sugar epitopes in all available hydroxyl groups has also been explored [56]. For *in vivo* applications, however, multivalent ligands based on partially deprotected carbohydrate scaffolds, may result in interesting biological advantages. Improved hydrophilicity and pharmacokinetics may be expected, for instance, as compared with peptidic, aromatic, or polymeric scaffolds [57].

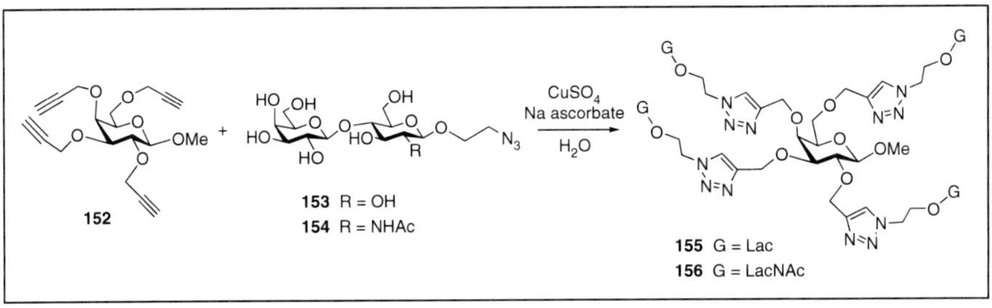

SCHEME 5.33 Galactose-centered multilactosides.

Carbohydrate-centered glycoclusters were prepared from methyl-2,3,4,6-tetra-O-propargyl β-D-galactopyranoside **152** and 2-azidoethyl lactoside **153** or 2-azidoethyl N-acetyllactosamine glycoside **154** (83% and 88% for **155** and **156**, respectively) (Scheme 5.33). The glycoclusters showed a 400-fold enhanced binding affinity to the plant lectin ricinus communis agglutinin I (RCA120) (minimum inhibition concentration 50 μM) relative to free lactose (20 mM) [56b].

Santoyo-Gonzalez and coworkers [58] reported the synthesis of multivalent neoglycoconjugates by clicking Man and Glc derivatives, in a sequential process, on monosaccharide and disaccharide scaffolds (**157–159**) (Scheme 5.34). The valency of the final product (**160–162**) as well as the α-Man/Glc ratio, could be determined by the number of alkyne and chloride substituents (as precursors of azides) on the scaffold. Also, this strategy allowed the introduction of linkers of different lengths (Scheme 5.35). The click reactions were performed using $(EtO)_3P.Cu(I)$ in toluene (70–90% yield).

The influence of the substitution pattern and the distance between sugar residues, as key features in multivalent interactions, was studied by evaluating the binding affinities of the products to Concanavalin A (Con A) lectin. Higher relative potencies per sugar moiety were observed, in both homo and hetero neoglycoconjugates, when the sugar residues were not directly connected to the 1,2,3-triazole ring through their anomeric positions.

In our laboratory, we developed a procedure to carry out the regioselective polyazidation of fully unprotected sugars (D-glucose, maltose, and maltotriose) using

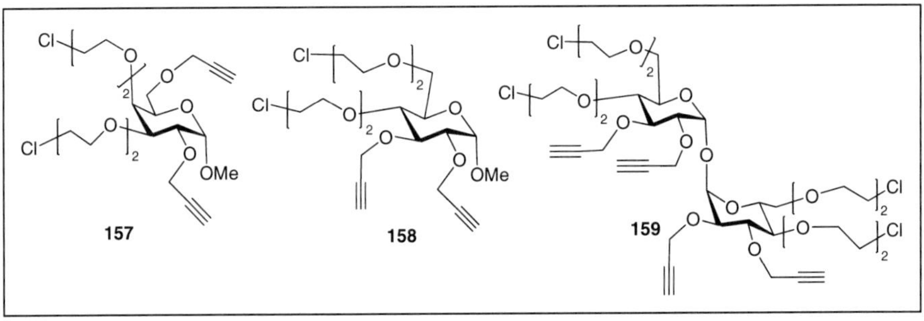

SCHEME 5.34 Mono-and disaccharide scaffolds.

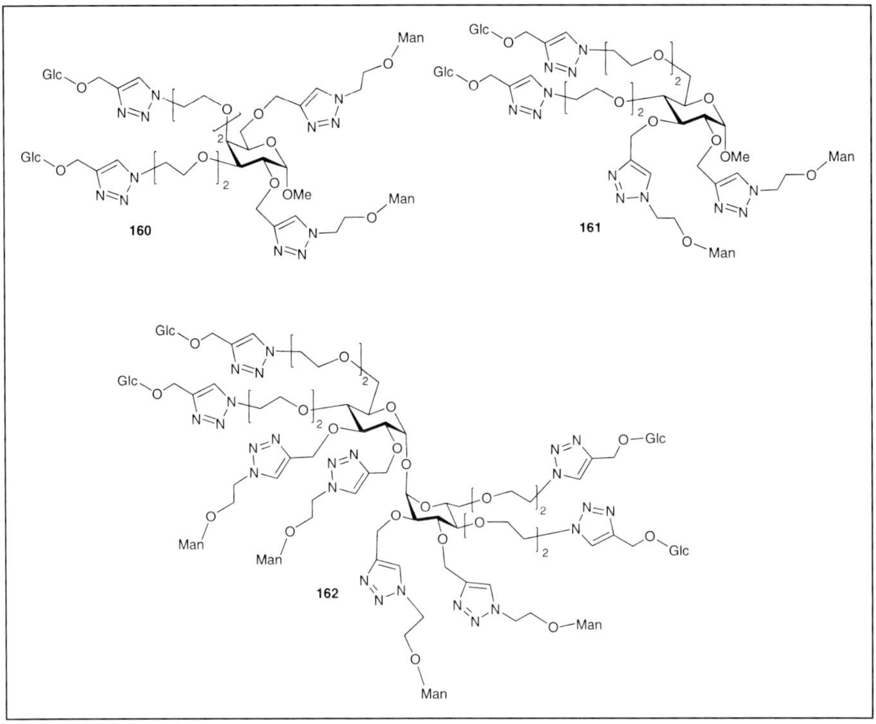

SCHEME 5.35 Multivalent neoglycoconjugates containing glucose and mannose units.

$PPh_3/CBr_4/NaN_3$ in dry DMF [59]. The number and the position of the azido groups in the final compound can be modulated by varying the relative proportions of reagents (Scheme 5.36). For example, when maltose (**163**) was treated with 2 equivalents PPh_3/CBr_4 and 10 equivalents NaN_3, the 6,6′-diazido derivative **164** was obtained after acetylation in 49% yield. Doubling the molar relationship of the abovementioned reactants led to the triazide **165** as the major final product (49% yield). We have shown that the resulting polyazido-derivatives are useful oligosaccharide scaffolds.

Therefore, as a first approach, a short and efficient strategy was developed for the synthesis of multi-mannosides [60]. This methodology involves four key steps: (i) direct regioselective azidation of the unprotected commercial saccharides; (ii) acetylation; (iii) linking of the mannosyl ligands by click chemistry; and (iv)

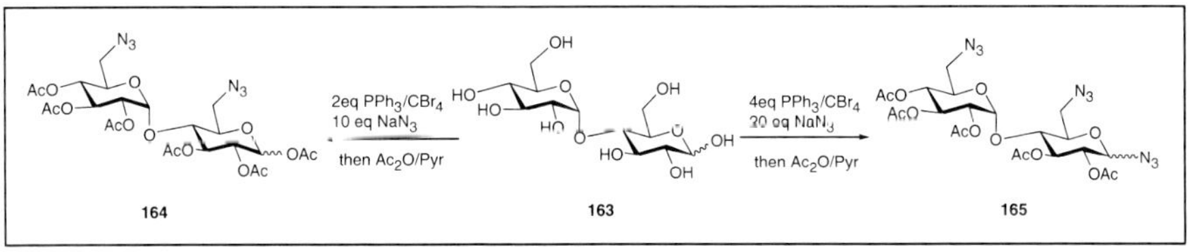

SCHEME 5.36 Direct azidation of unprotected sugars.

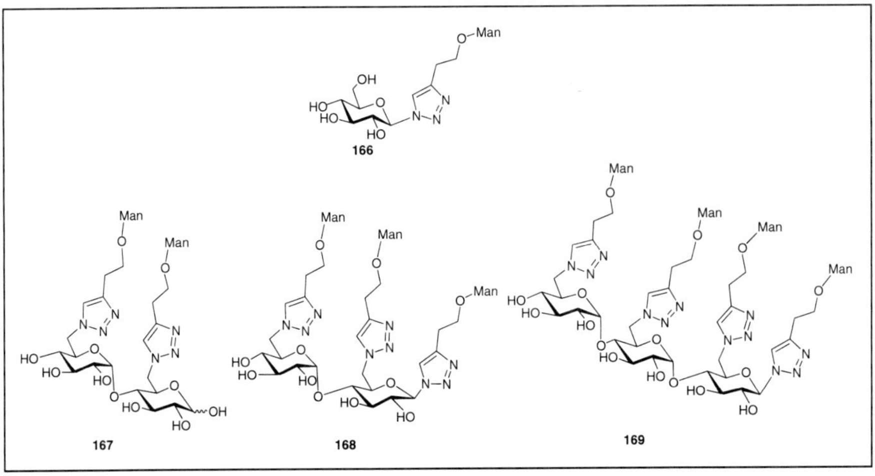

SCHEME 5.37 Multimannosides on carbohydrate scaffolds.

deacetylation. A new family of glycoclusters with a valency ranging from 1 to 4 were obtained (Scheme 5.37). Binding affinities towards Con A were investigated by ELLA. The synthetic multivalent compounds exhibited a remarkable cluster effect with a relative potency per mannoside residue ranging from 8.1 to 9.1 depending on the structures. Molecular dynamics performed on representative glycoconjugates revealed interesting structural features such as rigidity of the scaffold and highly flexible mannose counterparts.

We also synthesized a second family of glycoclusters based on carbohydrate scaffolds, bearing one to four lactoside epitopes linked through flexible oligoethyleneglycol (EG) spacers of different lengths (**170–173**) with yields ranging between 40% and 75% (Scheme 5.38). We demonstrated the relevance of the EG linkers lengths in the binding process by crosslinking-type assays. Alkynyl-armed lactosides were clicked onto azide-functionalized carbohydrate scaffolds. The solution-phase-binding affinities for galectin-1, galectin-3, and the lactose-binding plant lectin *Arachis hypogaea* (peanut) agglutinin were assessed by fluorescence polarization methods and ELLA, respectively. Besides, the crosslinking properties of the glycoclusters were specifically assessed by a two-site ELLA (sandwich assay). Molecular dynamic (MD) simulations were performed on the entire set of synthetic glycoconjugates to study their structural and dynamical features. Kd values of the multivalent lactosides ranged between 42 μM and 244 μM for galectin-1 while for galectin-3 the values ranged between 16 μM and 73 μM. Taking into account our results, the optimal distance of the EG-linker must be considered for an efficient binding process to the target receptor. This is particularly relevant in antiadhesive therapies, where a "natural" occurring binding phenomenon is to be inhibited by promoting an aggregation process involving a synthetic analog. On the other hand, shorter spacers could favor the "rebinding" mechanism if a single binding site is required, which is frequently the case for effector molecules [61].

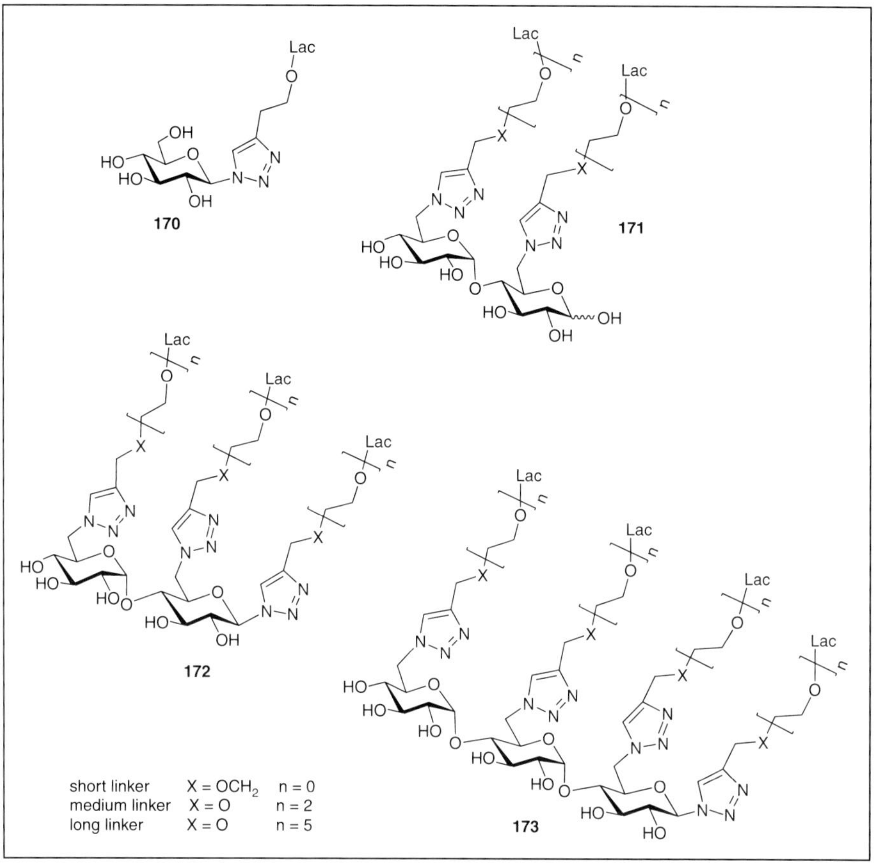

SCHEME 5.38 Multilactosides on carbohydrate scaffolds with different spacer length.

Urinary tract infections caused by uropathogenic *Escherichia coli* presents a serious communal and nosocomial health problem. The disease is initiated by the adhesion of bacteria to bladder cells, by means of the fimbriae, which express a mannose-binding adhesin called FimH. Heptyl α-D-mannoside (HM) is a strong inhibitor of this lectin at nanomolar concentrations. It was shown that this compound prevents the adhesion of type 1-piliated *E. coli* and reduces bacteria levels in a murine cystitis model [62]. A variety of multimeric heptyl mannosides (cf. **174**) with valencies ranging from one to four were synthesized using the strategy described above (Scheme 5.39). Biological evaluation of these multimannosides showed remarkable cluster effects in the bladder-binding inhibition assay [63].

As a continuation of our work, we envisaged a new set of multivalent HM ligands based on carbohydrate cores with structural valencies that range from 1 to 7 [64]. The strategy used to construct the regular hydrophilic structures involved the repetition of a critical glucoside fragment. Controlled acidic opening of a suitable functionalized

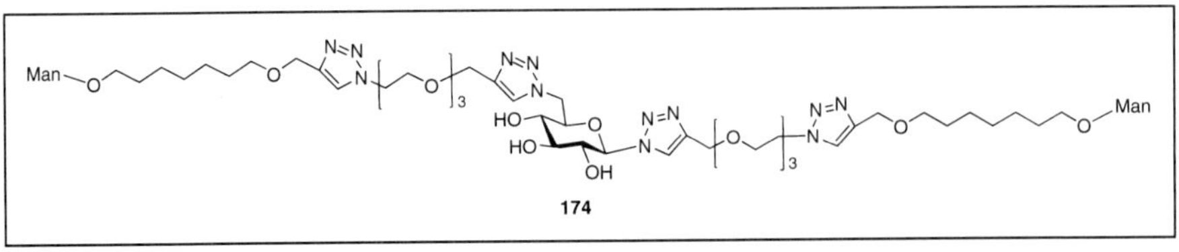

SCHEME 5.39 Multimeric heptyl mannosides for FimH inhibition.

heptaazido β-cyclodextrin [65] led to the maltoheptasaccharide scaffold **176** (Scheme 5.40). The azido-oligosaccharide was glycosylated with 9-fluorenylmethoxycarbonyl (Fmoc)-protected cysteamine. After CuAAC and deprotection (53–68% yield for the two steps) (Scheme 5.41), coupling of the multimers to a fluorescent label could be achieved through the amino group. A one-pot synthetic approach was developed to tether the ligands and the fluorescein isothiocyanate (FITC) probe to the scaffold simultaneously (75% yield) (Scheme 5.42). Isothermal calorimetry using the monomeric FimH lectin revealed affinities in the nanomolar range and saturation of all structurally available binding sites by the multivalent HM ligands. Direct titrations showed almost strict correlation of enthalpy–entropy compensation with the increasing valency of the ligand, whereas reverse titration calorimetry demonstrated negative cooperativity between the first and the second binding sites of the divalent heptyl mannoside. A multivalency effect was nevertheless observed by inhibiting the hemagglutination of type-1 piliated UTI89 *E. coli*, with a titer as low as 60 nm for the heptavalent HM ligand. An FITC-labeled HM trimer showed capture and crosslinking of living bacteria in solution, a phenomenon not previously described with low-valency ligands.

The "cluster effect" has been extensively studied for several lectin systems, but little is known about the influence of the multivalency on the activity of glycosidases

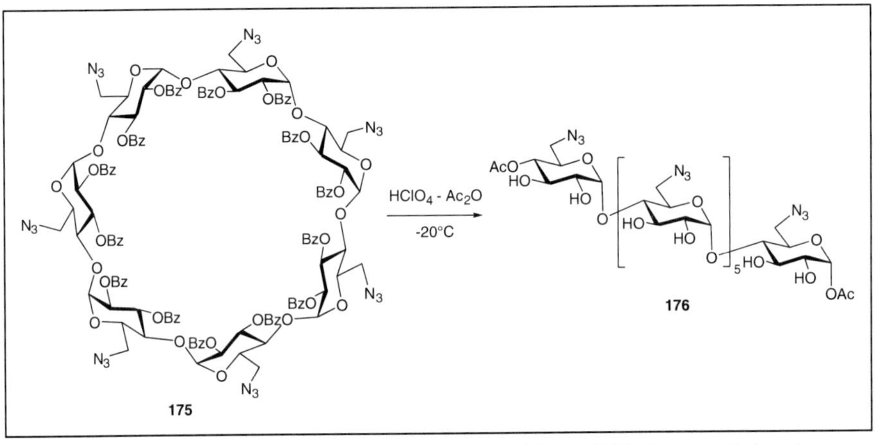

SCHEME 5.40 Synthesis of linear oligosaccharide scaffolds from cyclodextrin.

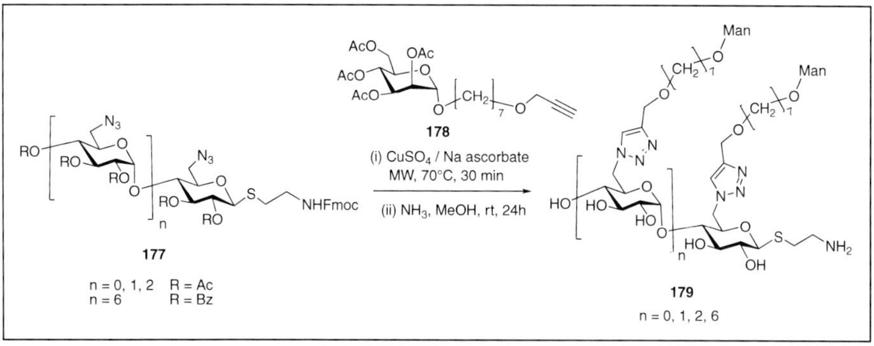

SCHEME 5.41 Synthesis of multimannosides on amino-functionalized oligosaccharide scaffolds.

and only a few examples have been reported [19b–c, 66]. When designing glycoclusters to study biological processes, specificity, affinity, and stability against glycolytic enzymes are important factors to be considered. In contrast to natural *O*-linked sugars, the *S*-carbohydrate mimetics are usually resistant to metabolic processes and are seen as potential precursors of promising carbohydrate-based therapeutics. Particularly, the resistance of glycoclusters to enzyme hydrolysis is crucial to circumvent degradation when they are internalized in the cell. The replacement of

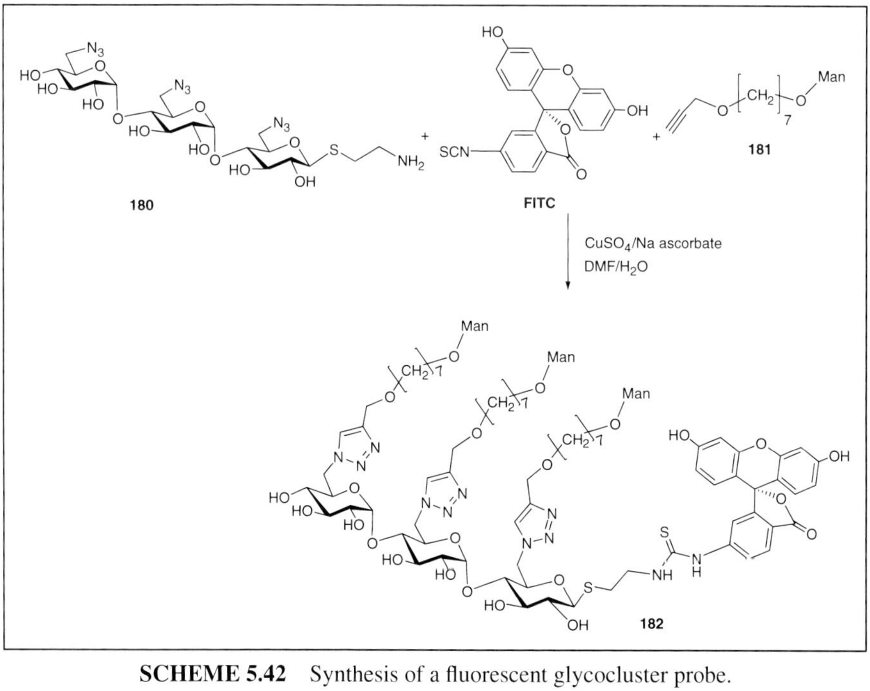

SCHEME 5.42 Synthesis of a fluorescent glycocluster probe.

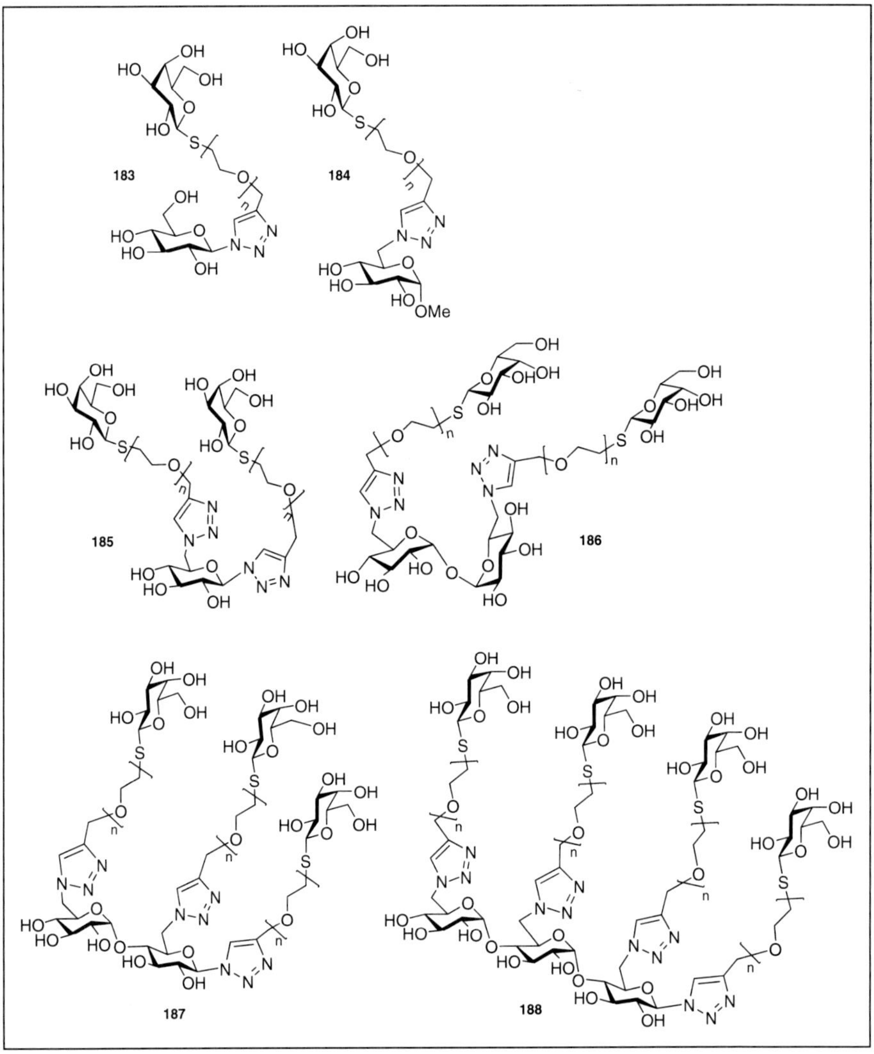

SCHEME 5.43 Multivalent S-galactosides on carbohydrate scaffolds.

the interglycosidic oxygen atom by a sulfur atom in glycoclusters could increase the resistance to hydrolytic degradation by enzymes and can even cause inhibition of this activity. Recently, we reported the synthesis of a family of glycoclusters based on carbohydrate scaffolds bearing one to four S-galactoside epitopes and EG spacers of different lengths (Scheme 5.43). The click reaction between thiogalactoside residues and azide oligosaccharide scaffolds was performed under microwave irradiation, in moderate to excellent yields, depending on the valencies of the desired products (72–100% for monovalents, 60–84% for divalents, 54–67% for trivalents and 50% for tetravalents) [67].

The K_i values against *E. coli* β-galactosidase showed that an increase of the linker length between the thiogalactose residue and the triazol, results in a decrease of the inhibitory activity for divalent derivatives. Moving from mono- to di-, tri- and tetravalent derivatives, increased inhibition is observed. However, this is probably due to a statistical effect, and no cluster or multivalency effect is observed for β-galactosidase inhibition.

5.6 CONCLUSION

The incorporation of an azide and/or an alkyne residue in oligosaccharidic fragments is the first step to access a new dimension of structural diversity in the field of glycoconjugates. Indeed, these new families of synthetic compounds contribute in a positive way to the variability already inherent to carbohydrates, giving rise to a great number of potential bioactive analogs.

Moreover, the presence of the triazol ring(s) could contribute to additional beneficial interactions between the aromatic heterocycle and the carbohydrate-binding molecules.

As shown here, the Huisgen cycloaddition between sugars has been used as a powerful "click" reaction for the synthesis of a wide spectrum of interesting analogs: oligosaccharide mimics, cyclodextrin-type macrocyclics, glycoclusters, "decorated" polysaccharides. There is no doubt that even more complex molecules, as biologically relevant active compounds and biomaterials, will be prepared in a near future.

REFERENCES

1. (a) Bertozzi, C. R.; Kiessling, L. L. *Science* **2001**, *291*, 2357–2364. (b) Danishefsky, S. J.; Allen, J. R. *Angew. Chem. Int. Ed.* **2000**, *39*, 836–863. (c) Ambrosi, M.; Cameron, N. R.; Davis, B. G. *Org. Biomol. Chem.* **2005**, *3*, 1593–1608. (d) Gabius, H. J.; André, S.; Kaltner, H.; Siebert, H.C. *Biochim. Biophys. Acta* **2002**, *1572*, 165–177. (e) Lis, H.; Sharon, N. *Chem. Rev.* **1998**, *98*, 637–674. (f) Rabinovich, G. A. *Cell. Death Differ.* **1999**, *6*, 711–722. (g) Kiessling, L. L.; Gestwicki, J. E.; Strong, L. E. *Angew. Chem. Int. Ed.* **2006**, *45*, 2348–2368.
2. (a) Hanessian, S.; Mascitti, V.; Rogel, O. *J. Org. Chem.* **2002**, *67*, 3346–3354. (b) Murphy, P. V.; Hubbard, R. E.; Manallack, D. T.; Wills, R. E.; Montana, J. G.; Taylor, R. J. K. *Bioorg. Med. Chem.* **1998**, *6*, 2421–2439.
3. Ochieng, J.; Furtack, V. *Glycoconj. J.* **2004**, *19*, 527–535.
4. Gabius, H. J. *Adv. Drug Deliv. Rev.* **2004**, *56*, 421–424.
5. Rabinovich, G. A.; Toscano, Marta. A.; Jackson, S. S.; Vasta, G. R. *Curr. Opin. Struct. Biol.* **2007**, *17*, 513–520.
6. Sharon, N. *Biochim. Biophys. Acta* **2006**, *1760*, 527–537.
7. (a) Huisgen, R. *Angew. Chem. Int. Ed. Engl.* **1963**, *2*, 565–598. (b) Huisgen, R. *Angew. Chem. Int. Ed. Engl.* **1963**, *2*, 633–645.
8. (a) Bock, V. D.; Hiemstra, H.; van Maarseveen, J. H. *Eur. J. Org. Chem. 2006*, 61–68. (b) Kolb, H. C.; Sharpless, K. B. *Drug Discov. Today* **2003**, *8*, 1128–1137. (c) Kolb, H. C.;

Finn, M. G.; Sharpless, K. B. *Angew. Chem. Int. Ed.* **2001**, *40*, 2004–2021. (d) Bräse, S.; Gil, C.; Knepper, K.; Zimmermann, V. *Angew. Chem. Int. Ed.* **2005**, *44,* 5188–5240. (e) St. Hilaire, P. M.; Todd, L. L.; Meldal, M.; Bock, K. *J. Am. Chem. Soc.* **1998**, *120*, 13312–13320. (f) Lin, C.H.; Shimazaki, M.; Wong, C.-H.; Koketsu, M.; Juneja, L. R.; Kim, M. *Bioorg. Med. Chem. Lett.* **1995**, *3*, 1625–1630. (g) Jiao, H.; Hindsgaul, O.; *J. Carbohydr. Chem.* **1999**, *18*, 499–513. (h) Tornøe, C. W.; Meldal, M. *Proc. Am. Peptide. Symp., San Diego* **2001**, 263–264.

9. Rostovtsev, V. V.; Green, L. G.; Fokin, V. V.; Sharpless, K. B. *Angew. Chem. Int. Ed.* **2002**, *41*, 2596–2599.
10. Tornøe, C. W.; Christensen, C.; Meldal, M. *J. Org. Chem.* **2002**, *67*, 3057–3064.
11. Pérez-Balderas, F.; Ortega-Muñoz, M.; Morales-Sanfrutos, J.; Hernández-Mateo, F.; Calvo-Flores, F. G.; Calvo-Asín, J. A.; Isac-García, J.; Santoyo-Gonzalez, F. *Org. Lett.* **2003**, *5*, 1951–1954.
12. (a) Agard, N. J.; Prescher, J. A.; Bertozzi, C. R. *J. Am. Chem. Soc.* **2004**, *126*, 15046–15047. (b) Poloukhtine, A. A.; Mbua, N. E.; Wolfert, M. A.; Boons, G. J.; Popik, V. V. *J. Am. Chem. Soc.* **2009**, *131*, 15769–15776.
13. Dedola, S.; Nepogodie, S. A.; Field, R. A. *Org. Biomol. Chem.* **2007**, *5*, 1006–1017.
14. Fazio, F.; Brian, M. C.; Blixt, O.; Paulson, J. C.; Wong, C.H. *J. Am. Chem. Soc.* **2002**, *124*, 14397–14402.
15. (a) Bodine, K. D.; Gin, D. Y.; Gin, M. S. *Org. Lett.* **2005**, *7*, 4479–4482.
15. (b) Bodine, K. D.; Gin, D. Y.; Gin, M. S. *J. Am. Chem. Soc.* **2004**, *126*, 1638–1639.
16. Chen, Q.; Yang, F.; Du, Y. *Carbohydr. Res.* **2005**, *340*, 2476–2482.
17. Hotha, S.; Kashyap, S. *J. Org. Chem.* **2006**, *71*, 364–365.
18. Ladmiral, V.; Mantovani, G.; Clarkson, G. J.; Cauet, S.; Irwin, J. L.; Haddleton, D. M. *J. Am. Chem. Soc.* **2006**, *128*, 4823–4830.
19. (a) Périon, R.; Ferrières, V.; García-Moreno, M. I.; Ortiz Mellet, C.; Duval, R.; García Fernández, J. M.; Plusquellec, D. *Tetrahedron* **2005**, *61*, 9118–9128. (b) Diot, J.; García-Moreno, M. I.; Gouin, S. G.; Ortiz Mellet, C.; Haupt, K.; Kovensky, J. *Org. Biomol. Chem.* **2009**, *13*, 357–363. (c) Compain, P.; Decroocq, C.; Iehl, J.; Holler, M.; Hazelard, D.; Barragán, T. M.; Ortiz Mellet, C.; Nierengarten, J.F. *Angew. Chem. Int. Ed.* **2010**, *49*, 5753–5756. (d) Diot, J. D.; Garcia Moreno, I.; Twigg, G.; Ortiz Mellet, C.; Haupt, K.; Butters, T. D.; Kovensky, J.; Gouin, S. G. *J. Org. Chem.* **2011**, *76*, 7757–7768.
20. Dalvie, D. K.; Kalgutkar, A. S.; Khojasteh-Bakht, S. C.; Obach, R. S.; O'Donnell, J. P. *Chem. Res. Toxicol.* **2002**, *15*, 269–299.
21. Horne, W. S.; Stout, C. D.; Ghadiri, M. R. *J. Am. Chem. Soc. 2003*, *125*, 9372–9376.
22. Horne, W. S.; Yadav, M. K.; Stout, C. D.; Ghadiri, M. R. *J. Am. Chem. Soc.* **2004**, *126*, 15366–15367.
23. Appukkutan, P.; Dehaen, W.; Fokin, V. V.; van der Eycken, E. *Org. Lett.* **2004**, *6*, 4223–4225.
24. (a) Davies, M. B. *Polyhedron* **1992**, *11*, 285–321. (b) Creutz, C. *Inorg. Chem.* **1981**, *20*, 4449–4452.
25. Chittaboina, S. Xie, F.; Wang, Q. *Tetrahedron Lett.* **2005**, *46*, 2331–2336.
26. Miner, P. L.; Wagner, T. R.; Norris, P. *Heterocycles* **2005**, *65*, 1035–1049.
27. Akula, R. A.; Temelkoff, D. P.; Artis, N. D.; Norris, P. *Heterocycles* **2004**, *63*, 2719–2725.
28. Wilkinson, B. L.; Bornaghi, L. F.; Poulsen, S.A.; Houston, T. A. *Tetrahedron* **2006**, *62*, 8115–8125.

29. Gouin, S. G.; Bultel, L.; Falentin, C.; Kovensky, J. *Eur. J. Org. Chem.* **2007**, 1160–1167.

30. Rajaganesh, R.; Jayakumar, J.; Sivaraj, C.; Raaman, N.; Das, T. M. *Carbohydr. Res.* **2010**, *345*, 1649–1657.

31. Anand, N.; Jaiswal, N.; Pandey, S. K.; Srivastava, A. K.; Tripathi, R. P. *Carbohydr. Res.* **2011**, *346*, 16–25.

32. Marra, A.; Vecchi, A.; Chiappe, C.; Melai, B.; Dondoni, A. *J. Org. Chem.* **2008**, *73*, 2458–2461.

33. Kolthoff, I. M.; Coetzeem, J. *J. Am. Chem. Soc.* **1957**, *79*, 1852–1858.

34. Weiwer, M.; Chen, C. C.; Kemp, M. M.; Linhardt, R. J. *Eur. J. Org. Chem.* **2009**, 2611–2620.

35. Hotha, S.; Kashyap, S. *J. Org. Chem.* **2006**, *71*, 364–367

36. Nepogodiev, S. A.; Dedola, S.; Marmuse, L.; de Oliveira, M. T.; Field, R. A. *Carbohydr. Res.* **2007**, *342*, 529–540.

37. Bera, S.; Linhardt, R. J. *J. Org. Chem.* **2011**, *76*, 3181–3193.

38. Hasegawa, T.; Umeda, M.; Numata, M.; Li, C.; Bae, A.H.; Fujisawa, T.; Haraguchi, S.; Sakurai, K.; Shinkai, S. *Carbohydr. Res.* **2006**, *341*, 35–40.

39. (a) Cheshev, P.; Marra, A.; Dondoni, A. *Org.Biomol. Chem. 2006*, *4*, 3225–3227. (b) Conte, M. L.; Marra, A.; Chambery, A.; Gurcha, S. S.; Besra, G. S.; Dondoni, A. *J. Org. Chem.* **2010**, *75*, 6326–6336.

40. Chabre, Y. M.; Roy, R. *Adv. Carbohydr. Chem. Biochem.* **2010**, *63*, 165–393.

41. Ortega-Muñoz, M.; Morales-Sanfrutos, J.; Perez-Balderas, F.; Hernandez-Mateo, F.; Giron-Gonzalez, M. D.; Sevillano-Tripero, N.; Salto-Gonzalez, R.; Santoyo-Gonzalez, F. *Org. Biomol. Chem.* **2007**, *5*, 2291–2301.

42. Chwalek, M.; Auzély, R.; Fort, S. *Org. Biomol. Chem.* **2009**, *7*, 1680–1688.

43. Hoffmann, B.; Bernet, B.; Vasella, A. *Helv.Chim. Acta* **2002**, *85*, 265–287.

44. Pietrzik, N.; Schmollinger, D.; Ziegler, T. *Beilstein J. Org. Chem.* **2008**, *4*, 30.

45. Campo, V. L.; Carvalho, I.; Silva, C. H. T. P.; Schenkman, S.; Hill, L.; Nepogodiev, S.; Field, R. A. *Chem. Sci.* **2010**, *1*, 507–514.

46. Muthana, S.; Yu, H.; Cao, H.; Cheng, J.; Chen, X. *J. Org. Chem.* **2009**, *74*, 2928–2936.

47. Lo Conte, M.; Grotto, D.; Chambery, A.; Dondoni, A.; Marra, A. *Chem. Commun.* **2011**, *47*, 1240–1242.

48. Vrasidas, I.; André, S.; Valentini, P.; Böck, C.; Lensch, M.; Kaltner, H.; Liskamp, R. M. J.; Gabius, H. J.; Pieters, R. J. *Org. Biomol. Chem.* **2003**, *1*, 803–810.

49. (a) Zhang, Z.; Merritt, E. A.; Ahn, M.; Roach, C.; Hou, Z.; Verlinde, C. L. M. J.; Hol, W. G. J.; Fan, E. *J. Am. Chem. Soc.* **2002**, *124*, 12991–12998. (b) Fan, E.; Zhang, Z.; Minke, W. E.; Hou, Z.; Verlinde, C. L. M. J.; Hol, W. G. J. *J. Am. Chem. Soc.* **2000**, *122*, 2663–2664.

50. Kitov, P. I.; Sadowska, J. M.; Mulvey, G.; Armstrong, G. D.; Ling, H.; Pannu, N. S.; Read, R. J.; Bundle, D. R. *Nature* **2000**, *403*, 669–672.

51. (a) Lundquist, J. J.; Toone, E. J. *Chem. Rev.* **2002**, *102*, 555–578. (b) Lee, R. T.; Lee, Y. C. *Glycoconjugate J.* **2000**, *17*, 543–551. (c) Mammen, M.; Choi, S. K.; Whitesides, G. M. *Angew. Chem. Int. Ed.* **1998**, *37*, 2754–2794. (d) Lee, Y. C.; Lee, R. T. *Acc. Chem. Res.* **1995**, *28*, 321–327.

52. Kiessling, L. L.; Gestwicki, J. E.; Strong, L. E. *Angew. Chem. Int. Ed.* **2006**, *45*, 2348–2368.

53. (a) Hashimoto, M.; Morimoto, M.; Saimoto, H.; Shigemasa, Y.; Sato, T. *Bioconjugate Chem.* **2006**, *17*, 309–316. (b) Nelson, A.; Stoddart, J. F. *Carbohydr. Res.* **2004**, *339*, 2069–2075. (c) Kalovidouris, S. A.; Blixt, O.; Nelson, A.; Vidal, S.; Turnbull, W. B.; Paulson, J. C.; Stoddart, J. F. *J. Org. Chem.* **2003**, *68*, 8485–8493. (d) Turnbull, W. B.; Kalovidouris, S. A.; Stoddart, J. F. *Chem. Eur. J.* **2002**, *8*, 2988–3000. (e) Turnbull, W. B.; Pease, A. R.; Stoddart, J. F. *ChemBioChem.* **2000**, *1*, 70–74. (f) Colonna, B.; Harding, V. D.; Nepogodiev, S. A.; Raymo, F. M.; Spencer, N.; Stoddart, J. F. *Chem. Eur. J.* **1998**, *4*, 1244–1254.

54. (a) Murphy, P. V.; Bradley, H.; Tosin, M.; Pitt, N.; Fitzpatrick, G. M.; Glass, W. K. *J. Org. Chem.* **2003**, *68*, 5692–5704. (b) Velasco-Torrijos, T.; Murphy, P. V. *Org. Lett.* **2004**, *6*, 3961–3964. (c) Tosin, M.; Gouin, S. G.; Murphy, P. V. *Org. Lett.* **2005**, *7*, 211–214. (d) Tosin, M.; Murphy, P. V. *J. Org. Chem.* **2005**, *70*, 4107–4117.

55. (a) Vargas-Berenguel, A.; Ortega-Caballero, F.; Casas-Solvas, J. M. *Mini-Rev. Org. Chem.* **2007**, *4*, 1–14. (b) Ortiz Mellet, C.; Defaye, J.; García Fernández, J. M. *J. Incl. Phenom. Macrocycl. Chem.* **2006**, *56*, 149–159. (c) Ortiz Mellet, C.; Defaye, J.; García Fernández, J. M. *Chem. Eur. J.* **2002**, *8*, 1982–1990.

56. (a) Sperling, O.; Fuchs, A.; Lindhorst, T. K. *Org. Biomol. Chem.* **2006**, *4*, 3913–3922.

56. (b) Gao, Y.; Eguchi, A.; Kakehi, K.; Lee, Y. C. *Bioorg. Med. Chem.* **2005**, *13*, 6151–6157.

57. (a) Haubner, R.; Wester, H.-J.; Burkhart, F.; Senekowitsch-Schmidtke, R.; Weber, W.; Goodman, S. L.; Kessler, H.; Schwaiger, M. *J. Nucl. Med.* **2001**, *42*, 326–336. (b) Beer, A. J.; Haubner, R.; Goebel, M.; Luderschmidt, S.; Spikler, M. E.; Wester, H.-J.; Weber, W. A.; Schwaiger, M. *J. Nucl. Med.* **2005**, *46*, 1333–1341.

58. Ortega-Muñoz, M.; Perez-Balderas, F.; Morales-Sanfrutos, J.; Hernández-Mateo, F.; Isac-García, J.; Santoyo-Gonzalez, F. *Eur. J. Org. Chem.* **2009**, 2454–2473.

59. Gouin, S. G.; Kovensky, J. *Tetrahedron Lett.* **2007**, *48*, 2875–2879.

60. Gouin, S. G.; Vanquelef, E.; García Fernandez, J. M.; Ortiz Mellet, C.; Dupradeau, F.Y.; Kovensky, J. *J. Org. Chem.* **2007**, *72*, 9032–9045.

61. Gouin, S. G.; García Fernández, J. M.; Vanquelef, E.; Dupradeau, F.Y.; Salomonsson, E.; Leffler, H.; Ortega-Muñoz, M.; Nilsson, U. J.; Kovensky, J.; *Chem. Bio. Chem.* **2010**, *11*, 1430–1442.

62. (a) Bouckaert, J.; Berglund, J.; Schembri, M.; De Genst, E.; Cools, L.; Wuhrer, M.; Hung, C. S.; Pinkner, J.; Slättegård, R.; Zavialov, A.; Choudhury, D.; Langermann, S.; Hultgren, S. J.; Wyns, L.; Klemm, P.; Oscarson, S.; Knight, S. D.; De Greve, H. *Mol. Microbiol.* **2005**, *55*, 441–455. (b) Wellens, A.; Garofalo, C.; Nguyen, H.; Van Gerven, N.; Slättegård, R.; Hernalsteens, P.; Wyns, L.; Oscarson, S.; De Greve, S.; Hultgren, S.; Bouckaert, J. *PLoS ONE.* **2008**, *3*, e2040.

63. Gouin, S. G.; Wellens, A.; Bouckaert, J.; Kovensky, J. *Chem. Med. Chem.* **2009**, *4*, 749–755.

64. Almant, M.; Moreau, V.; Kovensky, J.; Bouckaert, J.; Gouin, S. G. *Chem. Eur. J.* **2011**, *17*, 10029–10038.

65. Lesur, D.; Gassama, A.; Moreau, V.; Pilard, S.; Djedani-Pilard, F.; *Carbohydr. Res.* **2005**, *340*, 1225–1231.

66. Barrientos, A. G.; de la Fuente, J. M.; Jiménez, M.; Solís, D.; Cañada, F. J.; Martín-Lomas, M.; Penadés, S. *Carbohydr. Res.* **2009**, *344*, 1474–1478.

67. Cagnoni, A.; Varela, O.; Gouin, S. G.; Kovensky, J.; Uhrig, M. L. *J. Org. Chem.* **2011**, *76*, 3064–3077.

6

CLICK MULTIVALENT GLYCOMATERIALS: GLYCOCLUSTERS, GLYCODENDRIMERS, GLYCOPOLYMERS, HYBRID GLYCOMATERIALS, AND GLYCOSURFACES

Carmen Ortiz Mellet, Alejandro Méndez-Ardoy, and José Manuel García Fernández

6.1 INTRODUCTION

Carbohydrates at the cell surface (glycolipids, glycoproteins, proteoglycans)—forming the so-called glycocalyx—play key roles in the communication of cells with the external environment. Specific, reversible interactions of these carbohydrates with protein receptors (lectins) are at the origin of important biological and pathological processes including fertilization, bacterial, and viral infection or tumor metastasis, among others. Binding of lectins to carbohydrate ligands is typically characterized by high selectivities but low affinities, which is circumvented in nature by endowing the functional lectin molecules with two or more carbohydrate-binding sites and presenting the complementary glycotopes in multiple copies. The resulting global interaction involves several simultaneous binding events, resulting in associations that are significantly more efficient than the simple addition of individual recognition

Click Chemistry in Glycoscience: New Developments and Strategies, First Edition.
Edited by Zbigniew J. Witczak and Roman Bielski.

processes. This effect is known as the multivalent or cluster glycoside effect and is of utmost importance in glycobiology [1–4].

In order to study, control, interfere, or block carbohydrate–lectin interactions, accessing multivalent glycosystems with well-defined topologies is mandatory. Oligomerization of monovalent derivatives, a divergent approach, has been generally associated with the generation of polydisperse mixtures. The alternative convergent strategy based on glycocoating of functionalized scaffolds, has to face limitations derived from incomplete coupling reactions, resulting in structural defaults. Moreover, both routes have often required cumbersome protecting group strategies to avoid interference of coupling chemistry with the polyfunctionality of sugars. This scenario has abruptly changed in the last decade with the irruption of click chemistry [5], especially the archetypal click-type copper(I)-catalyzed azide–alkyne cycloaddition (CuAAC) reaction [6–8], and its implementation for bioconjugation purposes [9–12]. The bioorthogonal character of the functional groups at play and the unique property of CuAAC as a quantitative ligation reaction have boosted the design and synthesis of novel multivalent glycomaterials for fundamental studies on carbohydrate–protein interactions and therapeutic applications [13–17]. This chapter intends to provide an overview of the many classes of click multivalent glycoarchitectures that can be accessed through CuAAC, with emphasis on its extraordinary selectivity and performance. Due to space limitations, applications of CuAAC to the modification of carbohydrates with multivalent noncarbohydrate functional elements (e.g., elements for DNA [18–22] or toxin recognition [23, 24]) or to the preparation of hybrid glycomaterials not intended for multivalent recognition of the sugar units by protein receptors (e.g., hybrid glycomaterials for controlled drug release [25]) are not discussed here.

6.2 CLICK GLYCOCLUSTERS

Low-valency glycoclusters are fundamental tools to test the influence of architectural features on carbohydrate–protein interactions. The number of recognition elements, their separation distances and spatial orientation, the flexibility or rigidity of the system, and the presence of additional structural elements are parameters that can be systematically varied in these systems. Whereas achieving full control on glycocluster homogeneity maybe relatively easy for divalent or trivalent ligands built onto achiral cores, the difficulties on warranting monodispersity increase exponentially for higher valency platforms. The use of quantitative ligation methods then becomes essential to prevent the presence of under-substituted side products, often susceptible to positional isomerism in the reaction mixtures, which generally leads to unaffordable separation problems. A seminal report by Santoyo-González in 2003 [26], only one year after the discovery of the catalytic version of this reaction, already illustrated the tremendous potential of the CuAAC reaction to create multivalent glycodisplays. By using several per-*O*-acetylated α-D-mannopyranosides with terminal alkyne aglycons (**1**–**3**) and complementary mono-, di-, tri-, or hexafunctionalized azide cores (**4**–**9**), a broad series of multimannoside glycoclusters (**10**) was prepared in excellent

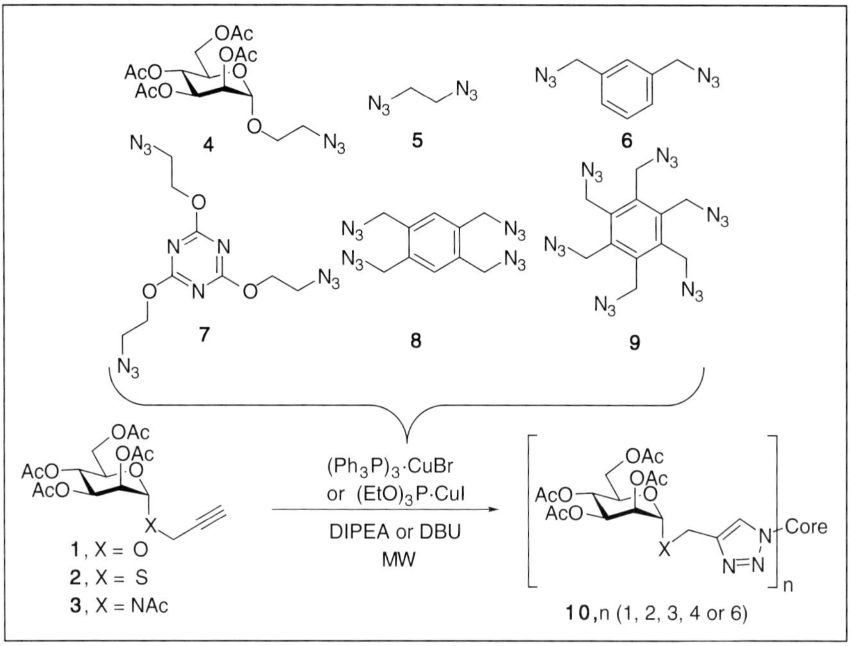

SCHEME 6.1 Synthesis of mulivalent click glycoconjugates onto aliphatic and aromatic platforms.

yields. The coupling reactions were conducted in toluene with $(Ph_3P)_3$,·CuBr, or $(EtO)_3P$·CuI as Cu(I) catalysts soluble in organic solvents, in the presence of diisopropylethylamine (DIPEA) or 1,8-diazabicyclo[5.4.0]undec-7-ene (DBU), and under microwave irradiation (MW; Scheme 6.1).

Several modifications of the reaction conditions have been proposed, such as the use of noncomplexed CuI [27], solid-supported Cu(I) catalysts [28, 29], unprotected carbohydrate precursors [30, 31], aqueous solvent mixtures [32], the production of Cu(I) in the reaction mixture by reduction of Cu(II) with sodium ascorbate (NaAsc) [33], and the generation of the azide reagent *in situ* from a fully unprotected sugar [34] or from an amine precursor by Cu(II)-catalyzed diazo transfer [35]. Nevertheless, the general scheme remains essentially the same. Click strategies have also been implemented to design glycosensors by clusterizing sugars onto electroactive (e.g., ferrocene) [36, 37], optically active (e.g., anthraquinone) [38], or fluorescent cores (e.g., tetraphenylethylene) [39]. Notably, the chemoselectivity and broad functional group tolerance of the click reaction conditions allows the use of bifunctional scaffolds (e.g., the α,α′-trehalose derivative **11**) that can be elaborated after a first click cycle (→**12**; by reaction of **11** with the mannose-derived azide **4**) to activate the adduct for a second coupling (→**13**), thereby accessing heteroglycoclusters with controlled arrangements of different sugar motifs (e.g., **15**; by reaction of **13** with the glucose-derived alkyne **14**) (Scheme 6.2) [40].

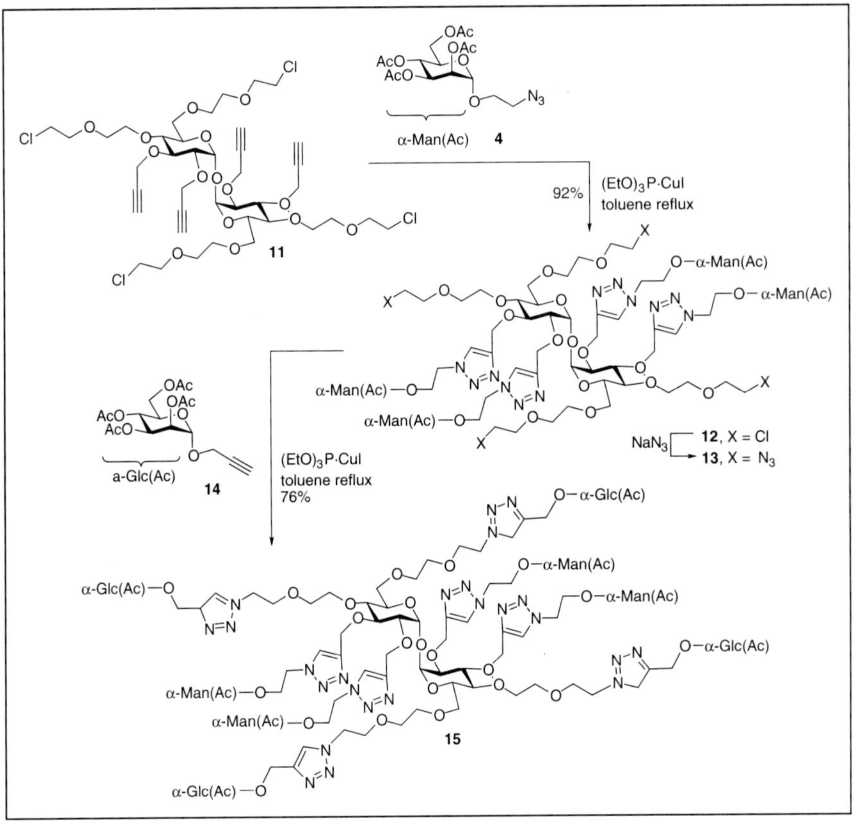

SCHEME 6.2 Synthesis of α,α′-trehalose-centred click heteroglycoclusters.

The above example illustrates the versatility of "clickable" carbohydrate cores to create well-defined architectures by taking advantage of configurational and conformational bias in combination with regioselective functionalization reactions [41]. Gouin, Kovensky, and coworkers have exploited this potential, in combination with the direct replacement of the primary or primary and anomeric hydroxyls in unprotected D-glucose, maltose, and maltotriose, to synthesize a battery of click mannopyranosyl [42] and lactosyl clusters [43] differing in valency and core–ligand distances. Molecular modeling revealed that the oligosaccharide scaffold features a relatively rigid platform providing defined orientations for the substituents, whereas the linkers can be modulated to provide sufficient flexibility for optimal lectin interactions. Most interestingly, the triazole segments can contribute to lectin-binding affinity in a lectin-dependent manner, for instance, being favorable in the case of the mannose-specific lectin concanavalin A (ConA) and detrimental or irrelevant in the case of the lactose/galactose-specific lectin peanut agglutinin (PNA). Other bioinspired clickable scaffolds with linear backbones used for the construction of glycoclusters include polypeptides [44], β-peptoids [45], and phosphodiester oligomers [46,47].

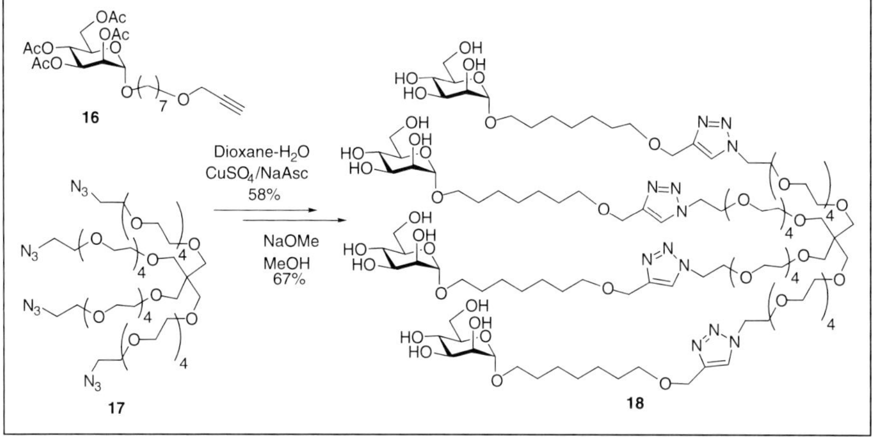

SCHEME 6.3 Synthesis of pentaerythritol-based click multimeric mannosides as *E. coli* antiadhesive agents.

In most cases, the structure of the linker between the recognizable sugar residues and the central core influences the binding of a multivalent ligand to a specific lectin in a generic manner by regulating intersaccharide distances and flexibility. However, in some cases, the aglyconic portion can interact with amino acids at the vicinity of the binding site, thereby reinforcing the association, which can be further enhanced by presenting the whole pattern in a multivalent manner. An outstanding example was reported by Gouin and coworkers by clicking the pentaerythritol-derived tetraazide **17** with alkyne **16**, containing the heptyl α-D-mannopyranoside motif previously identified as a preferred ligand for FimH lectin, situated at the fibrillum tip of uropathogenic *Escherichia coli* bacteria [48]. The resulting tetravalent cluster **18** inhibited bacterial bladder cell binding 64-fold more efficiently than the monomeric mannoside (Scheme 6.3).

In all the above-commented reports, the external carbohydrate units are linked either through the anomeric carbon to the nitrogen N-1 of the triazole ring (e.g., if a glycosyl azide is used as a reagent partner) or via an anomeric *O*-, *N*-, or *S*-atom to an N-1 (e.g., if the reaction involves an azide-armed glycoside) or to a C-4 triazole substituent (e.g., if the reaction involves an alkyne-armed glycoside) after clicking. Attachment of the sugar via the anomeric carbon directly to C-4 of the heterocycle (e.g., in **19**) or through a carbon bridge to N-1 (e.g., in **20**), implying alkyne- or azide-armed *C*-glycosides as reagents [49], as well as via other sugar ring atoms (e.g., in **21**) [50] has also been achieved (Fig. 6.1).

Macrocyclic platforms, particularly cyclooligosaccharides of the cyclomaltodextrin family (cyclodextrins, CDs), provide excellent opportunities to install a variety of functional elements with very precise spatial orientations and architectural arrangements [51]. Inhomogeneity problems typically associated to incomplete or side reactions have been largely overcome by the implementation of CuAAC ligation protocols. This was already exemplified in the original communication by

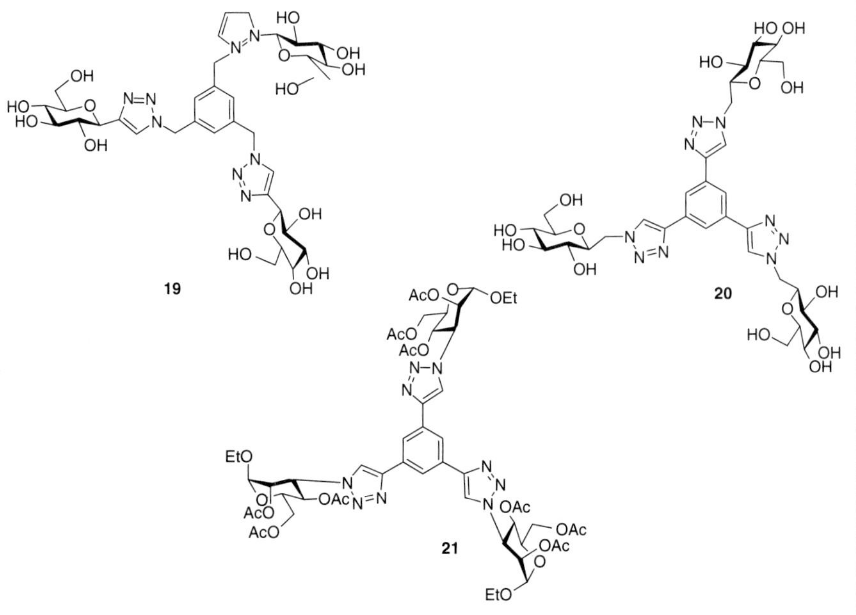

FIGURE 6.1 Examples of click glycoconjugates with *C*-linked and non-anomeric *N*-linked glycoligands.

Santoyo-González and coworkers with the preparation of click per-(C-6)-cyclomaltohepatose (β-cyclodextrin, βCD)-scaffolded hepta-*O*- and *S*-α-D-mannopyranosides [26]. In a latter report, the same group extended the application of the βCD platform to the preparation of analogs hepta-*O*-β-D-galactopyranosides and hepta-*O*-β-lactosides (**24**), as well as to fluorescently labeled galactopyranoside conjugates (**25**) obtained by coclicking the βCD heptaazide **22** with propargyl β-D-galactopyranoside and the propargyl dansyl derivative **23** (Scheme 6.4) [52].

The efficiency of the triazole-forming reaction was further demonstrated with the preparation or per-(*O*-2,*O*-3)-substituted, βCD-centered tetradecavalent galactosyl and lactosyl clusters. These compounds had the capability to promote cell adhesion and stimulation of monocytes, thereby mimicking the lipopolysaccharide (LPS) effects [52]. Other authors have exploited the capabilities of cyclomaltohexaose (α-cyclodextrin, αCD) to form polyrotaxanes to increment the functionality of mono-, di-, or triazido-substituted units by threading them on polyethylene glycol chains that are blocked at the ends with bulky stoppers before clicking the desired glycotopes [53]. Alternatively, monovalent click conjugates can be used to form pseudorotaxanes that are click-assembled in a later step to create multivalent polyrotaxanes [54]. Although intellectually very appealing, these strategies suffer from the intrinsically low yields of cyclodextrin rotaxanation.

Polyaromatic containers of the calixarene type and their relative resorcinarenes (resorcarenes) have been largely used for the construction and application of novel

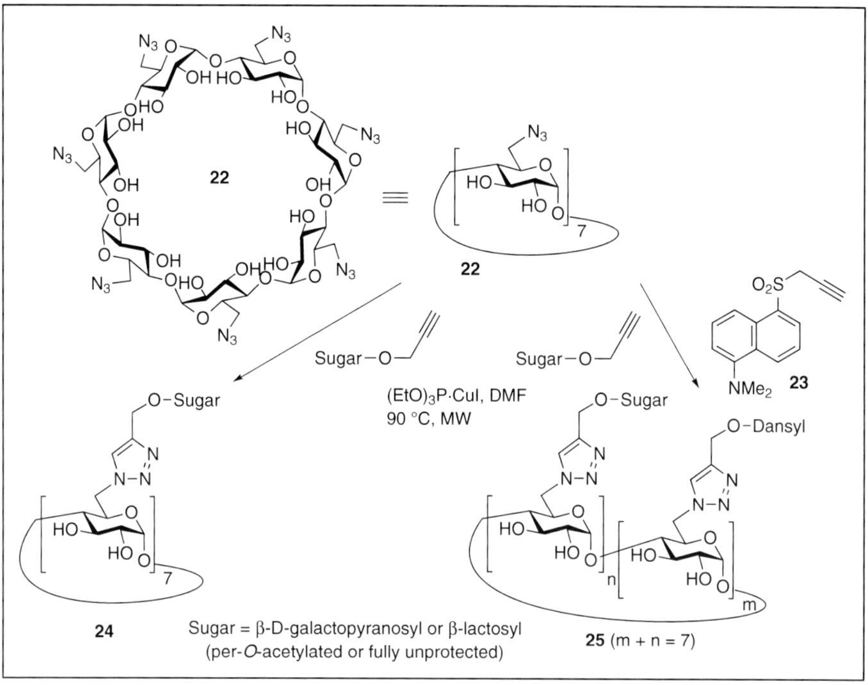

SCHEME 6.4 Synthesis of cyclodextrin-scaffolded click glycoclusters.

glycocluster architectures to explore multivalent interactions between carbohydrates and proteins [55], because of their nanometric three-dimensional rigid structure, commercial availability, and extremely rich preparative chemistry [56]. Dondoni, Marra, and coworkers pioneered the application of click chemistry for those channels, with the preparation of several tetra- and octavalent glycoclusters from calix[4]arene platforms [44, 57]. For instance, the sialic acid derivatives **30–32**, obtained by click coupling of the polyazides **26–28** with alkyne **29** followed by conventional deprotection (Scheme 6.5), were able to inhibit the viral infectivity by BK and influenza A viruses at submillimolar concentrations [57].

The possibility to fix the cone, the partial cone, and the 1,3-alternate conformation in calix[4]arenes bearing *tert*-butyl substituents at C-4 and propargyl groups at C-1 in the aromatic rings (**33–35**) was exploited by Vidal and coworkers for the preparation of a library of conformationally locked click glycoclusters differing in valency and topology [45, 58] (Fig. 6.2). Tetravalent galactosylated clusters bearing galactopyranosyl residues at both faces of the calixarene platform were found to be superior as ligands for the galactose-binding lectin PA-IL from the opportunistic bacteria *Pseudomonas aeruginosa* as compared with compounds with the same valency, but with monofacially oriented substituents [58]. In a recent report, Field and coworkers encountered the reverse situation for calix[4]arene-centered tetralactosyl click clusters when assayed for their tripanocidal activity, with monofacially oriented

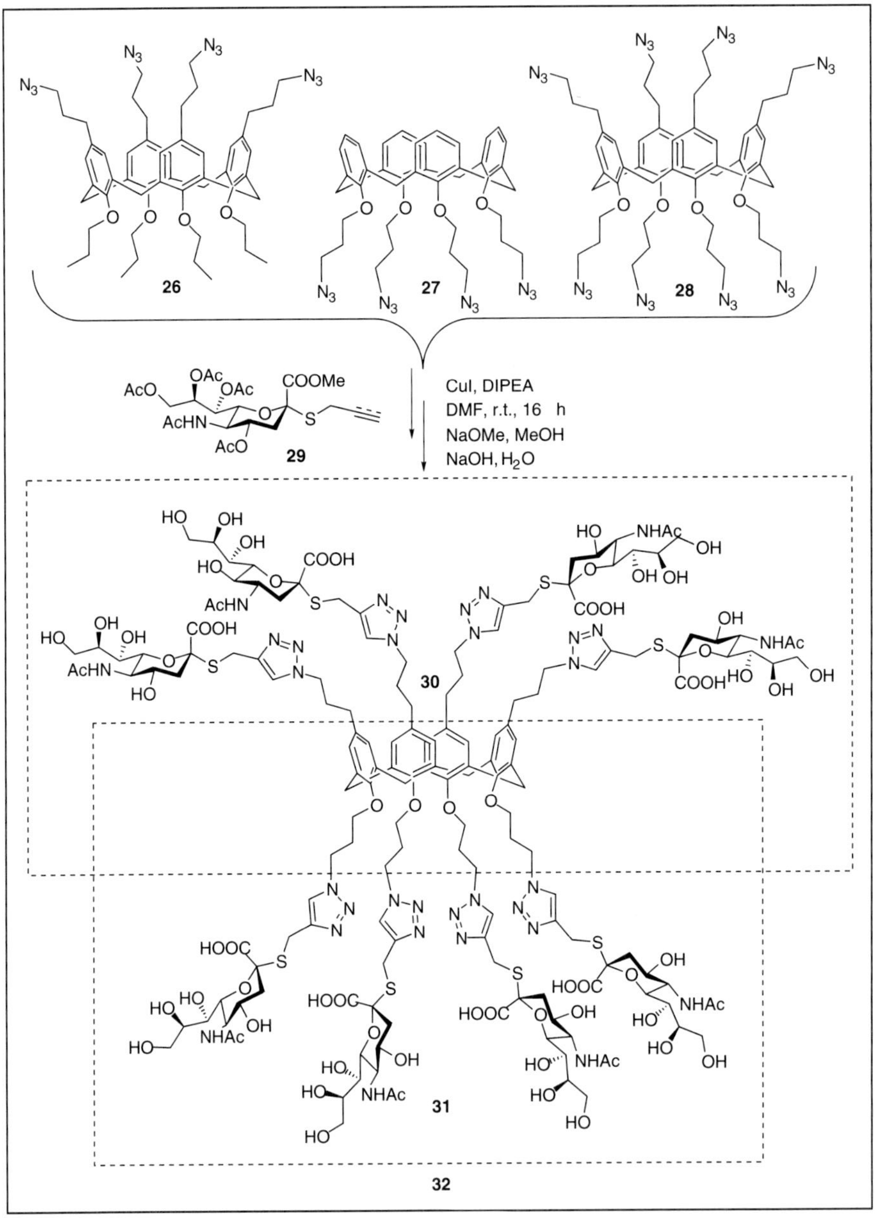

SCHEME 6.5 Synthesis of face-selective and dual-face modified calix[4]arene click clusters coated with neuraminic acid ligands.

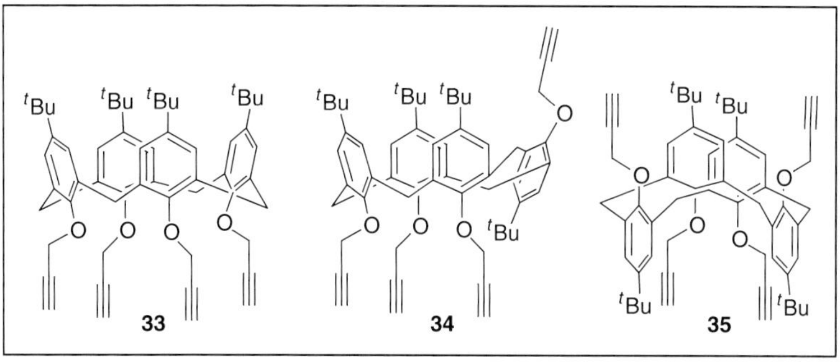

FIGURE 6.2 Structures of conformationally locked clickable calix[4]arenes in the cone, partial cone and 1,3-alternate conformations.

compounds being more efficient [59]. Altogether, these results provide a proof of concept of the importance of architectural features in the design of multivalent glycoligands for a given protein receptor. Further examples include the preparation of click glycoclusters from calix[6]arene [45] and resorcin[4]arene derivatives [60].

Other macrocyclic platforms well adapted to build glycoclusters with precise geometries through click strategies are porphyrins [45, 61, 62], cyclic peptoids [45], and cyclopeptides [63]. It is worth mentioning that multivalency mechanisms to enhance carbohydrate recognition are not restricted to lectins, but can also operate, for example, carbohydrate-directed antibodies. Thus, the cyclodecapeptide-templated oligomannose cluster **38**, assembled by CuAAC of the tetra-propargylamide derivative **36** and the azide-armed tetrasaccharide [Manα1,2Manα1,2Manα1,3Manα] **37** (Scheme 6.6), showed high-affinity binding to the HIV-neutralizing antibody 2G12, acting as a mimic of high-mannose *N*-glycans [64].

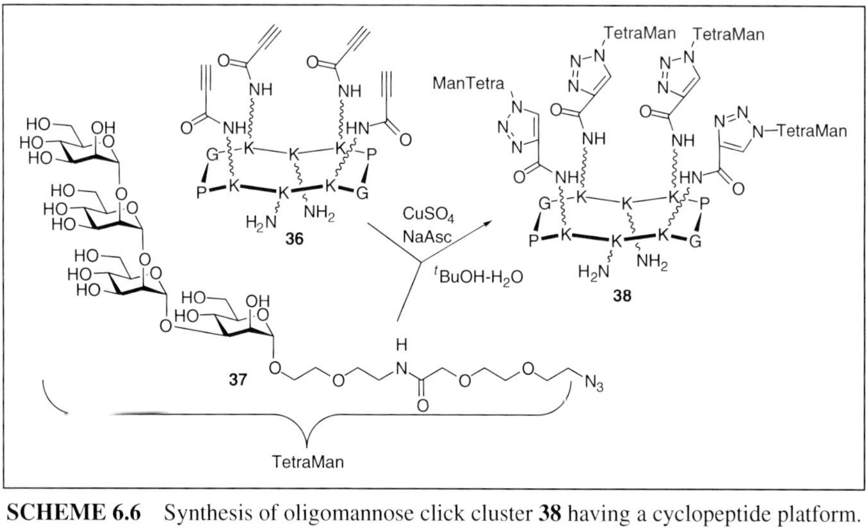

SCHEME 6.6 Synthesis of oligomannose click cluster **38** having a cyclopeptide platform.

6.3 CLICK GLYCODENDRIMERS

The high functional group's density at the chain ends of dendrimers, which can rapidly increase upon moving to higher generations, allows unprecedented control over molecular architecture, making these materials extremely attractive candidates for a variety of surface active applications. The suitability of dendrimeric scaffolds to create high-valency, yet monodisperse displays of sugar glycotopes has been broadly realized [65–67]. The need for chemical transformations which allow simple and facile preparations is, however, exacerbated when considering dendrimer glycocoating. In such circumstances, the CuAAC expresses all its power. The synthesis of the dendritic block copolymer **39** by Hawker, Sharpless, and coworkers illustrates the efficiency of this premier click reaction [68] (Fig. 6.3). The synthetic approach is based on 2,2-bis(hydroxymethyl)propionic acid (bis-MPA) multifunctional dendrimers. Two 7-dimethylaminocoumarin fluorescent dyes are first coupled at the focal point of the generation 4 (G4) acetal-protected dendritic core. The terminal hydroxyl groups are then liberated and propargylated to create an alkyne functional G4 dendrimer. In the last step, the 16 mannopyranosyl units are incorporated at the surface of the dendrimer by click reaction with the corresponding 2-azidoethyl glycoside without the use of protecting groups. The strategy based on focal–focal coupling of alkyne functional bis-MPA dendrimeric blocks was further extended by Boons and coworkers to the preparation of up to 32-valent β-D-galactopyranosyl glycodendrimers from two 16-valent moieties [69].

Click glycocoating of peripheral alkyne or azide functional dendrimers has become a preferred method in glycodendrimer synthesis [70–75]. Monodisperse glycoconstructs with up to 243 sugar residues have been reported, attesting for the supremacy of the CuAAC methodology in divergent schemes [76]. Moreover, the click strategy is compatible with sequential coupling cycles for dendrimer generation growth. This concept has been exploited by Morvan and coworkers to build oligonucleotide carbohydrate-centered glycocluster conjugates by implementing phosphoramidite and click chemistries [77]. The clickable G0 mannopyranosyl conjugate **40** was first coupled on the solid support with 3-azidopropyl-β-D-galactopyranoside (**41**) and the resulting G1 conjugate was activated for a next click cycle by reaction with the alkyne-armed phosphoramidite **42**. Coupling of **43** with the azide-armed galactopyranoside **44**, followed by cleavage from the solid support and deprotection of the peripheral sugar units afforded the 16-valent glycodendrimer **45** with a core that contains one mannose and four galactose motifs (Scheme 6.7).

Alternatively, CuAAC has been exploited to design click glycodendrons that are subsequently engaged in coupling reactions with appropriately functionalized dendritic cores in a divergent synthetic scheme. The final click glycodendron–core multiconjugation can be a next CuAAC (e.g., for **46**) [78,79] or any other coupling reaction providing the necessary efficiency. Examples on record include oxime (e.g., for **47**) [80] and thiourea-forming reactions (e.g., for **48**) [81] (Fig. 6.4). A complementary approach consisting in preforming amide-based glycodendrons bearing a focal azide group (e.g., **49**) for their coupling to an alkyne functional core (e.g., **50**) has also been put forward by Seeberger and coworkers to build Ru(II)-centered glycodendrimers (e.g., **51**; Scheme 6.8) [82].

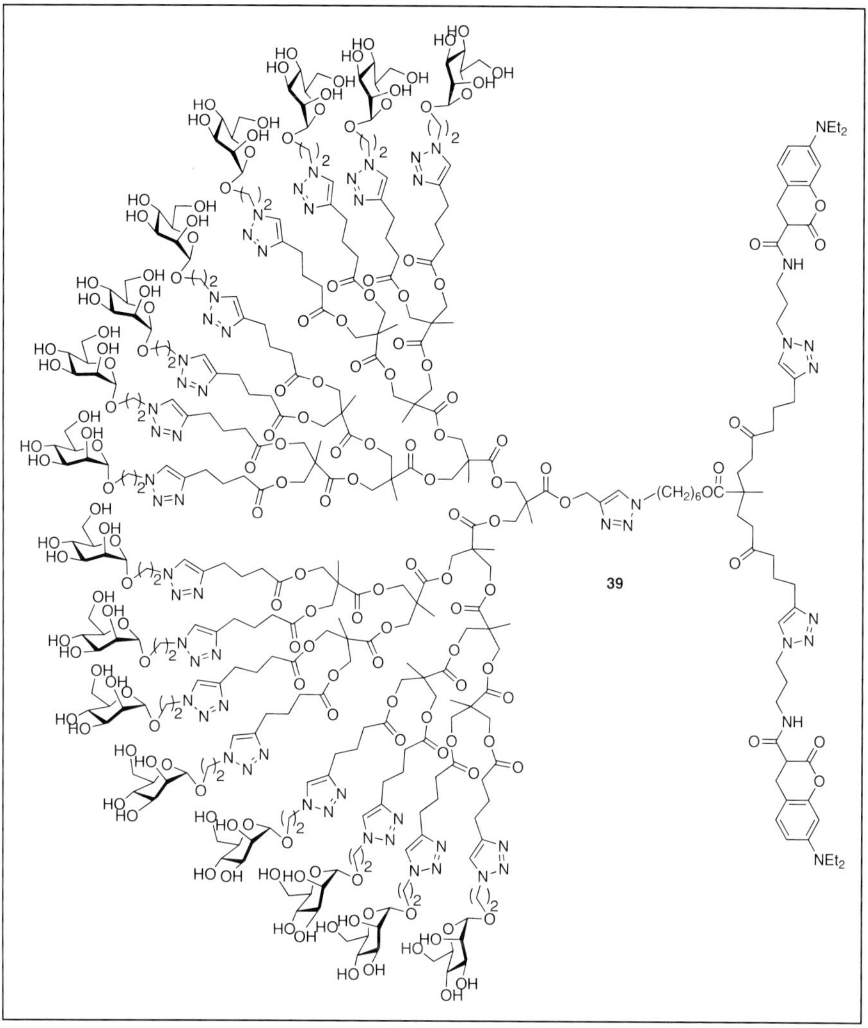

39

FIGURE 6.3 Bis(hydroxymethyl)propionic acid (bis-MPA)-based 16-valent click glycodendrimer incorporating two 7-dimethylaminocoumarin fluorescent dyes.

6.4 CLICK GLYCOPOLYMERS

Carbohydrate-containing synthetic polymers are a popular class of bioinspired materials that have been widely studied and used in multiple applications [83]. The carbohydrate moieties can be inserted in the polymer backbone to build glycooligomers or block copolymers that mimic the structure of polysaccharides. While there are several interesting reports on the use of CuAAC for the preparation of this type of biomaterials [84–87], they do not conform to the generally accepted notion of multivalency and will not be discussed here. This section will focus instead on synthetic

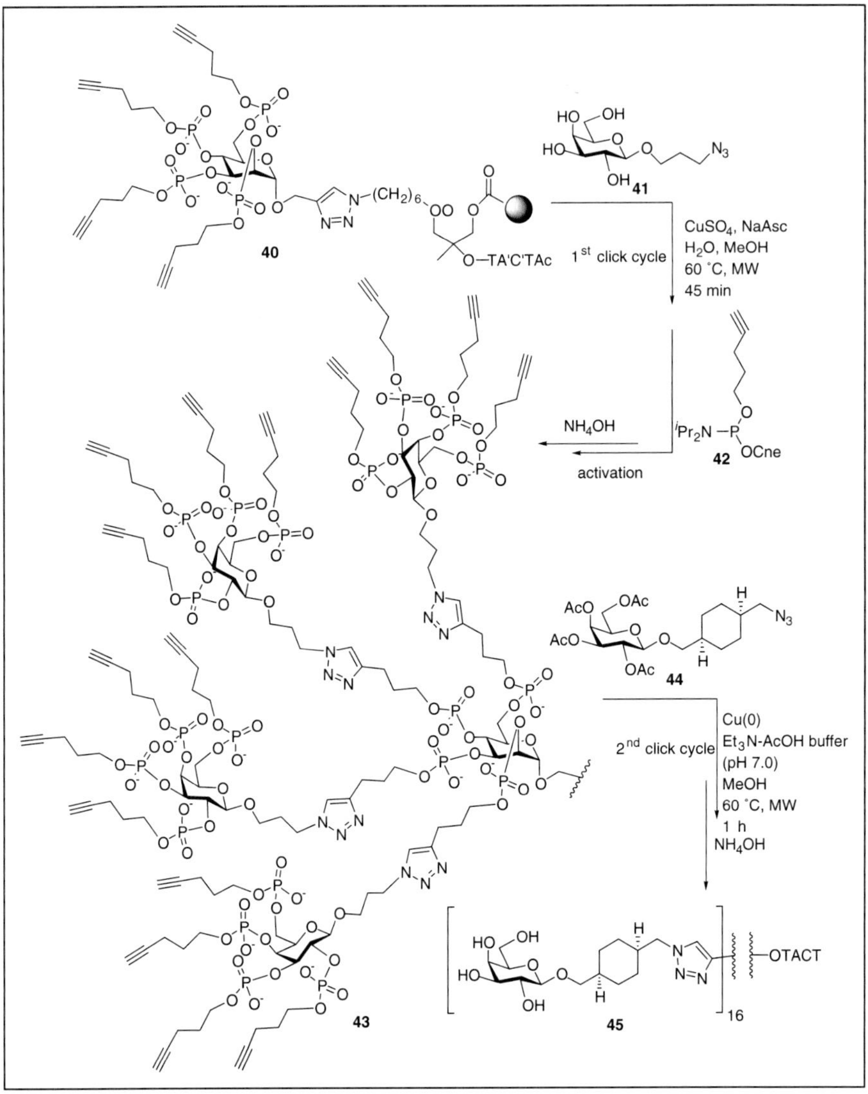

SCHEME 6.7 Synthesis of β-galatopyranosyl-coated glycodendrimers by phosphoramidite and CuAAC ligation chemistries.

polymeric macromolecules featuring pendant carbohydrate moieties as in the natural glycoproteins.

Until the irruption of the Cu(I)-catalyzed version of the Huisgen azide–alkyne cross-coupling in bioconjugation, it was generally accepted that the controlled polymerization of functional glycomonomers was by far the preferable strategy to access multivalent glycopolymers [88]. With few exceptions [89–93], post-glycosylation of pre-synthesized polymers does not afford 100% functionalization, producing defaults

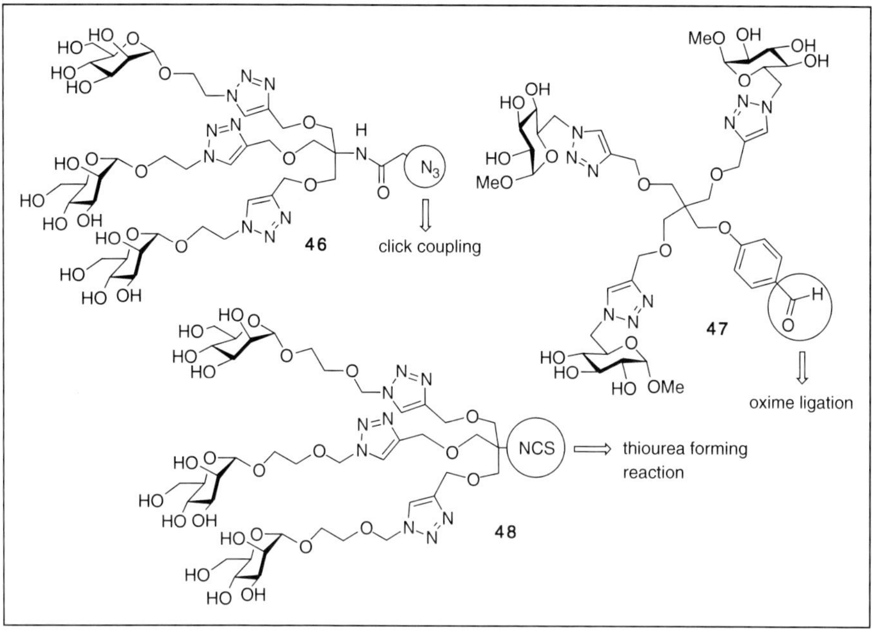

FIGURE 6.4 Examples of trivalent click glycodendrons.

within the chain that hamper systematic studies. The efficiency of CuAAC in post-polymerization modification approaches, generally affording quantitative yields of stable triazole-linked conjugates, drastically changed this scenario [94]. In a sense, click glyco-functionalization allows mimicking nature's manner of producing glycopeptides, which involves post-translational modification of pre-biosynthesized proteins. A wide range of glycosylated structures can thus be accessed from a reactive scaffold prepared from either a single or a limited number of monomers.

Because of the instability of organic azides, it was anticipated that the use of poly(alkynyl) polymers was preferable for CuAAC-based glycocoating strategies. To avoid the problems associated to radical transfer to the triple bond during radical polymerization of alkynyl monomers, Haddleton and coworkers [95] proposed the use of trimethylsilyl (TMS)-protected propargyl methacrylate precursors (e.g., **52**) and a benzyl α-bromoester (e.g., **53**) as polymerization initiator (→**54**). Clickable, well-defined alkyne polymers (e.g., **55**) were thus prepared to which glucosyl, galactosyl, and mannosyl residues were grafted via either a C-6 or an anomeric azide. The CuAAC reaction was conducted both with peracetylated or unprotected monosaccharides using $(PPh)_3CuBr$ as the catalyst in the presence of DIPEA, with no noticeable change in the polydispersity index (PDI) upon functionalization. The potential of the strategy for the synthesis of libraries of glycomaterials was further illustrated by "coclicking" 3-azidopropyl-α-D-mannopyranoside (**56**) and 3-azidopropyl-β-D-galactopyranoside (**57**) to the alkyne polymer (**55**) in different relative proportions. The resulting heteroglycopolymers (**58**) were assayed for their binding capacity to

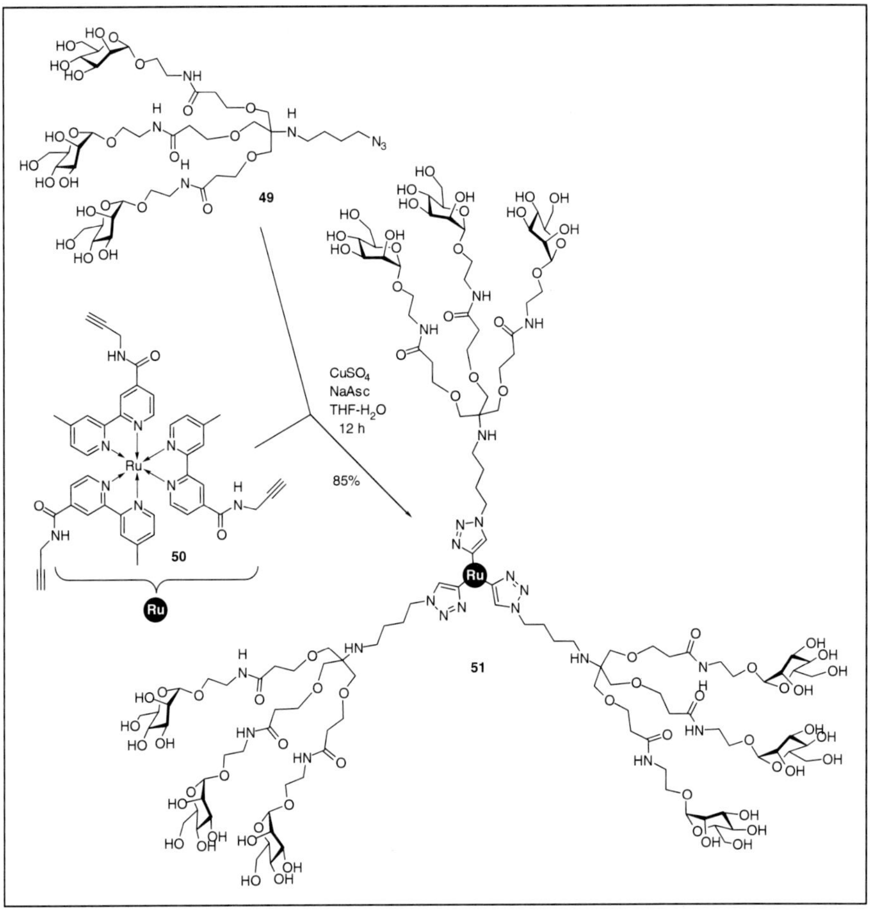

SCHEME 6.8 Synthesis of Ru(II)-centred click glycodendrimers.

the plant lectin ConA by turbidimetry and quantitative precipitation (Scheme 6.9). Interestingly, a 75:25 mannose:galactose ratio was found as efficient as the homogenously mannosylated polymer in clusterizing the lectin, implying a 1.5-fold higher efficiency in a mannose molar basis. Although a saturation effect cannot be discarded, synergic interactions might operate in this case, a phenomenon previously observed in highly dense heteroglycoclusters and termed heterocluster effect [96,97]. The versatility of the coclicking method allows the incorporation of further different functionalities in the polymer backbone. For instance, any of the sugar azides can be coclicked in combination with an azide-armed fluorescent probe to afford fluorescent glycopolymers that were used to quantitatively assess binding to an immobilized lectin through affinity HPLC.

The general strategy consisting in the click postfunctionalization of well-defined alkyne functional polymers with sugar azides has proven extremely powerful to

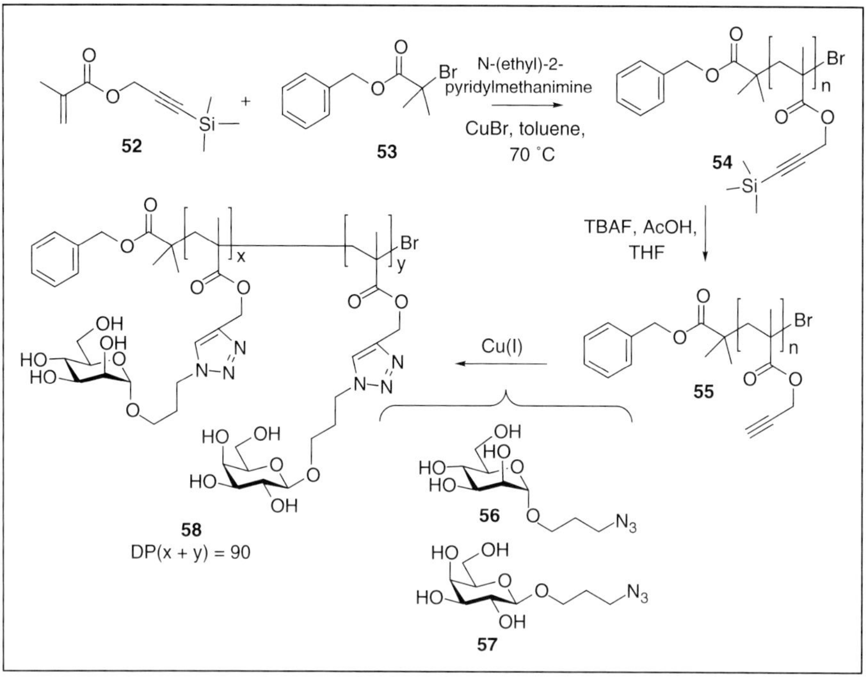

SCHEME 6.9 Synthesis of heteroglycopolymers through co-clicking strategies.

produce neoglycopolymers capable to interfere in carbohydrate–protein recognition events [98–104]. Mono-, di-, and trisaccharides with terminal D-mannose, D-galactose, and D-glucose motifs have been appended either directly through the anomeric position or through a spacer arm. It is worth highlighting that the presence of an anomeric triazole ring does not seem to compromise lectin recognition [98]. The possibility to coclick different functional elements onto the polymer backbone can be complemented with the incorporation of an orthogonal reactive functionality at the chain end for bioconjugation purposes. In a notable example, Haddleton and coworkers exploited the capability of novel maleimide-terminated glycopolymers to selectively react then with the free-cysteine residue of bovine serum albumin (BSA) to prepare neoglycopolymer–protein hybrid materials (Scheme 6.10) [105]. Copolymerization of **52** and the fluorescent rhodamine B derivative **59** in the presence of the maleimide-protected initiator **60** afforded the poly(propargyl methacrylate) **61,** which was later on coclicked with 2-azidoethyl α-D-mannopyranoside and β-D-galactopyranoside at different ratios. The resulting fluorescently labeled heteroglycopolymers were activated by a retro-Diels–Alder reaction to generate the maleimide moiety that was then conjugated to BSA (→**62**). Surface plasmon resonance (SPR)–binding studies showed clear and dose-dependent binding of the resulting neoglycopolymer–BSA conjugates to recombinant rat mannose-binding lectin (MBL), a key mammalian lectin of the immune system. Moreover, these biomaterials

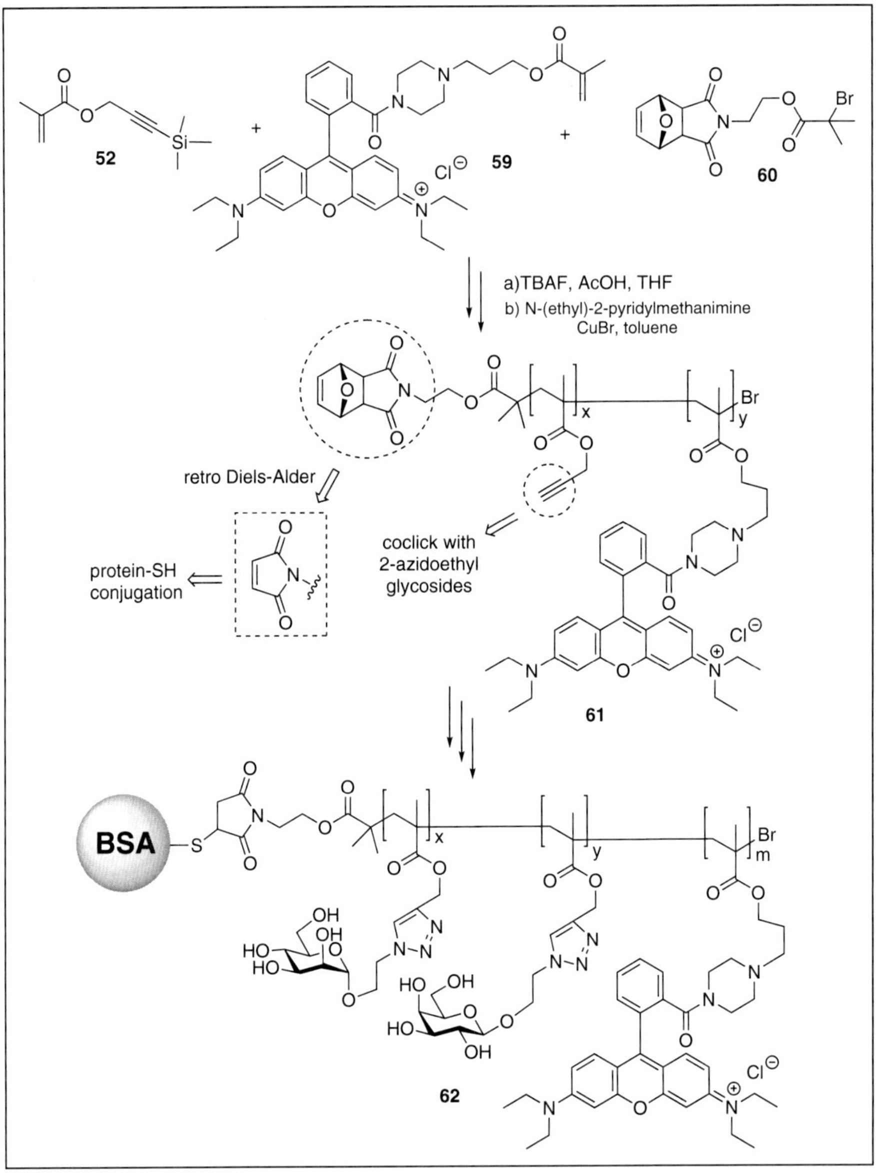

SCHEME 6.10 Synthesis of BSA-conjugated fluorescently-labelled click heteroglycopolymers.

were able to significantly activate the complement system via the lectin pathway, a powerful arm of the circulating immune system [105]. Other click glycopolymer applications explored include the inhibition of the interactions of a human dendritic cell-associated lectin and the viral envelope glycoprotein gp120, involved in HIV infection [106], the preparation of bead-immobilized glycopolymers for their

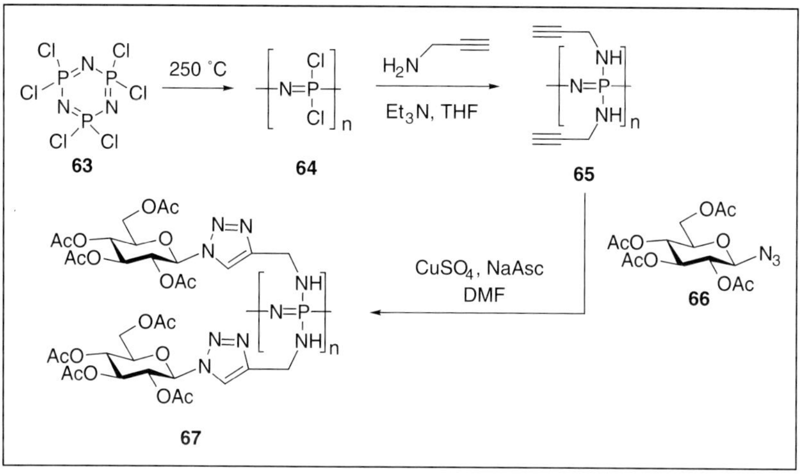

SCHEME 6.11 Synthesis of polyphosphazene click glycopolymers.

use in chemical sensing and affinity chromatography [107], and the elaboration of glycopolymer-coated microporous membranes for the isolation/purification or quantitative identification of carbohydrate-binding proteins [108].

Alkyne functional polymers suitable for click glycocoating have also been prepared by postmodification of preformed polymers bearing other reactive groups. Thus, Huang and coworkers reported the synthesis of polyphosphazene glycopolymers **67** through a reaction sequence that involved the nucleophilic substitution of chloro groups in poly(dichlorophosphazene) **64**, readily obtained by ring-opening polymerization of hexachlorocyclotriphosphazene (**63**), by propargylamine (→**65**) and subsequent CuAAC with peracetylated β-D-glucopyranosyl azide **66** (Scheme 6.11) [109].

Propargylation of hyperbranched polyglycerols (HPGs, **68**) has also been implemented to produce suitable scaffolds for click glycopolymer elaboration [110, 111]. The reaction is, however, incomplete and only a fraction of the hydroxyl groups is transformed into the corresponding propargyl ethers (**69**), which introduces some inhomogeneity in the structure. Nevertheless, high densities of sugar motifs could be achieved after subsequent click conjugation with the sugar azides **57** and **70**. The galactosylated glycopolymers (**71**, **72**) exhibited strong affinity inhibition toward leukocyte-derived selectins in an SPR-based competitive assay, in the nanomolar range, the sulfated glycopolymer (88% functionalization) being selective for L- and P-selectins [110]. Similarly, mannosylated glycopolymers exhibited strong affinity for the mannose-specific lectin ConA (Scheme 6.12) [111].

In spite of the potentially explosive character of organic azides, the reverse strategy consisting in the preparation of a polymeric azide and its click conjugation with acetylene-armed sugars has also proven very successful for the preparation of glycopolymers. The general approach implies either the polymerization of a halogen-bearing monomer or the incorporation of halogen at specific positions in a reactive

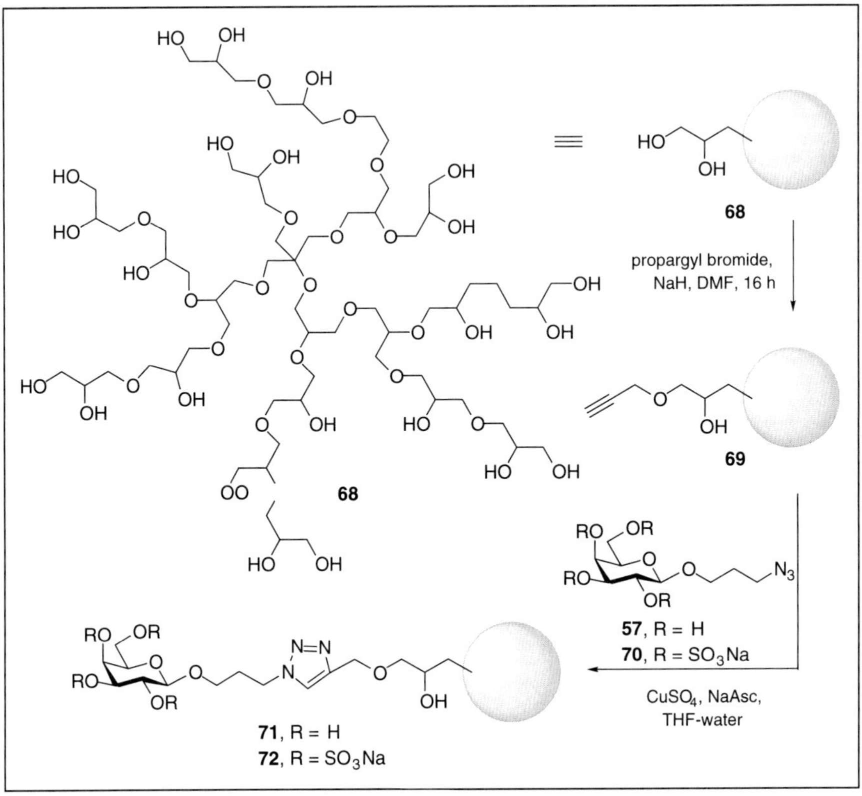

SCHEME 6.12 Synthesis of β-galactopyranosyl-coated hyperbranched hyperglycerol (HPG) dendrimers by CuAAC.

polymer, and the subsequent nucleophilic displacement of the halogen groups by azide anion. Poly[*p*-(azidomethyl)]styrene (**73**) [112], poly[2-azido-ε-caprolactone) (**74**) [113], and 6-azido-6-deoxy-curdlan (**75**) [114] have been prepared in this manner and used in multi-CuAAC reactions with propargyl glycosides derived from α-D-mannose, β-D-glucose, β-D-galactose, *N*-acetyl-β-D-glucosamine, β-maltose or β-lactose (Fig. 6.5). Coclick reactions are also possible, which has been exploited for

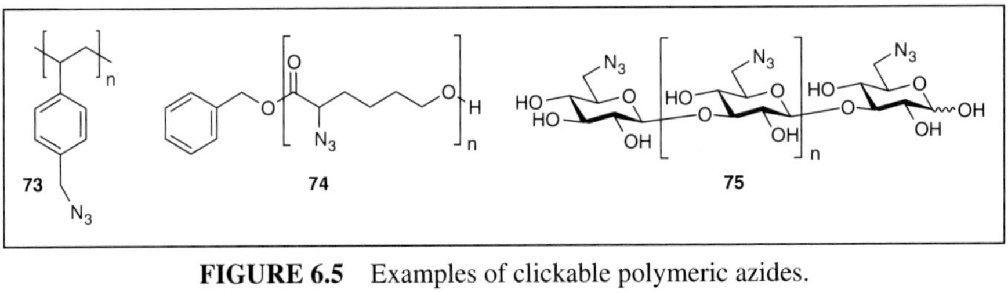

FIGURE 6.5 Examples of clickable polymeric azides.

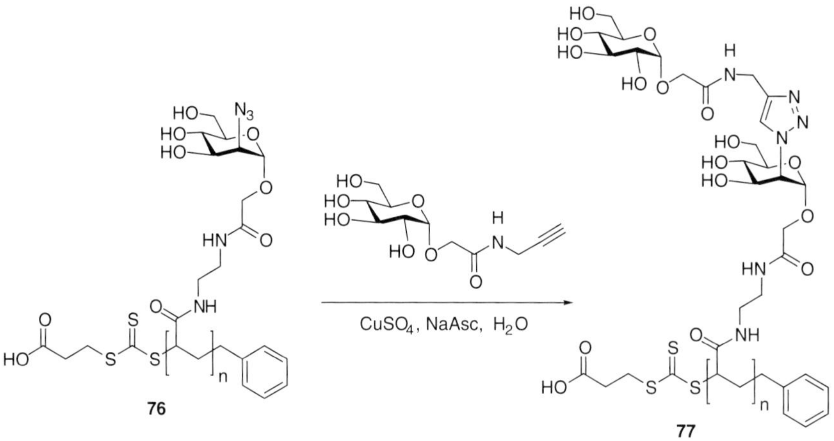

SCHEME 6.13 Click post-glycosylation of pre-synthesized glycopolymers.

the simultaneous incorporation of a sugar and a fluorescent dye [112]. Interestingly, the curdlan-mannopyranosyl neoglycopolymer was able to interact with polycytosine (polyC) in a similar manner as the native polysaccharide to form a triple-stranded stable macromolecular complex with two strands of the glycopolymer and one strand of polyC. These macromolecular complexes showed strong and specific affinity toward complementary lectins [114].

The above approach was implemented by Fleury and coworkers for the click post-glycosylation reaction of pre-synthesized glycopolymers bearing azide functionalities at the preinstalled sugar moieties [115]. Thus, they prepared the polyazide glycopolymer **76** by reversible addition–fragmentation chain transfer (RAFT) polymerization of *N*-(2-azido-2-deoxy-α-D-mannopyranosyl)acrylamide and subjected it to CuAAC with *N*-propargylamidomethyl-β-D-glucopyranoside, to give the corresponding neoglycopolymer with pseudodisaccharide pendant units **77** (Scheme 6.13).

A third general strategy to build triazole-linked multivalent glycopolymers takes advantage of the efficiency of the Cu(I)-catalyzed azide–alkyne click coupling to prepare monomers bearing polymerizable groups, such as vinyl [116] or 2-oxazoline groups [117], avoiding complex reactions and purification procedures often associated with carbohydrate monomer synthesis. Khiar and coworkers reported a remarkable application of this tactic to assemble polymeric "click glycoliposomes" by photopolymerization of click diacetylene monomers [118]. Propargylamide **79**, prepared from pentacosa-10,12-diynoic acid (PCDA, **78**) was clicked with sugar azidoglycosides with different spacers (e.g., **80**) to give diacetylenic glycoamphiphiles (e.g., **81**) that were inserted in PCDA liposomes before polymerization (→**82**), affording stable unilamellar vesicles (100–500 nm) whose color shifted from blue to red upon interaction with a specific lectin (Scheme 6.14).

Amphiphilic click glycopolymers can be used as precursors for the generation of self-assembled glyconanoparticles by nanoprecipitation [119, 120]. Zhao and

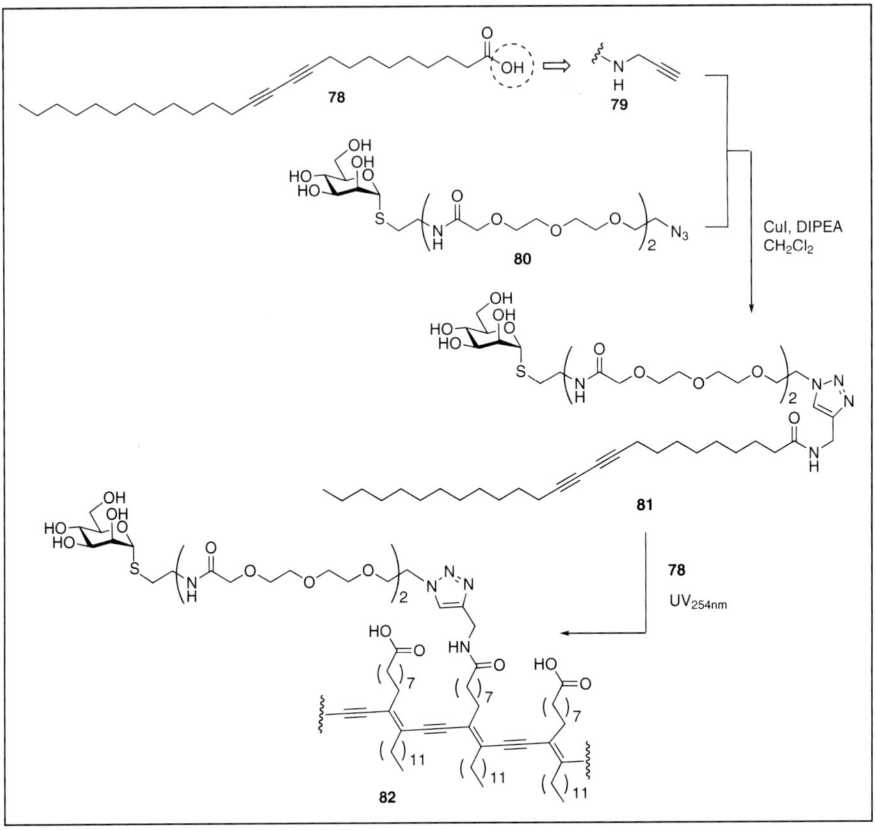

SCHEME 6.14 Click glycoliposomes obtained by photopolymerization of diacetylene monomers.

coworkers have reported an original alternative approach to obtain multivalent polymeric organic nanoparticles via surface clicking of alkynylated surfactant micelles [121]. The nanometric platform **85** is created by surface cross-linking of micelles obtained from the cationic surfactant **83** with diazido derivatives (e.g., **84**), using 1:1 molar ratios. The micelle then becomes stabilized while keeping multiple alkynyl groups at the surface, allowing further click glycodecoration (→**86**) (Scheme 6.15).

Fluorescent glycopolymers have attracted much interest as sensors for the detection and quantitation of biomolecules interacting with carbohydrates. Bunz and coworkers applied click glycocoating to alkyne functional poly(*p*-phenylethynylene) to obtain mannosyl-pendant conjugated polymers (e.g., **87**) (Fig. 6.6) that experienced fluorescence self-quenching induced by ConA and were able to aggregate *E. coli* [122, 123]. Han and coworkers exploited, instead, the advantages of CuAAC to synthesized glycofluorenyl monomers that were further polymerized, or the post-functionalization of azide functional polymers, to prepare fluorescent conjugated polyfluorenes with pendant sugars (e.g., **88**) (Fig. 6.6). Fluorescence quenching upon

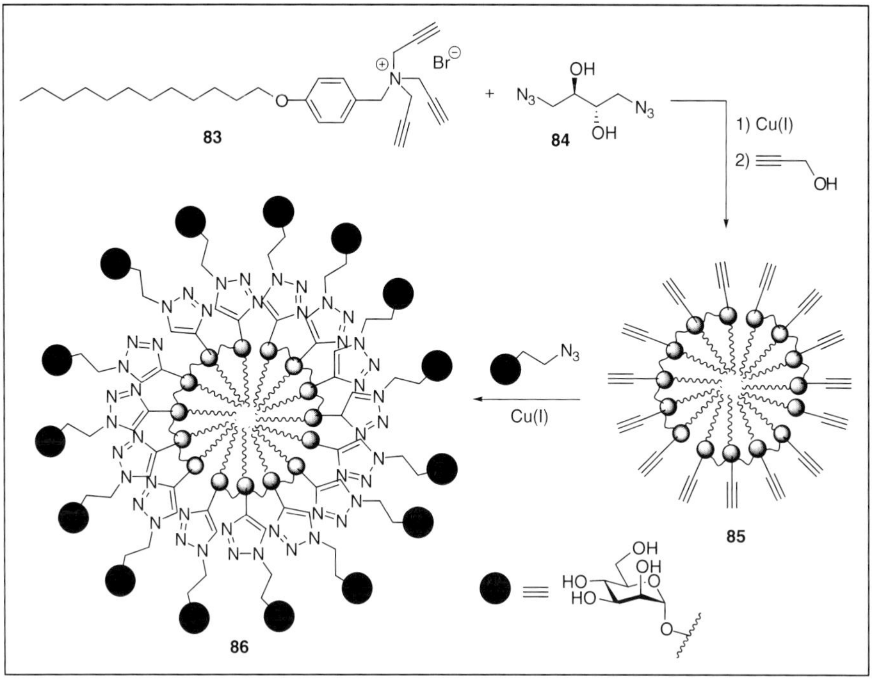

SCHEME 6.15 Click glycodecorated surfactant micelles.

recognition by receptor partners lets monitor carbohydrate–lectin [124], carbohydrate–DNA [125], carbohydrate–heparin [126], and carbohydrate–carbohydrate interactions [127].

Considering that the potential of glycopolymers to mimic natural glycoproteins has been the original motivation for the research developed in this field, it is not

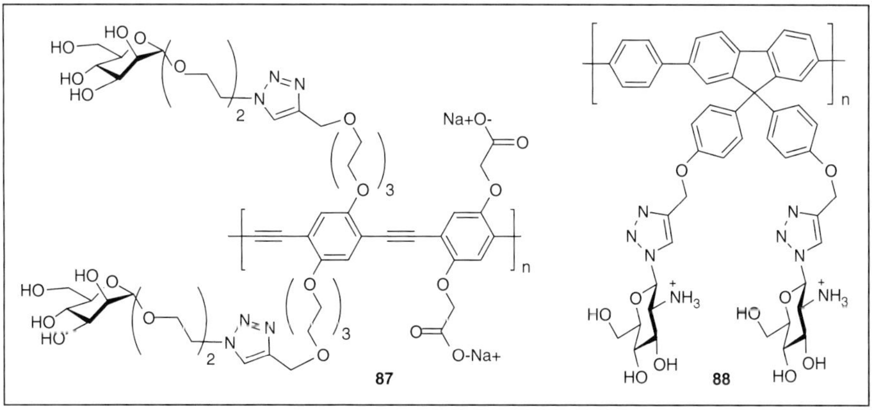

FIGURE 6.6 Examples of fluorescent click glycopolymers.

surprising that several groups have focused on the postfunctionalization of alkyne- [128–130] or azide- [131] functional oligo- and polypeptides by click glycoconjugation as a way to test the consequences of structural variations in lectin-binding properties. Particularly interesting is the possibility to implement cooperative interactions involving the polypeptide backbone and the appended carbohydrate ligands. For instance, Kiic and coworkers inserted propargylglycine residues in polypeptide sequences having affinity for the B_5 subunit of cholera toxin that were further modified with azido-functionalized galactopyranosides via CuAAC [132]. The results evidenced the importance of saccharide spacing, polypeptide chain extension, saccharide linker conformation, and the placement of charges on the polypeptide backbone on the inhibitory potencies of glycopeptide-based multivalent inhibitors toward this toxin.

The involvement of electrostatic interactions is particularly important for DNA complexation. Installation of bio-recognizable carbohydrates on cationic polymers offers a possibility to target nucleic acid material to specific cells through the formation of glycopolymer:DNA:lectin ternary complexes that can mediate cell internalization. Reineke and coworkers have designed polyamidoamine polycationic scaffolds bearing a controlled number of clickable alkyne sites by reaction of substituted malondichloride derivatives (**89**, **90**), with partially Boc-protected pentaethylenehexamine (**91**). Click conjugation with peracetylated azidoalkyl β-D-galactopyranosides (**92**) and final deprotection afforded polycationic glycopolymers (**93**) that efficiently compacted plasmid DNA (pDNA) into transfectious nanoparticles that were specifically recognized by the asialoprotein receptor at the cell surface of hepatocytes (Scheme 6.16) [133].

6.5 CLICK HYBRID GLYCOMATERIALS

The term hybrid materials here refers to chimerical composites that include two constituents amalgamated on the molecular scale, where one of these moieties is commonly inorganic and the other organic in nature. Hybrid materials represent one of the most fascinating progresses in materials sciences with applications in many branches, from biomedicine to organic solar cells [134]. Carbohydrates, as a prime class of biomolecules, have not remained apart from these developments. Taking into account the multifunctional and, quite often, fragile nature of saccharides, a major prerequisite for the preparation of hybrid glycomaterials is to proceed through bioorthogonal reaction conditions that combine mildness and efficiency. The CuAAC reaction perfectly meets these requirements, which has translated into an enormous impact in the field.

Among hybrid glycomaterials, glyconanomaterials, carbohydrate-decorated nanometric scale materials, are creating a high expectation in glycobiology. The merging between the glycosciences and the nanosciences (*glyconanotechnology*) provides new devices for more accurate fundamental studies and opens new and exciting opportunities in the biomedical field. Glyconanomaterials are creating a high expectation in areas such as diagnostics, therapy, drug-delivery-controlled drug release and, tissue

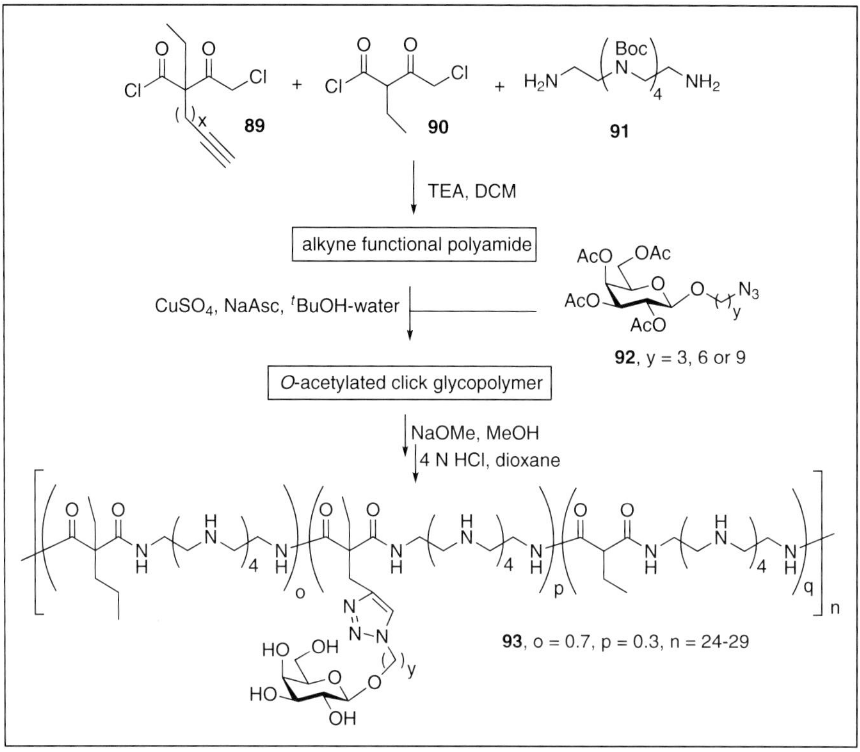

SCHEME 6.16 Polycationic galactosylated click glycopolymers for hepatocyte-specific gene delivery.

engineering, among others [135–137]. Critical to the performance of glyconanomaterials is the proper display of the relevant glycotopes, taking into consideration not only the nanoscaffold nature, but also the ligand density, the type and length of the linker, and coupling chemistry. The preparation of carbon, silicon, and metal-based hybrid glyconanomaterials have already benefited from the advantages of click chemistry in terms of efficiency and structure control. Significant examples of these families of click multivalent systems will be presented in the following paragraphs.

Fullerenes and carbon nanotubes (CNTs) [138], and their glycocoated derivatives [139, 140], are the focus of considerable interest because they can be used to test fundamental ideas about the roles of dimensionality and confinement in materials of greatly reduced size. In 2007, Nakamura and coworkers proposed the use of the pentaalkynyl C_{60} fullerene **94** as a platform for the construction of nanometer-scale molecular architectures presenting as many as 15 sugar moieties in C_5-symmetry via click chemistry [141]. The cycloaddition reaction with azide-armed mono- and oligosaccharides proceeded quantitatively under mild conditions without the need to protect the sugar hydroxyl groups. By using derivative **95**, incorporating the P^k trisaccharide moiety of the cell surface glycolipid globotriaosylceramide (Gb-3), the

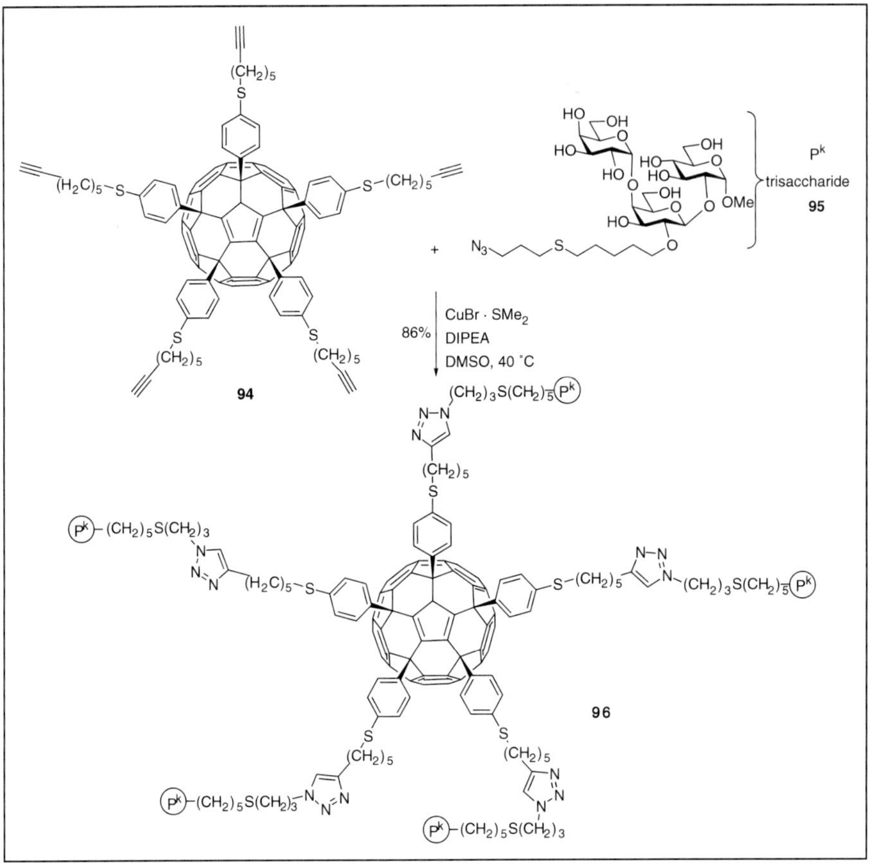

SCHEME 6.17 Synthesis of P^k trisaccharide-C_{60} fullerene pentavalent conjugate **96**.

pentavalent glycoconjugate **96**, exhibiting symmetry complementarity with Shiga-like toxins, was obtained (Scheme 6.17). The groups of Freitas and coworkers [142] and Nierengarten and coworkers [143] have further broadened the range of clickable fullerene platforms with the alkyne- and azide-functionalized C_{60} di- and hexaadducts **97–99** (Fig. 6.7), which were successfully coupled with a series of sugar derivatives bearing the complementary functionality to give di- or dodecavalent click glycofullerenes, respectively. Dodecagalactosyl derivatives exhibited a strong glycoside cluster effect against the PA-IL lectin from *P. aeruginosa* as determined by several binding evaluation techniques, including hemagglutination inhibition assay (HIA), enzyme-linked lectin assay (ELLA), and SPR.

In the case of CNTs, noncovalent functionalization, taking advantage of supramolecular π–π-stacking interactions, has been favored to append carbohydrate ligands at their surface, since in that manner the nanotube structure is preserved. Click chemistry has been put forward by several groups to facilitate the synthesis

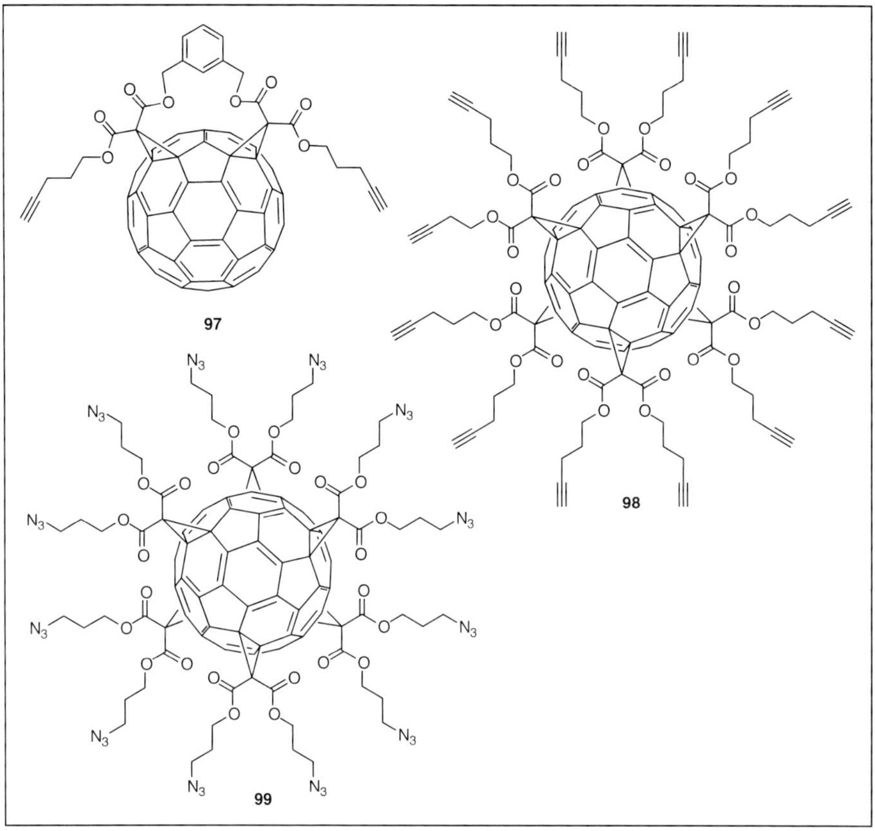

97

98

99

FIGURE 6.7 Representative clickable poly(propargyl)-functionalized fullerene scaffolds.

of the key polyaromatic platform glycoconjugates suitable for CNT surface coating. Bertozzi and coworkers [144] prepared octavalent mannose-, galactose-, and lactose-conjugated click glycodendrimers with a terminal pyrene-bearing segment at the focal point from the alkyne derivative **100**. Khiar and coworkers proposed, instead, the azide-armed butterfly-like tetrabenzo[*a,c,g,i*]fluorene polyaromatic platform **101** [145], whereas Star and coworkers used the tetraalkynyl porphyrine scaffold **102** [146]. In all cases, the resulting glycoconjugates self-organized on the CNT surface to form the corresponding multivalent click glyco-CNTs (Scheme 6.18).

Metal glyconanoparticles are probably the glyconanomaterials that have raised the highest expectations in nanomedicine [147–149]. Examples of click-chemistry-based glycocoating of either azide (TiO_2) [150] or alkyne functional metal nanoparticles (Au) [151] have been reported. Interestingly, ZnS/CdSe glyconanoparticles **103** (Fig. 6.8), obtained by thiol coupling of trivalent galactosyl click glycodendrons with preformed pyridine-capped nanoparticles were shown to be selectively uptake by lung cancer cells enriched with membrane-bound asialoprotein receptors [152].

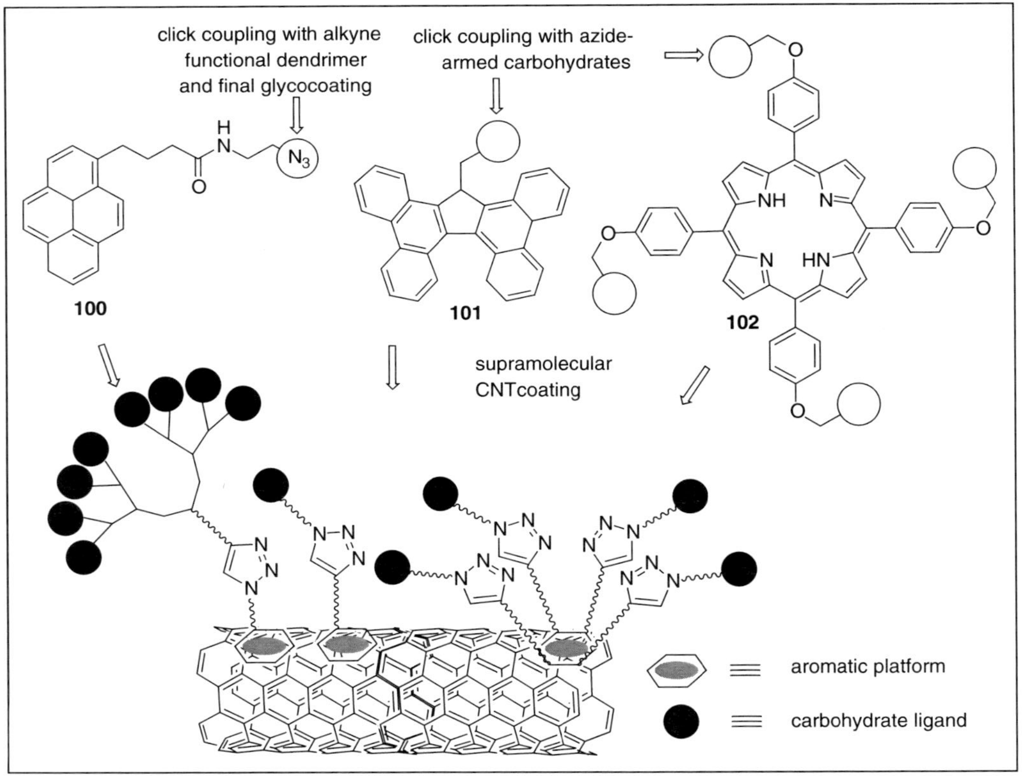

SCHEME 6.18 Synthesis of multivalent click glyco-carbon nanotubes (CNTs).

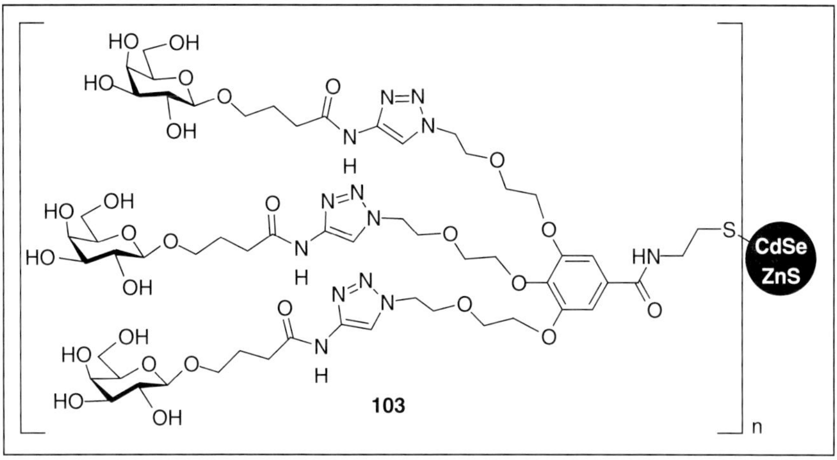

FIGURE 6.8 Glyconanoparticles obtained by thiol coupling of trivalent galactosyl click glycodendrons with ZnS/CdSe nanoparticles.

Noncovalent stabilization of nanoparticles by interaction with click glycodendrimers also has been achieved [153].

The groups of Chiara and coworkers [154] and Fessner and coworkers [155] have reported the preparation of a new family of octavalent cube-shaped click glyconanomaterials with an inorganic–organic hybrid core (**105**) based on the azide functional octahydrosilsesquioxane platform **104** (Scheme 6.19). Clickable azide-functionalized Qβ bacteriophage capsid particles have also been used as scaffolds for CuAAC with alkyne-armed sialic acid residues on the glycan chain of transferrin (Tf) to target the Tf receptors overexpressed on the surface of a variety of tumor cells [156].

Silica-based click hybrid glycomaterials represent one of the most important functional materials [157]. Click ligation of carbohydrates onto the surface of silica

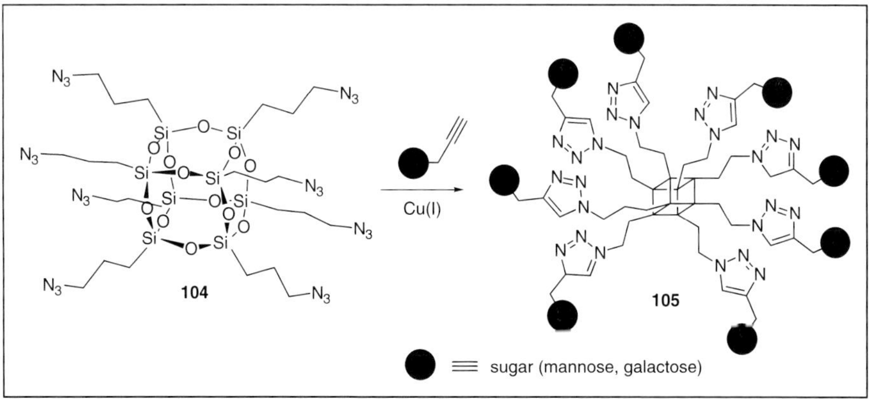

SCHEME 6.19 Synthesis of octahydrosilsesquioxane cube-shaped click glyconanomaterials.

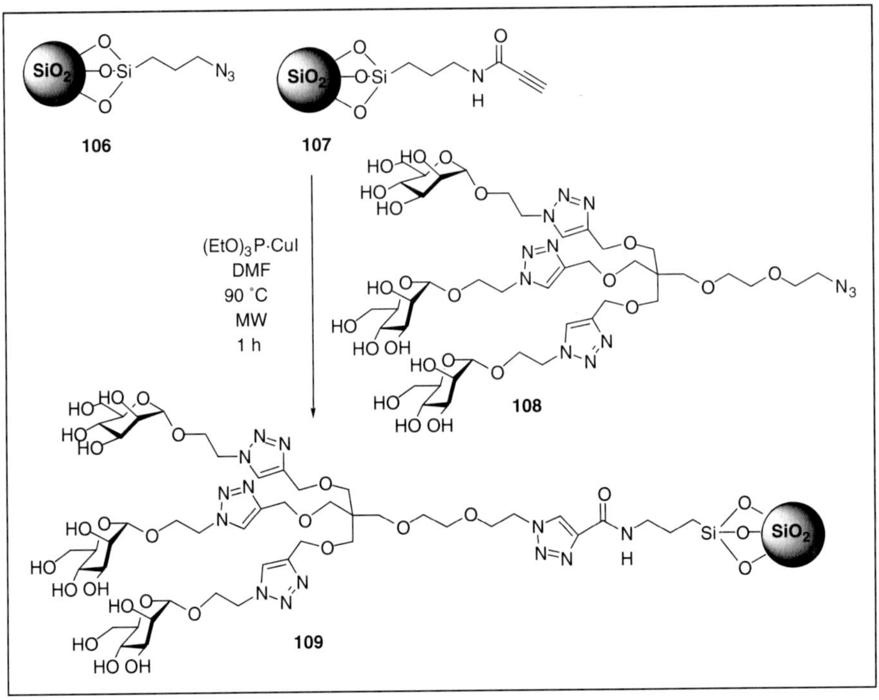

SCHEME 6.20 Elaboration of glycosilicas through click chemistry.

particles has been performed with the principal aim of developing new supports for chromatographic applications, especially for affinity chromatography. As a matter of fact, the carbohydrate residues arranged at the surface of silica particles form a multivalent display that benefits from the affinity enhancement toward complementary lectins derived from the cluster effect. The group of Santoyo-González was pioneer in the synthesis of tailored glycosilicas through click chemistry (e.g., **109**) by using either azide (**106**) or alkyne functional silicas (**107**) in combination with complementary sugar partners, including preformed glycodendrimers (**108**; Scheme 6.20) [158]. This general strategy was further adopted by other authors to bioactivate silica [159–162] or other inorganic materials such as granules of nanostructured hydroxyapatite [163] with carbohydrates.

6.6 CLICK GLYCOSURFACES AND GLYCOARRAYS

The analysis of the biomolecular interactions involving carbohydrates requires the development of robust and efficient sensors and high-throughput analysis methods. Several formats have been developed for this purpose that generally relay on glycoarray technology, that is, displaying carbohydrates onto a surface for mapping the carbohydrate–protein interactions The self-assembled monolayer (SAM) scaffold has appeared as one of the most promising model systems in this sense, because the glyco–SAMs offer extensive control over density and orientation of the sugar ligands and, more importantly, are amenable to real-time, label-free analysis methods such as SPR, quartz crystal microbalance (QCM), ellipsometry or atomic force microscopy (AFM). Glyco–SAMs can be prepared by direct sugar coating of preformed SAMs, from which SAMs of alkanethiolates on gold currently hold the best model system. CuAAC has once more proven its versatility by allowing a very efficient interphase coupling reaction ("click on SAM").

The groups of Wang and coworkers [164] and Lindhorst and coworkers [165] privileged the use of propargylamido-terminated thiolates, generated from either disulfides or thioacetate precursors, for the preparation of alkyne functional SAMs, whereas Miura and coworkers [166] favored propargyl-ether-terminated disulfides. These clickable SAMs were readily coupled with azido sugars, including mono-, di-, or trisaccharides or even azide-armed, preformed glycodendrons, using standard Cu(I)-catalyzed conditions. Co-assembling the alkyne-functionalized thiolate with a thiolate bearing irrelevant terminal groups was used to finely tune the surface carbohydrate density. The glyco-SAMs thus obtained proved to be potent platforms for high-throughput characterization of carbohydrate–lectin interactions. In a more sophisticated application, Miura and coworkers used a click 6-*O*-sulfo-*N*-acetylglucosamine (the most abundant structure of heparin) glycodendron-coated SAM (**112**), obtained from the azide-armed trivalent click conjugate **110** and the alkyne functional SAM **111,** by click on SAM (Scheme 6.21), to study the mechanism of amyloidosis of Alzheimer amyloid β peptide (1–42) (Aβ (1–42)) by AFM and SPR [167]. The results indicated that the multivalency of sugars was significant regarding both the morphology and the aggregation effects of Aβ (1–42). Other types of surfaces that have been made clickable by alkynylation in order to be functionalized with carbohydrates include silicon, quartz, and glass substrates [168], cellulose [169], polymeric surfaces (polystyrene, polyacrylamide, poly(ethylene glycol), poly(2-ethyl-2-oxazoline, polypropylene) [170], and hydrogel layers [171].

Several examples of preformation of click glycodendrons [172,173], glycoclusters [174], or glycooligomers [175] endowed with a suitable functional moiety for their attachment to a complementary surface have been reported and applied to the preparation of carbohydrate (micro)arrays displaying different sets of multivalent glycotopes. Alternatively, "clicking-on" approaches have also demonstrated their efficiency for such goals [176]. Wong and coworkers [177], in the very first report on the CuAAC reaction in the field of glycosciences, already presented a proof of this concept. Thus, *N*-tetradecyl propargylamide was noncovalently bound to the surface of microtiter plate wells and a series of azide-armed oligosaccharides was clicked at the interphase. Interestingly, the clicked oligosaccharides could be further transformed in the well by enzyme-assisted reactions. The "clicking-on" concept has been later extended by Bertozzi and coworkers to the microcontact printing of alkyne-functionalized glycopolymers on azide-functionalized silicon wafer chips [178].

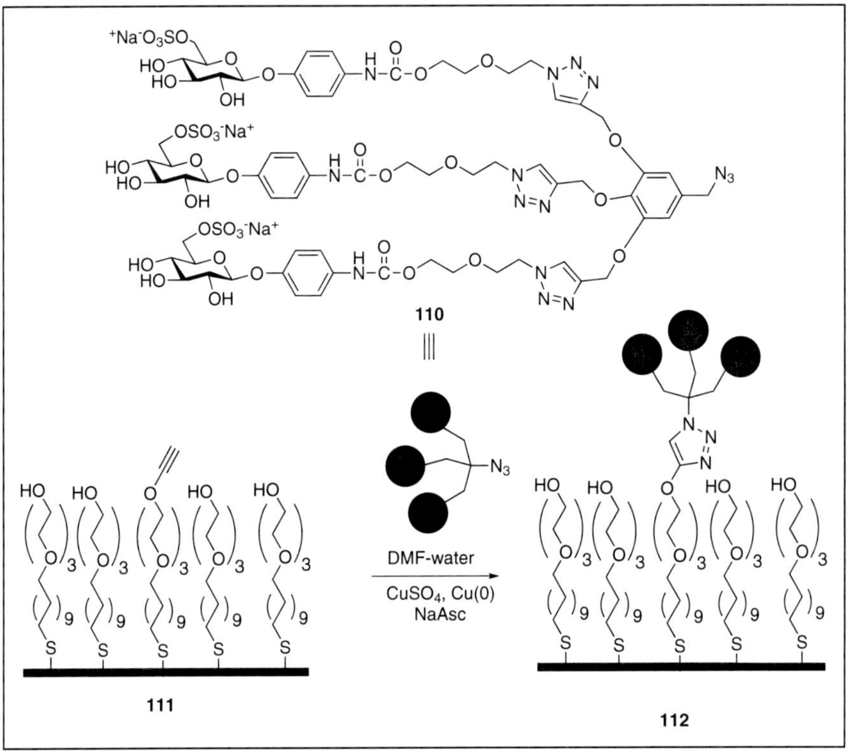

SCHEME 6.21 Click glycocoating of self-assembled monolayers (SAMs) for high-throughput characterization of carbohydrate-lectin interactions.

6.7 CLICK MULTIVALENT GLYCOMIMETICS

The carbohydrate-binding sites in lectins are very often superficially located, so that the bound ligand remains largely in contact with bulk solvent. On the other hand, lectin molecules tend to form oligomers and those oligomers can be expressed as clusters. All these are factors that contribute to the multivalent effect by facilitating sliding of multivalent conjugates over carbohydrate-binding sites, favoring chelate-type complexes, and promoting cross-linking and aggregation. The low binding affinities for monovalent ligands also favor fast on–off binding kinetics after increasing the local concentration through multivalent presentation, limiting the entropic penalty of the association. In principle, one could expect that some of those processes could also be at play in the interaction of carbohydrates with other receptor partners. Recently, the possible existence of a multivalent effect in the interaction of carbohydrates with carbohydrate-processing enzymes has attracted the attention of several groups. Should it operate, then one could hypothesize that a conceptually new type of enzyme inhibitors could be designed by displaying glycomimetic structures, capable of competitively competing with the substrate for the catalytic site, in multiple copies. For obvious reasons, click chemistry has been privileged to test this notion.

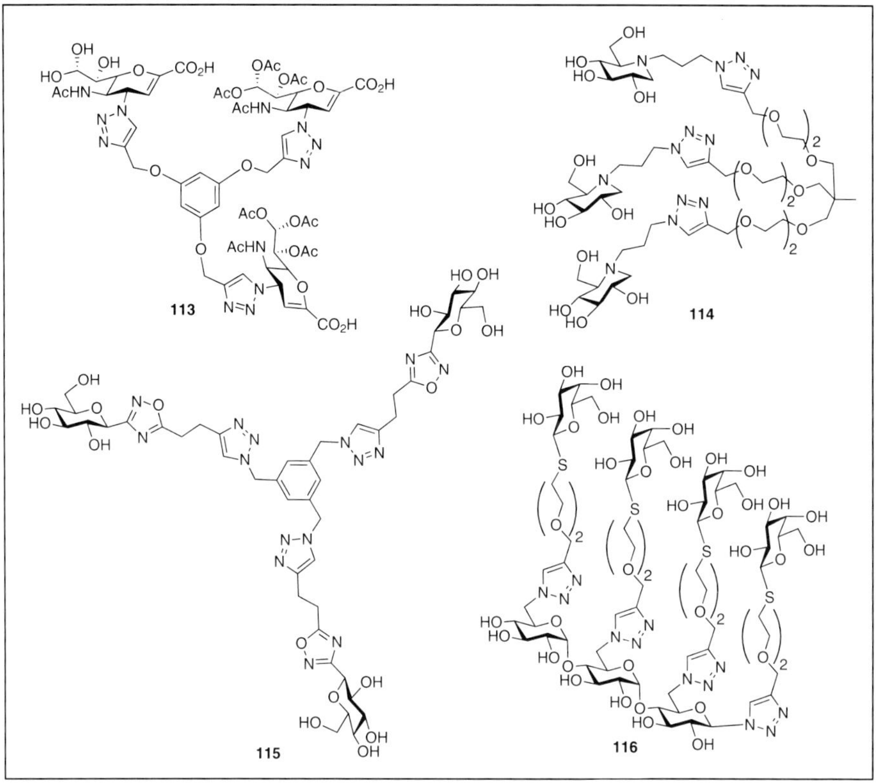

FIGURE 6.9 Low-valency glycomimetic clusters obtained by CuAAc.

In the light of experimental evidence pointing to a beneficial effect after presenting the neuraminic acid mimic zanamivir (Relenza®), an inhibitor of viral neuraminidase used in the treatment and prophylaxis of influenza, in multivalent manner [179, 180], Gervay-Hague and coworkers undertook the preparation of click trivalent conjugates built on aromatic scaffolds (e.g., **113**) (Fig. 6.9) [181]. Vidal and coworkers similarly applied click chemistry to the synthesis of trivalent derivatives containing the 3-*C*-glycosyl-1,2,4-oxadiazole motif (e.g., **115**) (Fig. 6.9) known to exhibit inhibitory activity toward glycogen phosphorylase, the enzyme that catalyses glycogen depolymerization [182]. No clear evidence on the existence of a multivalent effect was obtained in these cases; however, Gouin and coworkers also prepared trivalent and up to tetravalent click clusters incorporating the iminosugar 1-deoxynojirimycin (DNJ) (e.g., **114**) [183] or 1-thiogalactose units (e.g., **116**) (Fig. 6.9) [184] and assessed them for their glycosidase inhibitory activity. The overall conclusion was that no multivalent effect applies for enzymes acting on substrates that match the configurational pattern of the displayed glycomimetic (glucosidases and galactosidases, respectively). Interestingly, a modest but significant multivalent effect was observed for compound **114** when assayed against yeast α-mannosidase (2.4-fold in a molar basis).

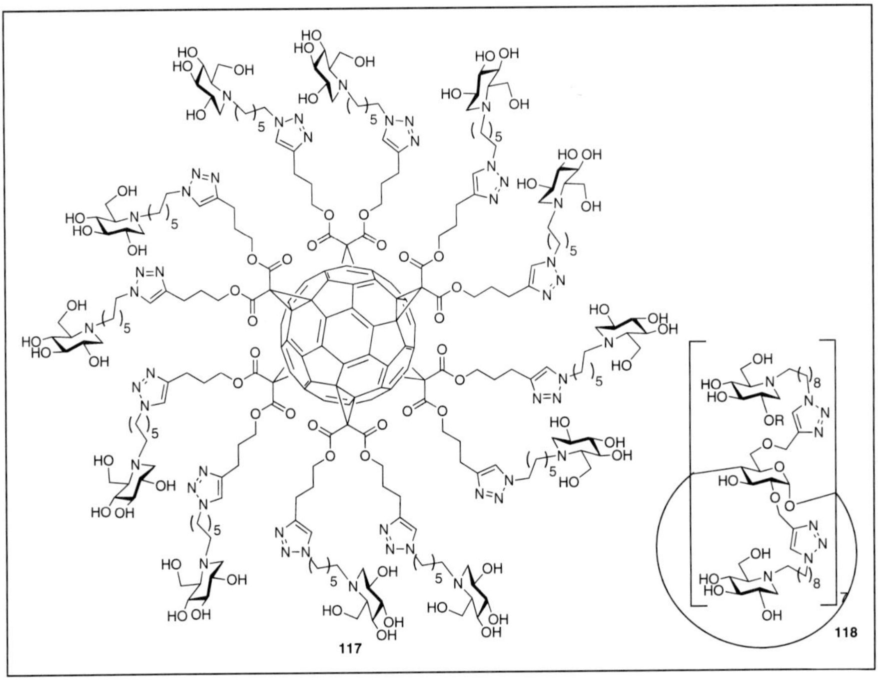

FIGURE 6.10 High-valency C_{60} fullerene and β-cyclodextrin-scaffolded click clusters exposing 1-deoxynojirimycin motifs.

The above result prompted a collaborative research by the groups of Nierengarten, Compain, and Ortiz Mellet to prepare high-valency click clusters exposing DNJ motifs based on C_{60} fullerene (i.e., **117**) (Fig. 6.10) [185] and β-cyclodextrin scaffolds (e.g., **118**) (Fig. 6.10) [186]. Amazing increases in the inhibition potency against yeast α-mannosidase, over 600-fold per DNJ unit in the case of the tetradeca-conjugate **118**, were achieved. Although the biological significance of multivalency in glycosidase inhibition is still unclear, this preliminary data represent a proof of concept and warrants further research in that direction. Click chemistry will be—with no doubt—instrumental for those channels.

REFERENCES

1. Reynolds, M.; Pérez, S. *C. R. Chimie.* **2011**, *14*, 74–95.
2. Mammen, M.; Choi, S. K.; Whitesides, G. M. *Angew. Chem. Int. Ed.* **1998**, *37*, 2754–2794.
3. Kiessling, L. L.; Gestwicki, J. E.; Strong, L. E. *Curr. Opin. Chem. Biol.* **2000**, *4*, 696–703.
4. Lundquist, J. J.; Toone, E. J. *Chem. Rev.* **2002**, *102*, 555–578.
5. Kolb, H. C.; Sharpless, K. B. *Drug Discov. Today.* **2003**, *8*, 1128–1137.

6. Meldal, M.; Tornøe, C. W. *Chem. Rev.* **2008**, *108*, 2952–3015.
7. Wood, W. J. L.; Patterson, A. W.; Tsuruoka, H.; Jain, R. K.; Ellman, J. A. *J. Am. Chem. Soc.* **2005**, *127*, 15521–15527.
8. Rostovtsev, V. V.; Green, L. G.; Fokin, V. V.; Sharpless, B. K. *Angew. Chem., Int. Ed.* **2002**, *41*, 2596–2599.
9. Dondoni, A. *Chem. Asian. J.* **2007**, *2*, 700–708.
10. Dirks, A. J.; Cornelissen, J. J. L. M.; van Delft, F. L.; van Hest, J. C. M.; Nolte, R. J. M.; Rowan, A. E.; Rutjes, F. P. J. T. *QSAR Comb. Sci.* **2007**, *26*, 1200–1210.
11. Nwe, K.; Brechbiel, M. W. *Cancer Biother. Radio.* **2009**, *24*, 289–302.
12. van Dijk, M.; Rijkers, D. T. S.; Liskamp, R. M. J.; van Nostrum, C. F.; Hennink, W. E. *Bioconjugate Chem.* **2009**, *20*, 2001–2016.
13. Pieters, R. J.; Rijkers, D. T. S.; Liskamp, R. M. J. *QSAR Comb. Sci.* **2007**, *26*, 1181–1190.
14. Dedola, S.; Nepogodiev, S. A.; Field, R. A. *Org. Biomol. Chem.* **2007**, *5*, 1006–1017.
15. Boltje, T. J.; Buskas, T.; Boons, G.J. *Nat. Chem.* **2009**, *1*, 611–622.
16. Lepenies, B.; Yin, J.; Seeberger, P. H. *Curr. Opin. Chem. Biol.* **2010**, *14*, 404–411.
17. Deniaud, D.; Julienne, K.; Gouin, S. *Org. Biomol. Chem.* **2011**, *9*, 966–979.
18. Ortiz Mellet, C.; Benito, J. M.; García Fernández, J. M. *Chem. Eur. J.* **2010**, *16*, 6728–6742.
19. Ortiz Mellet, C.; García Fernández, J. M.; Benito, J. M. *Chem. Soc. Rev.* **2011**, *40*, 1586–1608.
20. Méndez-Ardoy, A.; Gómez-García, M.; Ortiz Mellet, C.; Sevillano, N.; Girón, M. D.; Salto, R.; Santoyo-González, F.; García Fernández, J. M. *Org. Biomol. Chem.* **2009**, *7*, 2681–2684.
21. Méndez-Ardoy, A.; Guilloteau, N.; Di Giorgio, C.; Vierling, P.; Santoyo-González, F.; Ortiz Mellet, C.; García Fernández, J. M. *J. Org. Chem.* **2011**, *76*, 5882–5894.
22. Srinivassachari, S.; Fichter, K. M.; Reineke, T. M. *J. Am. Chem. Soc.* **2008**, *130*, 4618–4627.
23. Díaz-Moscoso, A.; Méndez-Ardoy, A.; Ortega-Caballero, F.; Benito, J. M.; Ortiz Mellet, C.; Defaye, J.; Robinson, T. M.; Yohannes, A.; Karginov, V. A.; García Fernández, J. M. *Chem. Med. Chem.* **2011**, *6*, 181–192.
24. Joshi, A.; Kate, S.; Poon, V.; Mondal, D.; Boggara, M. B.; Saraph, A.; Martin, J. T.; McAlpine, R.; Day, R.; Garcia, A. E.; Mogridge, J.; Kane, R. S. *Biomacromolecules* **2011**, *12*, 791–796.
25. Kim, H.; Kim, S.; Park, C.; Lee, H.; Park, H. J.; Kim, C. *Adv. Mater.* **2010**, *22*, 4280–4283.
26. Pérez-Balderas, F.; Ortega-Muñoz, M.; Morales-Sanfrutos, J.; Hernández-Mateo, F.; Calvo-Flores, F. G.; Calvo-Asín, J. A.; Isac-García, J.; Santoyo-González, F. *Org. Lett.* *2003*, *5*, 1951–1954.
27. Giguère, D.; Patnam, R.; Bellefleur, M.A.; St-Pierre, C.; Sato, S.; Roy, R. *Chem. Commun.* **2006**, 2379–2381.
28. Megía-Fernández, A.; Ortega-Muñoz, M.; López-Jaramillo, J.; Hernández-Mateo, F.; Santoyo-González, F. *Adv. Synth. Catal.* **2010**, *352*, 3306–3320.
29. Alix, A.; Chassaing, S.; Pale, P.; Sommer, J. *Tetrahedron* **2008**, *64*, 8922–8929.
30. Tejler, J.; Tullberg, E.; Frejd, T.; Leffler, H.; Nilsson, U. J. *Carbohydr. Res.* **2006**, *341*, 1353–1362.

31. Halila, S.; Manguian, M.; Fort, S.; Cottaz, S.; Hamaide, T.; Fleury, E.; Driguez, H. *Macromol. Chem. Phys.* **2008**, *209*, 1282–1290.

32. Lee, B.Y.; Park, S. R.; Jeon, H. B.; Kim, K. S. *Tetrahedron Lett.* **2006**, *47*, 5105–5109.

33. Gouin, S. G.; Bultel, L.; Falentin, C.; Kovensky, J. *Eur. J. Org. Chem.* **2007**, 1160–1167.

34. Chittaboina, S.; Xie, F.; Wang, Q. *Tetrahedron Lett.* **2005**, *46*, 2331–2336.

35. Beckmann, H. S. G.; Wittmann, V. *Org. Lett.* **2007**, *9*, 1–7.

36. Casas-Solvas, J. M.; Vargas-Berenguel, A.; Capitán-Vallvey, L. F.; Santoyo-González, F. *Org. Lett.* **2004**, *6*, 3687–3690.

37. Casas-Solvas, J. M.; Ortiz Salmerón, E.; Jiménez-Martínez, J. J.; García Fuentes, L.; Capitán-Vallvey, L. F.; Santoyo-González, F.; Vargas-Berenguel, A. *Chem. Eur. J.* **2009**, *15*, 710–725.

38. Zhang, Y. J.; He, X. P.; Hua, M.; Li, Z.; Shi, X. X.; Chen, G. R. *Dyes. Pigments* **2011**, *88*, 391–395.

39. Hu, X. M.; Chen, Q.; Wang, J. X.; Cheng, Q. Y.; Yan, C. G.; Cao, J.; He, Y. J.; Han, B. H. *Chem. Asian J.* **2011**, *6*, 2376–2381.

40. Ortega-Muñoz, M.; Pérez-Balderas, F.; Morales-Sanfrutos, J.; Hernández-Mateo, F.; Isac-García, J.; Santoyo-González, F. *Eur. J. Org. Chem.* **2009**, 2454–2473.

41. Marmuse, L.; Nepogodiev, S. A.; Field, R. A. *Org. Biomol. Chem.* **2005**, *3*, 2225–2227.

42. Gouin, S. G.; Vanquelef, E.; García Fernández, J. M.; Ortiz Mellet, C.; Dupradeau, F.Y.; Kovensky, J. *J. Org. Chem.* **2007**, *72*, 9032–9045.

43. Gouin, S. G.; García Fernández, J. M.; Vanquelef, E.; Dupradeau, F.Y.; Salomonsson, E.; Leffler, H.; Ortega-Muñoz, M.; Nilsson, U. J.; Kovensky, J.; *ChemBioChem.* **2010**, *11*, 1430–1442.

44. Wan, Q.; Chen, J.; Chen, G.; Danishefsky, S. *J. Org. Chem.* **2006**, *71*, 8244–8249.

45. Cecioni, S.; Faure, S.; Darbost, U.; Bonnamour, I.; Parrot-López, H.; Roy, O.; Taillefumier, C.; Wimmerov, M.; Praly, J.P.; Imberty, A.; Vidal, S. *Chem. Eur. J.* **2011**, *17*, 2146–2159.

46. Morvan, F.; Meyer, A.; Jochum, A.; Sabin, C.; Chevolot, Y.; Imberty, A.; Praly, J. P.; Vasseur, J. J.; Souteyrand, E.; Vidal, S. *Bioconjugate Chem.* **2007**, *18*, 1637–1643.

47. Pourceau, G.; Meyer, A.; Vasseur, J. J.; Morvan, F. *J. Org. Chem.* **2009**, *74*, 1218–1222.

48. Gouin, S. G.; Wellens, A.; Bouckaert, J.; Kovensky, J. *ChemMedChem* **2009**, *4*, 749–755.

49. Dondoni, A.; Marra, A. *J. Org. Chem.* **2006**, *71*, 7546–7557.

50. de Oliveira, R. N.; Sinou, D.; Srivastava, R. M. *J. Carbohydr. Chem.* **2006**, *25*, 407–425.

51. Vargas-Berenguel, A.; Ortega-Caballero, F.; Casas-Solvas, J. M. *Mini-Rev. Org. Chem.* **2007**, *4*, 1–14.

52. Ortega-Muñoz, M.; Morales-Sanfrutos, J.; Pérez-Balderas, F.; Hernández-Mateo, F.; Dolores Girón-González, M.; Sevillano-Tripero, N.; Salto-González, R.; Santoyo-González, F. *Org. Biomol. Chem.* **2007**, *5*, 2291–2301.

53. Hyun, H.; Yui, N. *Macromol. Rapid Commun.* **2011**, *32*, 326–331.

54. Chwalek, M.; Auzély, R.; Fort, S. *Org. Biomol. Chem.* **2009**, *7*, 1680–1688.

55. Aoyama, Y.; Matsuda, Y.; Chuleeraruk, J.; Nishiyama, K.; Fujimoto, T.; Fujimoto, K.; Shimizu, T.; Hayashida, O. *Pure Appl. Chem.* **1998**, *70*, 2379–2384.

56. Rudkevich, D. M. *Bull. Chem. Soc. Jpn.* **2002**, *75*, 393–413.
57. Marra, A.; Moni, L.; Pazzi, D.; Corallini, A.; Bridi, D.; Dondoni, A. *Org. Biomol. Chem.* **2008**, *6*, 1396–1409.
58. Cecioni, S.; Lalor, R.; Blanchard, B.; Praly, J.-P.; Imberty, A.; Matthews, S. E.; Vidal, S. *Chem. Eur. J.* **2009**, *15*, 13232–13240.
59. Galante, E.; Geraci, C.; Sciuto, S.; Campo, V. L.; Carvalho, I.; Sesti-Costa, R.; Guedes, P. M. M.; Silva, J. S.; Hill, L.; Nepogodiev, S. A.; Field, R. A. *Tetrahedron* **2011**, *67*, 5902–5912.
60. Soomro, Z. H.; Cecioni, S.; Blanchard, H.; Praly, J.P.; Imberty, A.; Vidal, S.; Matthews, S. E. *Org. Biomol. Chem.* **2011**, *9*, 6587–6597.
61. Locos, O. B.; Heindl, C. C.; Corral, A.; Senge, M. O.; Scanlan, E. M. *Eur. J. Org. Chem.* **2010**, 1026–1028.
62. Garcia, G.; Naud-Martin, D.; Carrez, D.; Croisy, A.; Maillard, P. *Tetrahedron* **2011**, *67*, 4924–4932.
63. Chen, Y. X.; Zhao, L.; Huang, Z. P.; Zhao, Y. F.; Li, Y. M. *Bioorg. Med. Chem. Lett.* **2009**, *19*, 3775–3778.
64. Wang, J.; Li, H.; Zou, G.; Wang, Xi, L. *Org. Biomol. Chem.* **2007**, *5*, 1529–1540.
65. Cheng, Y.; Zhao, L.; Li, Y.; Zu, T. *Chem. Soc. Rev.* **2011**, *40*, 2673–2703.
66. Jaroslav, S.; Petr, N.; Jan, J. *Amino Acids* **2011**, *40*, 301–370.
67. Yoann, M. C.; Roy, R. *Adv. Carbohydr. Chem. Biochem.* **2010**, *63*, 165–393.
68. Wu, P.; Malkoch, M.; Hunt, J. N.; Vestberg, R.; Kaltgrad, E.; Finn, M. G.; Fokin, V. V.; Sharpless, K. B.; Hawker, C. J. *Chem. Commun.* **2005**, 5775–5777.
69. Ledin, P. A.; Friscourt, F.; Guo, J.; Boons, G.J. *Chem. Eur. J.* **2011**, *17*, 839–846.
70. Joosten, J. A. F.; Tholen, N. T. H.; El Maate, F. A.; Brouwer, A. J.; van Esse, G. W.; Rijkers, D. T. S.; Liskamp, R. M. J.; Pieters, R. J. *Eur. J. Org. Chem.* **2005**, 3182–3185.
71. Branderhorst, H. M.; Kooij, R.; Salminen, A.; Jongeneel, L. H.; Arnusch, C. J.; Liskamp, R. M. J.; Finne, J.; Pieters, R. J. *Org. Biomol. Chem.* **2008**, *6*, 1425–1434.
72. Fernández-Megia, E.; Correa, J.; Rodríguez-Meizoso, I.; Riguera, R. *Macromolecules* **2006**, *39*, 2113–2120.
73. Touaibia, M.; Shiao, T. C.; Papadopoulos, A.; Vaucher, J.; Wang, Q.; Benhamioud, K.; Roy, R. *Chem. Commun.* **2007**, *4*, 380–382.
74. Touaibia, M.; Roy, R. *J. Org. Chem.* **2008**, *73*, 9292–9302.
75. Touaibia, M.; Wellens, A.; Shiao, T. C.; Wang, Q.; Sirois, S.; Bouckaert, J.; Roy, R. *ChemMedChem* **2007**, *2*, 1190–1201.
76. Camponovo, J.; Hadad, C.; Ruiz, J.; Cloutet, E.; Gatard, S.; Muzart, J.; Bouquillon, S.; Astruc, D. *J. Org. Chem.* **2009**, *74*, 5071–5074.
77. Pourceau, G.; Meyer, A.; Chevolot, Y.; Souteyrand, E.; Vasseur, J. J.; Morvan, F. *Bioconjugate. Chem.* **2010**, *21*, 1520–1529.
78. Chabre, Y. M.; Brisebois, P. P.; Abbassi, L.; Kerr, S. C.; Fahy, J. V.; Marcotte, I.; Roy, R. *J. Org. Chem.* **2011**, *76*, 724–727.
79. Chabre, Y. M.; Giguère, D.; Blanchard, B.; Rodrigue, J.; Rocheleau, S.; Neault, M.; Rauthu, S.; Papadopoulos, A.; Arnold, A. A.; Imberty, A.; Roy, R. *Chem. Eur. J.* **2011**, *17*, 6545–6562.

80. Karskela, M.; Helkearo, M.; Virta, P.; Lönnberg, H. *Bioconjugate. Chem.* **2010**, *21*, 748–755.

81. Schlick, K. H.; Morgan, J. R.; Weiel, J. J.; Kelsey, M. S.; Cloninger, M. J. *Bioorg. Med. Chem. Lett.* **2011**, *21*, 5078–5083.

82. Kikkeri, R.; Liu, X.; Adibekian, A.; Tsaiab, Y.H.; Seeberger, P. H. *Chem. Commun.* **2010**, *46*, 2197–2199.

83. Ting, S. R. S.; Chen, G.; Stenzel, M. H. *Polym. Chem.* **2010**, *1*, 1392–1412.

84. Schatz, C.; Lecommandoux, S. *Macromol. Rapid Commun.* **2010**, *31*, 1664–1684.

85. Srinivasachari, S.; Liu, Y.; Zhang, G.; Prevette, L.; Reineke, T. M. *J. Am. Chem. Soc.* **2006**, *128*, 8176–8184.

86. Srinivasachari, S.; Liu, Y.; Prevette, L. E.; Reineke, T. M. *Biomaterials.* **2007**, *28*, 2885–2898.

87. Eissa, A. M.; Khosravi, E. *Eur. Polym. J.* **2011**, *47*, 61–69.

88. Spain, S. G.; Gibson, M. I.; Cameron, N. R. *J. Polym. Sci. Pol. Chem.* **2007**, *45*, 2059–2072.

89. Takasu, A.; Itou, H.; Takada, M.; Inai, Y.; Hirabayashi, T. *Polymer.* **2002**, *43*, 227–231.

90. Takasu, A.; Takada, M.; Itou, H.; Hirabayashi, T.; Kinoshita, T. *Biomacromolecules.* **2004**, *5*, 1029–1037.

91. Sigal, G. B.; Mammen, M.; Dahmann, G.; Whitesides, G. M. *J. Am. Chem. Soc.* **1996**, *118*, 3789–3800.

92. Gestwicki, J. E.; Strong, L. E.; Borchardt, S. L.; Cairo, C. W.; Schnoes, A. M.; Kiessling, L. L. *Bioorg. Med. Chem.* **2001**, *9*, 2387–2393.

93. Uzawa, H.; Ito, H.; Izumi, M.; Tokuhisa, H.; Taguchi, K.; Minoura, N. *Tetrahedron* **2005**, *61*, 5895–5905.

94. Slavin, S.; Burns, J.; Haddleton, D. M.; Becer, C. R. *Eur. Polym. J.* **2011**, *47*, 435–446.

95. Ladmiral, V.; Mantovani, G.; Clarkson, Guy J.; Cauet, S.; Irwin, J. L.; Haddleton, D. M. *J. Am. Chem. Soc.* **2006**, *128*, 4823–4830.

96. Gómez-García, M.; Benito, J. M.; Rodríguez-Lucena, D.; Yu, J.-X.; Chmurski, K.; Ortiz Mellet, C.; Gutiérrez Gallego, R.; Maestre, A.; Defaye, J.; García Fernández, J. M. *J. Am. Chem. Soc.* **2005**, *127*, 7970–7971.

97. Gómez-García, M.; Benito, J. M.; Gutiérrez Gallego, R.; Maestre, A.; Ortiz Mellet, C.; García Fernández, J. M.; Jiménez Blanco, J. L. *Org. Biomol. Chem.* **2010**, *8*, 1849–1860.

98. Geng, J.; Lindqvist, J.; Mantovani, G.; Chen, G.; Sayers, C. T.; Clarkson, G. J.; Haddleton, D. M. *QSAR Comb. Sci.* **2007**, *26*, 1220–1228.

99. Nurmi, L.; Lindqvist, J.; Randev, R.; Syrett, J.; Haddleton, D. M. *Chem. Commun.* **2009**, 2727–2729.

100. Vinson, N.; Gou, Y.; Becer, C. R.; Haddleton, D. M.; Gibson, M. I. *Polym. Chem.* **2011**, *2*, 107–113.

101. Otman, O.; Boullanger, P.; Drockenmuller, E.; Hamaide, T. *Beilstein J. Org. Chem.* **2010**, *6*, 58. DOI:10.3762/bjoc.6.58.

102. Kumar, J.; McDowall, L.; Cheng, G.; Stenzel, M. H. *Polym. Chem.* **2011**, *2*, 1879–1886.

103. Semsarilar, M.; Ladmiral, V.; Perrier, S. *Macromolecules.* **2010**, *43*, 1438–1443.

104. Ott, C.; Easton, C. D.; Gengenbach, T. R.; McArthur, S. L.; Gunatillake, P. A. *Polym. Chem.* **2011**, *2*, 2782–2784. DOI: 10.1039/c1py00412c.

105. Geng, J.; Mantovani, G.; Tao, L.; Nicolas, J.; Chen, G.; Wallis, R.; Mitchell, D. A.; Johnson, B. R. G.; Evans, S. D.; Haddleton, D. M. *J. Am. Chem. Soc.* **2007**, *129*, 15156–15163.

106. Becer, C. R.; Gibson, M. I.; Geng, J.; Ilyas, R.; Wallis, R.; Mitchell, D. A.; Haddleton, D. M. *J. Am. Chem. Soc.* **2010**, *132*, 15130–15132.

107. Chen, G.; Tao, L.; Mantovani, G.; Geng, J.; Nystrolm, D.; Haddleton, D. M. *Macromolecules.* **2007**, *40*, 7513–7520.

108. Wang, C.; Wu, J.; Xu, Z.K. *Macromol. Rapid. Commun.* **2010**, *31*, 1078–1082.

109. Huang, X.; Huang, X. J.; Yu, A. G.; Wang, C.; Dai, Z. W.; Xu, Z. K. *Macromol. Chem. Phys.* **2011**, *212*, 272–277.

110. Papp, I.; Dernedde, J.; Enders, S.; Haag, R. *Chem. Commun.* **2008**, 5851–5853.

111. Papp, I.; Dernedde, J.; Enders, S.; Riese, S. B.; Shiao, T. C.; Roy, R.; Haag, R. *ChemBioChem.* **2011**, *12*, 1075–1083.

112. Ortega-Muñoz, M.; Morales-Sanfrutos, J.; Pérez-Balderas, F.; Hernández-Mateo, F.; Girón-González, M. D.; Sevillano-Tripero, N.; Salto-González, R.; Santoyo-González, F. *Org. Biomol. Chem.* **2007**, *5*, 2291–2301.

113. Xu, N.; Wang, R.; Du, F.S.; Li, Z.C. *J. Polym. Sci. Pol. Chem.* **2009**, *47*, 3583–3594.

114. Hasegawa, T.; Numata, M.; Okumura, S.; Kimura, T.; Sakurai, K.; Shinkai, S. *Org. Biomol. Chem.* **2007**, *5*, 2404–2412.

115. Abdelkader, O.; Moebs-Sánchez, S.; Queneau, Y.; Bernard, J.; Fleury, E. *J. Polym. Sci. Pol. Chem.* **2011**, *49*, 1309–1318.

116. Hetzer, M.; Chen, G.; Barner-Kowollik, C.; Stenzel, M. H. *Macromol. Biosci.* **2010**, *10*, 119–126.

117. Kempe, K.; Weber, C.; Babiuch, K.; Gottschaldt, M.; Hoogenboom, R.; Schubert, U. S. *Biomacromolecules* **2011**, *12*, 2591–2600.

118. Pernía Leal, M.; Assali, M.; Fernández, I.; Khiar, N. *Chem. Eur. J.* **2011**, *17*, 1826–1826.

119. Wang, R.; Chen, G.-T.; Du, F.-S.; Li, Z.-C. *Colloid. Surface. B.* **2011**, *85*, 56–62.

120. Jubeli, E.; Moine, L.; Barrat, G. *J. Polym. Sci. Pol. Chem.* **2010**, *48*, 3178–3187.

121. Zhang, S.; Zhao, Y. *Macromolecules* **2010**, *43*, 4020–4022.

122. Phillips, R. L.; Kim, I.-B.; Tolbert, L. M.; Bunz, U. H. F. *J. Am. Chem. Soc.* **2008**, *130*, 6952–6954.

123. Phillips, R. L.; Kim, I.B.; Carson, B. E.; Tidbeck, B.; Bai, Y.; Lowary, T. L.; Tolbert, L. M.; Bunz, U. H. F. *Macromolecules* **2008**, *41*, 7316–7320.

124. Chen, Q.; Han, B.H. *J. Polym. Sci. Pol. Chem.* **2009**, *47*, 2948–2957.

125. Chen, Q.; Cheng, Q. Y.; Zhao, Y. C.; Han, B. H. *Macromol. Rapid Commun.* **2009**, *30*, 1651–1655.

126. Chen, Q.; Cui, Y.; Cao, J.; Han, B.H. *Polymer.* **2011**, *52*, 383–390.

127. Chen, Q.; Cui, Y.; Zhang, T. L.; Cao, J.; Han, B. H. *Biomacromolecules* **2010**, *11*, 13–19.

128. Barra, M.; Roy, O.; Traïkia, M.; Taillefumier, C. *Org. Biomol. Chem.* **2010**, *8*, 2941–2955.

129. Xiao, C.; Zhao, C.; He, P.; Tang, Z.; Chen, X.; Jing, X. *Macromol. Rapid Commun.* **2010**, *31*, 991–997.

130. Huang, J.; Habraken, G.; Audouin, F.; Heise, A. *Macromolecules* **2010**, *43*, 6050–6057.

131. Tang, H.; Zhang, D. *Biomacromolecules* **2010**, *11*, 1585–1592.

132. Maheshwari, R.; Levenson, E. A.; Kiick, K. L. *Macromol. Biosci.* **2010**, *10*, 68–81.

133. Lee, C.C.; Grandinetti, G.; McLendon, P. M.; Reineke, T. M. *Macromol. Biosci.* **2010**, *20*, 585–598.
134. *Hybrid materials: Synthesis, Characterization, and Applications*, ed. G. Kickelbick, Wiley: Weinheim, **2007**.
135. Wang, X.; Ramström, O.; Yan, M. *Adv. Mater.* **2010**, *22*, 1946–1953.
136. El-Boubbou, K.; Huang, X. *Curr. Med. Chem.* **2011**, *18*, 2060–2078.
137. García Fernández, J. M.; Ortiz Mellet, C.; Defaye, J. *J. Incl. Phenom. Macrocycl. Chem.* **2006**, *56*, 149–159.
138. Guldi, D. M.; Rahman, G. M. A.; Sgobba, V.; Ehli, C. *Chem. Soc. Rev.* **2006**, *35*, 471–487.
139. Gorityala, B. K.; Ma, J.; Wang, X.; Chen, P.; Liu, X.W. *Chem. Soc. Rev.* **2010**, *39*, 2925–2934.
140. Hirsch, A. *Pure. Appl. Chem.* **2008**, *80*, 571–587.
141. Isobe, H.; Cho, K.; Solin, N.; Werz, D. B.; Seeberger, P. H.; Nakamura, E. *Org. Lett.* **2007**, *9*, 4611–4614.
142. Pereira, G. R.; Santos, L. J.; Luduvico, I.; Alves, R. B.; P de Freitas, R. *Tetrahedron Lett.* **2010**, *51*, 1022–1025.
143. Nierengarten, J.-F.; Iehl, J.; Oerthel, V.; Holler, M.; Illescas, B. M.; Muñoz, A.; Martín, N.; Rojo, J.; Sánchez-Navarro, M.; Cecioni, S.; Vidal, S.; Buffet, K.; Durka, M.; Vincent, S. P. *Chem. Commun.* **2010**, *46*, 3860–3862.
144. Wu, P.; Chen, X.; Hu, N.; Tam, U. C.; Blixt, O.; Zettl, A.; Bertozzi, C. R. *Angew. Chem. Int. Ed.* **2008**, *47*, 5022–5025.
145. Assali, M.; Pernía Leal, M.; Fernández, I.; Romero-Gomez, P.; Baati, R.; Khiar, N.; *Nano. Res.* **2010**, *3*, 764–767.
146. Vedala, H.; Chen, Y.; Cecioni, S.; Imberty, A.; Vidal, S.; Star, A. *Nano. Lett.* **2011**, *11*, 170–175.
147. García, I.; Marradi, M.; Penadés, S.; *Nanomedicine.* **2010**, *5*, 777–792.
148. Marradi, M.; Martín-Lomas, M.; Penadés, S. *Adv. Carbohydr. Chem. Biochem.* **2010**, *64*, 211–290.
149. Dong, C.M. *Comb. Chem. High T. Scr.* **2011**, *14*, 173–181.
150. Moni, L.; Rossetti, S.; Scoponi, M.; Marra, A.; Dondoni, A. *Chem. Commun.* **2010**, *46*, 475–477.
151. Zhang, L.; Wei, G.; Du, Y. *Carbohydr. Res.* **2009**, *344*, 2083–2087.
152. Chen, C. T.; Munot, Y. S.; Salunke, S. B.; Wang, Y. C.; Lin, R. K.; Lin, C. C.; Chen, C. C.; Liu, Y. H. *Adv. Funct. Mater.* **2008**, *18*, 527–540.
153. Gatard, S.; Liang, L.; Salmon, L.; Ruiz, J.; Astruc, D.; Bouquillon, S. *Tetrahedron Lett.* **2011**, *52*, 1842–1846.
154. Trastoy, B.; Pérez-Ojeda, M. E.; Sastre, R.; Chiara, J. L. *Chem. Eur. J.* **2010**, *16*, 3833–3841.
155. Heyl, D.; Rikowski, E.; Hoffmann, R. C.; Schneider, J. J.; Fessner, W. D. *Chem. Eur. J.* **2010**, *16*, 5544–5548.
156. Banerjee, D.; Liu, A. P.; Voss, N. R.; Schmid, S. L.; Finn, M. G. *ChemBioChem.* **2010**, *11*, 1273–1279.
157. Santoyo-González, F.; Hernández-Mateo, F. *Chem. Soc. Rev.* **2009**, *38*, 3449–3462.

158. Ortega-Muñoz, M.; López-Jaramillo, J.; Hernández-Mateo, F.; Santoyo-González, F. *Adv. Synth. Catal.* **2006**, *348*, 2410–2420.

159. Yu, L.; Li, X.; Guo, Z.; Zhang, X.; Liang, X. *Chem. Eur. J.* **2009**, *15*, 12618–12626.

160. Guo, Z. M.; Lei, A. W.; Zhang, Y. P.; Xu, Q.; Xue, X. Y.; Zhang, F. F.; Liang, X. M. *Chem. Commun.* **2007**, 2491–2493.

161. Malvi, B.; Sarkar, B. R.; Pati, D.; Mathew, R.; Ajithkumar, T. G.; Gupta, S. S. *J. Mater. Chem.* **2009**, *19*, 1409–1416.

162. Huang, H.; Jin, Y.; Xue, M.; Yu, L.; Fu, Q.; Ke, Y.; Chu, C.; Liang, X. *Chem. Commun.* **2009**, 6973–6975.

163. Russo, L.; Landi, E.; Tampieri, A.; Natalello, A.; Doglia, S. M.; Gabrielli, L.; Cipolla, L.; Nicotra, F. *Carbohydr. Res.* **2011**, *346*, 1546–1568.

164. Zhang, Y.; Luo, S.; Tang, Y.; Yu, L.; Hou, K.-Y.; Cheng, J.-P.; Zeng, X.; Wang, P. G. *Anal. Chem.* **2006**, *78*, 2001–2008.

165. Kleinert, M.; Winkler, T.; Terfort, A.; Lindhorst, T. K. *Org. Biomol. Chem.* **2008**, *6*, 2118–2132.

166. Fukuda, T.; Onogi, S.; Miura, Y. *Thin. Solid. Films.* **2009**, *518*, 880–888.

167. Fukuda, T.; Matsumoto, E.; Onogi, S.; Miura, Y. *Bioconjugate Chem.* **2010**, *21*, 1079–1086.

168. Matsumoto, E.; Yamauchi, T.; Fukuda, T.; Miura, Y. *Sci. Technol. Adv. Mater.* **2009**, *10*, DOI: 10.1088/1468-6996/10/3/034605.

169. Montañez, M. I.; Hed, Y.; Utsel, S.; Ropponen, J.; Malmström, E.; Wågberg, L.; Hult, A.; Malkoch, M. *Biomacromolecules* **2011**, *12*, 2114–2125.

170. Norberg, O.; Deng, L.; Aastrup, T.; Yan, M.; Ramstrom, O. *Anal. Chem.* **2011**, *83*, 1000–1007.

171. Anders, T.; Adamiak, K.; Keul, H.; Elling, L.; Möller, M. *Macromol. Biosci.* **2011**, *11*, 1201–1210.

172. Branderhorst, H. M.; Ruijtenbeek, R.; Liskamp, R. M. J.; Pieters, R. J. *ChemBioChem.* **2008**, *9*, 1836–1844.

173. Pera, N. P.; Branderhorst, H. M.; Kooij, R.; Maierhofer, C.; van der Kaaden, M.; Liskamp, R. M. J.; Wittmann, V.; Ruijtenbeek, R.; Pieters, R. J.; *ChemBioChem.* **2010**, *11*, 1896–1904.

174. Moni, L.; Pourceau, G.; Zhang, J.; Meyer, A.; Vidal, S.; Souteyrand, E.; Dondoni, A.; Morvan, F.; Chevolot, Y.; Vasseur, J.-J.; Marra, A. *ChemBioChem.* **2009**, *10*, 1369–1378.

175. Chevolot, Y.; Zhang, J.; Meyer, A.; Goudot, A.; Rouanet, S.; Vidal, S.; Pourceau, G.; Cloarec, J. P.; Praly, J. P.; Souteyrand, E.; Vasseurb, J. J.; Morvan, F. *Chem. Commun.* **2011**, *47*, 8826–8828.

176. Hanson, S. R.; Greenberg, W. A.; Wong, C. H. *QSAR Comb. Sci.* **2007**, *26*, 1243–1252.

177. Fazio, F.; Bryan, M. C.; Blixt, O.; Paulson, J. C.; Wong, C. H. *J. Am. Chem. Soc.* **2002**, *124*, 14397–14402.

178. Godla, K.; Rabuka, D.; Nam, K. T.; Bertozzi, C. R. *Angew. Chem. Int. Ed.* **2009**, *48*, 4973–4976.

179. MacDonald, S. J. F.; Cameron, R.; Demaine, D. A.; Fenton, R. J.; Foster, G.; Gower, D.; Hamblin, J. N.; Hamilton, S.; Hart, G. J.; Hill, A. P.; Inglis, G. G. A.; Jin, B.; Jones, H. T.; McConnell, D.; McKimm-Breschkin, J.; Mills, G.; Nguyen, V.; Owens, I. J.; Parry,

N.; Shanahan, S. E.; Smith, D.; Watson, K. G.; Wu, W.Y.; Tucker, S. P. *J. Med. Chem.* **2005**, *48*, 2964–2971.

180. Watson, K. G.; Cameron, R.; Fenton, R. J.; Gower, D.; Hamilton, S.; Jin, B.; Krippner, G. Y.; Luttick, A.; McConnell, D.; MacDonald, S. J. F.; Mason, A. M.; Nguyen, V.; Tucker, S. P.; Wu, W. Y. *Bioorg. Med. Chem. Lett.* **2004**, *14*, 1589–1592.
181. Lu, Y.; Gervay-Hague, J. *Carbohydr. Res.* **2007**, *342*, 1636–1650.
182. Cecioni, S.; Argintaru, O.A.; Docsa, T.; Gergely, P.; Praly, J. P.; Vidal, S. *New J. Chem.* **2009**, *33*, 148–156.
183. Diot, J.; García-Moreno, M. I.; Gouin, S. G.; Ortiz Mellet, C.; Haupt, K.; Kovensky, J.; *Org. Biomol. Chem.* **2009**, *7*, 357–363.
184. Cagnoni, A. J.; Varela, O.; Gouin, S. G.; Kovensky, J.; Uhrig, M. L. *J. Org. Chem.* **2011**, *76*, 3064–3077.
185. Compain, P.; Decroocq, C.; Iehl, J.; Holler, M.; Hazelard, D.; Mena Barragán, T.; Ortiz Mellet, C.; Nierengarten, J.-F. *Angew. Chem. Int. Ed.* **2010**, *49*, 1–5.
186. Decroocq, C.; Rodríguez-Lucena, D.; Russo, V.; Mena Barragán, T.; Ortiz Mellet, C.; Compain, P. *Chem. Eur. J.* **2011**, *17*, 13825–13831.

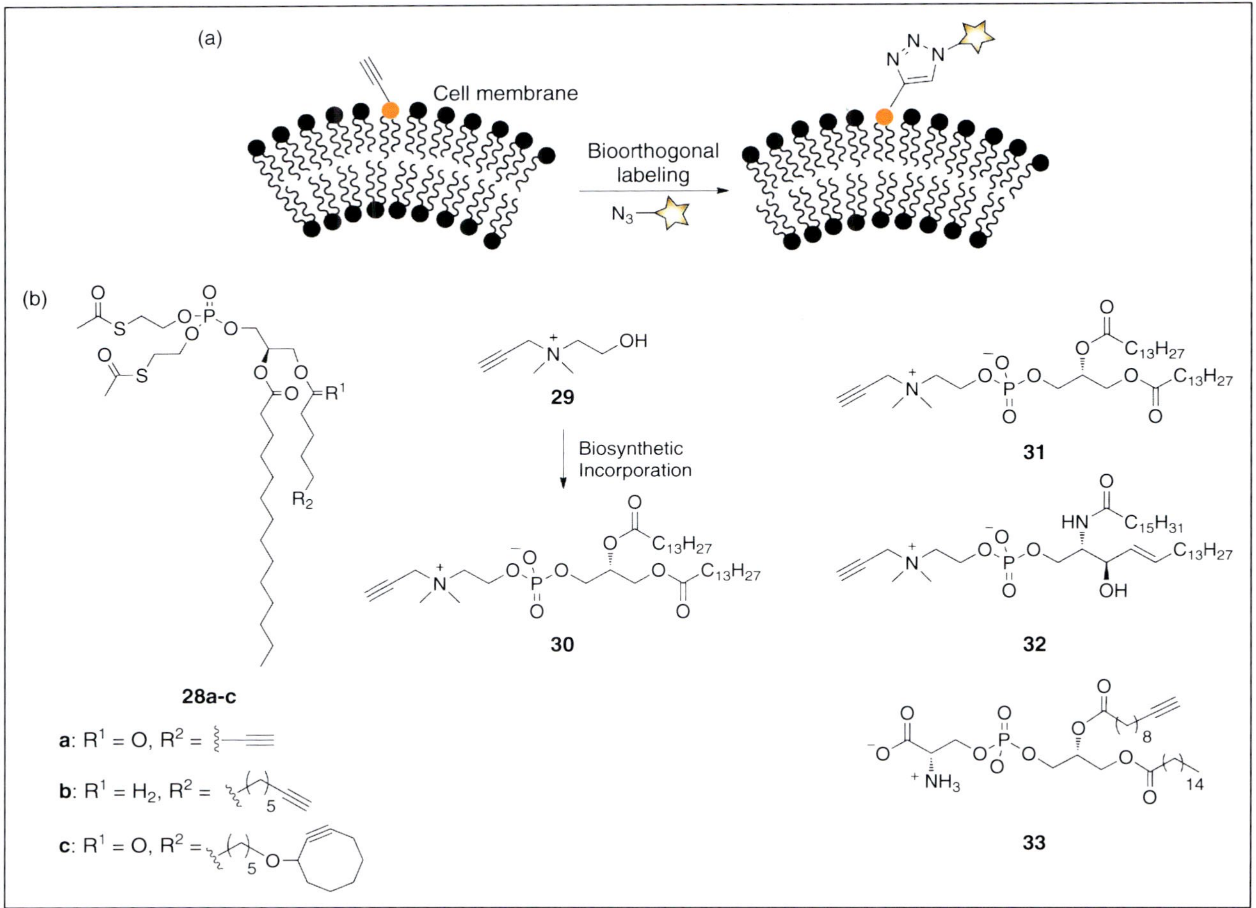

FIGURE 4.8 Strategies for in situ labeling of tagged lipid analogs. (a) Cartoon depicting bioorthogonal labeling of lipids in complex environment. (b) Structures that have been developed for *in situ* labeling studies.

Click Chemistry in Glycoscience: New Developments and Strategies, First Edition.
Edited by Zbigniew J. Witczak and Roman Bielski.

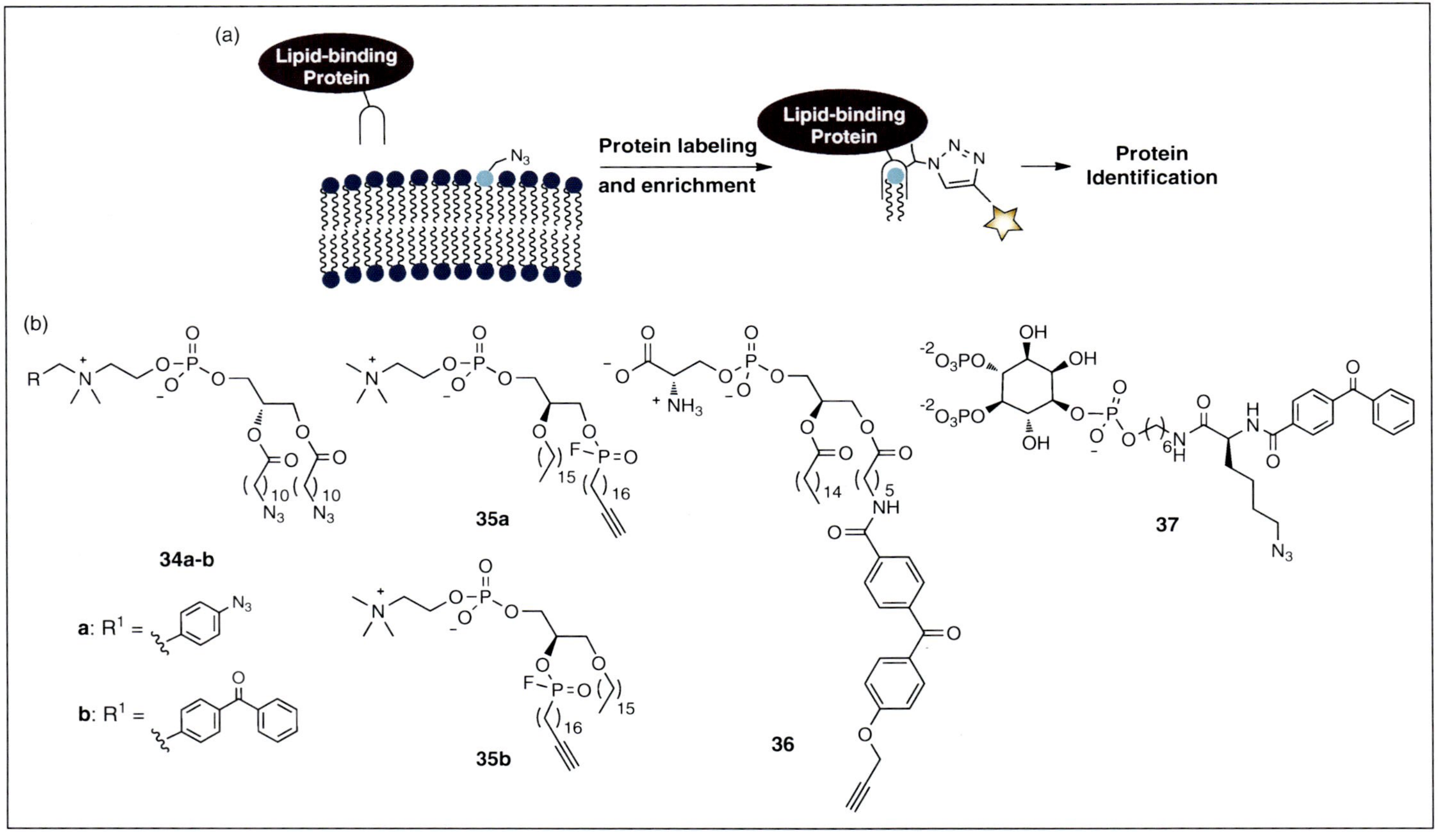

FIGURE 4.9 The development of activity-based lipid probes. (a) Cartoon depicting the application of lipid activity probes for the labeling and identification of target proteins. (b) Examples of lipid activity probes that have been reported.

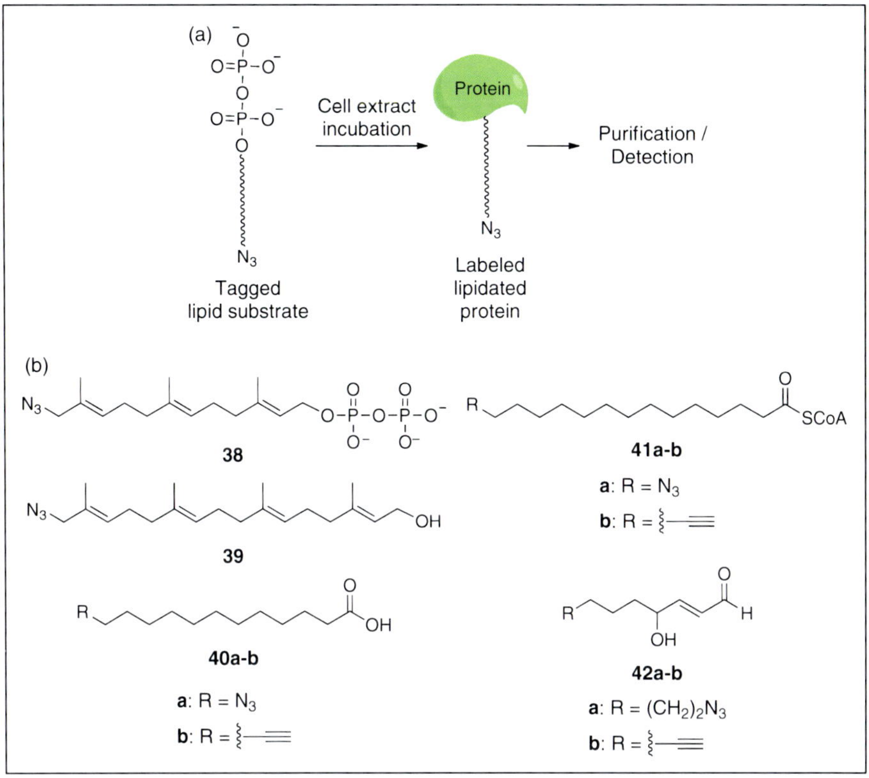

FIGURE 4.10 Investigation of covalent protein lipidation using bioorthogonal labeling. (a) Cartoon depicting the covalent labeling of target using azide-tagged lipid substrate analogs. (b) Examples of structures used to infiltrate covalent protein modification using tagged lipid substrate derivatives.

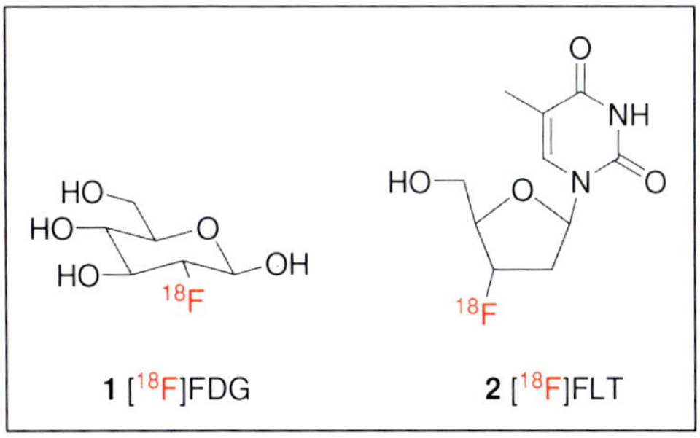

FIGURE 7.1 Two labeled carbohydrates used in PET.

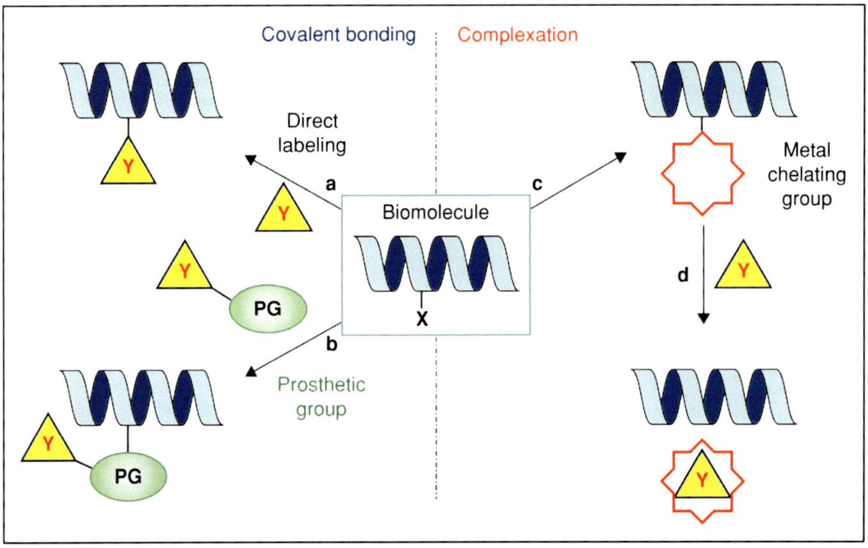

FIGURE 7.2 Strategies for the construction of radiotracers.

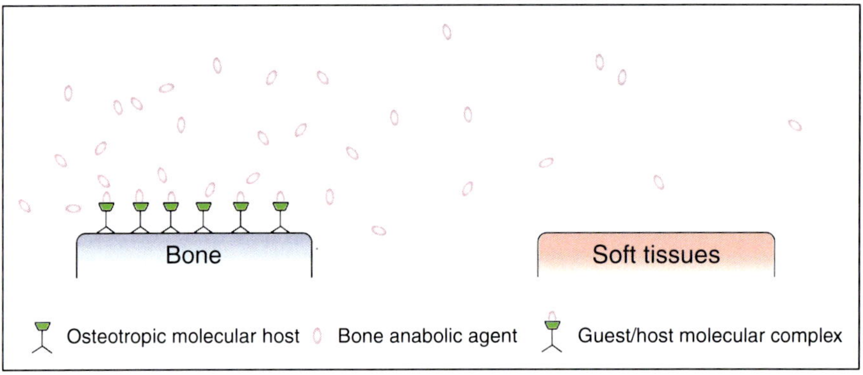

FIGURE 11.1 The proposed bone anabolic mechanism of ALN-β-CD as a molecular sequestrant.

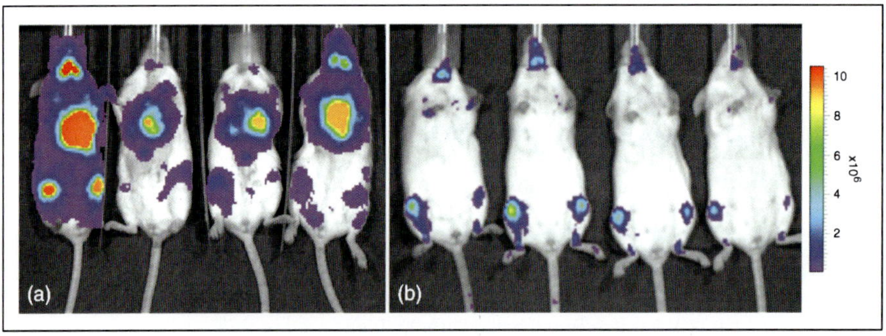

FIGURE 11.2 The early identification of breast cancer bone metastasis lesion by NIR dye-labeled osteotropic polyrotaxane. (a) Luciferase activity associated with breast cancer cells was mainly found at knee joints (bone metastatic sites) and heart (due to intracardial injection of Luciferase-expressing breast cancer cells); (b) High signal intensity of NIR-dye-labeled osteotropic polyrotaxane was found at knee joints but not the heart, which identifies these positions as cancer bone metastasis.

7

TOWARD IMAGING GLYCOTOOLS BY CLICK COUPLING

YVES CHAPLEUR, CHRISTINE VALA, FRANÇOISE CHRÉTIEN, AND SANDRINE LAMANDÉ-LANGLE

7.1 INTRODUCTION

In the last century, medical imaging has become an invaluable tool in medicine. Since the first X-ray image recording of Mrs Roentgen's hand in 1895, acquisition of human body images considerably helped physicians to make a diagnosis. In addition to the pioneering X-ray examination, several other techniques are now at disposition such as ultrasonic imaging, magnetic resonance and nuclear imaging techniques. Taken together, these techniques allowed the emergence of a new field in medicine, the so-called molecular imaging. This new multidisciplinary field at the interface of molecular biology and imaging allows the visualization of cellular functions or the follow-up of biological processes at the molecular level without interference.

Applications of this technique are increasing every day, ranging from cancer, neurological disorders and cardiovascular disease diagnosis to treatment evaluation and improvement at the preclinical and clinical stages. In contrast to medical imaging, molecular imaging makes use of biomarkers as target- or process-specific probes. The chemical interaction between the biomarker and its environment induces localized molecular changes at the origin of image alteration. These noticeable differences allow the visualization of fine modifications paving the way to early diagnosis and fast pharmacological developments. As a technology able to explore and measure the chemistry of life in a noninvasive manner, this new science undoubtedly strongly contributes to the development of biomedicine. Not surprisingly, molecular imaging

Click Chemistry in Glycoscience: New Developments and Strategies, First Edition.
Edited by Zbigniew J. Witczak and Roman Bielski.

was listed as one of the "Ten emerging technologies that will change the world" by the MIT 2003 Technology Review [1].

Because nuclear medicine uses radioelements, this discipline is central to the development of molecular imaging. The two main techniques used in nuclear medicine, single-photon emission computerized tomography (SPECT) and positron emitter tomography (PET) rely on the use of radiotracers able to interact with living organisms to provide images. Coupled with X-ray computed tomography, the latter provide high-resolution images useful as diagnostic or therapeutic follow-up tools. It is clear that the progresses in PET imaging are related to available radiotracers. Thus, the search for new efficient radiotracers is of tremendous importance and constitutes an interesting playground for chemists [2].

7.2 NUCLEAR IMAGING AND RADIOTRACERS

A radiotracer can be defined as a biologically active compound associated with a radioisotope with a short half-life. The radioisotope can be directly linked to the active compound or complexed by means of a suitable complexing molecule. The principle of imaging is to detect the decay of the radioisotope of the radiotracer.

Radiotracers useful in SPECT imaging are gamma-emitting radioelements like technetium-99m (^{99m}Tc), iodine-123 (^{123}I), indium-111 (^{111}In), or gallium-67 (^{67}Ga). SPECT imaging is acquired with a gamma camera as multiple 2-D images from several angles. A computed tomographic reconstruction is then carried to yield 3-D datasets which are used to show thin slices of the body.

PET radiotracers incorporate one of the β^+-emitters like fluoride-18 (^{18}F), carbon-11 (^{11}C), nitrogen-13 (^{13}N), oxygen-15 (^{15}O), copper-64 (^{64}Cu) or gallium-68 (^{68}Ga). All these elements decay, producing a positron which encounters an electron in tissues. This so-called annihilation reaction produces the two photons of high energy (511 keV) which are detected by the camera allowing the acquisition of 3-D images constructed by computer analysis. This three dimensional image construction is often accomplished with the aid of CT X-ray scan performed at the same time.

Working with radioelements induces some constraints among which the half-life ($t_{1/2}$) of the element is prominent. For these reasons only a few radioelements are useful in imaging. For SPECT, ^{99m}Tc $t_{1/2}$ 6 hours, ^{123}I $t_{1/2}$ 13 hours, ^{111}In $t_{1/2}$ 67 hours and ^{67}Ga $t_{1/2}$ 3.3 days are used in radiotracers. For PET, ^{18}F $t_{1/2}$ 109 minutes, ^{11}C $t_{1/2}$ 20.4 minutes, ^{68}Ga $t_{1/2}$ 67.7 minutes, ^{13}N $t_{1/2}$ <10 minutes, ^{15}O $t_{1/2}$ 122.24 seconds, ^{64}Cu $t_{1/2}$ 12.7 hours can be used. Because of radioelement physical characteristics, (decay energy and $t_{1/2}$), only a few are routinely usable. ^{18}F, ^{11}C and more recently ^{64}C have found applications, ^{18}F being the most extensively used in clinics and in research and development as well.

Until recently, only a few radiotracers were available in the clinics and 2-deoxy-2-[^{18}F]fluoro-D-glucose (FDG) (**1**), the first radiotracer with approval, is still the workhorse in PET imaging. This carbohydrate derivative is used for tumor imaging and for a lot of applications. As glucose, FDG is taken up by cells using GLUT transporters and accumulates in highly metabolic brain, kidney, and cancer cells

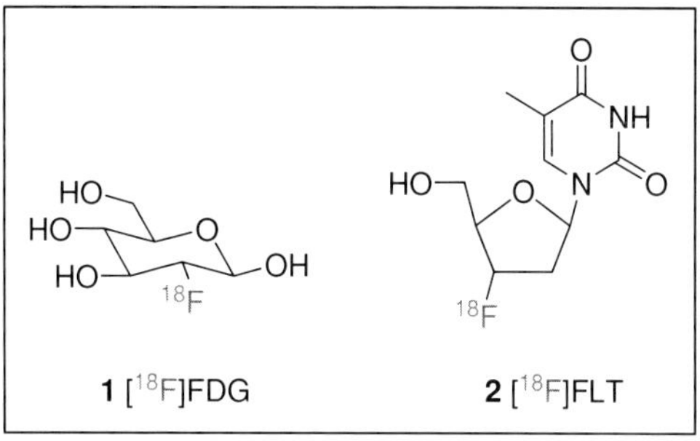

FIGURE 7.1 Two labeled carbohydrates used in PET. (See color plate section.)

and phosphorylation prevents glucose from being released. Since the 2′ hydroxyl group is absent in FDG, it is not further metabolized in cells and accumulates. Other radiotracers are now in use such as 3′-deoxy 3′-[^{18}F]fluorothymidine (FLT) (**2**) or fluorocholine (FC) both for tumor detection applications (Figure 7.1).

FDG as well as FLT and FC have been used in a number of clinical studies but have some drawbacks and are limited to some pathologies. Major advances in PET imaging will undoubtedly come from new, more sophisticated radiotracers specially designed for a given application. The design of a new radiotracer comprises the selection of an appropriate target and of a molecule able to interact with the target, two choices relevant to biomedicine and molecular biology. The determination of the radioelement and of its place in the molecule is the second step in the radiotracer design, relevant to radiochemistry. Finally, the synthetic plan toward the radiotracer synthesis is relevant to organic chemistry.

As shown in Figure 7.2, two main strategies are available for the introduction of the radioelement either by formation of a covalent bond (for fluorine or carbon, paths a and b) or by the formation of a complex (for technetium, gallium, indium, or copper, path c and d). Regarding the formation of covalent bonds shown on the left–hand side of Figure 7.2, there is again two ways to introduce the radioelement, either directly, by modification of the active molecule (direct labeling path a, Fig. 7.2), or by coupling the bioactive molecule to a small molecule bearing the radioelement. This small molecule is called a prosthetic group. The coupling reaction (path b Fig. 7.2) is then crucial to the synthesis. Given the half-life of the radioelement, the labeling reaction must be a fast reaction, ideally performed within 10–20 minutes. When using a prosthetic group, its labeling and the coupling reaction should take place within the same time range. Moreover, the coupling reaction must be efficient in order to be sure that the totality of the prosthetic group is coupled to the active molecule, that is, all the radioactivity is located on the expected radiotracer, ideally without the need to extensively purify the radiotracer. As already mentioned, radiotracers of the future will certainly be elaborated around complex biomolecules like peptides, proteins, carbohydrates, or nucleic acids, all of them being rather sensitive compounds [3–5]. The use of prosthetic groups should be the method of choice to label these biomolecules. One current challenge in the field of radiochemistry is to design new

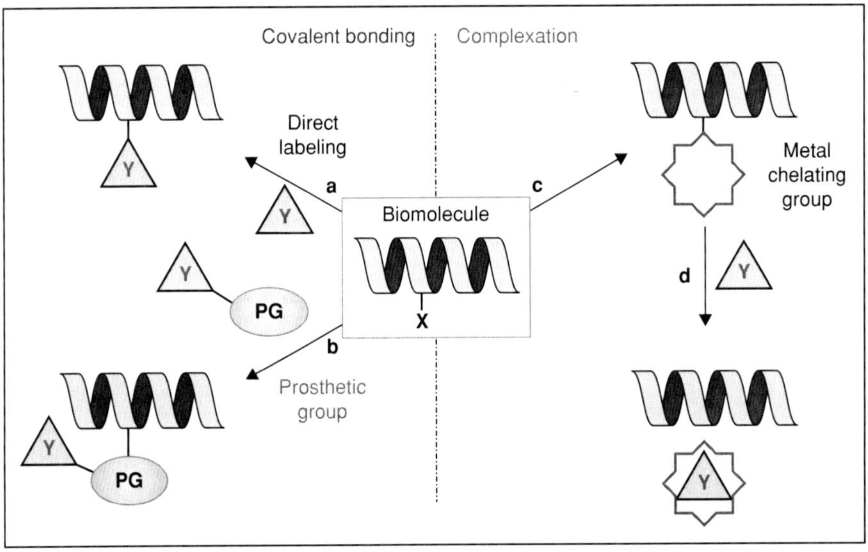

FIGURE 7.2 Strategies for the construction of radiotracers. (See color plate section.)

prosthetic groups and to design new efficient coupling reactions of the latter to a given biomolecule. On the one hand, the so-called "click chemistry" technologies [6] would be of tremendous interest to provide valuable solutions to the coupling problem. On the other hand, there is a long-standing interest in the use of sugars in the elaboration of radiotracers and labeled complex carbohydrates themselves are interesting tools for imaging [7]. Some reviews dealing with the use of click reactions in radiochemistry [8–12] or in carbohydrate chemistry [13–15] have been published. The aim of this chapter is to summarize different achievements recently reported which made use of the two concepts for the elaboration of carbohydrate-using imaging tools. It is important to note that the carbohydrate moiety can be used in different ways: as the carrier of the radioelement (^{18}F), as a key element to improve the biokinetics of the radiotracer or to participate in the chelation structure for radiometals (^{99m}Tc, etc.). In all cases, a robust and specific ligation method is sought for. This chapter will be arranged according to the type of ligation or click reaction used (triazole, oxime formation, etc.) and then to the type of compounds prepared (peptides, carbohydrates, etc.)

7.3 TRIAZOLE-CONTAINING RADIOTRACERS

Since its revival in 2001, the Huisgen cycloaddition of an azide and an alkyne, now catalyzed by Cu(I) [16, 17], has become the leading reaction in the "click chemistry" approach to achieve chemical ligation of two molecules [18]. Under the new catalytic conditions, this triazole-forming reaction proceed readily, usually within hours, in a number of solvents including aqueous medium at room temperature or above. As seen from above, synthesis of radiotracers must be rapid, selective, and must preserve functional groups on the bioactive molecule. The copper-catalyzed azide–alkyne cycloaddition (CuAAC) obeys to these requirements as azide or alkyne groups are rather inert to other functional groups encountered in peptides, proteins, carbohydrates, or nucleic acids. For example, the azide group is recognized as bio-orthogonal [19, 20], meaning that it is inert towards metabolic reactions occurring in living systems. This stability is a major advantage in radiochemistry where chemoselective ligation would be the rule. On the other hand, any cycloaddition reaction is a no-byproduct reaction as it is stoichiometric in each component avoiding purification of labeled compounds. Finally, the CuAAC reaction provides a single 1,4-triazole in high yields [17] and this triazole is stable under physiological conditions. It has also been shown that the triazole moiety is capable of hydrogen-bond forming comparable to that of amide bonds [21, 22]. Since 2002, many groups have explored this opportunity particularly in carbohydrate chemistry [23, 24]. The ability of triazole to be involved in metal complexes, an important construction in molecular imaging, also triggered the interest in this heterocycle [25]. Taken together, this makes CuAAC a very attractive solution for the chemical ligation between a prosthetic group and the bioactive molecule component of a radiotracer [26–34].

Various receptors are overexpressed in tumors and can be used to visualize these tumors with the appropriately labeled peptidic ligand. Peptides are often delicate entities which cannot be labeled directly with positron emitters like ^{18}F. Conjugation of the peptide with a more robust labeled moiety using CuAAC is thus a useful strategy. When this robust molecule to label, i.e. the prosthetic group, is a sugar derivative, the net result is the construction of a neoglycopeptide. In addition, it is now well established that sugar conjugation to a radiotracer largely improves its pharmacological properties by reducing lipophilicity, plasma protein binding, hepatic uptake, hepatobiliary transfer or unspecific uptake, and increasing tumor uptake and renal excretion [35–39].

It is important to keep in mind that in a search for radiotracers, the chemist needs to prepare first a molecule containing a ^{19}F atom, the so-called cold reference from a precursor. This compound is needed for biological studies and will serve as a reference for the synthesis of the hot compound. In most cases, the precursor will undergo the same reaction with ^{18}F in the radiochemical step. Together with the synthesis of labeled compounds, some examples of cold references and precursor syntheses will be given.

We will first examine the use of labeled carbohydrates as prosthetic groups to be linked with amino acids and peptides by triazole formation.

7.3.1 Conjugation of Labeled Carbohydrates

[^{18}F]FDG is routinely prepared by nucleophilic substitution of 2-*O*-triflate mannopyranose with [^{18}F]F^{-}. Starting from this point, Maschauer and Prante exampli-fied the preparation of different mannopyranose triflates suitable for CuAAC, that is, bearing an azide or alkyne function (Scheme 7.1) [40]. Starting from

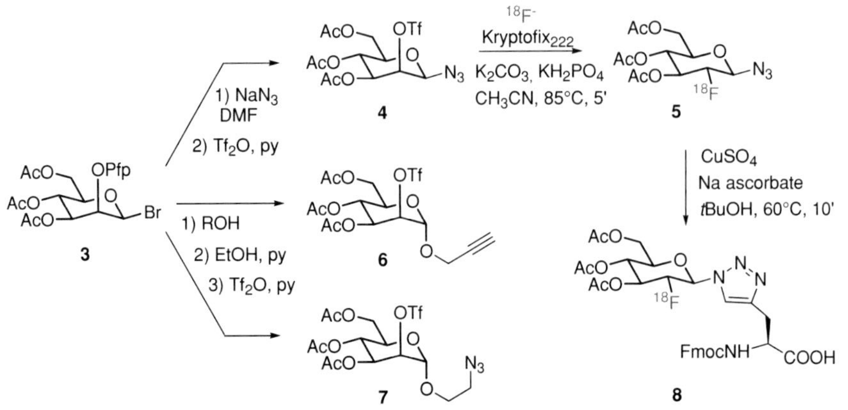

SCHEME 7.1 Coupling a labeled sugar azide with a propagyl amino acid using CuAAC.

2-*O*-pentafluoropropionyl acetobromomannose **3**, different 2-*O*-triflates were elaborated by glycosylation with azide, propargyl alcohol and azidoethanol leading to **4, 6,** and **7** respectively.

These triflates were submitted to labeling conditions ($[^{18}F]F^-$ / $Kryptofix_{222}$ in acetonitrile at 85°C for 5 minutes). Not unexpectedly, only the β-azido triflate **4** gave a useful 71% radiochemical yield of **5**. This compound was successfully coupled with fluorenylmethyloxycarbonyl (Fmoc) propargyl glycine using CuSO4, sodium ascorbate in *t*BuOH at 60°C for 10 minutes with 60–75% yields to give **8**. It is interesting to note the failure of compounds **6** and **7** as well as the α-azido anomer of **4** to produce significant yields of fluorinated derivatives. This should be attributed to a competition between the entering fluoride nucleophile and the participation of the anomeric group [41,42].

Our group is involved in a program aiming at the use of carbohydrates for peptide labeling. This approach cumulates two advantages: labeling the sugar moiety is an easily automated, well-established process in FDG synthesis and the presence of the sugar moiety on the final radiotracer could improve the pharmacokinetics of the latter. In connection with this program, we investigated the labeling of **12** [43]. As shown in Scheme 7.2, using a double inversion at C-2 and C-4 developed by Dong [44,45], the synthesis of **13** was achieved in 5 steps from galactoside **9**. This known compound **9** [46] was selectively acetylated in two steps in one pot by way of stannylidene formation followed by acetylation to give **10** in 65% yield. The bis-2,4-*O*-triflate was then formed and treated with tetra *N*-butylammonium acetate at room temperature in acetonitrile to give the *gluco* triflate **11** in 46% yield. The latter was treated with tetra *N*-butyl ammonium nitrite to give the *manno* derivative **12** in 47% yield. Interestingly this compound was easily fluorinated by treatment with diethylaminosulfur trifluoride (DAST) at 50°C giving **14** in 68% yield. The corresponding α-anomer of **12** was not fluorinated at position 2 by the same reaction, but at the anomeric position with migration of the azido ethyl group at C-2. Finally, ^{18}F fluorination of **12**, carried

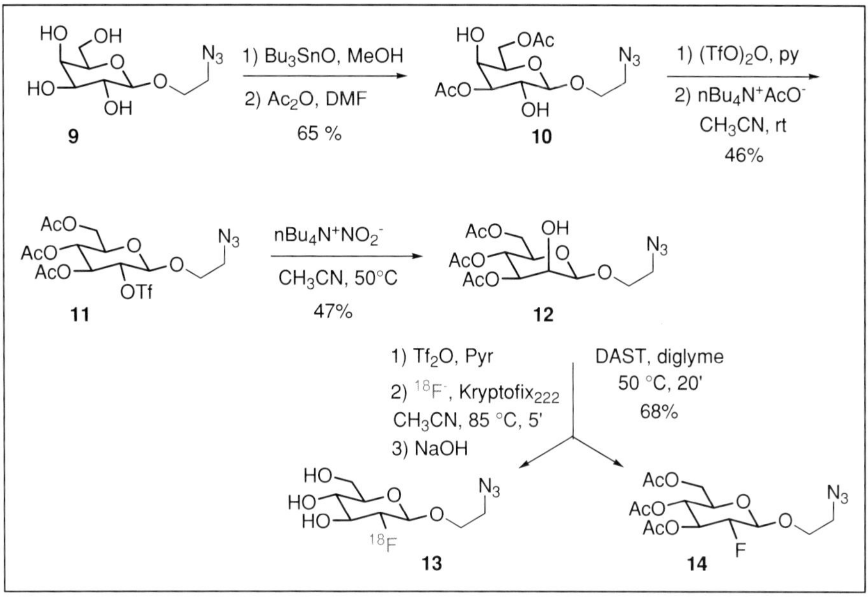

SCHEME 7.2 Elaboration of [2-azidoethyl]-2-fluoro glucosides.

out using [^{18}F]F^- and K_{222} followed by removal of acetyl groups, led to **13** in 38% radiochemical yield. Not unexpectedly, it is very important to start from β-anomers to achieve fluorination at C-2 of 2-*O*-triflyl-*manno* derivatives in significant yields.

Following their preliminary study, Prante et al. developed the [^{18}F]-labeling of alkyne-bearing peptides with concomitant glycosylation (Scheme 7.3) [47]. The alkyne function was introduced by coupling commercially available L-propargylglycine with the peptide of interest. The method was applied to neurotensin (NT), a peptidic ligand of neurotensin receptor 1 (NTR-1), a receptor overexpressed in pancreatic tumor cells. Radioactive ligands of this receptor should be useful in colorectal cancer imaging. Different peptidic sequences designed from the NT(8-13) hexapeptide were used as a starting point. As an example, several peptides bearing a triple bond were reacted with [^{18}F]fluoroglucosyl azide **15** prepared like [^{18}F]FDG starting from the azidomannosyl triflate **4** via compound **5**. Four labeled glycopeptides like **16** were prepared in 70–80% radiochemical yield by heating the two compounds in PBS (10% ethanol) at 60°C for 20 minutes in the presence of $Cu(SO_4)$ and sodium ascorbate. Interestingly, the glucose-conjugated c(RGDf) **17**, prepared in 20% yield within 70 minutes, had an excellent *in vivo* metabolic stability, fast tumor accumulation and compared well with the clinically used [^{18}F]galacto-RGD [48].

It is clear that most of the ligation methods to connect a peptide to a prosthetic group need some modifications of the peptide which could be a limiting step. Our group recently explored a new approach for the direct derivatization of peptides with an alkyne residue taking advantage of the unique reactivity of thiol [49]. The thiol nucleophilicity of cysteine-containing peptide has not been widely used till now. We were able to selectively introduce a propargyl group on the cysteine residue

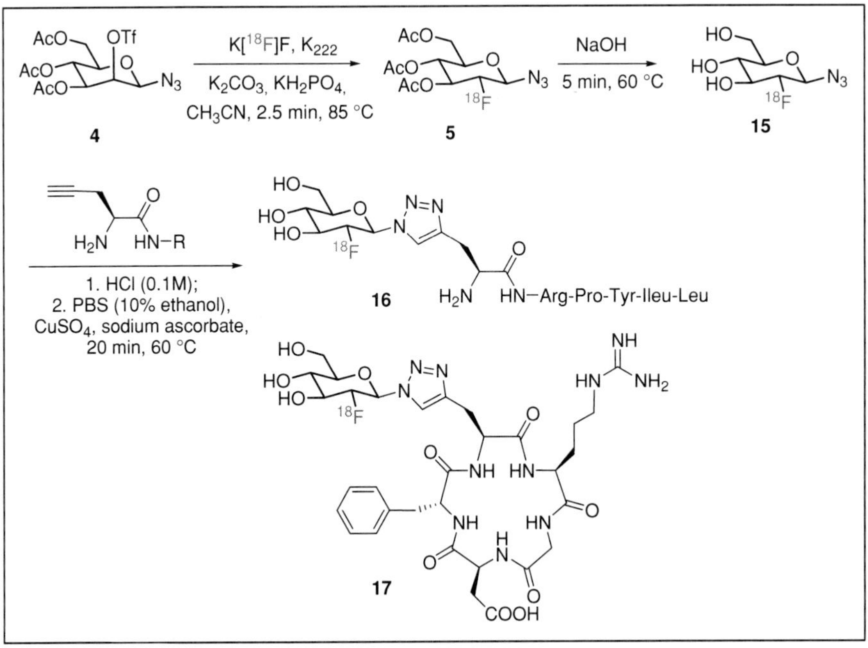

SCHEME 7.3 Labeling pentapeptides with 1-azido-2-fluoro glucose using CuAAC.

of several native peptides by simple treatment with propargyl bromide in aqueous basic solution ($Ba(OH)_2$ or NH_4OH) for 2 hours at 0°C. This methodology has been successfully applied to glutathione as a model peptide and to the tetrapeptide RGDC (Arg-Gly-Asp-Cys) (Scheme 7.4). The CuAAC coupling with different azido sugars in water at room temperature in the presence of copper acetate and sodium ascorbate proceeded uneventfully. In the meantime, we have developed new fluoro sugar prosthetic groups. We took advantage of the high reactivity of the primary position to introduce a fluorine atom at C-6 using 6-*O*-triflyl derivatives like **20**. The latter were prepared from the azido sugars **18** by a classical sequence of protection–deprotection using the trityl derivatives **19**. The labeling sequence with $[^{18}F]F^-$ has been applied in the *gluco* and *manno* series with radiochemical yields of 63% and 64% respectively. The CuAAC sequence was carried out with the *gluco* derivative **21** and the propargylated glutathione **22** using classical conditions ($Cu(OAc)_2$, sodium ascorbate, water, 60°C, 12 minutes) to provide the labeled glutathione derivative **23**. This shows that 6-deoxy-6-fluoro sugars are good candidates as prosthetic groups.

7.3.2 Glycoconjugation of Radiotracers

As mentioned above, sugar conjugation to a radiotracer strongly modifies the pharmacological properties by reducing lipophilicity, plasma protein binding, hepatic uptake, hepatobiliary transfer or unspecific uptake, and increasing tumor uptake and

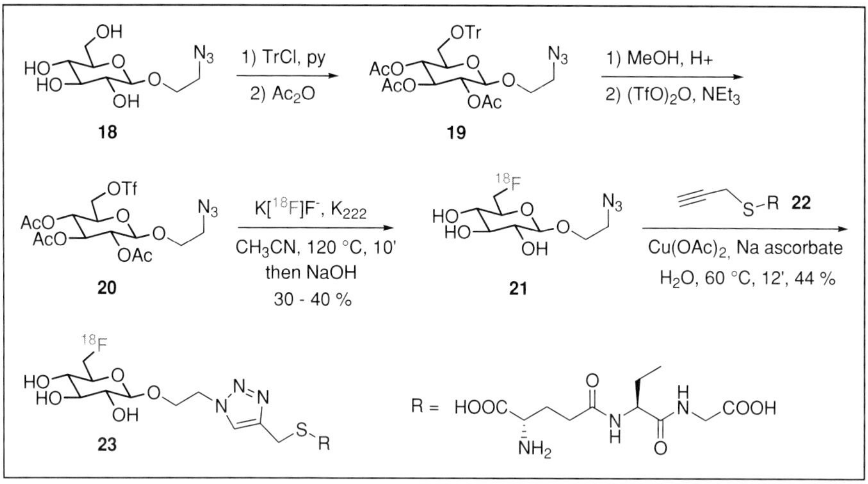

SCHEME 7.4 Elaboration of [2-azidoethyl]-6-fluoro glucosides and coupling with S-propargylcysteine.

renal excretion. Thus, many radiotracers have been conjugated to sugars by different ways including the CuAAC.

*7.3.2.1 **Radiotracers with Covalent Labeling*** The short Arg-Gly-Asp (RGD) peptide is a famous sequence found in proteins of the extracellular matrix such as laminin, fibrinogen, vitronectin involved in the adhesion between cells and extracellular matrix. This sequence is recognized by integrins, heterodimeric transmembrane proteins, which play a major role in signaling in and out of cells. The vitronectin receptor or $\alpha_v\beta_3$ integrin is a key receptor expressed on endothelial cells which modulates migration and survival during angiogenesis. This receptor is also expressed in tumor cells and plays a role in metastasis. Imaging angiogenesis or tumor cells with a suitable labeled ligand of $\alpha_v\beta_3$ integrin has become a challenging area in cancer and cardiovascular diseases diagnosis. Several efficient ligands of $\alpha_v\beta_3$ integrin have been disclosed and used to prepare imaging tools [29, 39, 48, 50–52].

In a study aimed at defining the influence of the carbohydrate moiety of active peptide glycoconjugates, Rutjes et al. reported the use of the triazolo glycoamino acid **26** (Scheme 7.5) [53]. The latter was obtained by CuAAC between the known 2,3,4,6-tetra-*O*-acetyl-glucosyl azide **24** [54] and N-Fmoc-propargyl glycine **25** under standard conditions, that is, $Cu(OAc)_2$, sodium ascorbate, *t*BuOH, H_2O [55]. The free glycoamino acid **26** was involved in a solid phase synthesis of the cyclic peptide c(RGDy-NTGA) **27**. In order to evaluate the validity of the triazole ring as a surrogate of the amide bond, the analog **29** with an amide linkage was prepared from 2,3,4,6-tetra-*O*-acetyl-glucosyl amine and aspartic acid. An improved chemical stability of the triazole ring with respect to the amide bond was found. The triazole ring was also found compatible with either basic or acidic conditions. Both compounds

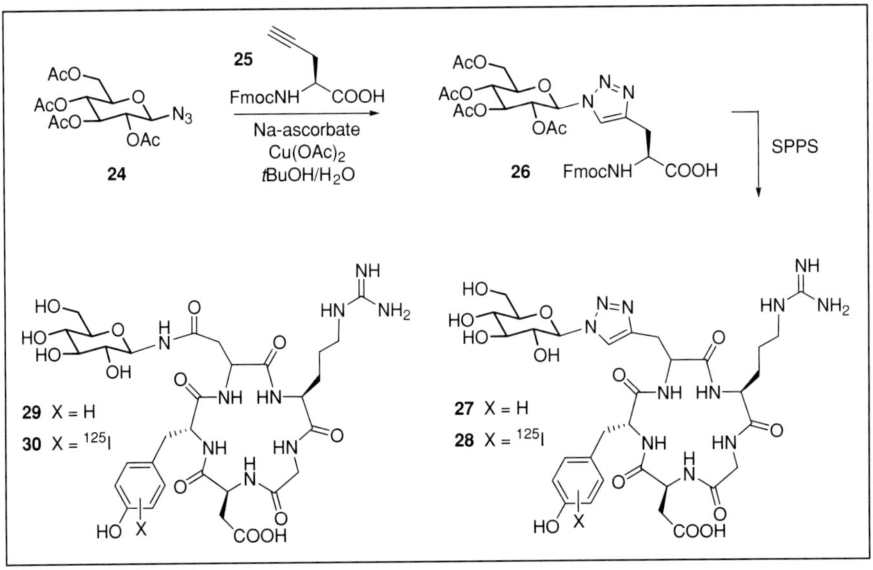

SCHEME 7.5 Synthesis of glycosylated cRGD peptide using CuAAC.

were tested as ligands of human $\alpha_v\beta_3$ integrin in a solid-phase assay by competition binding with radioactive of ^{111}In-1,4,7,10-tetraazacyclododecane-1,4,7,10-tetraacetic acid (DOTA)-E-[*c*(RGDfK)] as the reference compound. IC_{50} of 144nM for **27** and 238 nM for **29** were measured, in the same range of the Kessler's peptide *c*(RGDfV) [56] at 65 nM, indicating that side chain modification had only limited effect on the affinity. Imaging tools were built from **27** and **29** by standard iodination of the tyrosine residue with ^{125}I. It has been shown that the radiolabeled compounds **28** and **30** accumulate in tumors and that this accumulation is mediated by $\alpha_v\beta_3$ integrin. Furthermore, compound **27** with the triazole ring showed improved pharmacological properties with reduced liver and higher tumor uptakes. This study confirms the nondeleterious effect of glycosylation via triazole ring formation.

7.3.2.2 ***Radiotracers Based on Metal Chelates*** Metal complexes are key compounds in molecular imaging because of the use of radiometals in SPECT (^{99m}Tc, ^{111}In, ^{186}Re, etc.) or in PET (^{64}Cu), these elements being included in complexing moieties often made of nitrogen heterocycles or macrocycles. In this regard, the triazole ring is interesting because it could be involved in part in cooperation with other groups in such metal complexes. In 2006, the group of Schibli introduced the "click to chelate" approach in which the synthesis of the complexing moiety and its complexation is achieved in a single step making use of the CuAAC method (Scheme 7.6) [57]. This approach has been tested on two azido sugars, 3′-azido thymidine **31** and 1-azido-1-deoxy-β-D-galactopyranose **32** reacting with propargyl–glycine in aqueous solutions at 100°C for 30 minutes in the presence of $Cu(OAc)_2$ and sodium

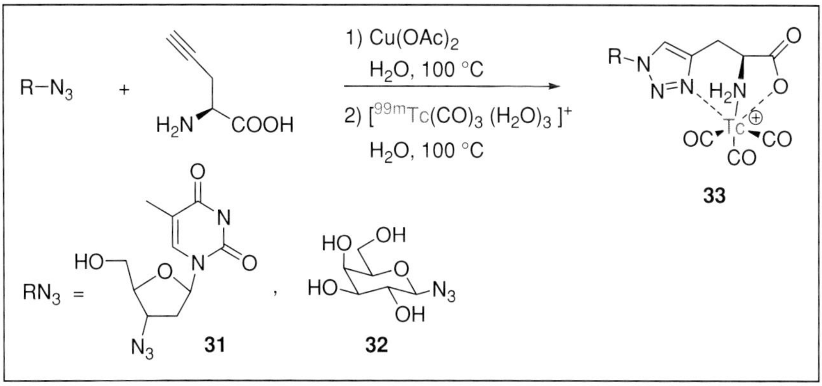

SCHEME 7.6 Synthesis of ^{99m}Tc complexes based on thymidyl- and glucosyl-triazolo-alanine.

ascorbate. [^{99m}Tc(OH$_2$)$_3$-(CO)$_3$] was then added and the mixtures heated for 30 minutes to provide the expected complexes **33**.

Using a similar approach, carbohydrate-based technetium complexes have been prepared (Scheme 7.7) [58]. Glucosyl azide **34** reacted with different pyridine derivatives like **35** as metal chelator in the presence of copper sulfate and sodium ascorbate, the reaction being advantageously activated under microwave irradiation. The triazole **36** was then complexed with technetium in the usual way to give **37**. This complex was stable *in vitro* in the presence of histidine and *in vivo*. This study also revealed that the cycloaddition reaction can be performed on rhenium precomplexed alkynes.

The use of di-triazolyl radiometal chelate has been studied by Mindt et al. (Scheme 7.8) [59]. Starting from a bis-propargylated glycinamide derivative **38** with one disarmed propargyl group, sequential CuAAC reactions with two different azides were performed to give rise first to **39**. Hydrolysis of the disarming silyl group gave

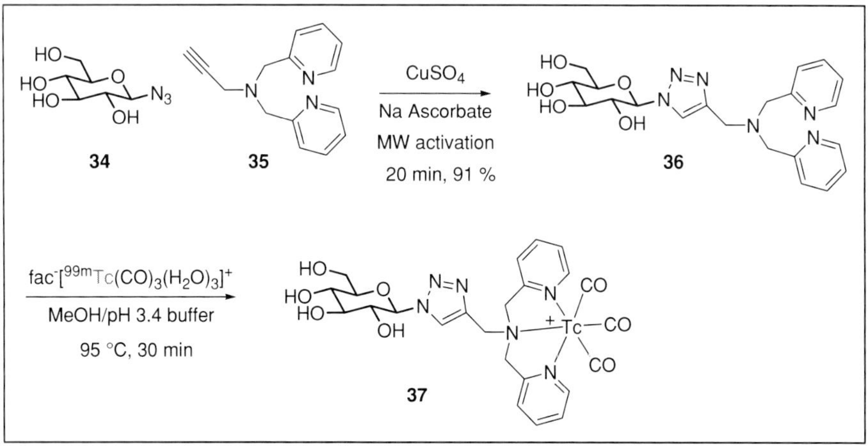

SCHEME 7.7 Synthesis of ^{99m}Tc complexes based on glucosyl-triazolo-bipyridine.

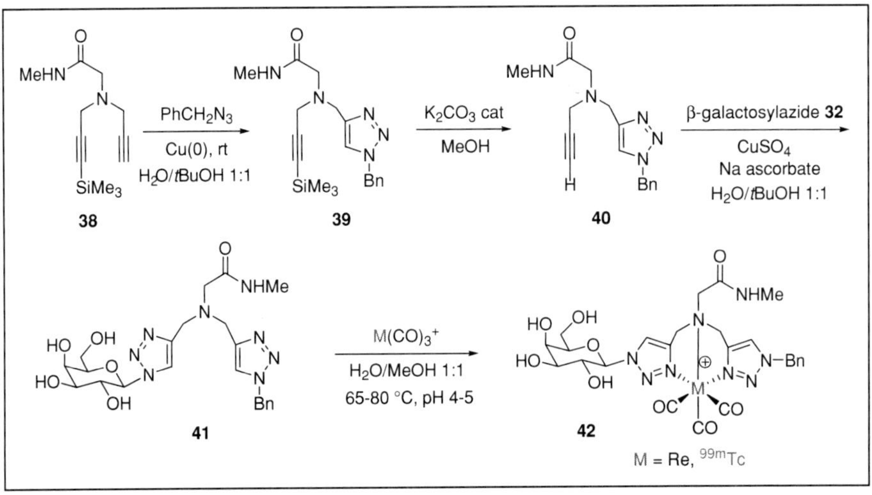

SCHEME 7.8 Synthesis of ^{99m}Tc complexes based on glucosylated bis-triazolo-amine.

40 subsequently transformed into the triazole **41** by reaction with the galactosylazide **32** in 86% yield. This sequence can also be performed in one pot in 80% yield. Treatment of **41** with rhenium or technetium carbonyl complexes ($[^{99m}Tc(CO)_3(H_2O)_3]^+$, phosphate-buffered saline [PBS] buffer pH 5.5, 80°C–100°C, 20–30 minutes) led to complexes **42** in good yield (68% with Re).

In an extension of the "click to chelate" approach, thymidine has been labeled at N-3 position by ^{99m}Tc using the complexing properties of the triazole linker (Scheme 7.9) [60]. Thus compound **45** prepared from thymidine **43** via the bromide

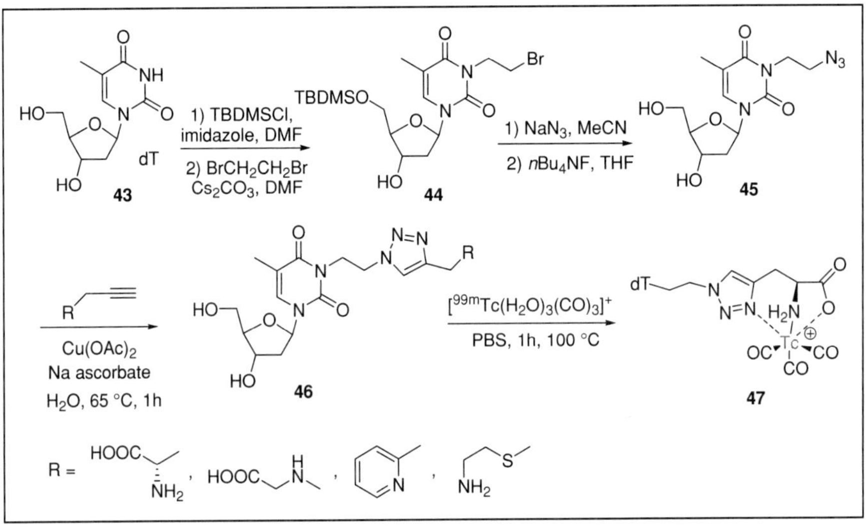

SCHEME 7.9 ^{99m}Tc labeled thymidine.

44 was reacted with different amino-acid-derived alkynes and the resulting triazoles **46** were then complexed with ^{99m}Tc to produce the expected radioactive probes **47** (obtained from propargyl glycine as the alkyne component). These compounds were further transformed into the 5′-monophosphate and tested as substrates of human cytosolic thymidine kinase (hTK1), a key enzyme in tumor proliferation.

7.3.3 Carbohydrate Labeling

The search for surrogates of [^{18}F]FDG and the labeling of biologically significant and/or complex carbohydrates such as nucleoside or oligonucleotide is very active and the efficiency of CuAAC has stimulated some researches along these lines which are reported in this section.

7.3.3.1 Analogs of 2-Deoxy-2-[18F]fluoro-Glucose [^{18}F]FDG is widely used in all PET centers and its automated synthesis is now well established. Nevertheless, the search for new labeled sugars as surrogate of [^{18}F]FDG is still a challenge. As shown in Scheme 7.10, the sugar does not carry the label but is coupled to a fluorinated prosthetic group, taking advantage of CuAAC. Choe et al. constructed the triazolo glucose derivative **50** [61]. Two methods of ^{18}F introduction have been investigated. The first one is based on the elaboration of the triazole **48** obtained by CuAAC of protected β-azido glucose **24** with homopropargyl alcohol. Subsequent tosylation followed by nucleophilic displacement with $nBu_4N[^{18}F]F$ in acetonitrile/tBuOH at 100°C for 20 minutes gave the labeled derivative **50** after deacetylation. The second

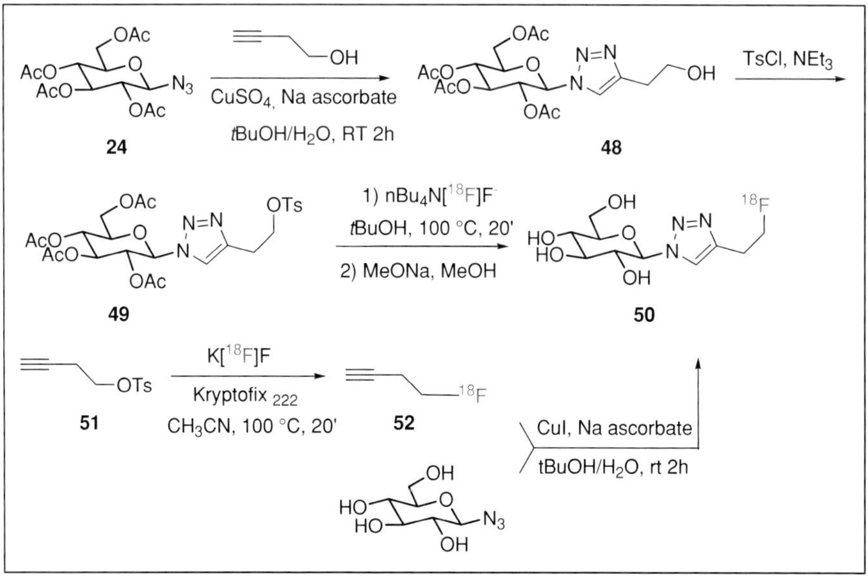

SCHEME 7.10 Synthesis of ^{18}F labeled triazolyl-glucose.

method involved the CuAAC between the free β-azido glucose and the homopropargyl [^{18}F]fluoride **52** prepared from the corresponding tosylate **51** by treatment with K[^{18}F]F, K_{222} in acetonitrile at 100°C for 20 minutes. The biological evaluation of compound **50** showed a good *in vitro* stability but was not a substrate of hexokinase, the enzyme responsible for the phosphorylation of [^{18}F]FDG, for example. Moreover, in contrast to [^{18}F]FDG, compound **50** was not taken up by human colon carcinoma cells limiting its use in PET.

7.3.3.2 Nucleosides Among labeled carbohydrates, 3′-deoxy-3′-[^{18}F]fluorothymidine (FLT) is currently the most widely used proliferation imaging agent in PET. The uptake of this tracer is dependent on the presence of thymidine kinase-1 (TK1), an enzyme implicated in the salvage pathway of nucleotides. In cells FLT is monophosphorylated by TK1 and this phosphate accumulates in the cell. TK1 activity can be upregulated by treatment and FLT could be of interest in the treatment assessment of tumor response to treatment. However, high FLT uptake in the liver and bone marrow limits its use in cancer diagnostic and prompted the search for new analogs [62, 63].

New thymidine derivatives have recently been prepared taking advantage of the CuAAC reaction (Scheme 7.11). New analogs of FLT have been described in the patent literature [64]. The synthesis has recently been improved and the compound used in testings [65]. The known uridine derivative **53** [66] was submitted to CuAAC with 2-[^{18}F]fluoroethylazide to provide the labeled thymidine derivative **54** in about 75% radiochemical yield. The reaction was carried out in the presence of copper iodide and ascorbic acid in water at 80°C for 10 minutes. Compound **54** was stable in human plasma but is taken up by myocardis suggesting *in vivo* metabolism. It should be a useful probe for tumor cell proliferation.

The CuAAC reaction has been developed for the labeling of glucose and 3-azidothymidine as model reactions. As shown in Scheme 7.12, Chi and coworkers proposed the labeled acetylenic derivative **56** as a prosthetic group [67]. This compound, prepared from mesylate **55**, reacted in good yields with acetylated or free azidoglucose, to provide the expected adducts **58** within 10 minutes. Classic reaction conditions,

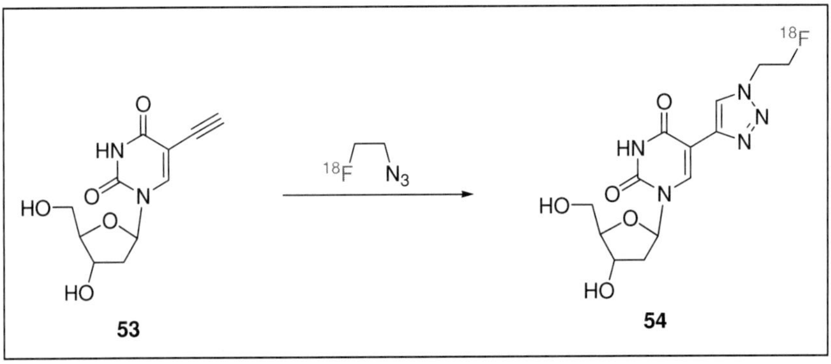

SCHEME 7.11 Synthesis of ^{18}F labeled triazolyl-thymidine.

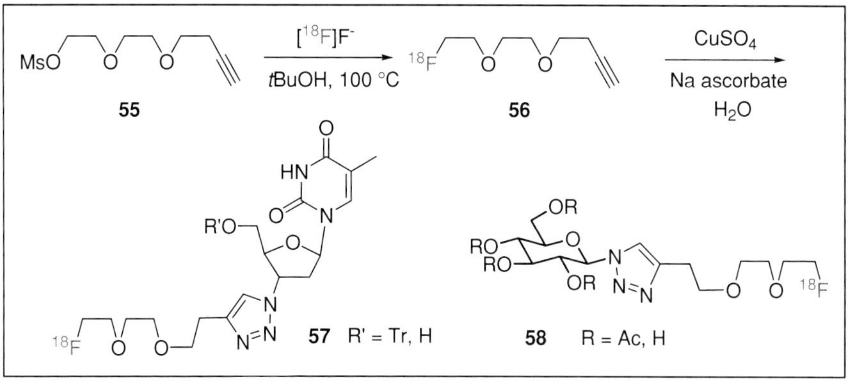

SCHEME 7.12 Synthesis of ^{18}F labeled triazolyl-glucose and triazolyl-thymidine.

that is, $CuSO_4$ and sodium ascorbate in *t*BuOH/H_2O mixture were used. AZT and 5′-*O*-trityl AZT reacted under the same conditions to give rise to triazoles **57**. Labeled derivatives of AZT obtained by CuAAC have also been patented [64].

7.3.3.3 ***Oligonucleotides*** Oligonucleotide-based probes are promising tools in molecular imaging, for example, to detect the expression of specific genes in a noninvasive way. Labeling oligonucleotides is still a challenging problem due to the complexity of these molecules. Click chemistry can help to solve the problem by the use of suitable prosthetic groups. Several ways to label nucleotides have been investigated taking advantage of CuAAC. For example, thymidine has been elaborated to several derivatives having an alkyne function at position 5′ or an azide function at position 3′ (Scheme 7.13) [68]. Thymidine was protected at 3′ position and oxidized at 5′ to give aldehyde **59**. The latter was elaborated to azide **60** and to alkyne **62**. Both compounds have been used to anchor prosthetic groups bearing an iodoaryl function suitable for ^{11}C labeling as shown in structures **61** and **63**. Compounds **61**, **62,** and **63** have been transformed into the corresponding phosphoramidites which were used for oligonucleotide synthesis. The protected thymidine derivative **64** was reacted under classical conditions with benzylazide **65** bearing a di-*t*-butylsilyl group at the para position to give **66**. The silicon group of **66** is suitable for ^{18}F labeling by simple treatment with [^{18}F]F^- [69–72].

As shown in Scheme 7.14, the labeling using this approach has been studied in detail in a subsequent paper [73]. For example, 5′-azido thymidine **67** was coupled with the propargyl ether **68** to give **69** in very high yield at high temperature with the help of microwave irradiation under classical catalysis by copper sulfate and sodium ascorbate. The labeled fluoride **70** was obtained from **69** in good radiochemical yield (34% not decay corrected). Compound **69** was further elaborated to the phosphoramidite **71** under classical conditions and engaged in an automated deoxyoligonucleotide synthesis to provide **72**. [^{18}F]-fluorination of **72** as above provided

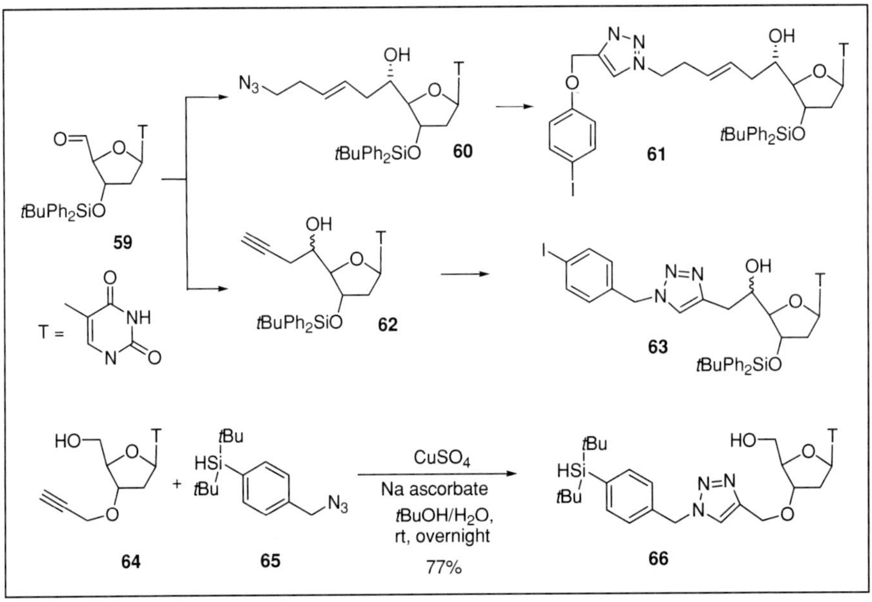

SCHEME 7.13 Synthesis of 3′- and 6′- triazolo-thymidine derivatives suitable for iodination and fluorination.

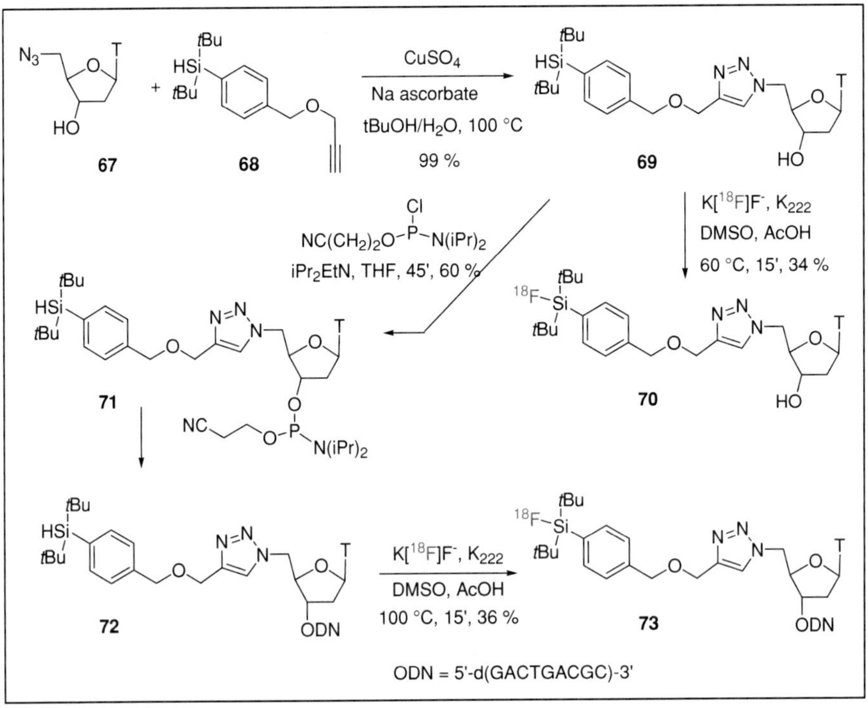

SCHEME 7.14 ^{18}F-fluorination of 3′- and 6′-triazolo-thymidine derivatives.

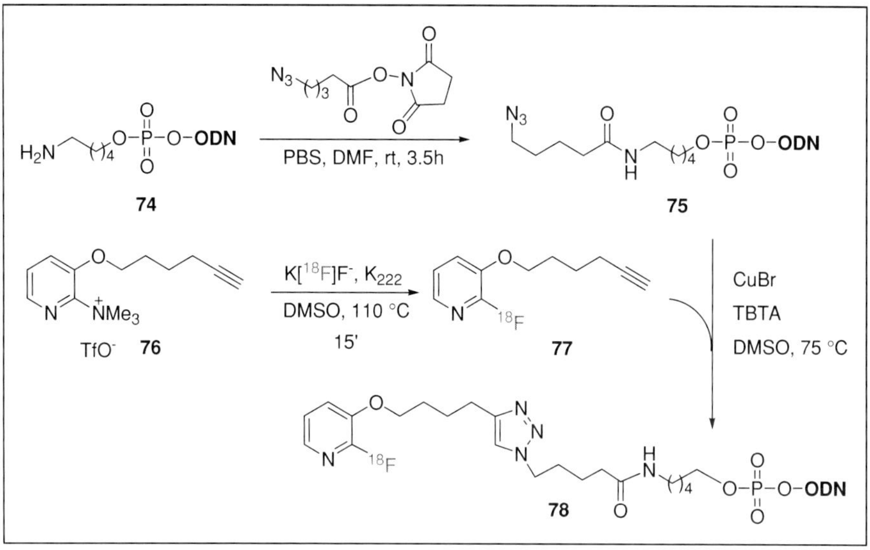

SCHEME 7.15 CuAAC in the labeling of oligodeoxy nucleotides using pyridinyl prosthetic groups.

labeled **73** in not decay corrected 36% radiochemical yield. The stability of these compounds was not reported.

Inkster et al. used their prosthetic group 2-[^{18}F]fluoro-3-(hex-5-ynyloxy)pyridine ([^{18}F]FPy5yne) **77** [74] obtained from ammonium salt **76** (Scheme 7.15). The 5′-phosphate of the oligodeoxynucleotide, esterified with a 5-aminohexanol **74**, was coupled on the amino function with succinimidyl 5-azido valerate to produce the azido oligonucleotide **75**. The CuAAC between **77** and 5′-azide-modified antisense DNA sequence was performed in the presence of copper bromide and tris-(triazolyl)benzylamine (TBTA) at 75°C giving the ^{18}F-labeled oligonucleotide **78** with a radiochemical yield of 24.6 (decay corrected) in 276 minutes [75].

Among oligonucleotides, small interfering RNAs (siRNA) are important compounds because of their possible use as therapeutic agents by inhibition of gene expression. The development of these compounds as drugs is impaired by the poor current knowledge of their pharmacokinetics. Molecular imaging using siRNA may help this therapeutic approach [76]. Recently, the Luxen group reported the first labeling of siRNA relying on the CuAAC reaction (Scheme 7.16) [77]. The single-stranded oligonucleotidic sequence **79** was synthesized on a solid support equipped with a linker bearing an alkyne function. Oligonucleotide synthesis of a single-stranded RNA provided **80** which was elaborated to the siRNA **81**. The key CuAAC was performed with two prosthetic groups **82** and **84** prepared according to a new method [78]. The cycloaddition was performed with large excess of sodium ascorbate (160 equiv) and Cu(II)TBTA complex (20 equiv) in DMSO/H_2O/*t*BuOH mixture. The final radiochemical yields of labeled siRNA **83** and **85** were about 15% and 35% respectively.

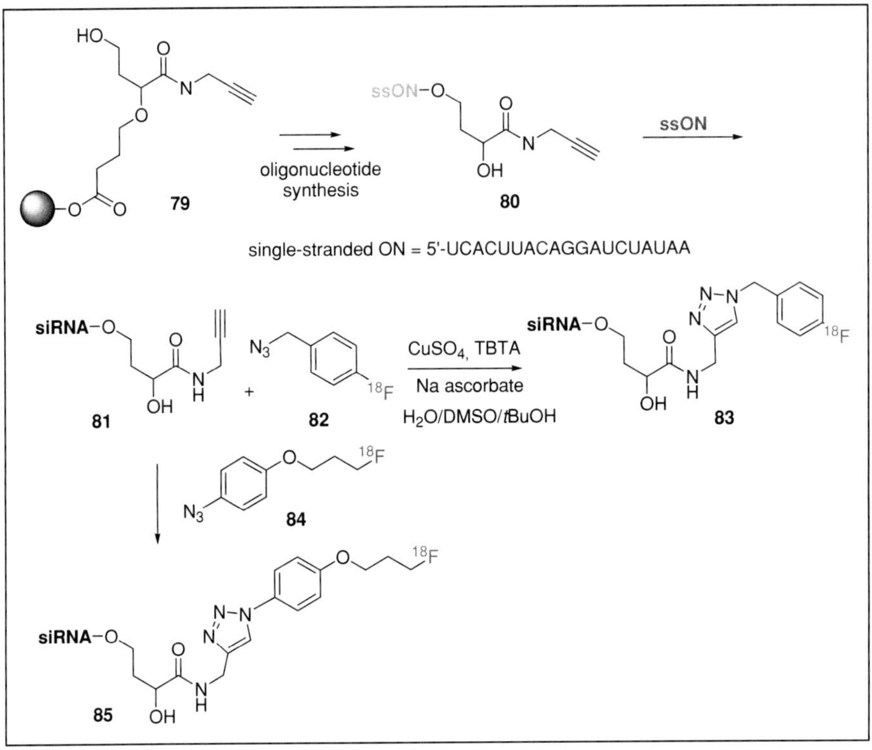

SCHEME 7.16 CuAAC in the labeling of siRNA using aryl- and benzyl-azides as prosthetic groups.

7.4 OTHER METHODS

Several methods relevant of click chemistry and specific to carbohydrate chemistry have been investigated to couple a sugar and a biomolecule suitable to provide a radiotracer. The unique reactivity of the anomeric center is a key element in these researches. The aldehydic form has been used in the Maillard reaction and oxime formation whereas the reactivity of the anomeric alcohol has been used in the glycosylation reaction. Thiosugars have also been used for unsymmetrical disulfides formation.

7.4.1 Maillard Reaction

Radiotracers based on somatostatin receptor ligands are useful for cancer imaging of somatostatin receptor (SR) positive neuroendocrine tumors. Octreotide is a cyclic eight amino acid peptide which binds to the SRs overexpressed on the surface of neuroendocrine tumors. Several octreotide analogs have been developed as potential radiotracers [79, 80]. Carbohydration of octreotide has been investigated as a means to reduce lipophilicity of the peptide and to limit its elimination by the liver.

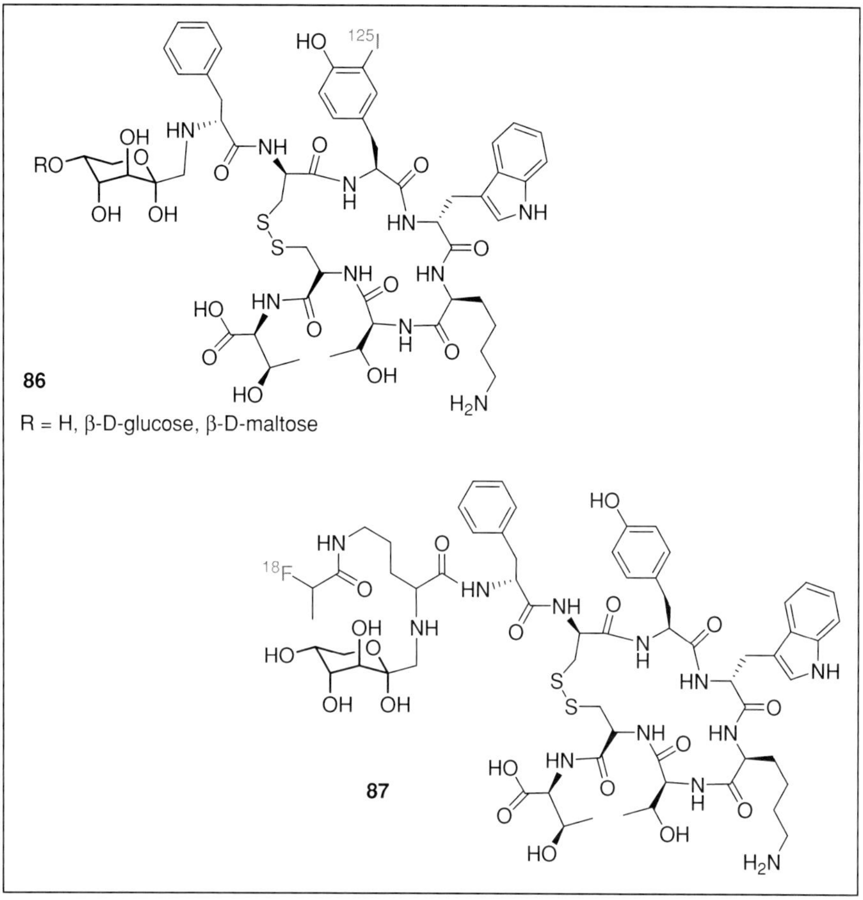

SCHEME 7.17 Glycosylation of labeled octreotide using the Maillard reaction as a click coupling.

As shown in Scheme 7.17, Wester et al. used the Maillard reaction and the involved Amadori rearrangement to introduce a sugar moiety to octreotide derivatives [35,36]. Using a method reported previously [81], glucose, maltose, or maltotriose in excess were reacted with the amino group of *N*-terminal phenylalanine of Lys^5(Boc)-Tyr^3-Thr^8octreotide analog in a MeOH/acetic acid mixture (95/5) at 60°C. However, prolonged reaction times of 24–48 hours were needed. After Boc removal, the adduct was subsequently radiolabeled with ^{125}I on the tyrosine residue to provide radiotracers **86** suitable for imaging [37,38]. A slightly different approach to reach ^{18}F-labeled tracers has been used [35, 82]. A Lys0 Tyr3 octreotide was glycosylated at the NH of lysine using glucose. The resulting fructosyl amine was then reacted at the ω-terminus of lysine to introduce a [^{18}F]fluoropropionyl residue giving the radiotracer **87** suitable for the imaging of SR-positive tumors.

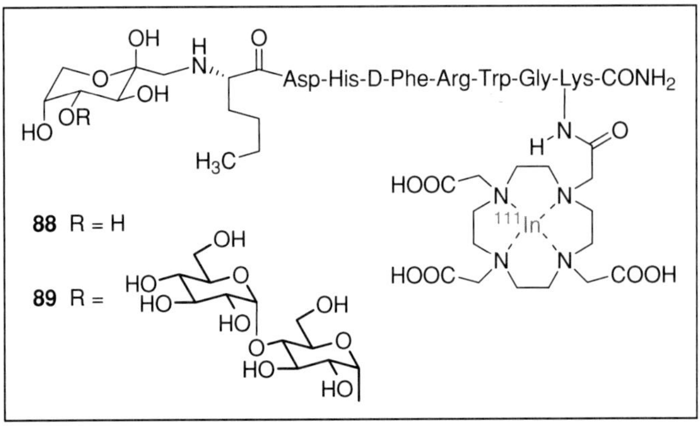

SCHEME 7.18 Glycosylation of ^{111}In labeled heptapetide using the Maillard reaction as a click coupling.

The Maillard reaction has been used for the glycosylation of α-melanocyte-stimulating hormone (MSH) analogs in order to improve the biokinetics of the radiotracer based on ^{111}In complexation by DOTA (Scheme 7.18) [83]. Interestingly, the *N*-terminus of a heptapeptide derived from α-MSH was glycosylated on resin at the end of the peptide synthesis using ten equivalent of glucose in a mixture of MeOH/AcOH (95:5) at 60°C for 72 hours. Cleavage from the resin and coupling with DOTA yielded after deprotection and complexation with ^{111}In, the radioligand **88**. The glycosylation with maltotriose giving **89** was also reported. Other glycopeptides have been prepared by glycosylation of the Lys residue of the same peptide under Maillard conditions. The glycosylated peptides **88** and **89** showed increased kidney reuptake and retention.

NT is a tridecapeptide localized in both the central nervous system and in the gastrointestinal tract acting as a neurotransmitter and a hormone. One subtype of the receptor of NT, the NTR1 receptors, is highly overexpressed on pancreatic carcinoma cells. NT-based radiotracers are thus interesting tools to develop. In the search for neurotensin-based radiotracers, Maes and Tourwé prepared glycosylated peptides equipped with a $^{99m}Tc(CO)_3$ chelator using Amadori rearrangement. On model peptides, glycosylation occurred mainly at the ε amino group of the lysine residue but the secondary amine present in the histidine residue of the chelator could also be glycosylated [84].

Bombesin is a tetradecapeptide isolated from a frog skin which binds to gastrin-releasing peptide receptor, a receptor overexpressed in human prostate, breast, colon, and small-cell lung carcinoma. Bombesin-based radiotracers have been prepared by coupling this peptide with glucose and further complexation of the adduct with ^{99m}Tc [85–87].

7.4.2 Oxime Formation

Oxime formation is a useful rapid and traceless reaction which has found numerous developments in chemical ligation of peptides and sugars [88] and in

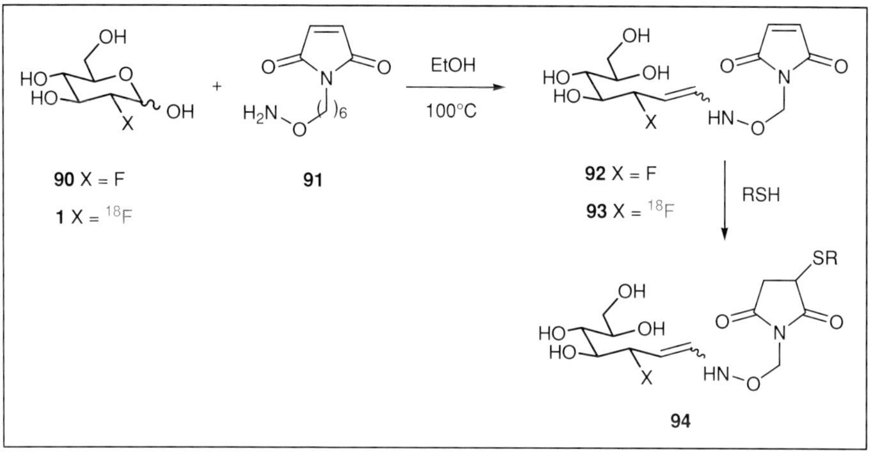

SCHEME 7.19 Oxime formation as a click coupling between a sugar and maleimide.

radiochemistry [89, 90]. Wuest and coworkers proposed oxime formation as a click reaction to construct an FDG-based prosthetic group (Scheme 7.19) [91]. As [^{18}F]FDG is available in all PET centers, it is obvious to use it as a precursor to elaborate reactive species able to label peptides or proteins. The cold reference was obtained by heating equimolecular amounts of free FDG **90** with *N*-(6-aminooxy) hexyl maleimide **91** in a EtOH/H_2O mixture (4:1) to give the oxime **92**. Purification on a Sep-Pack cartridge gave pure **92** as an open-chain mixture of *Z* and *E* isomers as shown by proton nuclear magnetic resonance spectroscopy ^{1}H NMR. The ^{18}F-labeled compound **93** was obtained from [^{18}F]FDG **1** and maleimide **91** used in excess. Heating the mixture at 100°C for 15 minutes in 80% ethanol gave 45–69% radiochemical yields in radiochemical purity up to 95%. The labeled oximes **93** have been used to achieve another click reaction, that is, thiol 1,4-addition on the activated double bond of the succinimide moiety to produce radiotracers **94** where R can be glutathione or annexin A5.

The same approach has been used by Gambhir et al. using [^{18}F]FDG and RGD-containing peptides (Scheme 7.20) [92]. In this case, the oxime **95** was directly built from [^{18}F]FDG **1** reacting with the aminooxy function introduced at the N-terminus of an RGD sequence by amide bond formation. Incubation of [^{18}F]FDG with the peptide 16% ethanol in saline (120 μL) in the presence of 0.4% TFA at 100°C for 30 minutes gave the conjugate **96** (*Z/E* mixture) in 27.5 radiochemical yield. Conjugation to c(RGDyK) proceeded the same way in 41.4% radiochemical yield. The method has also been used for the synthesis of a [^{18}F]FDG-c(RGDfK) conjugate **97** by direct reaction with the Boc-protected aminooxy function grafted at the ε position of a lysine residue (Scheme 7.20) [93].

The Wuest group also investigated this direct method to several aminooxy derivatized peptides [94]. It was concluded that the method works well with small peptides although it requires large amounts of peptide and tedious separation of the glucose–peptide conjugate also formed in the ligation reaction.

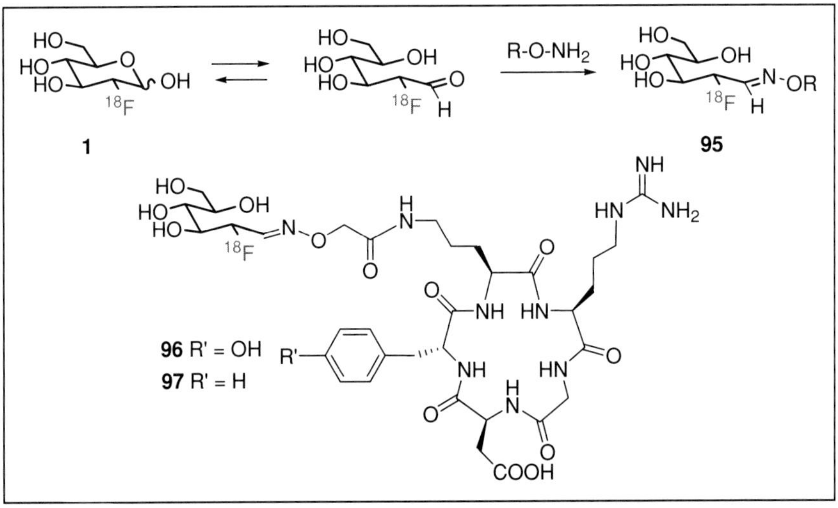

SCHEME 7.20 Oxime formation as a click coupling between FDG and a cyclic RGD peptide.

7.4.3 Glycosylation

Although not yet considered as a click chemistry method, glycosylation of sugar has been envisioned in radiochemistry. The construction of glucosylated 2-nitroimidazole has been reported [95]. 2-nitroimidazole derivatives are considered as markers of hypoxia. Under hypoxia, they are reduced and the resulting species bind to macromolecules of cells and then accumulate in hypoxic cells. This cell status is crucial in cancer treatment, oxygen-deprived tumor cells being less reactive to treatment, and the noninvasive detection of hypoxia is still a challenge. In a practical way, the *N*-glycosylation of imidazole with cold tetra-*O*-acetyl-2-deoxy-2-fluoro-glucose was achieved in 36% yield in acetonitrile at 60°C for 3 hours in the presence of $SnCl_4$ and $Hg(CN)_2$. When using the ^{18}F analog **98**, a higher yield of **99** (80%) was obtained in 60 minutes. (Scheme 7.21)

The direct glycosylation of the free hydroxyl of serine or threonine by the widely used acetylated [^{18}F]FDG **98** under the Lewis acid catalyst has been reported

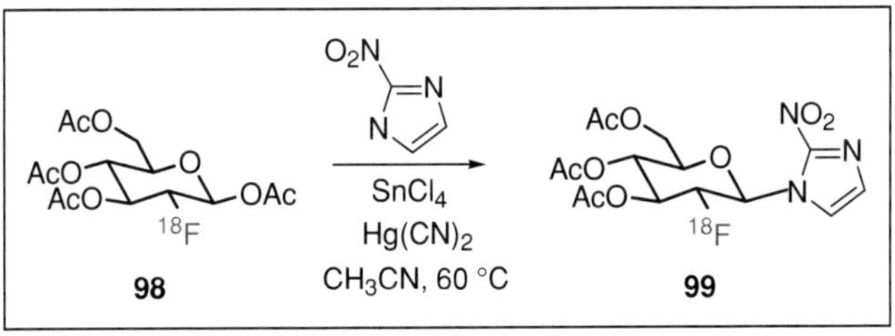

SCHEME 7.21 Glycosylation as a click coupling reaction between FDG and a nitroimidazole.

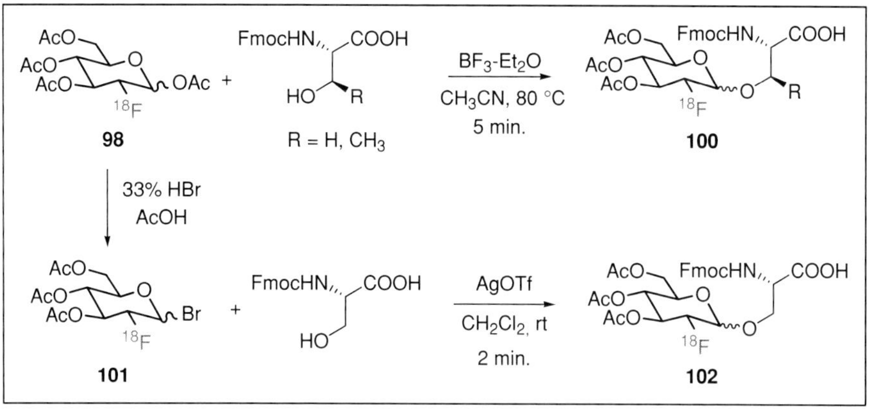

SCHEME 7.22 Glycosylation as a click coupling reaction between FDG and protected serine or threonine derivatives.

(Scheme 7.22) [96]. The best Lewis acid was BF_3-Et_2O in acetonitrile as the best solvent. It is interesting to note that the reaction was performed at 80°C and proceeded in good radiochemical yield to give **100**. *N*-glycosylation of lysine amino group cannot be achieved this way.

In a comparative study, the same group proposed the use of Koenigs–Knorr conditions [97]. Acetylated [^{18}F]FDG **98** was converted into the corresponding bromide **101** by treatment with 33% HBr/AcOH for 5 minutes. Glycosylation of Fmoc-Ser-OH was then achieved using silver triflate as the promoter in dichloromethane to give **102** in about 21% radiochemical yield together with the corresponding glycosyl ester. It is interesting to note that the reaction proceeded in 67% radiochemical yield with Z-Ser-OBn. From these results, it is clear that this methodology cannot be applied to peptide having unprotected carboxylic residues.

7.4.4 Disulfide Formation

Dissymmetrical disulfide bond formation introduced by Davis [98] may be regarded as a click reaction as it proceeds rapidly and with high selectivity in aqueous medium. A good example of the ligation of activated [^{18}F]FDG thioglycosides with cysteine containing peptides has been reported (Scheme 7.23) [99]. On treatment of **98** under classical bromination conditions, the radioactive bromide **101** was formed in 20 minutes in about 65% radiochemical yield. The phenylthiosulfonate formation was studied in detail and the use of a mixture of acetonitrile/DMF (4:1) at 70°C for 20 minutes gave the best corrected radiochemical yield (33%) in thioglycoside **103**. The coupling reaction with c(RGDfC) peptide proceeded rapidly in 50 mM tris buffer (pH 7.7) at room temperature for 15 minutes to give **104** in up to 95% radiochemical yield. This conjugate was found to be stable in human serum at 37°C.

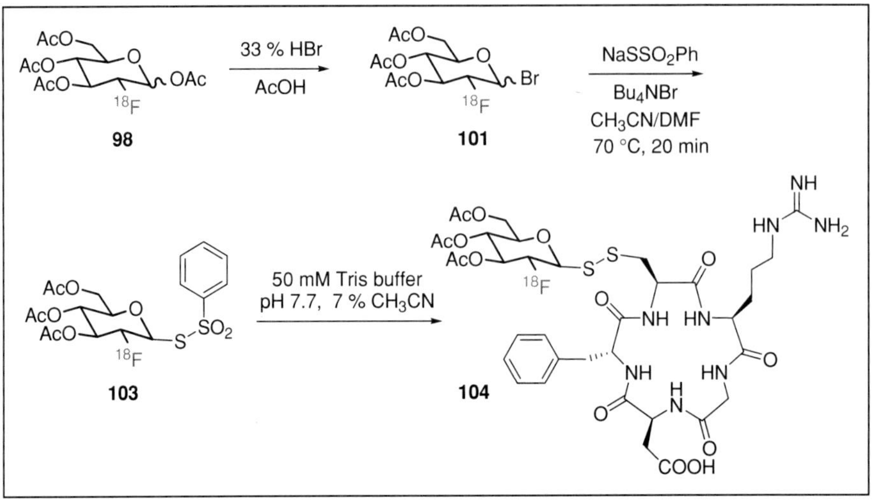

SCHEME 7.23 S-S click coupling reaction between FDG and a cyclic RGD peptide.

7.5 CONCLUSION AND OUTLOOK

Click chemistry is a significant advance in the elaboration of radiotracers where time and final product purification are important constraints. CuAAC, a fast and almost total reaction, has attracted considerable interest in the field of radiochemistry. It is a unique way to assemble a biomolecule and a carbohydrate either as a radioactive prosthetic group or as a helpful component to modify the pharmacokinetics of the radiotracer. This reaction gave high chemical and radiochemical yields and will certainly find several applications in the near future of molecular imaging. If the introduction of azide or alkyne functions on carbohydrates and nucleosides is well documented, thanks to a large body of work performed in the last five decades, the modification of peptide or nucleotide is still a challenging process. There is no doubt that efforts should be done to find efficient methods of derivatization of these rather sensitive molecules. Carbohydrates are also suitable for specific ligation methods relying on the presence of the anomeric hydroxyl or aldehydic function. Some interesting applications have been proposed but this new aspect of click chemistry must be thoroughly studied in the near future to add new ligation methods to the toolbox of radiochemists.

REFERENCES

1. http://www.technologyreview.com/Infotech/13060/.
2. Ametamey, S. M.; Honer, Schubiger, M. P. A. *Chem. Rev.* **2008**, *108*, 1501–1516.
3. Schirrmacher, R.; Wangler, C.; Schirrmacher, E. *Mini-Rev. Org. Chem.* **2007**, *4*, 317–329.

4. Mather, S. *Mol. Biosyst.* **2007**, *3*, 30–35.
5. Olberg, D.; Hjelstuen, O. *Curr. Top. Med. Chem.* **2010**, *10*, 1669–1679.
6. Kolb, H. C.; Finn, M. G.; Sharpless, K. B. *Angew. Chem. Int. Ed.* **2001**, *40*, 2004–2021.
7. Beuthien-Baumann, B.; Hamacher, K.; Oberdorfer, F.; Steinbach, J. *Carbohydr. Res.* **2000**, *327*, 107–118.
8. Mamat, C.; Ramenda, T.; Wuest, F. R. *Mini-Rev. Org. Chem.* **2009**, *6*, 21–34.
9. Glaser, M.; Robins, E. G. *J. Label. Compd. Radiopharm.* **2009**, *52*, 407–414.
10. Lei, M.; Tian, M.; Zhang, H. *Curr. Med. Imaging Rev.* **2010**, *6*, 33–41.
11. Walsh, J. C.; Kolb, H. *Chimia.* **2010**, *64*, 29–33.
12. Wängler, C.; Schirrmacher, R.; Bartenstein, P.; Wängler, B. *Curr. Med. Chem.* **2010**, *17*, 1092–1116.
13. Santoyo-Gonzalez, F.; Hernandez-Mateo, F. *Top. Heterocycl. Chem.* **2007**, *7*, 133–177.
14. Pieters, R. J.; Rijkers, D. T. S.; Liskamp, R. M. J. *QSAR Comb. Sci.* **2007**, *26*, 1181–1190.
15. Wilkinson, B. L.; Bornaghi, L. F.; Houston, T. A.; Poulsen, S.A. *Click Chemistry in Carbohydrate Based Drug Development and Glycobiology*, ed. S. P. Kaplan, Nova Science Publishers, New York, **2007**; pp. 57–102.
16. Tornøe, C. W.; Christensen, C.; Meldal, M. *J. Org. Chem.* **2002**, *67*, 3057–3064.
17. Rostovtsev, V. V.; Green, L. G.; Fokin, V. V.; Sharpless, K. B. *Angew. Chem. Int. Ed.* **2002**, *41*, 2596–2599.
18. Meldal, M.; Tornøe, C. W. *Chem. Rev.* **2008**, *108*, 2952–3015.
19. Saxon, E.; Bertozzi, C. R. *Science.* **2000**, *287*, 2007–2010.
20. van Swieten, P. F.; Leeuwenburgh, M. A.; Kessler, B. M.; Overkleeft, H. S. *Org. Biomol. Chem.* **2005**, *3*, 20–27.
21. Horne, W. S.; Stout, C. D.; Ghadiri, M. R. *J. Am. Chem. Soc.* **2003**, *125*, 9372–9376.
22. Bock, V. D.; Speijer, D.; Hiemstra, H.; van Maarseveen, J. H. *Org. Biomol. Chem.* **2007**, *5*, 971–975.
23. Miller, N.; Williams, G. M.; Brimble, M. A. *Int. J. Pept. Res. Ther.* **2010**, *16*, 125–132.
24. Slavin, S.; Burns, J.; Haddleton, D. M.; Becer, C. R. *Eur. Polymer J.* **2011**, *47*, 435–446.
25. Struthers, H.; Mindt, T. L.; Schibli, R. *Dalton Trans.* **2010**, *39*, 675–696.
26. Marik, J.; Sutcliffe, J. L. *Tetrahedron Lett.* **2006**, *47*, 6681–6684.
27. Ramenda, T.; Bergmann, R.; Wuest, F. *Lett. Drug Des. Discovery* **2007**, *4*, 279–285.
28. Glaser, M.; Arstad, E. *Bioconjugate Chem.* **2007**, *18*, 989–993.
29. Li, Z.-B.; Wu, Z.; Chen, K.; Chin, F. T.; Chen, X. *Bioconjugate Chem.* **2007**, *18*, 1987–1994.
30. Hausner, S. H.; Marik, J.; Gagnon, M. K. J.; Sutcliffe, J. L. *J. Med. Chem.* **2008**, *51*, 5901–5904.
31. Ross, T. L.; Honer, M.; Lam, P. Y. H.; Mindt, T. L.; Groehn, V.; Schibli, R.; Schubiger, P. A.; Ametamey, S. M. *Bioconjugate Chem.* **2008**, *19*, 2462–2470.
32. Thonon, D.; Kech, C.; Paris, J.; Lemaire, C.; Luxen, A. *Bioconjugate Chem.* **2009**, *20*, 817–823.
33. McConathy, J.; Zhou, D.; Shockley, S. E.; Jones, L. A.; Griffin, E. A.; Lee, H.; Adams, S. J.; Mach, R. H. *Mol. Imaging.* **2010**, *9*, 329–342.

34. Campbell-Verduyn, L. S.; Mirfeizi, L.; Schoonen, A. K.; Dierckx, R. A.; Elsinga, P. H.; Feringa, B. L. *Angew. Chem. Int. Ed.* **2011**, *50*, 11117–11120.
35. Wester, H. J.; Schottelius, M.; Scheidhauer, K.; Meisetschlager, G.; Herz, M.; Rau, F. C.; Reubi, J. C.; Schwaiger, M. *Eur. J. Nucl. Med. Mol. Imaging.* **2003**, *30*, 117–122.
36. Leisner, M.; Kessler, H.; Schwaiger, M.; Wester, H. J. *J. Label. Compd. Radiopharm.* **1999**, *42* (Suppl. 1), 549–551.
37. Wester, H.-J.; Schottelius, M.; Scheidhauer, K.; Reubi, J. C.; Wolf, I.; Schwaiger, M. *Eur. J. Nucl. Med. Mol. Imag.* **2002**, *29*, 28–38.
38. Schottelius, M.; Wester, H.-J.; Reubi, J. C.; Senekowitsch-Schmidtke, R.; Schwaiger, M. *Bioconjugate Chem.* **2002**, *13*, 1021–1030.
39. Haubner, R.; Wester, H. J.; Burkhart, F.; Senekowitsch-Schmidtke, R.; Weber, W.; Goodman, S. L.; Kessler, H.; Schwaiger, M. *J. Nucl. Med.* **2001**, *42*, 326–336.
40. Maschauer, S.; Prante, O. *Carbohydr. Res.* **2009**. *344*, 753–761.
41. Tsuchiya, T. *Adv. Carbohydr. Chem. Biochem.* **1990**, *48*, 91–277.
42. Nicolaou, K. C.; Ladduwahetty, T.; Randall, J. L.; Chucholowski, A. *J. Am. Chem. Soc.* **1986**, *108*, 2466–2467.
43. Vala, C. *Synform.* **2009**, A3–A4.
44. Dong, H.; Pei, Z.; Bystroem, S.; Ramstroem, O. *J. Org. Chem.* **2007**, *72*, 1499–1502.
45. Dong, H.; Pei, Z.; Angelin, M.; Bystroem, S.; Ramstroem, O. *J. Org. Chem.* **2007**, *72*, 3694–3701.
46. Fazio, F.; Bryan, M. C.; Blixt, O.; Paulson, J. C.; Wong, C.-H. *J. Am. Chem. Soc.* **2002**, *124*, 14397–14402.
47. Maschauer, S.; Einsiedel, J.; Haubner, R.; Hocke, C.; Ocker, M.; Haubner, H.; Kuwert, T.; Gmeiner, P.; Prante, O. *Angew. Chem. Int. Ed.* **2010**, *49*, 976–979.
48. Haubner, R.; Kuhnast, B.; Mang, C.; Weber, W. A.; Kessler, H.; Wester, H. J.; Schwaiger, M. *Bioconjugate Chem.* **2004**, *15*, 61–69.
49. Vala, C.; Chrétien, F.; Balentova, E.; Lamandé-Langle, S.; Chapleur, Y. *Tetrahedron Lett.* **2011**, *5*, 17–20.
50. Haubner, R.; Wester, H.; Reuning, U.; Senekowitsch-Schmidtke, R.; Diefenbach, B.; Kessler, H.; Stocklin, G.; Schwaiger, M. *J. Nucl. Med.* **1999**, *40*, 1061–1071.
51. Haubner, R.; Wester, H. J. *Curr. Pharm. Des.* **2004**, *10*, 1439–1455.
52. Schottelius, M.; Laufer, B.; Kessler, H.; Wester, H.-J. *Acc. Chem. Res.* **2009**, *42*, 969–980.
53. Kuijpers, B. H. M.; Groothuys, S.; Soede, A. C.; Laverman, P.; Boerman, O. C.; van Delft, F. L.; Rutjes, F. P. J. T. *Bioconjugate Chem.* **2007**, *18*, 1847–1854.
54. Shiozaki, M.; Mochizuki, T.; Hanzawa, H.; Haruyama, H. *Carbohydr. Res.* **1996**, *288*, 99–108.
55. Kuijpers, B. H. M.; Groothuys, S.; Keereweer, A. B. R.; Quaedflieg, P. J. L. M.; Blaauw, R. H.; van Delft Floris, L.; Rutjes, F. P. J. T. *Org. Lett.* **2004**, *6*, 3123–3126.
56. Gurrath, M.; Müller, G.; Kessler, H.; Aumailley, M.; Timpl, R. *Eur. J. Biochem.* **1992**, *210*, 911–921.
57. Mindt, T. L.; Struthers, H.; Brans, L.; Anguelov, T.; Schweinsberg, C.; Maes, V.; Tourwe, D.; Schibli, R. *J. Am. Chem. Soc.* **2006**, *128*, 15096–15097.
58. Benoist, E.; Coulais, Y.; Almant, M.; Kovensky, J.; Moreau, V.; Lesur, D.; Artigau, M.; Picard, C.; Galaup, C.; Gouin, S. G. *Carbohydr. Res.* **2011**, *346*, 26–34.

59. Mindt, T. L.; Schweinsberg, C.; Brans, L.; Hagenbach, A.; Abram, U.; Tourwe, D.; Garcia-Garayoa, E.; Schibli, R. *ChemMedChem.* **2009**, *4*, 529–539.

60. Struthers, H.; Spingler, B.; Mindt, T. L.; Schibli, R. *Chem. Eur. J.* **2008**, *14*, 6173–6183.

61. Kim, D. H.; Choe, Y. S.; Jung, K. H.; Lee, K. H.; Choi, J. Y.; Choi, Y.; Kim, B. T. *Arch. Pharm. Res.* **2008**, *31*, 587–593.

62. Dimitrakopoulou-Strauss, A.; Strauss, L. G. *Eur. J. Nucl. Med. Mol. Imaging* **2008**, *35*, 523–526.

63. Barwick, T.; Bencherif, B.; Mountz, J.; Avril, N. *Nucl. Med. Commun.* **2009**, *30*, 908–917.

64. Kolb, H. C.; Walsh, J. C.; Chen, K. WO 2006116629, 2006 Apr 27.

65. Ackermann, U.; O'Keefe, G.; Lee, S.-T.; Rigopoulos, A.; Cartwright, G.; Sachinidis, J. I.; Scott, A. M.; Tochon-Danguy, H. J. *J. Label. Compd. Radiopharm.* **2011**, *54*, 260–266.

66. Perman, J.; Sharma, R. A.; Bobek, M. *Tetrahedron Lett.* **1976**, 2427–2430.

67. Sirion, U.; Kim, H. J.; Lee, J. H.; Seo, J. W.; Lee, B. S.; Lee, S. J.; Oh, S. J.; Chi, D. Y. *Tetrahedron Lett.* **2007**, *48*, 3953–3957.

68. James, D.; Escudier, J.-M.; Amigues, E.; Schulz, J.; Vitry, C.; Bordenave, T.; Szlosek-Pinaud, M.; Fouquet, E. *Tetrahedron Lett.* **2010**, *51*, 1230–1232.

69. Mu, L.; Hohne, A.; Schubiger, P. A.; Ametamey, S. M.; Graham, K.; Cyr, J. E.; Dinkelborg, L.; Stellfeld, T.; Srinivasan, A.; Voigtmann, U.; Klar, U. *Angew. Chem. Int. Ed.* **2008**, *47*, 4922–4925.

70. Hohne, A; Yu, L; Mu, L; Reiher, M; Voigtmann, U; Klar, U; Graham, K; Schubiger, P. A; Ametamey, Simon M. *Chem. Eur. J.* **2009**, *15*, 3736–3743.

71. Kostikov, A. P.; Iovkova, L.; Chin, J.; Schirrmacher, E.; Wängler, B.; Wängler, C.; Jurkschat, K.; Cosa, G.; Schirrmacher, R. *J. Fluorine Chem.* **2011**, *132*, 27–34.

72. Balentova, E.; Collet, C.; Lamandé-Langle, S.; Chrétien, F.; Thonon, D.; Aerts, J.; Lemaire, C.; Luxen, A.; Chapleur, Y. *J. Fluorine Chem.* **2011**, *132*, 250–257.

73. Schulz, J.; Vimont, D.; Bordenave, T.; James, D.; Escudier, J.-M.; Allard, M.; Szlosek-Pinaud, M.; Fouquet, E. *Chem. Eur. J.* **2011**, *17*, 3096–3100.

74. Inkster, J. A. H.; Guérin, B.; Ruth, T. J.; Adam, M. J. *J. Label. Compd. Radiopharm.* **2008**, *51*, 444–452.

75. Inkster, J. A.; Adam, M. J.; Storr, T.; Ruth, T. J. *Nucleosides, Nucleotides. Nucleic Acids.* **2009**, *28*, 1131–1143.

76. Tavitian, B. *Gut.* **2003**, *52* (Suppl. 4), iv40–iv47.

77. Mercier, F.; Paris, J.; Kaisin, G.; Thonon, D.; Flagothier, J.; Teller, N.; Lemaire, C.; Luxen, A. *Bioconjugate Chem.* **2011**, *22*, 108–114.

78. Lemaire, C. F.; Aerts, J. J.; Voccia, S.; Libert, L. C.; Mercier, F.; Goblet, D.; Plenevaux, A. R.; Luxen, A. J. *Angew. Chem. Int. Ed.* **2010**, *49*, 3161–3164.

79. Wester, H.-J.; Schottelius, M.; Poethko, T.; Bruus-Jensen, K.; Schwaiger, M. *Cancer Biother. Radiopharm.* **2004**, *19*, 231–244.

80. Iddon, L.; Leyton, J.; Indrevoll, B.; Glaser, M.; Robins, E. G.; George, A. J. T.; Cuthbertson, A.; Luthra, S. K.; Aboagye, E. O. *Bioorg. Med. Chem. Lett.* **2011**, *21*, 3122–3127.

81. Albert, R.; Marbach, P.; Bauer, W.; Briner, U.; Fricker, G.; Bruns, C.; Pless, J. *Life Sci.* **1993**, *53*, 517–525.

82. Schottelius, M.; Poethko, T.; Herz, M.; Reubi, J.-C.; Kessler, H.; Schwaiger, M.; Wester, H.-J. *Clin. Cancer Res.* **2004**, *10*, 3593–3606.

83. Bapst, J.-P.; Calame, M.; Tanner, H.; Eberle, A. N. *Bioconjugate Chem.* **2009**, *20*, 984–993.
84. Maes, V.; Tourwe, D. *Int. J. Pept. Res. Ther.* **2006**, *12*, 197–202.
85. Brans, L.; Maes, V.; Garcia-Garayoa, E.; Schweinsberg, C.; Daepp, S.; Blauenstein, P.; Schubiger, P. A.; Schibli, R.; Tourwe, D. A. *Chem. Biol. Drug Des.* **2008**, *72*, 496–506.
86. Schweinsberg, C.; Maes, V.; Brans, L.; Blauenstein, P.; Tourwe, D. A.; Schubiger, P. A.; Schibli, R.; Garcia Garayoa, E. *Bioconjugate Chem.* **2008**, *19*, 2432–2439.
87. Maes, V.; Brans, L.; Schweinsberg, C.; Garcia-Garayoa, E.; Blauenstein, P.; Schubiger, P. A.; Tourwe, D. *Adv. Exp. Med. Biol.* **2009**, *611*, 409–410.
88. Renaudet, O.; Dumy, P. *Org. Lett.* **2003**, *5*, 243–246.
89. Poethko, T.; Schottelius, M.; Thumshirn, G.; Hersel, U.; Herz, M.; Henriksen, G.; Kessler, H.; Schwaiger, M.; Wester, H. J. *J. Nucl. Med.* **2004**, *45*, 892–902.
90. Schottelius, M.; Berger, S.; Poethko, T.; Schwaiger, M.; Wester, H. J. *Bioconjugate Chem.* **2008**, *19*, 1256–1268.
91. Wuest, F.; Berndt, M.; Bergmann, R.; van den Hoff, J.; Pietzsch, J. *Bioconjugate Chem.* **2008**, *19*, 1202–1210.
92. Namavari, M.; Cheng, Z.; Zhang, R.; De, A.; Levi, J.; Hoerner, J. K.; Yaghoubi, S. S.; Syud, F. A.; Gambhir, S. S. *Bioconjugate Chem.* **2009**, *20*, 432–436.
93. Hultsch, C.; Schottelius, M.; Auernheimer, J.; Alke, A.; Wester, H. J. *Eur. J. Nucl. Med. Mol. Imaging* **2009**, *36*, 1469–1474.
94. Wuest, F.; Hultsch, C.; Berndt, M.; Bergmann, R. *Bioorg. Med. Chem. Lett.* **2009**, *19*, 5426–5428.
95. Patt, M.; Sorger, D.; Scheunemann, M.; Stöcklin, G. *Appl. Radiat. Isot.* **2002**, *57*, 705–712.
96. Maschauer, S.; Pischetsrieder, M.; Kuwert, T.; Prante, O. *J. Label. Compd. Radiopharm.* **2005**, *48*, 701–719.
97. Maschauer, S.; Kuwert, T.; Prante, O. *J. Label. Compd. Radiopharm.* **2006**, *49*, 101–108.
98. Gamblin, D. P.; Garnier, P.; Ward, S. J.; Oldham, N. J.; Fairbanks, A. J.; Davis, B. G. *Org. Biomol. Chem.* **2003**, *1*, 3642–3644.
99. Prante, O.; Einsiedel, J.; Haubner, R.; Gmeiner, P.; Wester, H.-J.; Kuwert, T.; Maschauer, S. *Bioconjugate Chem.* **2007**, *18*, 254–262.

8

BIOORTHOGONAL REACTIONS FOR LABELING GLYCOCONJUGATES

Frédéric Friscourt and Geert-Jan Boons

8.1 INTRODUCTION

There is a growing appreciation that posttranslational modifications, such as protein glycosylation, dramatically increase polypeptide complexity and function [1–10]. For example, almost all cell surface and secreted proteins are modified by covalently linked carbohydrate moieties, and the glycan structures on these glycoproteins have been implicated as essential mediators in processes such as protein folding, cell signaling, fertilization, embryogenesis, neuronal development, hormone activity, and the proliferation of cells and their organization into specific tissues. In addition, overwhelming data supports the relevance of glycosylation in pathogen recognition, inflammation, innate immune responses, and the development of autoimmune diseases and cancer. The importance of protein glycosylation is also underscored by the developmental abnormalities observed in a growing number of human disorders known as Congenital Disorders of Glycosylation caused by defects in the glycosylation machinery.

Almost all naturally occurring protein glycosylations can be classified as either *N*-glycosides whereby *N*-acetyl glucosamine is linked to the amide side chain of an asparagine, or as *O*-glycosides whereby a saccharide is linked to the hydroxyl of serine, threonine, or tyrosine. The biosynthesis of *N*-linked oligosaccharides occurs in the endoplasmic reticulum (ER) and Golgi complex. In the ER, a dolichol-linked $Glc_3Man_9GlcNAc_2$ oligosaccharide precursor is biosynthesized and transferred *en bloc* to an Asn-X-Ser/Thr sequon on newly synthesized polypeptides through the

Click Chemistry in Glycoscience: New Developments and Strategies, First Edition.
Edited by Zbigniew J. Witczak and Roman Bielski.

action of the multi-subunit oligosaccharide transferase complex [11, 12]. Subsequent trimming and processing of the transferred oligosaccharide results in a $Man_3GlcNAc_2$ core structure, which is transported to the medial stacks of the Golgi complex where maturation of the oligosaccharide gives rise to extreme structural diversity [13–16]. This complexity of *N*-glycan structures is largely based on the cell-specific expression of a collection of glycosyl transferases that specify the extension of oligosaccharide structures onto the trimmed $Man_3GlcNAc_2$ core structure. The switch from structural uniformity in the ER to diversification in the Golgi complex coincides with a marked change in glycan function. In the early secretory pathway, the glycans have a common role in the promotion of protein folding, quality control, and certain sorting events. This is in contrast to their roles in the Golgi complex, in which they are modified to perform a wide spectrum of functions displayed by the mature glycoproteins.

Despite their importance, in most cases the precise roles that glycans play in biological systems is not well understood because of their underpinning complexity. Over the past decade, integrated approaches have been developed, which are broadly termed glycomics, which are aimed at unraveling structure–function relationships of complex carbohydrates. Key glycomics technologies include mass spectrometric profiling of glycan structures isolated from cells and tissues [17–19], glyco-gene microarray technology for measuring the expression levels of glycoenzymes and glycan-binding proteins and screening for glycan–protein interactions using glycan and lectin array technologies [20–25]. The diverse data sets generated by the use of these technologies are beginning to provide an understanding of the fundamental structure–function relationships of glycans. Critical components that enable this process are bioinformatics platforms that store, integrate, process, and disseminate the data in a meaningful way [26–28].

The use of tandem MS for glycomics is driven by the need to obtain structural information of glycans present in serum/plasma or tissue samples to understand metabolic or disease processes and discover new biomarkers [17–19]. Strategies for acquisition and interpretation of multistage MS have been most fully developed for permethylated glycans. The advantage to this approach is that tandem mass spectrometric dissociation of a glycosidic bond leaves a site that lacks a methyl group that is clearly indicated by mass. It is possible to differentiate some types of positional isomers based on the formation of specific product ion types. However, the assignment of glycan structures is very challenging due to the isobaric nature of glycans, that is, different structures with the identical molecular weights.

Glycoarray technology is a key tool for glycomics that has as a distinct advantage that only minute amounts of precious oligosaccharides are required while allowing fast, quantitative, systematic identification and characterization of carbohydrate–protein interactions [20–25]. In addition, a glycoarray format presents glycans in a multivalent fashion, which is often an important requirement for high affinity binding. In particular, the glycan array developed by the consortium of functional glycomics (CFG) has found wide utility in profiling interactions with carbohydrate binding proteins, growth factors, pathogen- and cancer-induced antibodies, viruses, and bacteria [26, 27]. The saccharides of this array are prepared by chemoenzymatic approaches and comprise the most common terminal saccharide motifs found on *N*- and *O*-linked saccharides.

An emerging technology that makes it possible to isolate and visualize glycans exploits promiscuity of the biosynthetic machinery, which makes it possible to incorporate monosaccharides that have a unique chemical functionality (the reporter) into glycoconjugates of living cells. The chemical reporter of these glycans can then be reacted with a probe linked to a complementary bio-orthogonal functional group. The mutually selective chemical reactivity of the two functional groups ensures that only the metabolically labeled glycans are targeted for reaction. Bertozzi and coworkers have coined this technology "bio-orthogonal chemical reporter" strategy, which is the topic of this review [29–35].

8.2 KETONE-BASED BIOORTHOGONAL LABELING OF GLYCOCONJUGATES

In the late eighties, it was found that sialyltransferases were remarkably permissive for substrates bearing certain functional groups and, for example, even bulky fluorophores appended to CMP-Neu5Ac could be biosynthetically incorporated into glycans [36, 37]. These modified nucleotide sugars were, however, highly polar and hence not membrane permeable and therefore could not readily be employed to label oligosaccharides of living cells.

Reutter and coworkers demonstrated that synthetic *N*-acyl-modified D-mannosamines can be taken up by cells and efficiently metabolized to the respective *N*-acyl-modified neuraminic acids, which can then be incorporated into cell surface glycoconjugates [38]. For example, derivatives such as *N*-propanoyl, *N*-butanoyl, *N*-pentanoyl, *N*-hexanoyl, and *N*-crotonoyl can be metabolized by the promiscuous sialic acid biosynthetic pathway with an efficiency of 10–85%. The use of this technology has revealed the unexpected functions of the *N*-acyl side chain of sialic acids in biological processes such as host-pathogen interactions, cell adhesion and proliferation, and immune responses [39–42].

An important conceptual advance was the discovery that monosaccharide analogs having a unique chemical functionality can also be employed in metabolic glycoengineering. The attraction of the use of such derivatives is the ability to perform chemoselective ligations on surfaces of living cells [43]. A stringent requirement of these reactions is the need of two participating functional groups to react under physiological conditions without interference with coexisting functionalities [44].

It has been shown that the ketone-containing ManNLev can be incorporated into the sialic acid pathway and then employed for bioorthogonal reactions [45]. Although aldehydes and ketones are intracellular biosynthetic precursors, they are absent from cell surfaces and hence in this environment can serve as unique chemical reporters. For example, keto sugars expressed on cell surfaces can be reacted with aminooxy and hydrazide probes to form oximes and hydrazones, respectively [45, 46]. In this way, various tags such as Flag peptides, biotin, and fluorescent or fluorogenic molecules have been linked to the surface of living cells. The attraction of these conjugation approaches is that the tagged glycans can be enriched for further analysis by, for example, mass spectrometry, detected by flow cytometry for quantitative studies, or visualized by confocal microscopy to study localization and trafficking.

It has been found that hydrazones hydrolyze when internalized into vesicular compartments where the pH is relatively low. This chemical lability can be exploited for drug [47], gene [48], or toxin delivery [45]. It can also be employed for loading cell with an MRI contrast agent [49]. On the other hand, oximes are much more acid resistant and can be recycled after internalization [47].

The use of ketones as chemical reporter has a number of drawbacks such as the requirement to perform the ligation reaction under mild acidic conditions and a limited possibility to extend the technology to a wide range of sugar derivatives. Paulson and coworkers have pioneered an alternative approach for introducing aldehydes into cell-surface sialic acid residues by mild periodate oxidation followed by an aniline catalyzed reaction with aminooxy-biotin to give biotinylated glycoconjugates (Scheme 8.1). Importantly, it has been found that aniline accelerates oxime formation by approximately 10-fold, and furthermore can be performed under neutral conditions [50, 51]. Reagents to implement the periodate oxidation aniline catalyzed oxime ligation (PAL) are commercially available. Furthermore, the use of aniline makes it possible to reduce the amount of aminooxy reagent by 10- to 50-fold compared to what typically is employed in uncatalyzed reactions.

Selective oxidation of sialic acids has also been employed for enrichment of glycoproteins for analysis by mass spectrometry. In the most advanced approach,

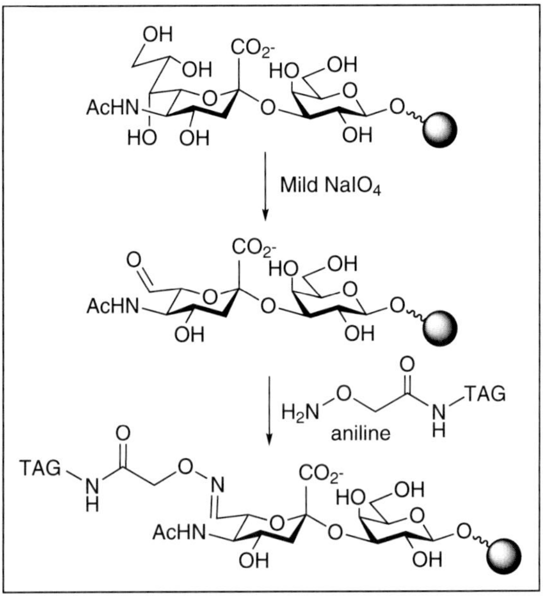

SCHEME 8.1 Covalent labeling of sialylated glycoconjugates by mild periodate cleavage followed by aniline mediated oxime ligation. The tag can be biotin or a fluorescent probe.

sialylated glycoproteins were selectively periodate-oxidized, captured on hydrazide beads, trypsinized and the released peptides analyzed by mass spectrometry for protein identification [52]. Next, the remaining glycopeptides were released by acid hydrolysis of sialic acid glycosidic bonds and the released compounds analyzed by MS for glycan identification. It is to be expected that this approach will be more facile when aniline is employed to catalyze the oxime formation.

8.3 AZIDO- AND ALKYNE-BASED STRATEGIES FOR BIOORTHOGONAL LABELING OF GLYCOCONJUGATES

Azides, which are extremely rare in biological systems, are emerging as a powerful chemical reporter [30–33, 53, 54]. The azido group is small and therefore only minimally perturbs a substrate. Furthermore, a number of attractive reaction partners for azides have been developed that are compatible with biological systems.

A number of modified monosaccharides have been employed to introduce azides into cell surface glycoconjugates (Fig. 8.1) [33]. In this respect, sialic acid-containing glycans have been labeled by feeding cell *N*-azidoacetylmannosamine (ManNAz) [44], which is an analog of the biosynthetic precursor *N*-acetylmannosamine (ManNAc) or by employing *N*-azidoacetylneuraminic acid (SiaNAz) [55] or 9-azido-*N*-acetylneuraminic acid [56]. Mucin-type glycans have been labeled with azido functions using *N*-azidoacetylgalactosamine (GalNAz) [57]. This unnatural sugar is

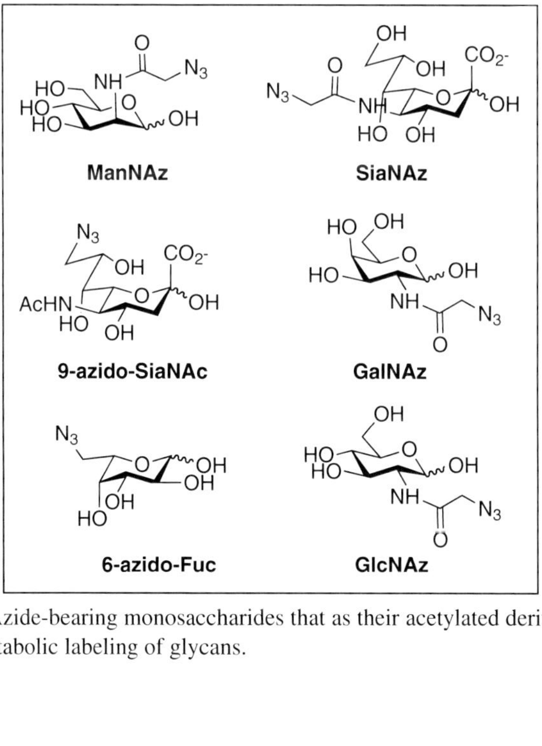

FIGURE 8.1 Azide-bearing monosaccharides that as their acetylated derivatives have been employed for metabolic labeling of glycans.

processed by the GalNAc salvage pathway to form the intermediate uridine diphospho UDP-GalNAz, which is utilized by a family of polypeptide GalNAc transferases in the Golgi compartment, resulting in the formation of mucins that have a GalNAz moiety.

Modified fucose derivatives have been employed as a chemical reporter [58–60]. In this case, the salvage pathway converts the fucose analogs into the corresponding GDP-fucose, which is then employed by fucosyltransferases for the biosynthesis of glycoconjugates. It has been found that the salvage pathway and fucosyltransferases tolerate modifications at C-6 of fucose. 6-Azido-6-deoxy-fucose was the first sugar to be investigated as a chemical reporter, however, it was found that this derivative exhibited some cytotoxicity. Fortunately, 6-alkynylfucose (FucAl) is accepted by the fucose salvage pathway and exhibits lower toxicity. The alkyne group provides a bioorthogonal reporter that can be chemoselectively labeled with azido-containing probes using Cu^{I}-catalyzed $[3+2]$ cycloadditions, which will be described in detail below [58–60]. It has also been found that 4-pentynoyl mannosamine (Ac_4ManNAl) is metabolically incorporated into cultured cells and mice with greater efficiency than Ac_4ManNAz [61]. This result underscores the sensitivity of sialic acid biosynthetic enzymes to subtle differences in the *N*-acyl structures of ManNAc analogs.

O-GlcNAc-modified proteins, which occur in the cytosol and nucleus, have also been labeled with bioorthogonal chemical reporters by using per-*O*-acetylated *N*-azidoacetylglucosamine (Ac_4GlcNAz) [62]. This azidosugar is modified by the GlcNAc salvage pathway enzymes to form UDP-GlcNAz, which is used as a substrate by the cytosolic *O*-GlcNAc transferase. More recently, it was shown that GalNAz can be converted by endogenous mammalian biosynthetic enzymes to UDP-GalNAz and then epimerized to UDP-*N*-GlcNAz. *O*-GlcNAc transferase can then append the azidosugar onto its native substrates, which can then be detected by covalent labeling using azide-reactive chemical probes [63]. This approach was employed for affinity-purification and identification of a large number of *O*-GlcNAc-modified proteins.

Another attractive approach to study *O*-GlcNAc modification of proteins exploits an engineered galactosyltransferase enzyme to selectively label *O*-GlcNAc proteins with an azide-biotin tag [64, 65]. The tag permits enrichment of low-abundance *O*-GlcNAc species from complex mixtures and localization of the modification to short amino acid sequences. Using this approach, changes in *O*-GlcNAc glycosylation on several proteins involved in the regulation of transcription and mRNA translocation were detected. Also, it provided evidence that *O*-GlcNAc glycosylation is dynamically modulated by excitatory stimulation of the brain *in vivo*.

Glycoconjugates containing azido or alkyne moieties can also be introduced in living systems by enzymatic transformations, and, for example, Wu and coworkers [66] took advantage of a mutant *Helicobacter pylori* α(1,3)-fucosyltransferase, which converts LacNAc to the Le^{X} trisaccharide by transferring a fucose residue from GDP-fucose to the C-3 hydroxyl of the GlcNAc in the LacNAc unit. This enzyme exhibits promiscuity for GDP-fucose, and analogs containing a C-6 azido or alkyne moiety at fucose are also well tolerated by this enzyme thereby providing opportunities for installing these moieties in LacNAc-containing glycoconjugates. Lec2 cells, which express complex and hybrid *N*-glycans mainly terminating in LacNAc residues, could be labeled with alkynyl tags by treating them with α(1,3)FucT and GDP-6-alkynylfucose (GDP-FucAl), and detected by BTTES-promoted CuAAC [67].

8.4 STAUDINGER LIGATION

The Bertozzi group disclosed the first example of the use of azide as a chemical reporter by exploiting a modified Staudinger reaction. The Staudinger reaction between an azide and a phosphine proceeds smoothly in water to give a primary amine and the corresponding phosphine oxide (Scheme 8.2a) [68]. To avoid hydrolysis of the aza-ylide intermediate of the Staudinger reaction, a phosphine reagent was designed that contains a strategically placed ester functionality, which serves as an electrophilic trap to capture the nucleophilic aza-ylide intermediate by an intramolecular cyclization, leading to the formation of a stable amide bond (Scheme 8.2b) [44]. The usefulness of this reagent for bioorthogonal chemistry was demonstrated by metabolic labeling of Jurkat cells with $Ac_4ManNAz$ (20 mM) for 3 days to install azido-containing cell surface glycoconjugates. The latter moieties could be reacted with a biotinylated phosphine reagent and the resulting reaction products could be visualized by staining with fluorescein isothiocyanate-labeled avidin (FITC-avidin). The azido-labeled cells showed a marked increase in fluorescence intensity when reacted with phosphine (1 mM) compared to cells that were not exposed to

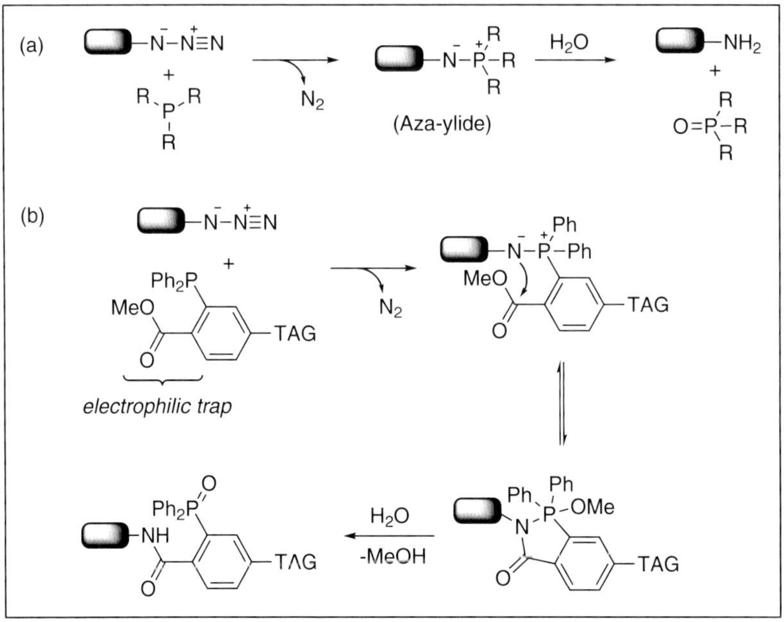

SCHEME 8.2 (a) Staudinger reaction between azides and phosphines; (b) Staudinger ligation of azides with triarylphosphines.

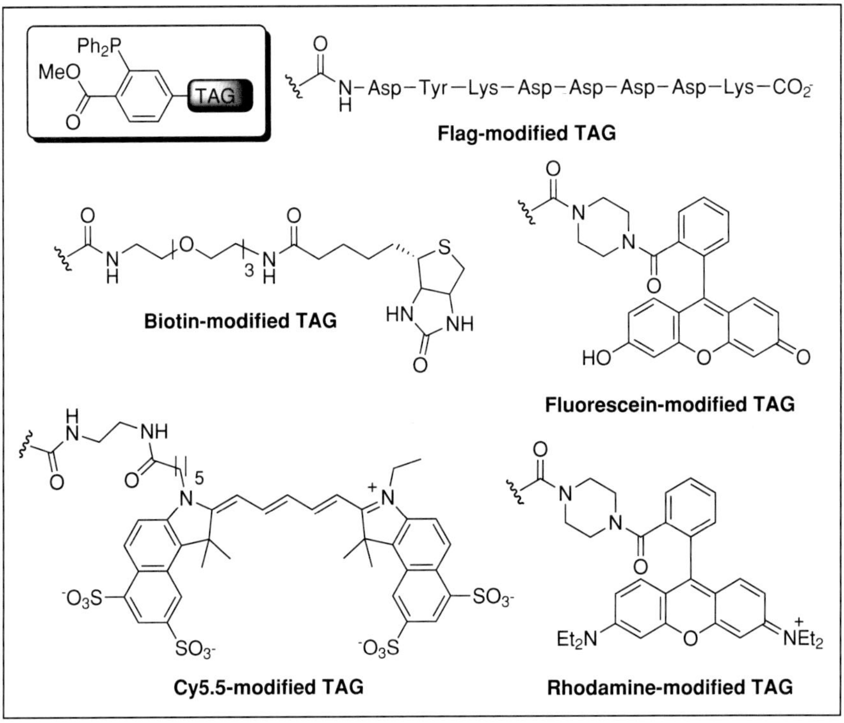

FIGURE 8.2 Various triarylphosphine-TAG conjugates for Staudinger Ligation.

ManNAz. The Staudinger ligation protocol was extended to C57BL/6 mice which were administered intraperitoneally daily doses of Ac_4ManNAz (300 mg.kg^{-1}) for 7 days [69]. Next, a phosphine probe comprising a Flag peptide was administered (Phos–Flag, 16 μmol), and 90 minutes later, the mice were euthanized, splenocytes isolated, and cell-surface Flag epitopes probed by FITC–anti-Flag antibody. As expected, only splenocytes isolated from mice treated with Ac_4ManNAz and Phos–Flag displayed a significant increase in fluorescence intensity, indicating a successful Staudinger ligation *in vivo*. However, the labeling intensity was significantly reduced when splenocytes were harvested after prolonged exposure to Phos–Flag, (12 hours) and no significant labeling was detected after 24 hours introduction time. The loss of detection was attributed to degradation of the Flag peptide, and this problem could be addressed by employing fluorescently labeled phosphine probes (Fig. 8.2) [70]. In particular, the near-infrared emitter Cy5.5–phosphine conjugate was found to be superior compared to fluorescein– and rhodamine–phosphine conjugates, which was attributed to a greater solubility allowing a more efficient removal of excess phosphine conjugate thereby increasing the signal-to-noise ratio.

The Staudinger ligation has been extended to a phosphine that becomes fluorescent upon reaction with azides (Scheme 8.3) [71]. In this case, the lone pair of electrons of the phosphine moiety quenches the excited state of the fluorophore,

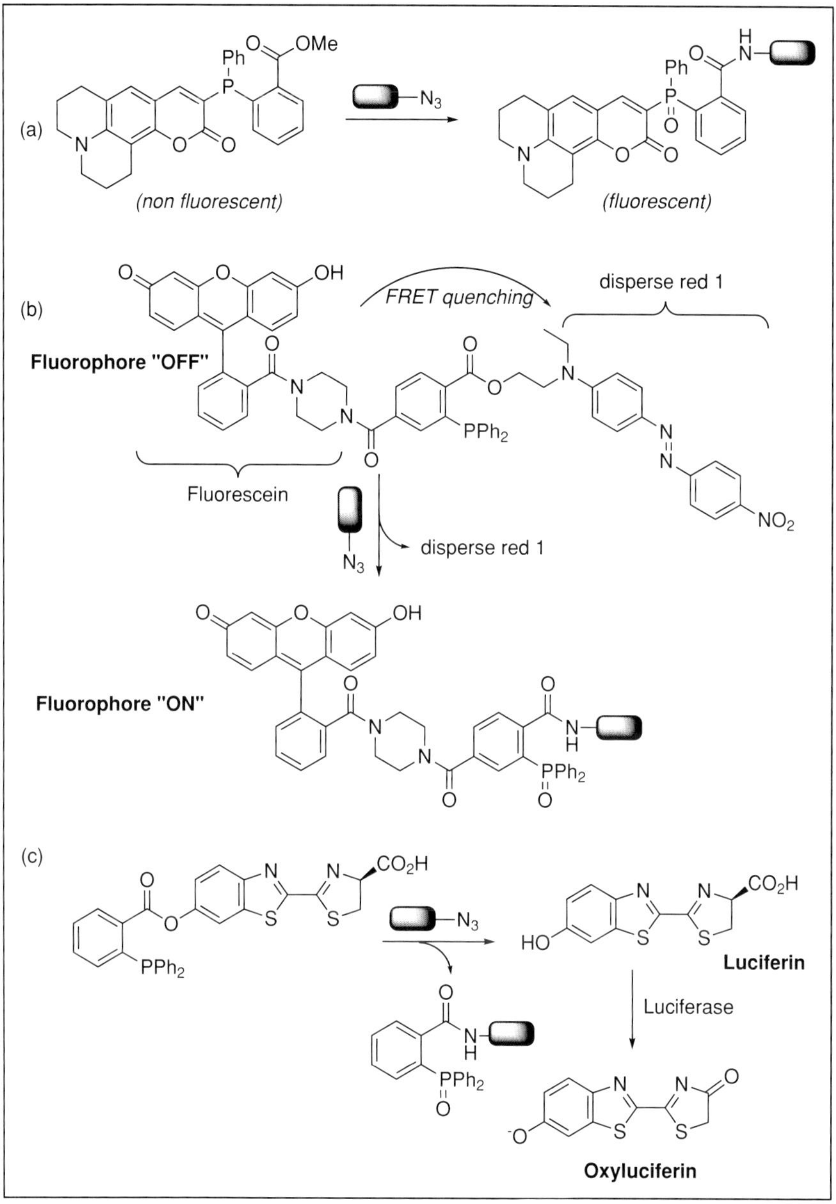

SCHEME 8.3 Fluorogenic and bioluminescent triarylphosphines for Staudinger Ligation.

rendering the reagent non-fluorescent. Staudinger ligation results in oxidation of the phosphine thereby enhancing fluorescence intensity by 60-fold. Despite attractive features of fluorogenic phosphines, this class of reagents is prone to nonspecific air oxidation resulting in high fluorescence background in cell-based experiments. To address this side effect, a more elaborate FRET-based fluorogenic phosphine was developed [72], which comprises a phosphine-tethered fluorophore (fluorescein) that is quenched intramolecularly by an ester-linked Förster resonance energy transfer (FRET) quencher disperse red 1. Upon Staudinger ligation, the ester is cleaved, resulting in the release of the quencher, leading to a concomitant fluorescence increase of the labeled molecule.

While fluorogenic phosphines have partially solved the problem of low signal-to-noise ratios, the issue of tissue autofluorescence has complicated efforts to image glycans of animals. In this respect, bioluminescence imaging (BLI) is gaining appreciation for real-time animal imaging, since the total absence of tissue luminescence confers negligible background. Recently, a phosphine–luciferin conjugate was reported for real-time BLI of azidosugars of live cells [73]. Upon Staudinger ligation, ester bound luciferin is released which can then freely diffuse into the cell where it is converted into the bioluminescent oxyluciferin by luciferase.

8.5 Cu^I-CATALYZED AZIDE–ALKYNE CYCLOADDITIONS (CuAAC)

The independent discovery by Sharpless [74] and Meldal [75] that the Huisgen's 1,3-dipolar cycloaddition of azides and alkynes can be catalyzed by Cu^I has revolutionized the field of bioorthogonal ligation. The resulting 1,4-disubstituted triazole adducts are formed in high yields with exclusive regioselectivity and have exceptional high functional group tolerance making it possible to modify biomolecules such as proteins, carbohydrates, lipids, and oligonucleotides. The rate of cycloaddition can be further accelerated by employing Cu^I-stabilizing ligands such as tris[(1-benzyl-1*H*-1,2,3-triazol-4-yl)methyl]amine (TBTA) [76] and the more polar tris(3-hydroxypropyltriazolylmethyl)amine (THPTA) [77].

Despite many attractive features of CuAAC, the cytotoxicity of Cu^I hampers applications to living systems. To address this issue, a series of water-soluble chelating ligands have been developed that make the CuAAC more biocompatible (Fig. 8.3) [78–82]. In particular, 2-4-(bis1-tert-butyl-1*H*-1,2,3-triazol-4-yl)methylamino(methyl-1*H*-1,2,3-triazol-1-yl)ethanesulfonic acid (BTTES) exhibits an appropriate balance between reactivity and solubility and is effective in promoting the CuAAC in aqueous media [78]. While its bulky, *t*-Bu groups enhance the rate of cycloaddition (three times faster than TBTA), the sulfate moiety minimizes cell permeability of the coordinated copper. BTTES-Cu^I-catalyzed click reactions have been successfully applied for the rapid detection of alkyne-modified sialylated glycans of live cell surface with robust labeling achieved within one minute [78]. The biocompatibility of the BTTES-Cu^I system was further demonstrated by *in vivo* imaging of fucosylated glycans dynamic changes during zebrafish early development. However, in this study some toxicity was observed and some of the treated embryos exhibited minor developmental defects [78]. To overcome this shortcoming, Wu and coworkers developed BTTAA [80], which is a carboxylic acid analog of BTTES, and the use of this reagent did not lead to developmental defects. These new chelating ligands are paving the way for rapid, noninvasive imaging of biomolecules in living organisms.

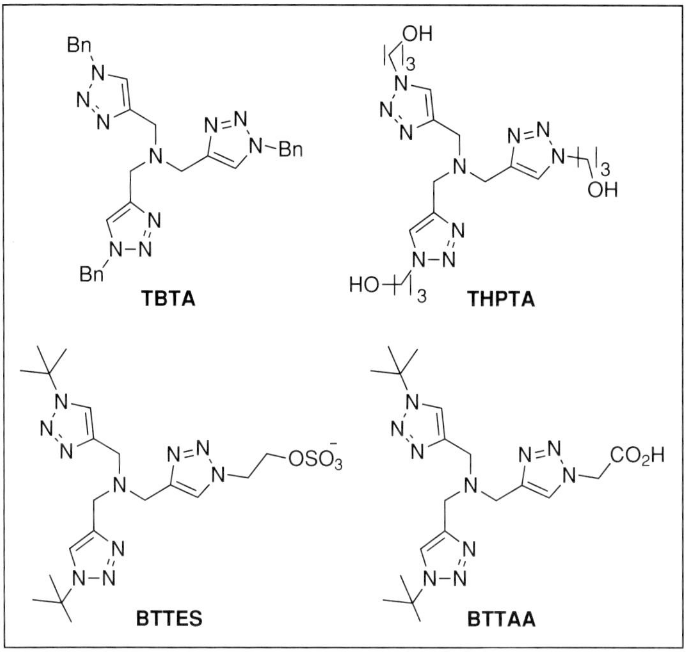

FIGURE 8.3 Various chelating ligands for biocompatible CuAAC.

Click-activated fluorogenic probes have been developed to address the problem of background labeling (Scheme 8.4) [59]. In this approach, the click product generates a strong detectable signal whereas the starting materials remain traceless. The fluorogenic probe design was based on 4-amino-1,8-naphthalimide, which absorbs light in the visible region and emits at long wavelengths (λ_{max} 540–550 nm) [59]. It was postulated that the fluorescent signal of such a compound could be modulated by the formation of a triazole ring because electron-donating groups at the C-4 position were known to strongly affect fluorescent properties. Indeed, a 1,8-naphthalimide derivative, which has an azido moiety at the C-4 position, has a very low fluorescence

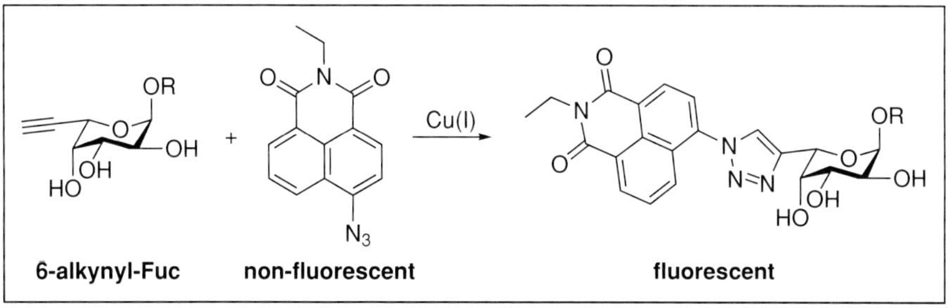

SCHEME 8.4 Fluorescence of 4-azido-1,8-naphthalimide is quenched by the azido moiety. Upon formation of a triazole, the derivative becomes fluorescent.

intensity. However, the emission maximum of the click product is at 422 nm with a quantum yield of 0.29 when excited at 357 nm.

Metabolic labeling with alkynyl containing monosaccharides has also been employed for glycoprotein identification and glycan mapping [83]. In this approach, glycoconjugates of cells are metabolically labeled with an alkynyl sugar probe, which can be selectively reacted with a biotin azide reagent using a Cu^{I}-catalyzed $[3+2]$ azide–alkyne cycloaddition. Introduction of a biotin moiety permits enrichment of glycoproteins by affinity capture with immobilized streptavidin. Tandem protein ID and glycan site mapping were carried out on-bead. Thus, first, non-glycosylated peptides were harvested by tryptic digestion, allowing protein identification. The remaining resin bound glycopeptides were analyzed by treatment with *N*-glycosidase F (PNGase), which hydrolyzes the amide bond between the biotinylated glycan and Asn residue of the bound peptide. The shift from Asn to Asp at formerly glycosylated sites can be identified by a differential modification of $+1$ Da on Asn in SEQUEST searches of MS data. The approach, which was applied to analyze the sialylated *N*-linked glycoproteins of prostate cancer cells, offers a method to identify proteins and map glycosylation sites of glycoproteins that carry a specific monosaccharide such as sialic acid. However, it does not enable the elucidation of the nature of the glycan at a particular glycosylation site.

8.6 STRAIN PROMOTED AZIDE–ALKYNE CYCLOADDITIONS (SPAAC)

Alkynes can be activated by ring strain and for example, cyclooctynes undergo $[3+2]$ cycloadditions with azides at ambient temperature without the need for a metal catalyst (Fig. 8.4) [84–86]. The increased reactivity of cyclooctynes has been attributed to ground-state destabilizing caused by the triple bond, however, this model has been challenged with the introduction of the concept of dipole distortion [87–89]. The strain-promoted cycloaddition has been used for labeling biomolecules without observable cytotoxicity [85]; however, the slow rate of reaction has limited the scope of the approach [90]. Fortunately, it has been found that appending electron-withdrawing groups to the cyclooctyne ring can increase rates of strain-promoted cycloadditions. For example, reactions of difluorinated cyclooctyne (DIFO) with azides proceed 60 times faster compared to an unsubstituted cyclooctyne [91]. The rate enhancement by the *gem*-difluoro substitution has been attributed to the LUMO-lowering effect of the fluoro substituent [91, 92]. DIFO linked to Alexa Fluor has been successfully employed to investigate the dynamics of glycan trafficking and it was found that after incubation for 1 hour, labeled glycans colocalize with markers for endosomes and Golgi. DIFO–fluorophore conjugates have been used for imaging azido-labeled biomolecules within complex biological systems such as living cells [91], and zebrafish embryos [93] with low background fluorescence. Background labeling has, however, been observed when samples were analyzed by Western blotting, possibly due to nonspecific hydrophobic interactions or as yet uncharacterized reactions with protein functions [33]. Therefore, a more hydrophilic azacyclooctyne [94], which could be prepared in nine chemical steps from a glucose derivative, was

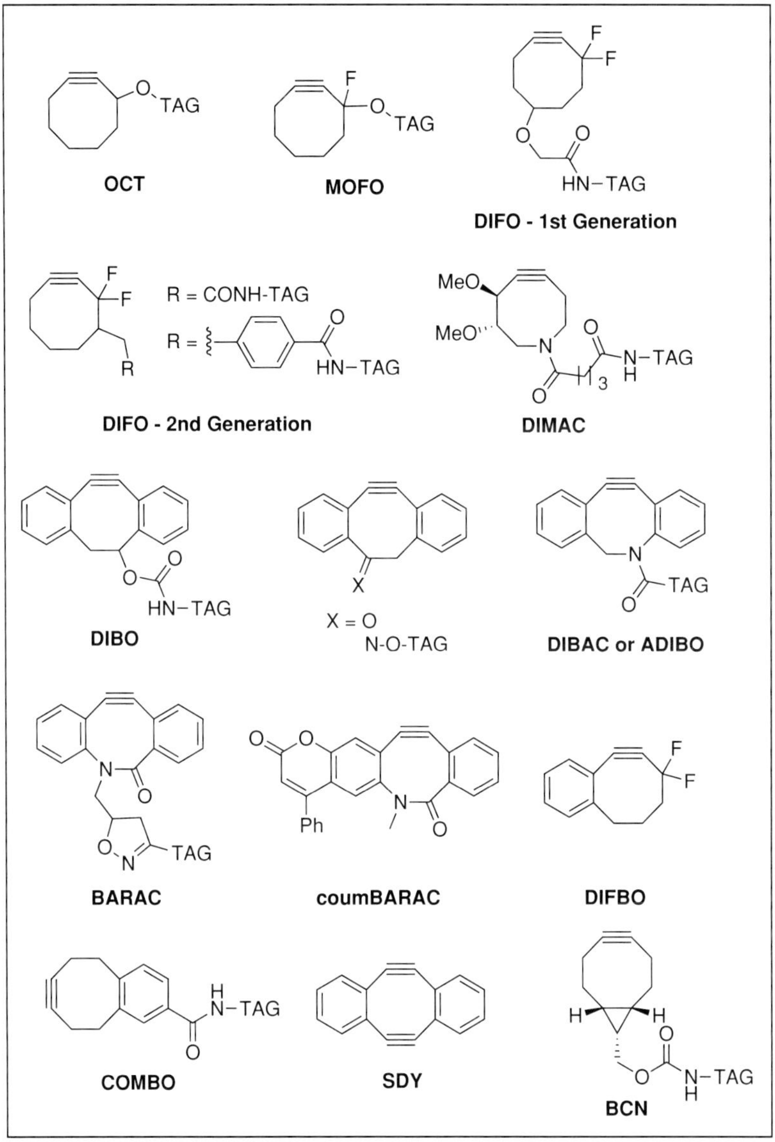

FIGURE 8.4 Various cyclooctynes for Strain-Promoted Alkyne–Azide Cycloaddition.

developed for SPAAC with azido-labeled proteins and cells. Interestingly, its superior polarity and water solubility reduced nonspecific binding, thereby improving the sensitivity of azide detection.

Recently, we reported that derivatives of 4-dibenzocyclooctynol (DIBO), (Fig. 8.4) react exceptionally fast in the absence of a metal catalyst with azido-containing saccharides and amino acids and can be employed for visualizing glycoconjugates of living cells that are metabolically labeled with azido-containing monosaccharides

[95]. The attraction of DIBO is that its synthesis is extremely simple, it is nontoxic and its hydroxyl can be employed to attach a variety of probes. Furthermore, it was found that the rate of cycloaddition of DIBO can be enhanced by modifying the eight-membered ring. For example, oxidation of the hydroxyl moiety to ketone resulted in the compound that exhibits a fivefold increase in the rate of cycloaddition [96]. Keto-DIBO as well as its oxime analog possessed additional interesting properties and it was found that the cyclooctynes were fluorescent whereas this was not the case for the corresponding triazole products. This type of fluoro-switches can, for example, be exploited for monitoring reactions in real time.

Several cyclooctynes have been described that exhibit even faster rates of cycloaddition by introducing sp^2-hybridized ring-atoms. For example, the groups of van Delft [97] and Popik [98] independently reported a more reactive DIBO derivative that has a nitrogen atom within the cyclooctyne ring. The cycloaddition reaction between aza-dibenzocyclooctyne (DIBAC or ADIBO) and benzyl azide was found to be five times faster compare to DIBO $k_{DIBAC} \approx 0.3\ M^{-1}.s^{-1}$; $k_{DIBO} \approx 0.057\ M^{-1}.s^{-1}$. DIBAC was also successfully employed for protein modification and surface functionalization. Further enhancement of the reaction rate $k \approx 0.96\ M^{-1}.s^{-1}$ was achieved by using BARAC [99], which is a dibenzo-fused lactam. Due to its high reactivity, very low concentration (250 nM) of BARAC-Fluor (BARAC conjugated to fluorescein) was required to image living cells metabolically labeled with azido-sialic acid. In an attempt to render BARAC fluorogenic, it was modified by a coumarin moiety [100]. Although increase in fluorescence was observed upon reaction with an azide, the low fluorescence quantum yield $\Phi_F \approx 0.04$ of the triazole product, combined with a relatively high-energy wavelength of excitation ($\approx$300 nm) makes coumBARAC unsuitable for biological imaging.

Recent DFT calculations by Goddard [101] have suggested that although the aryl rings of DIBO promote the rate of cycloaddition by increasing the ring strain, the effect is somewhat tempered by unfavorable steric interactions between the approaching azide and the ortho-hydrogen atom of the aryl rings. Therefore, it was anticipated that removal of an aryl ring of DIBO could lead to compounds with enhanced rates of cycloaddition. Indeed, Bertozzi and coworkers demonstrated that DIFBO [102], which is a hybrid of DIBO and DIFO, was found to be highly reactive and underwent spontaneous homotrimerization in solution. For analytical purposes, DIFBO was ingeniously stabilized as an inclusion complex with β-cyclodextrin, enabling long-term storage and used as a reagent for rapid cycloaddition with azides ($k \approx 0.22\ M^{-1}.s^{-1}$). Unfortunately, to date no biological applications have been reported, most probably due to its high instability. Furthermore, a monobenzylated cyclooctyne (COMBO) [103] has been designed that exhibits similar reaction kinetics ($k \approx 0.23\ M^{-1}.s^{-1}$) as DIFBO, but is chemically stable at room temperature over a 2-day period, but quickly degrades at 37°C (30% degradation after 9 hours). To underscore its stability within complex biological environment, a COMBO-fluorescein conjugate was successfully used for imaging azido-labeled sialic acids present on the surface of living cells.

Although increasing the number of sp^2-like centers within the cyclooctyne ring has proved to be a successful strategy to enhance the rate of cycloaddition, it renders

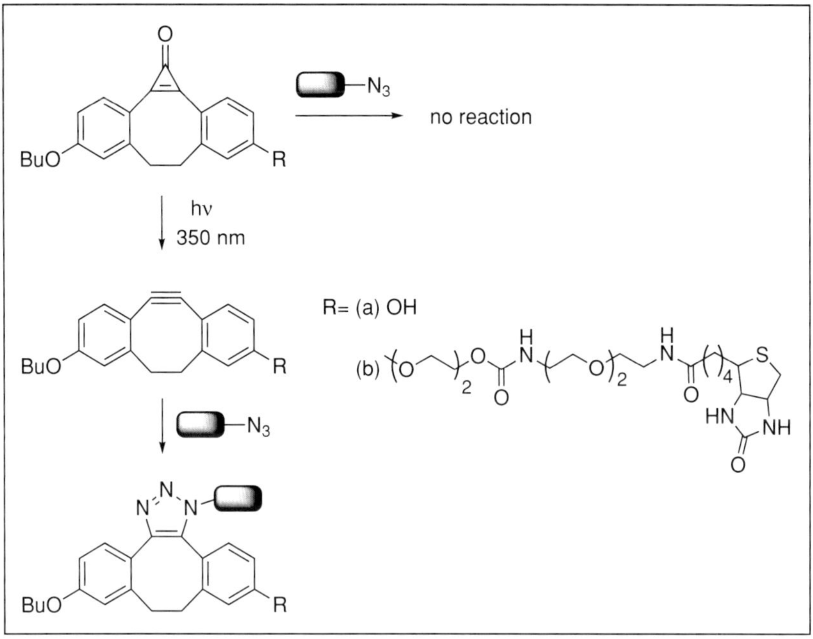

SCHEME 8.5 Photoactivated metal-free click reaction.

cyclooctyne more lipophilic, making it more susceptible to nonspecifically interact with proteins. To address this issue, van Delft and coworkers designed a cyclooctyne that has a 5,6-fused-cyclopropane at the cyclooctyne ring [104]. Bicyclo[6.1.0]nonyne (BCN) has the advantages to be easy to prepare (3-step synthesis with 61% overall yield), highly reactive towards azides ($k \approx 0.14\ M^{-1}.s^{-1}$) and symmetrical.

We have further extended the DIBO-based bioorthogonal reporter strategy by developing a photochemically triggered click reaction (Scheme 8.5) [105]. Such an approach provides opportunities for spatial and temporal control of labeling of the target substrates. In fact, photochemical release or generation of an active molecule is a widely employed strategy to deliver bioactive compounds to addressable target sites in a time-controlled manner [106–109]. To achieve this goal, we have explored photochemical generation of reactive dibenzocyclooctynes. In this respect, it is known that single [110–115] or two-photon [116] excitation of cyclopropenones results in the formation of corresponding acetylenes. Photochemical decarbonylation of thermally stable diaryl-substituted cyclopropenones is especially efficient ($\Phi = 0.2 - 1$) and produces alkynes in a quantitative yield [114]. This reaction is also extremely fast and is complete within a few hundred picoseconds after excitation [117, 118]. Thus, we anticipated that cyclopropenone derivatives would not react with azides under ambient conditions in the dark but efficiently produce reactive dibenzocyclooctynes upon irradiation. Indeed, the latter type of compound could be employed for labeling of living cells modified with azido-containing cell surface saccharides.

8.7 STRAIN-PROMOTED CYCLOADDITIONS WITH NITRONES, NITRILE OXIDES, AND DIAZOCARBONYL DERIVATIVES

Although metal-free cycloadditions of cyclooctynes and azides to give stable 1,2,3-triazoles have found wide utility in chemical biology and material sciences, there is an urgent need for faster and more versatile bioorthogonal reactions. It has been found that nitrones [119–122], nitrile oxides [123,124], and diazocarbonyl derivatives [124] undergo facile 1,3-dipolar cycloadditions with cyclooctynes.

Nitrones can readily be prepared by condensation of appropriate aldehydes with *N*-methylhydroxylamine. Cycloadditions of these derivatives with dibenzylcyclooctynol in a mixture of acetonitrile and water gave the corresponding oxazolines in high yield (Table 8.1, entries 1–5) [120]. It was found that reactions with α-carboxy nitrones (entries 4 and 5) proceeded 18- and 32-fold faster, than a similar cycloaddition with benzyl azide. The usefulness of this reaction for bioorthogonal chemistry was demonstrated by PEGylation of peptides and proteins. For example, the *N*-terminal serine of the chemokine interleukin-8 (IL-8) was oxidized with sodium periodate to rapidly generate and aldehyde, which was treated with *p*-MeOPhSH to reduce iodate species. Next, *N*-methylhydroxylamine and *p*-anisidine were added to the mixture leading to rapid formation of a nitrone as judged by mass spectrometric analysis. Finally, addition of PEG_{2000}-modified dibenzylcyclooctyne (25 eq.) led to a single protein with a mass corresponding to the PEGylated conjugate. The *p*-MeOPhSH treatment was essential otherwise *N*-methylhydroxylamine was rapidly converted into nitrosomethane dimer $(MeNO)_2$ by oxidation with iodate formed in the previous step. Furthermore, *p*-anisidine greatly enhanced the rate of nitrone formation, probably *via* a similar mechanism as recently described for the condensation of aldehydes with hydroxylamines to give oximes.

It was also found that cycloadditions of DIBO with diazocarbonyl derivatives exhibited similar kinetics compared to azides whereas the reaction rates of cycloadditions with nitrile oxides were much faster (Table 8.1, entries 6–11) [124]. In the case of nitrile oxides, substituents exerted only small influence on the reaction rate and it appeared that strongly electron-withdrawing substituents, such as a nitro group, somewhat increase the rate of reaction. Nitrile oxides can conveniently be prepared by direct oxidation of the corresponding oximes with [bisacetoxyiodo]benzene (BAIB) and these conditions made it possible to perform oxime formation, oxidation, and cycloaddition as a one-pot procedure. This sequence of reactions was, for example, employed to functionalize the anomeric center of carbohydrates with various tags. Cycloadditions of cyclooctynes with nitrile oxides have also been used for labeling glycoproteins by mild oxidation of sialic acid moieties with $NaIO_4$ to form a C-7 aldehyde, which upon treatment with hydroxylamine gave an oxime that could be oxidized to a nitrile oxide for reaction with derivatives of DIBO. The attraction of this strategy is that tags can be installed into glycoproteins by stable isoxazoles linkages and for example the glycoprotein fetuin could be biotinylated by employing a DIBO-biotin derivative. It has also been shown that oximes and azides provide an orthogonal pair of functional groups for sequential metal-free click reactions and this feature makes it possible to multi-functionalize biomolecules and materials by a simple synthetic procedure that does not require toxic metal catalysts.

TABLE 8.1 Rate Constants and Yields of Cycloaddition Reactions between DIBO and Various Dipoles

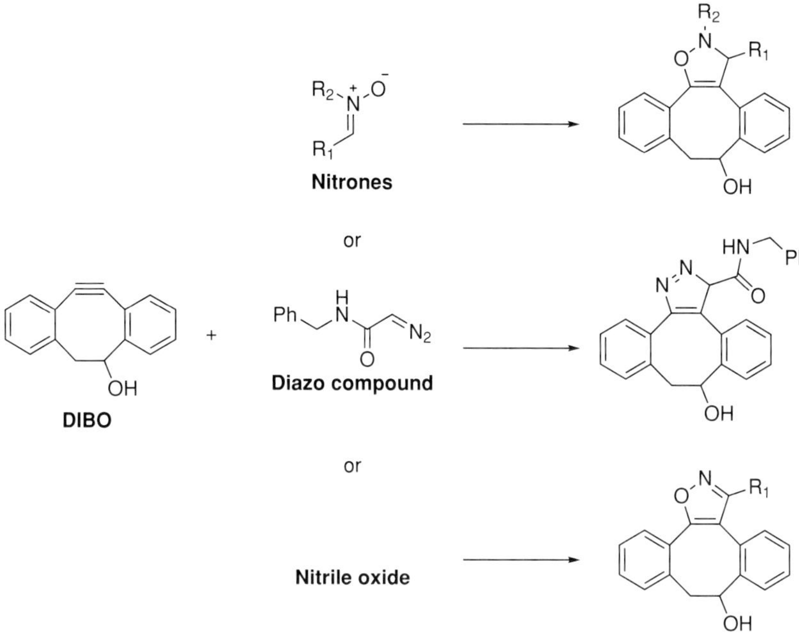

Entry	Dipole		k ($M^{-1}.s^{-1}$)	Yield (%)
1		R_1=Ph; R_2=Me	0.013	95
2		R_1=$(CH_2)_2$Ph; R_2=Me	0.032	80
3	*Nitrone*	R_1=Ph; R_2=Ph	0.2	89
4		R_1=CO_2Et; R_2=Me	3.9	92
5		R_1=C(O)NHBn; R_2=Me	2.2	93
6	*Diazo*		0.031	92
7		R_3=Ph	3.38	93
8		R_3=4-MeOPh	2.15	89
9	*Nitrile oxide*	R_3=4-NO_2Ph	8.47	93
10		R_3=4-FPh	3.99	90
11		R_3=Ph$(CH_2)_2$	1.38	90

8.8 METABOLIC LABELING WITH PHOTOACTIVATABLE CROSSLINKING SUGARS

Photoactivatable crosslinking reagents such as aryl azides and diazirine have been employed to covalently link protein complexes for further analysis. Inspired by these chemical reactions, metabolic labeling strategies have been developed that employ monosaccharides modified by photoactivatable crosslinking moiety. The use of such a reagent makes it possible to analyze glycoconjugate–protein complexes.

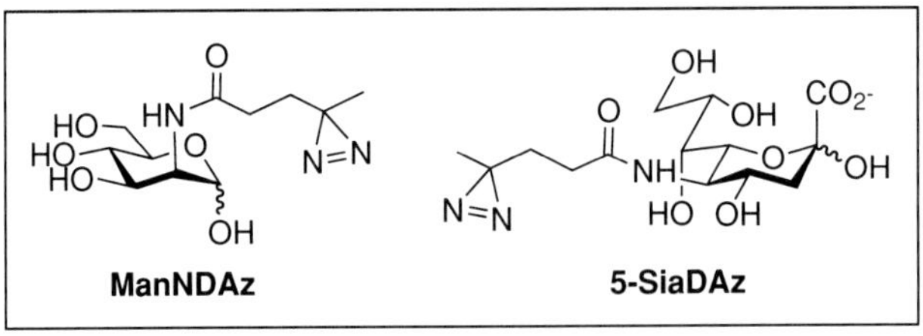

FIGURE 8.5 Photoactivatable crosslinking sugars.

Paulson and coworkers reported an attractive example of the use of a photo-activated aryl azide group to uncover *cis*-glycoprotein interactions on the surface of B-cells [125]. In this study, cells were metabolically labeled with sialic acid modified at C-9 with a photo-activated aryl azide group.

Inspired by the successful use of photo-activated aryl azide group, the Kohler group developed monosaccharides containing diazirine crosslinkers (Fig. 8.5) for metabolic incorporation onto cell surfaces, and demonstrated the utility of this approach by detecting multimerization of a cell surface glycoprotein [126]. The diazirine crosslinker was selected because its small size was expected to minimize interference with carbohydrate–protein interactions. It was shown that peracetylated 5-SiaDAz and ManNDAz could be metabolically incorporated into cellular glycoproteins. The latter was demonstrated by employing a cell line (KD20) that is impaired in sialic acid biosynthesis by a lack of UDP-GlcNAc 2-epimerase that converts UDP-GlcNAc to ManNAc. Incubation of the cells with the sugar derivatives resulted in cell surface expression of sialic acid as demonstrated by flow cytometry using FITC-labeled *Sambucus nigra* agglutinin (FITC-SNA). As expected, untreated cells display low FITC-SNA binding. The photoreactive sugars could be used to capture a carbohydrate-mediated interaction by sialic acid-dependent multimerization of CD22 when irradiated with 350 nm light for 20 minutes and their contents analyzed by Western blot using an anti-CD22 antibody.

8.9 METABOLIC LABELING WITH THIOL-CONTAINING SUGARS

Thiol group have been installed into the glycocalyx by metabolic labeling with thioglycolyl analogs of ManNAc (Ac_5ManNTGc) [127]. Thiols are common within cells but are not found in the glycocalyx. Surprisingly, by locating this versatile functional group on the outer periphery of normally nonadhesive human Jurkat cells, we obtained spontaneous cell–cell clustering and attachment to complementary maleimide-derivatized substrates. When analyzed in human embryoid body–derived (hEBD) stem cells, Ac_5ManNTGc induced catenin expression and altered cell morphology, consistent with neuronal differentiation. Notably, these effects were modulated by the growth substrate of the cells, with a stronger response observed on a gold surface than on glass. These results establish a novel approach for tissue engineering and offer a means to influence stem cell fates.

8.10 CONCLUDING REMARKS

Bioorthogonal reactions are beginning to open new avenues for the investigation of biological processes in which carbohydrates are involved. Attention needs to be focused on possible perturbations in oligosaccharide structures arising from metabolic labeling with monosaccharides. In this respect, labeling experiments generally use relatively high concentration of monosaccharide derivative, which may perturb the metabolic flux, which in turn may influence glycoconjugate biosynthesis. Furthermore, monosaccharides are generally fed as acylated derivatives and it has been suggested that the acyl moieties may influence biological process [31]. Despite attractive features of currently reported probes and reagents, there is a need for new bioorthogonal functional groups that exhibit faster kinetics and minimal background reactions. Computational chemistry has been employed to rationalize reaction mechanism of bioorthogonal reactions and this insight is being exploited for the design of new reagents [101].

REFERENCES

1. Varki, A. *Glycobiology* **1993**, *3*, 97–130.
2. Dwek, R. A. *Chem. Rev.* **1996**, *96*, 683–720.
3. Bertozzi, C. R.; Kiessling, L. L. *Science* **2001**, *291*, 2357–2364.
4. Rudd, P. M.; Elliott, T.; Cresswell, P.; Wilson, I. A.; Dwek, R. A. *Science* **2001**, *291*, 2370–2376.
5. Ritchie, G. E.; Moffatt, B. E.; Sim, R. B.; Morgan, B. P.; Dwek, R. A.; Rudd, P. M. *Chem. Rev.* **2002**, *102*, 305–319.
6. Roth, J. *Chem. Rev.* **2002**, *102*, 285–303.
7. Zachara, N. E.; Hart, G. W. *Chem. Rev.* **2002**, *102*, 431–438.
8. Schmidt, M. A.; Riley, L. W.; Benz, I. *Trends Microbiol.* **2003**, *11*, 554–561.
9. Bucior, I.; Burger, M. M. *Curr. Op. Struct. Biol.* **2004**, *14*, 631–637.
10. Kleene, R.; Schachner, M. *Nat. Rev. Neurosci.* **2004**, *5*, 195–208.
11. Dempski, R. E.; Imperiali, B. *Curr. Opin. Chem. Biol.* **2002**, *6*, 844–850.
12. Weerapana, E.; Imperiali, B. *Glycobiology* **2006**, *16*, 91R–101R.
13. Helenius, A.; Aebi, M. *Science* **2001**, *291*, 2364–2369.
14. Ellgaard, L.; Helenius, A. *Nat. Rev. Mol. Cell Biol.* **2003**, *4*, 181–191.
15. Helenius, A.; Aebi, M. *Annu. Rev. Biochem.* **2004**, *73*, 1019–1049.
16. Mast, S. W.; Moremen, K. W. *Methods Enzymol.* **2006**, *415*, 31–46.
17. Zaia, J. *Mass Spectrom. Rev.* **2004**, *23*, 161–227.
18. Haslam, S. M.; North, S. J.; Dell, A. *Curr. Opin. Struct. Biol.* **2006**, *16*, 584–591.
19. Nana, K.; Satsuki, I.; Noritaka, H.; Akira, H.; Daisuke, T.; Teruhide, Y. *Trends Glycosci. Glycotech.* **2008**, *20*, 97–116.
20. Feizi, T.; Fazio, F.; Chai, W. C.; Wong, C. H. *Curr. Opin. Struct. Biol.* **2003**, *13*, 637–645.
21. Alvarez, R. A.; Blixt, O. *Methods Enzymol.* **2006**, *415*, 292–310.

22. de Paz, J. L.; Horlacher, T.; Seeberger, P. H. *Methods Enzymol.* **2006**, *415*, 269–292.
23. Miyamoto, S. *Curr. Opin. Mol. Ther.* **2006**, *8*, 507–513.
24. Paulson, J. C.; Blixt, O.; Collins, B. E. *Nat. Chem. Biol.* **2006**, *2*, 238–248.
25. Hirabayashi, J. *J. Biochem.* **2008**, *144*, 139–147.
26. Raman, R.; Raguram, S.; Venkataraman, G.; Paulson, J. C.; Sasisekharan, R. *Nat. Methods* **2005**, *2*, 817–824.
27. Raman, R.; Venkataraman, M.; Ramakrishnan, S.; Lang, W.; Raguram, S.; Sasisekharan, R. *Glycobiology* **2006**, *16*, 82R–90R.
28. Aoki-Kinoshita, K. F. *Exp. Opin. Drug Discov.* **2008**, *3*, 877–890.
29. Prescher, J. A.; Bertozzi, C. R. *Cell* **2006**, *126*, 851–854.
30. Du, J.; Meledeo, M. A.; Wang, Z. Y.; Khanna, H. S.; Paruchuri, V. D. P.; Yarema, K. J. *Glycobiology* **2009**, *19*, 1382–1401.
31. Elmouelhi, N.; Aich, U.; Paruchuri, V. D. P.; Meledeo, M. A.; Campbell, C. T.; Wang, J. J.; Srinivas, R.; Khanna, H. S.; Yarema, K. J. *J. Med. Chem.* **2009**, *52*, 2515–2530.
32. Laughlin, S. T.; Bertozzi, C. R. *Proc. Natl. Acad. Sci. U. S. A.* **2009**, *106*, 12–17.
33. Sletten, E. M.; Bertozzi, C. R. *Angew. Chem. Int. Ed.* **2009**, *48*, 6974–6998.
34. Jewett, J. C.; Bertozzi, C. R. *Chem. Soc. Rev.* **2010**, *39*, 1272–1279.
35. Hao, Z.; Hong, S.; Chen, X.; Chen, P. R. *Acc. Chem. Res.* **2011**, *44*, 742–751.
36. Gross, H. J.; Rose, U.; Krause, J. M.; Paulson, J. C.; Schmid, K.; Feeney, R. E.; Brossmer, R. *Biochemistry* **1989**, *28*, 7386–7392.
37. Gross, H. J.; Brossmer, R. *Glycoconj. J.* **1995**, *12*, 739–746.
38. Keppler, O. T.; Horstkorte, R.; Pawlita, M.; Schmidts, C.; Reutter, W. *Glycobiology* **2001**, *11*, 11R–18R.
39. Hayrinen, J.; Jennings, H.; Raff, H. V.; Rougon, G.; Hanai, N.; Gerardyschahn, R.; Finne, J. *J. Infect. Dis.* **1995**, *171*, 1481–1490.
40. Liu, C.; Gelius, E.; Liu, G.; Steiner, H.; Dziarski, R. *J. Biol. Chem.* **2000**, *275*, 24490–24499.
41. Wu, J.; Guo, Z. *Bioconjug. Chem.* **2006**, *17*, 1537–1544.
42. Wang, Q. L.; Ekanayaka, S. A.; Wu, J.; Zhang, J. P.; Guo, Z. W. *Bioconj. Chem.* **2008**, *19*, 2060–2067.
43. Lemieux, G. A.; Bertozzi, C. R. *Trends Biotechnol.* **1998**, *16*, 506–513.
44. Saxon, E.; Bertozzi, C. R. *Science* **2000**, *287*, 2007–2010.
45. Mahal, L. K.; Yarema, K. J.; Bertozzi, C. R. *Science* **1997**, *276*, 1125–1128.
46. Hang, H. C.; Bertozzi, C. R. *J. Am. Chem. Soc.* **2001**, *123*, 1242–1243.
47. Nauman, D. A.; Bertozzi, C. R. *Biochim. Biophys. Acta* **2001**, *1568*, 147–154.
48. Lee, J. H.; Baker, T. J.; Mahal, L. K.; Zabner, J.; Bertozzi, C. R.; Wiemer, D. F.; Welsh, M. J. *J. Biol. Chem.* **1999**, *274*, 21878–21884.
49. Lemieux, G. A.; Yarema, K. J.; Jacobs, C. L.; Bertozzi, C. R. *J. Am. Chem. Soc.* **1999**, *121*, 4278–4279.
50. Kohler, J. J. *ChemBioChem.* **2009**, *10*, 2147–2150.
51. Zeng, Y.; Ramya, T. N. C.; Dirksen, A.; Dawson, P. E.; Paulson, J. C. *Nat. Methods* **2009**, *6*, 207–209.
52. Nilsson, J.; Ruetschi, U.; Halim, A.; Hesse, C.; Carlsohn, E.; Brinkmalm, G.; Larson, G. *Nat. Methods* **2009**, *6*, 809–811.

53. Akira, S. *J. Biol. Chem.* **2003**, *278*, 38105–38108.
54. Hanson, S. R.; Greenberg, W. A.; Wong, C. H. *QSAR Comb. Sci* **2007**, *26*, 1243–1252.
55. Luchansky, S. J.; Goon, S.; Bertozzi, C. R. *ChemBioChem.* **2004**, *5*, 371–374.
56. Kosa, R. E.; Gross, H. J. *Biochem. Biophys. Res. Commun.* **1993**, *190*, 914–920.
57. Hang, H. C.; Yu, C.; Kato, D. L.; Bertozzi, C. R. *Proc. Natl. Acad. Sci. USA* **2003**, *100*, 14846–14851.
58. Rabuka, D.; Hubbard, S. C.; Laughlin, S. T.; Argade, S. P.; Bertozzi, C. R. *J. Am. Chem. Soc.* **2006**, *128*, 12078–12079.
59. Sawa, M.; Hsu, T. L.; Itoh, T.; Sugiyama, M.; Hanson, S. R.; Vogt, P. K.; Wong, C. H. *Proc. Natl. Acad. Sci.* **2006**, *103*, 12371–12376.
60. Hsu, T. L.; Hanson, S. R.; Kishikawa, K.; Wang, S. K.; Sawa, M.; Wong, C. H. *Proc. Natl. Acad. Sci. USA* **2007**, *104*, 2614–2619.
61. Chang, P. V.; Chen, X.; Smyrniotis, C.; Xenakis, A.; Hu, T. S.; Bertozzi, C. R.; Wu, P. *Angew. Chem. Int. Ed.* **2009**, *48*, 4030–4033.
62. Vocadlo, D. J.; Hang, H. C.; Kim, E. J.; Hanover, J. A.; Bertozzi, C. R. *Proc. Natl. Acad. Sci. USA* **2003**, *100*, 9116–9121.
63. Boyce, M.; Carrico, I. S.; Ganguli, A. S.; Yu, S.–H.; Hangauer, M. J.; Hubbard, S. C.; Kohler, J. J.; Bertozzi, C. R. *Proc. Natl. Acad. Sci. USA* **2011**, *108*, 3141–3146.
64. Khidekel, N.; Ficarro, S. B.; Clark, P. M.; Bryan, M. C.; Swaney, D. L.; Rexach, J. E.; Sun, Y. E.; Coon, J. J.; Peters, E. C.; Hsieh–Wilson, L. C. *Nat. Chem. Biol.* **2007**, *3*, 339–348.
65. Clark, P. M.; Dweck, J. F.; Mason, D. E.; Hart, C. R.; Buck, S. B.; Peters, E. C.; Agnew, B. J.; Hsieh-Wilson, L. C. *J. Am. Chem. Soc.* **2008**, *130*, 11576–11577.
66. Wang, W.; Hu, T.; Frantom, P. A.; Zheng, T.; Gerwe, B.; del Amo, D. S.; Garret, S.; Seidel, R. D., III; Wu, P. *Proc. Natl. Acad. Sci. USA* **2009**, *106*, 16096–16101.
67. Zheng, T.; Jiang, H.; Gros, M.; del Amo, D. S.; Sundaram, S.; Lauvau, G.; Marlow, F.; Liu, Y.; Stanley, P.; Wu, P. *Angew. Chem. Int. Ed.* **2011**, *50*, 4113–4118.
68. Staudinger, H.; Meyer, J. *Helv. Chim. Acta* **1919**, *2*, 635–646.
69. Prescher, J. A.; Dube, D. H.; Bertozzi, C. R. *Nature* **2004**, *430*, 873–877.
70. Chang, P. V.; Prescher, J.; Hangauer, M. J.; Bertozzi, C. R. *J. Am. Chem. Soc.* **2007**, *129*, 8400–8401.
71. Lemieux, G.; De Graffenried, C. L.; Bertozzi, C. R. *J. Am. Chem. Soc.* **2003**, *125*, 4708–4709.
72. Hangauer, M. J.; Bertozzi, C. R. *Angew. Chem. Int. Ed.* **2008**, *47*, 2394–2397.
73. Cohen, A. S.; Dubikovskaya, E.; Rush, J. S.; Bertozzi, C. R. *J. Am. Chem. Soc.* **2010**, *132*, 8563–8565.
74. Rostovtsev, V. V.; Green, L. G.; Fokin, V. V.; Sharpless, K. B. *Angew. Chem. Int. Ed.* **2002**, *41*, 2596–2599.
75. Tornoe, C. W.; Christensen, C.; Meldal, M. *J. Org. Chem.* **2002**, *67*, 3057–3064.
76. Chan, T. R.; Hilgraf, R.; Sharpless, K. B.; Fokin, V. V. *Org. Lett.* **2004**, *6*, 2853–2855.
77. Hong, V.; Steinmetz, N. F.; Manchester, M.; Finn, M. G. *Bioconjugate Chem.* **2010**, *21*, 1912–1916.
78. Soriano Del Amo, D.; Wang, W.; Jiang, H.; Besanceney, C.; Yan, A. C.; Levy, M.; Liu, Y.; Marlow, F. L.; Wu, P. *J. Am. Chem. Soc.* **2010**, *132*, 16893–16899.

79. Besanceney–Webler, C.; Jiang, H.; Wang, W.; Baughn, A. D.; Wu, P. *Bioorg. Med. Chem. Lett.* **2011**, *21*, 4989–4992.

80. Besanceney-Webler, C.; Jiang, H.; Zheng, T.; Feng, L.; Soriano del Amo, D.; Wang, W.; Klivansky, L. M.; Marlow, F. L.; Liu, Y.; Wu, P. *Angew. Chem. Int. Ed.* **2011**, *50*, 8051–8056.

81. Kennedy, D. C.; McKay, C. S.; Legault, M. C. B.; Danielson, D. C.; Blake, J.; Pegoraro, A. F.; Stolow, A.; Mester, Z.; Pezacki, J. P. *J. Am. Chem. Soc.* **2011**, *133*, 17993–18001.

82. Wang, W.; Hong, S.; Tran, A.; Jiang, H.; Triano, R.; Liu, Y.; Chen, X.; Wu, P. *Chem. Asian J.* **2011**, *6*, 2796–2802.

83. Hanson, S. R.; Hsu, T. L.; Weerapana, E.; Kishikawa, K.; Simon, G. M.; Cravatt, B. F.; Wong, C. H. *J. Am. Chem. Soc.* **2007**, *129*, 7266–7267.

84. Turner, R. B.; Goebel, P.; Mallon, B. J.; Jarrett, A. D. *J. Am. Chem. Soc.* **1972**, *95*, 790–792.

85. Agard, N. J.; Prescher, J. A.; Bertozzi, C. R. *J. Am. Chem. Soc.* **2004**, *126*, 15046–15047.

86. Debets, M. F.; Van Berkel, S. S.; Dommerholt, J.; Dirks, A. J.; Rutjes, F. P. J. T.; Van Delft, F. L. *Acc. Chem. Res.* **2011**, *44*, 805–815.

87. Ess, D. H.; Jones, G. O.; Houk, K. N. *Org. Lett.* **2008**, *10*, 1633–1636.

88. Bach, R. D. *J. Am. Chem. Soc.* **2009**, *131*, 5233–5243.

89. Schoenebeck, F.; Ess, D. H.; Jones, G. O.; Houk, K. N. *J. Am. Chem. Soc.* **2009**, *131*, 8121–8133.

90. Agard, N. J.; Baskin, J. M.; Prescher, J. A.; Lo, A.; Bertozzi, C. R. *ACS Chem. Biol.* **2006**, *1*, 644–648.

91. Baskin, J. M.; Prescher, J. A.; Laughlin, S. T.; Agard, N. J.; Chang, P. V.; Miller, I. A.; Lo, A.; Codelli, J. A.; Bertozzi, C. R. *Proc. Natl. Acad. Sci. USA* **2007**, *104*, 16793–16797.

92. Codelli, J. A.; Baskin, J. M.; Agard, N. J.; Bertozzi, C. R. *J. Am. Chem. Soc.* **2008**, *130*, 11486–11493.

93. Laughlin, S. T.; Baskin, J. M.; Amacher, S. L.; Bertozzi, C. R. *Science* **2008**, *320*, 664–667.

94. Sletten, E. M.; Bertozzi, C. R. *Org. Lett.* **2008**, *10*, 3097–3099.

95. Ning, X. H.; Guo, J.; Wolfert, M. A.; Boons, G. J. *Angew. Chem. Int. Ed.* **2008**, *47*, 2253–2255.

96. Mbua, N. E.; Guo, J.; Wolfert, M. A.; Steet, R.; Boons, G. J. *ChemBioChem* **2011**, *12*, 1912–1921.

97. Debets, M. F.; van Berkel, S. S.; Schoffelen, S.; Rutjes, F. P. J. T.; van Hest, J. C. M.; van Delft, F. L. *Chem. Commun.* **2010**, *46*, 97–99.

98. Kuzmin, A.; Poloukhtine, A.; Wolfert, M. A.; Popik, V. V. *Bioconjug. Chem.* **2010**, *21*, 2076–2085.

99. Jewett, J. C.; Sletten, E. M.; Bertozzi, C. R. *J. Am. Chem. Soc.* **2010**, *132*, 3688–3690.

100. Jewett, J. C.; Bertozzi, C. R. *Org. Lett.* **2011**, *13*, 5937–5939.

101. Chenoweth, K.; Chenoweth, D.; Goddard, W. A. *Org. Biomol. Chem.* **2009**, *7*, 5255–5258.

102. Sletten, E. M.; Nakamura, H.; Jewett, J. C.; Bertozzi, C. R. *J. Am. Chem. Soc.* **2010**, *132*, 11799–11805.

103. Varga, B. R.; Kállay, M.; Hegyi, K.; Béni, S.; Kele, P. *Chem. Eur. J.* **2012**, *18*, 822–828.

104. Dommerholt, J.; Schmidt, S.; Temming, R.; Hendriks, L. J.; Rutjes, F. P.; van Hest, J. C.; Lefeber, D. J.; Friedl, P.; van Delft, F. L. *Angew. Chem. Int. Ed.* **2010**, *49*, 9422–9425.

105. Poloukhtine, A. A.; Mbua, N. E.; Wolfert, M. A.; Boons, G. J.; Popik, V. V. *J. Am. Chem. Soc.* **2009**, *131*, 15769–15776.

106. Pelliccioli, A. P.; Wirz, J. *Photochem. Photobiol. Sci.* **2002**, *1*, 441–458.

107. Mayer, G.; Heckel, A. *Angew. Chem. Int. Ed.* **2006**, *45*, 4900–4921.

108. Ellis–Davies, G. C. R. *Nat. Methods* **2007**, *4*, 619–628.

109. Song, W.; Wang, Y.; Qu, J.; Lin, Q. *J. Am. Chem. Soc.* **2008**, *130*, 9654–9655.

110. Chapman, O. L.; Gano, J.; West, P. R.; Regitz, M.; Maas, G. *J. Am. Chem. Soc.* **1981**, *103*, 7033–7036.

111. Dehmlow, E. V.; Neuhaus, R.; Schell, H. G. *Chem. Ber.* **1988**, *121*, 569–571.

112. Murata, S.; Yamamoto, T.; Tomioka, H. *J. Am. Chem. Soc.* **1993**, *115*, 4013–4023.

113. Chiang, Y.; Kresge, A. J.; Paine, S. W.; Popik, V. V. *J. Phys. Org. Chem.* **1996**, *9*, 361–370.

114. Poloukhtine, A.; Popik, V. V. *J. Org. Chem.* **2003**, *68*, 7833–7840.

115. Kuzmanich, G.; Natarajan, A.; Chin, K. K.; Veerman, M.; Mortko, C. J.; Garcia-Garibay, M. A. *J. Am. Chem. Soc.* **2008**, *130*, 1140–1141.

116. Urdabayev, N. K.; Poloukhtine, A.; Popik, V. V. *Chem. Commun.* **2006**, 454–456.

117. Takeuchi, S.; Tahara, T. *J. Chem. Phys.* **2004**, *120*, 4768–4776.

118. Poloukhtine, A.; Popik, V. V. *J. Phys. Chem. A* **2006**, *110*, 1749–1757.

119. McKay, C. S.; Moran, J.; Pezacki, J. P. *Chem. Commun.* **2010**, *46*, 931–933.

120. Ning, X.; Temming, R. P.; Dommerholt, J.; Guo, J.; Ania, D. B.; Debets, M. F.; Wolfert, M. A.; Boons, G. J.; van Delft, F. L. *Angew. Chem. Int. Ed.* **2010**, *49*, 3065–3068.

121. McKay, C. S.; Blake, J. A.; Cheng, J.; Danielson, D. C.; Pezacki, J. P. *Chem. Commun.* **2011**, *47*, 10040–10042.

122. Colombo, M.; Sommaruga, S.; Mazzucchelli, S.; Polito, L.; Verderio, P.; Galeffi, P.; Corsi, F.; Tortora, P.; Prosperi, D. *Angew. Chem. Int. Ed.* **2012**, *51*, 496–499.

123. Jawalekar, A. M.; Reubsaet, E.; Rutjes, F. P. J. T.; van Delft, F. L. *Chem. Commun.* **2011**, *47*, 3198–3200.

124. Sanders, B. C.; Friscourt, F.; Ledin, P. A.; Mbua, N. E.; Arumugam, S.; Guo, J.; Boltje, T. J.; Popik, V. V.; Boons, G. J. *J. Am. Chem. Soc.* **2011**, *133*, 949–957.

125. Han, S.; Collins, B. E.; Bengtson, P.; Paulson, J. C. *Nat. Chem. Biol.* **2005**, *1*, 93–97.

126. Tanaka, Y.; Kohler, J. J. *J. Am. Chem. Soc.* **2008**, *130*, 3278–3279.

127. Sampathkumar, S. G.; Li, A. V.; Jones, M. B.; Sun, Z. H.; Yarema, K. J. *Nat. Chem. Biol.* **2006**, *2*, 149–152.

9

"SWEET" SUCROSE MACROCYCLES VIA A "CLICK CHEMISTRY" ROUTE

MYKHAYLO A. POTOPNYK AND SŁAWOMIR JAROSZ

9.1 INTRODUCTION

The term "click chemistry" was introduced by Sharpless in 2001 to describe the processes which allow the connection of two molecules (or fragments) in a predictable way, providing well-defined products in high yields. These reactions should be characterized by high atom economy, application of readily available reagents (or their convenient precursors), be easy to perform, and should be rather insensitive to environmental factors such as oxygen and water. All these parameters would allow the connection of complex molecules through a relatively simple linker [1]. The 1,3-dipolar cycloaddition of azides **1** and acetylenes **2**, known as Huisgen reaction [2, 3] (Fig. 9.1), is one of the most convenient processes for the preparation of 5-membered heterocyclic compounds. This process is easy to perform, has high atom economy, and—moreover—the starting materials (azide and acetylene) are usually rather easy to prepare. However, because it gives more than one product (1,4- and 1,5-adduct; the ratio of them depends on the nature of the substituents R' and R"), it does not fulfill all the requirements of the "click chemistry" philosophy.

In 2002, Sharpless introduced a *small* modification to this process relying on an application of Cu(I) species as catalyst which resulted in the formation of the 1,4-adduct exclusively [4]. At the same time, a paper by Meldal [5] describing the same phenomenon appeared.

The Cu(I) species are provided either directly (as Cu(I), CuOAc, etc.) or generated *in situ* (from $CuSO_4$ and sodium ascorbate) [6, 7]. More complex species such as

Click Chemistry in Glycoscience: New Developments and Strategies, First Edition.
Edited by Zbigniew J. Witczak and Roman Bielski.

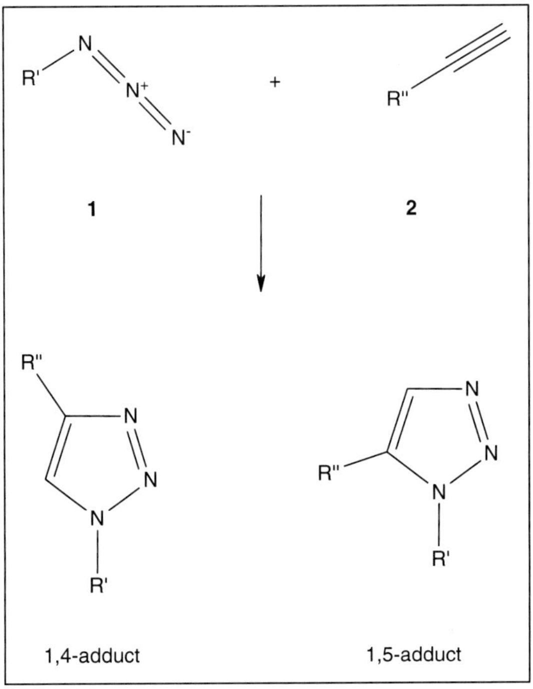

FIGURE 9.1 The general scheme of Huisgen reaction.

Cu(I)-modified zeolites (offering stabilized Cu(I) species which may exclude the formation of undesirable alkyne homocoupling products [8]) can be successfully applied, also for complex products such as carbohydrate derivative **3** (Fig. 9.2) [9].

The click chemistry approach quickly became a standard methodology of coupling, in a predictable way, two molecules (fragments) in various fields of organic,

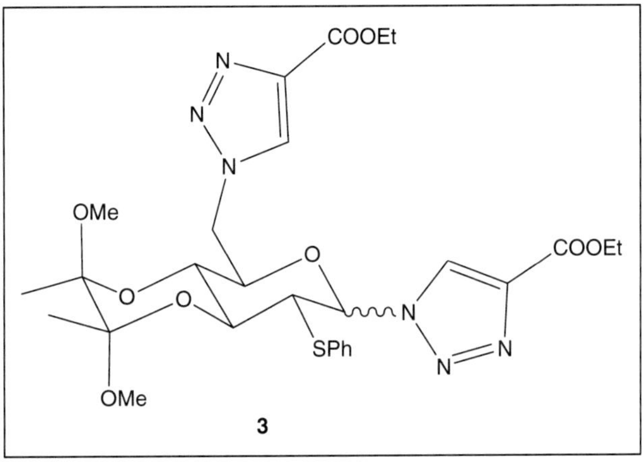

FIGURE 9.2 Double functionalized sugar block by "click" approach.

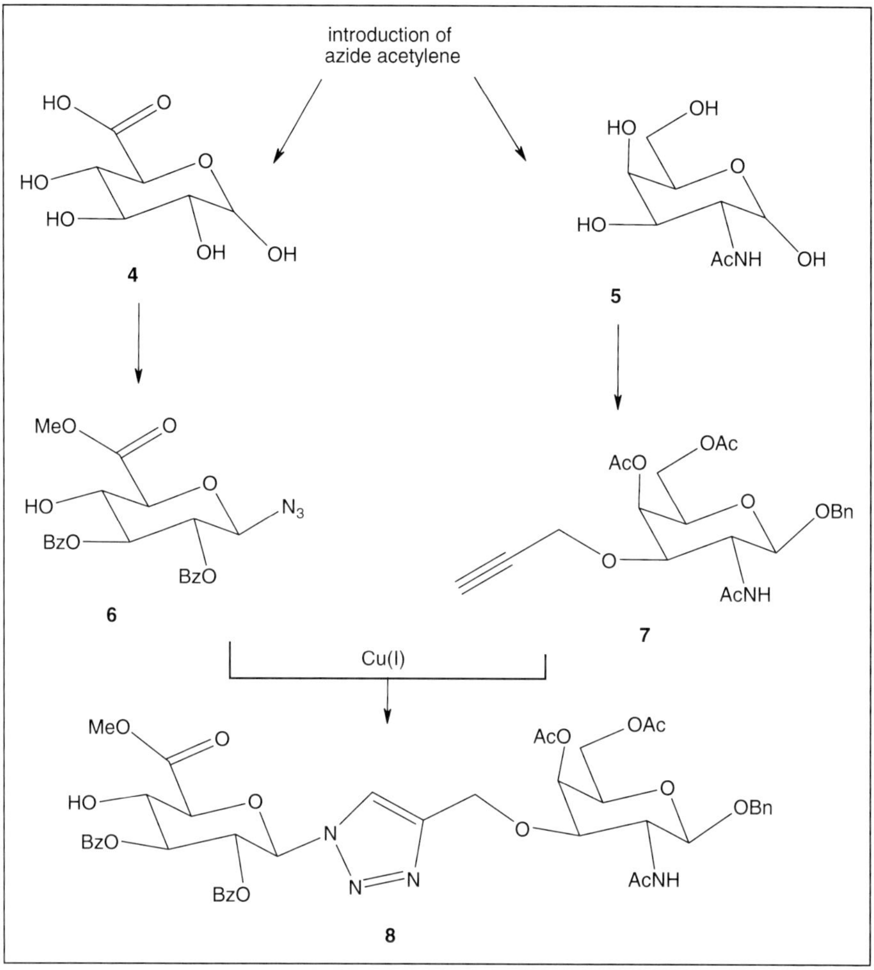

FIGURE 9.3 The coupling of two sugar molecules via click chemistry.

bioorganic, and medicinal chemistry [7,10–22]. One of these fields is sugar chemistry. Sugars are very convenient, "graceful" objects not only because of their biological activity or usefulness as components of the "chiral pool" but also they are useful materials for testing new concepts of synthesis. Moreover, the basic chemistry of simple sugars is well known, which allows the introduction of the reactive azide functionality at the precisely designed position of the carbohydrate molecule. This may be illustrated by the example shown in Figure 9.3.

Standard carbohydrate methodology (involving proper protection–deprotection) allows the introduction of the azide group onto the C1 of the gluconic acid (**4** → **6**) and the acetylene at the C3 of the galactosamine part (**5** → **7**). Coupling of these two activated molecules by click methodology provided the building block **8**, used further to prepare unnatural heparosan and chondroitin [23].

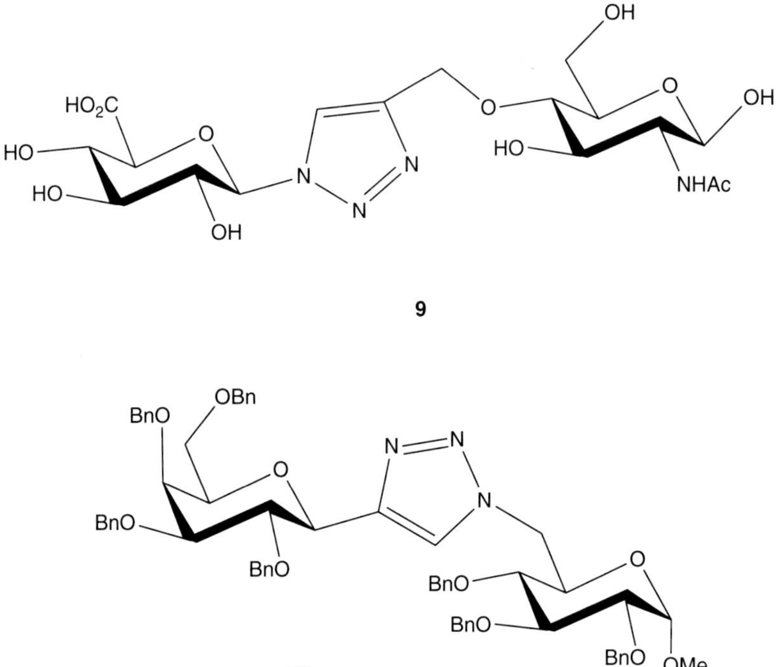

FIGURE 9.4 Disaccharide mimics available from simple sugars by click approach.

Similarly, other disaccharide mimics such as **9** [23] or **10** [24] (Fig. 9.4) were prepared. The azide and acetylene functionalities can be introduced at the desired position of carbohydrate, which makes possible not only convenient synthesis of natural important carbohydrate mimics but also allows for planning the structure of various analogs.

"Click" analogs of important higher sugars such as sialic acid were also prepared in a relatively simple way. For example, the azide **11** can be converted either into the dimer (by reaction with acetylenic derivative of sialic acid) or, via the "click" intermediate **12**, to dendrimeric derivatives (Fig. 9.5) [25].

Application of this approach to oligosaccharides is rather limited, although this methodology provides an easy and rather simple way to biologically interesting molecules obtained by relatively easy coupling of disaccharides (or even oligosaccharides). For example, compounds **16** and **17** (representing a new class of glycoconjugate benzene sulfonamides) were easily prepared and further applied in biological tests against Human Isozymes hCA I, II, and IX (Fig. 9.6) [26].

Dendrimeric structures containing the galabiose and galatriose units were also prepared (such as **18** and **19** in Fig. 9.7) [27].

Carbohydrates (mono-, oligo- or cyclic, e.g., cyclodextrins) are often used as stationary phases in the HPLC technique. The combination of polar stationary phases

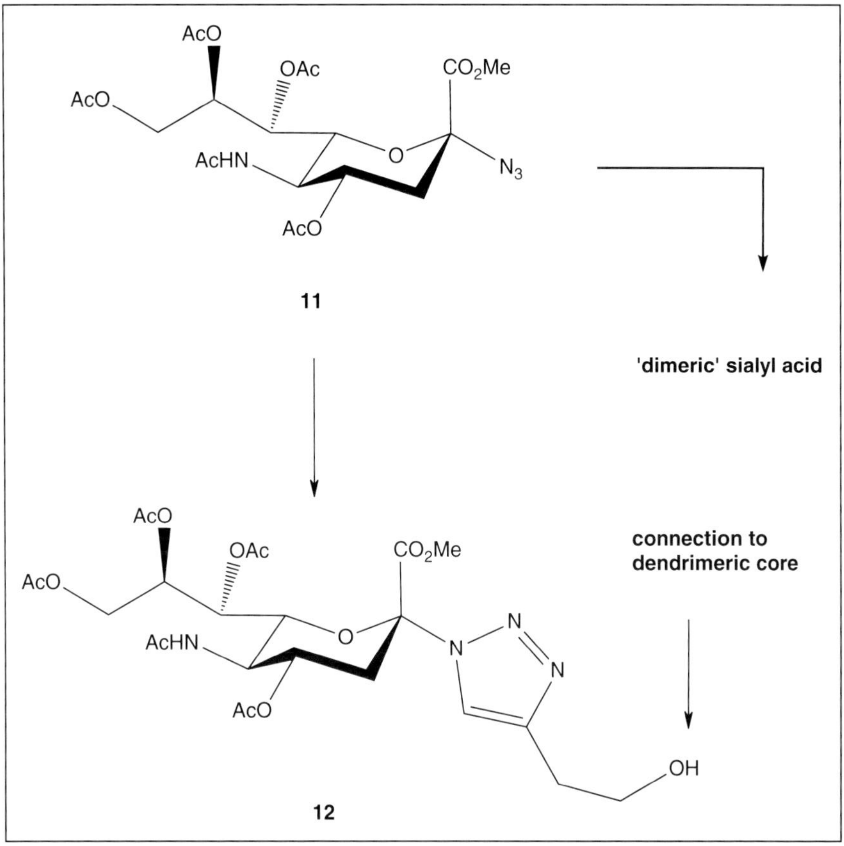

FIGURE 9.5 The sialic acid mimics prepared by click approach.

(sugars) with aqueous/organic mobile phases allows for the separation of very polar compounds. The sugars can be deposited on silica also using the click chemistry approach as shown in Figure 9.8 [28].

The click approach is a convenient methodology for the preparation of macrocyclic compounds, also those built of sugar units. For example, derivatives **23** and **24**, in which glucose or glucosamine are engaged in the macrocyclic skeleton, are available (Fig. 9.9) [29].

Recently, we used the click approach for the preparation of the macrocyclic derivatives with sucrose scaffold.

Sucrose (β-D-fructofuranosyl α-D-glucopyranoside) (**25**), one of the most common disaccharides occurring in nature, is available in amounts of more than 160 million tons per year [30]. Although most of it is consumed on the food market, still a few percent is placed *out of this market.* No wonder, therefore, that "overproduced" sucrose is a subject of interest in chemical laboratories as a starting material for the preparation of fine chemicals [31–36], bio-degradable polymers, or surfactants [37–40]. This cheap raw material is very demanding to work with, because of its very

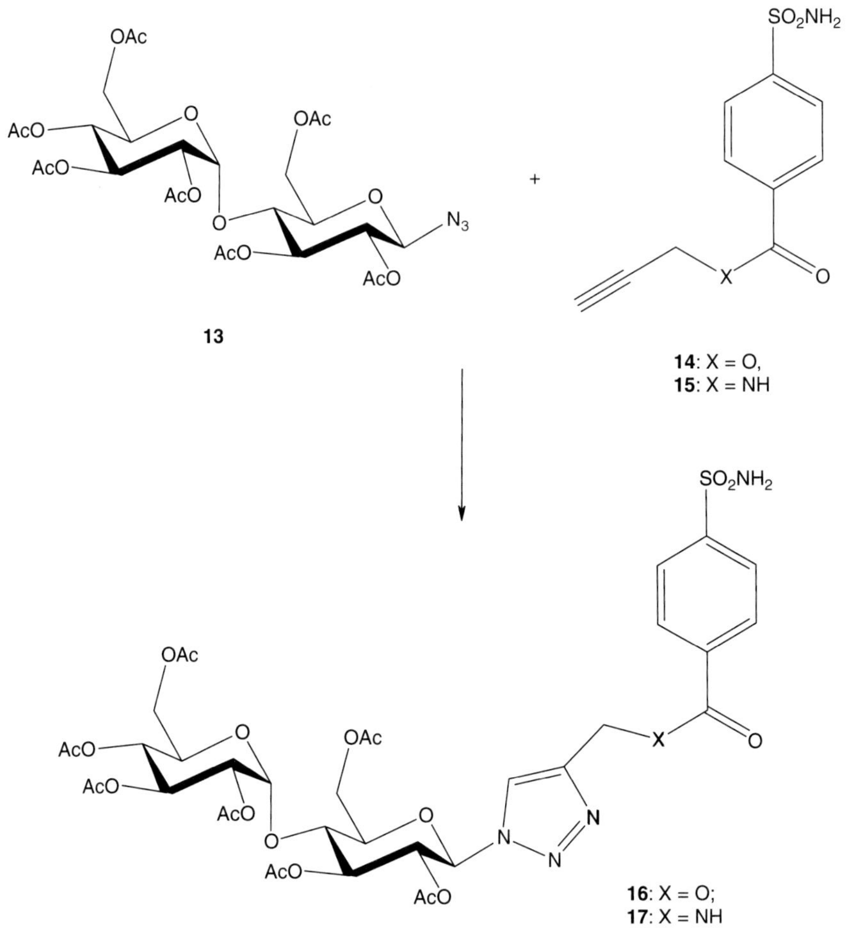

FIGURE 9.6 Glycoconjugate benzene sulfonamides prepared by click approach.

poor solubility in most organic solvents, presence of eight hydroxyl groups which are difficult to differentiate, and high sensitivity of the glycosidic bond in acidic media.

As a part of an ongoing program, we have elaborated a convenient route to 1′,2,3,3′,4,4′-hexa-*O*-benzylsucrose [41] (**26**), useful starting material for the preparation of derivatives modified at the terminal (C6,C6′) positions. Diol **26** also seemed to be a convenient starting material for the preparation of crown ether analogs with sucrose scaffold (Fig. 9.10) [42].

Such compounds possess interesting complexing properties and may be used as chiral receptors for the recognition of quaternary ammonium salts (e.g., derived from amines, amino acids). We have shown that some of them display interesting enantioselectivity toward α-phenylethyl-ammonium chloride. For example, compound **29** showed remarkable enantioselectivity toward the *S*-enantiomer ($K_S/K_R = 1.5$),

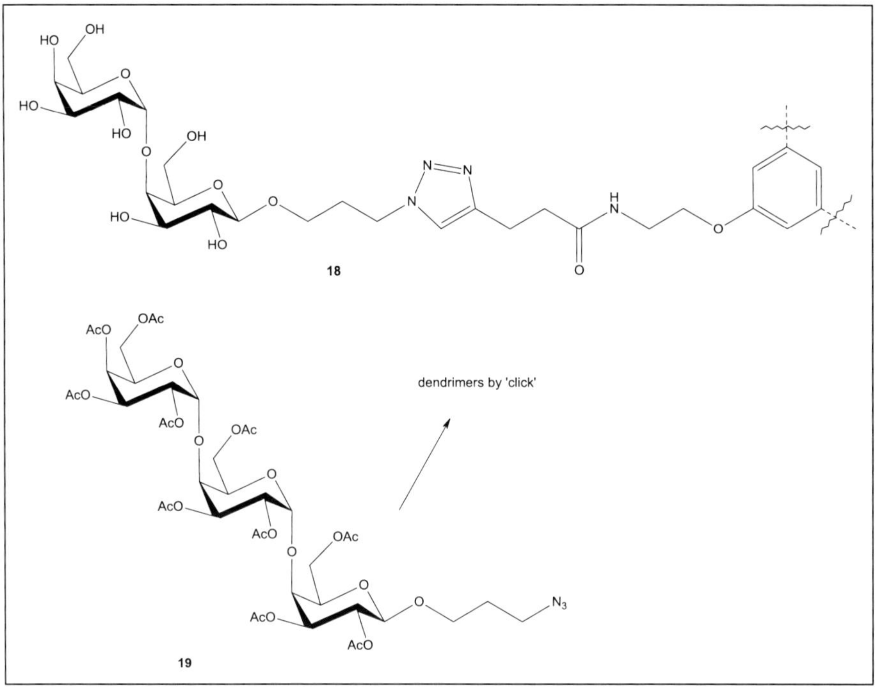

FIGURE 9.7 Preparation of dendrimeric structures containing galabiose and galatriose units.

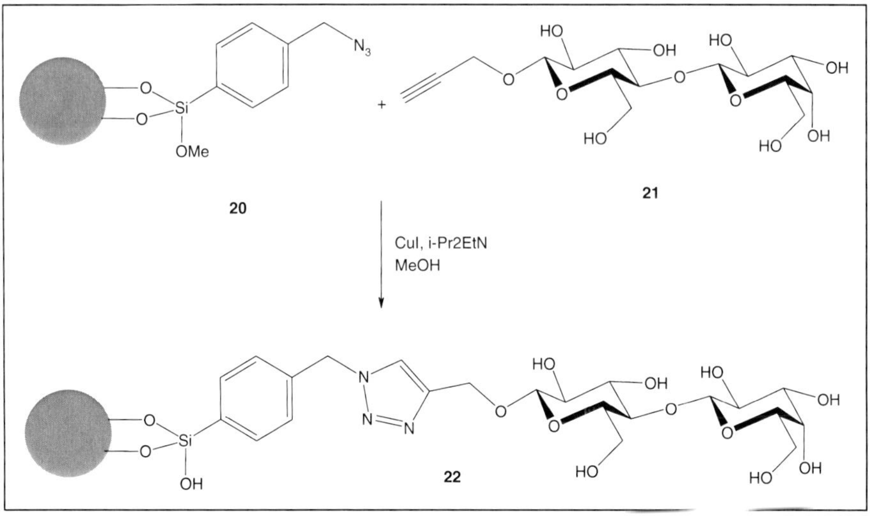

FIGURE 9.8 Preparation of stationary phases for HPLC modified by disaccharide units by click approach.

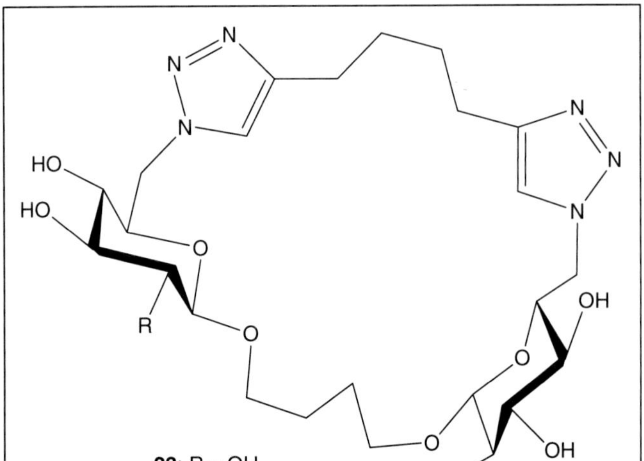

23: R = OH
24: R = NH2

FIGURE 9.9 Macrocycles **23** and **24** available by click approach from simple sugars.

while receptor **30** was much more selective and complexed only one *S*-enantiomer (Fig. 9.11) [43].

The complexing properties of the sucrose receptors might be eventually improved for more symmetrical derivatives, i.e., those with the C2-symmetry; one of the most convenient ways of the preparation of such receptors seemed to be the coupling of two sucrose units by Sharpless "click" approach. As already mentioned, the click methodology is predictable, relatively efficient and cheap, and might be a convenient alternative to other methods of the synthesis of large ring compounds such as classical

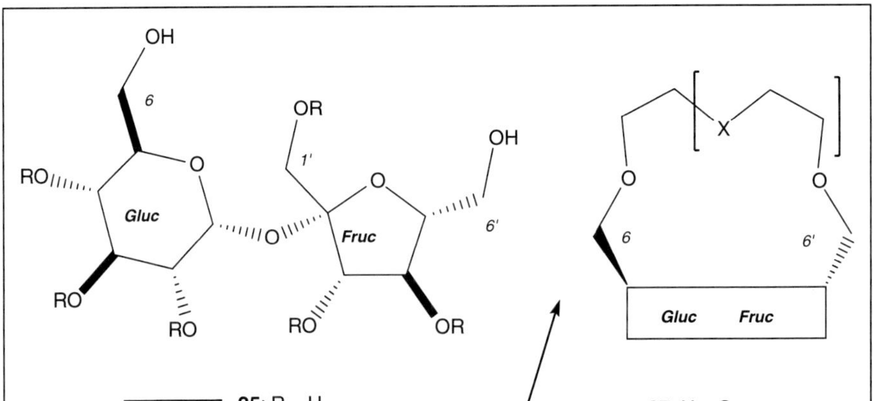

25: R = H
26: R = Bn
27. X = O
28. X = NBn

FIGURE 9.10 Crown ether analogs based on sucrose.

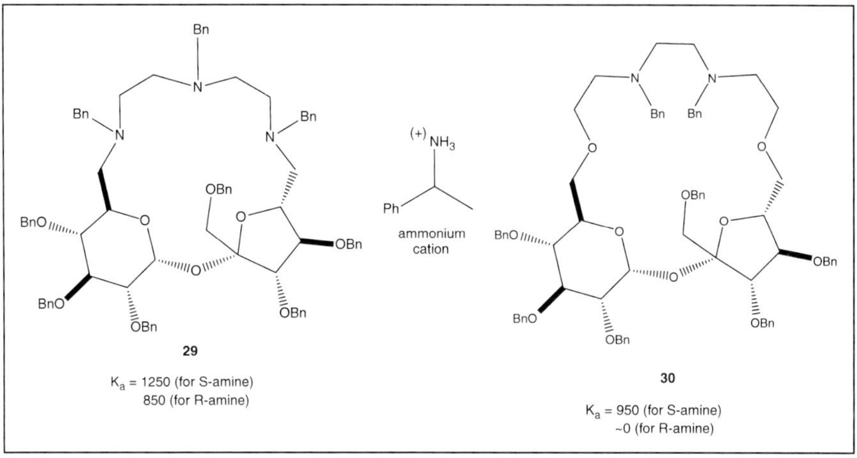

FIGURE 9.11 Complexation of phenylethylammonium cation by sucrose macrocycles.

high-dilution [44], high-pressure techniques [45], metal-templated synthesis of crown ethers [46], or metal-templated macrocyclizations [47].

The sucrose azidoacetylene **33**—chosen as starting material—was prepared readily taking advantage of the very high affinity of the "fructose end" (C-6′) toward silylation (Fig. 9.12) [48].

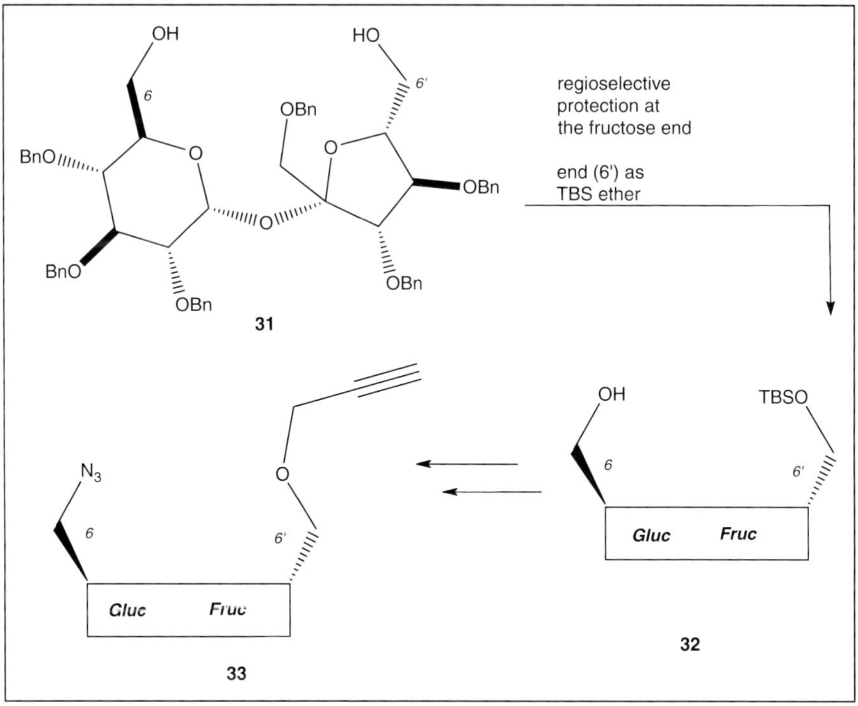

FIGURE 9.12 Preparation of the sucrose azido-acetylene for the synthesis of dimeric sucroses by click approach.

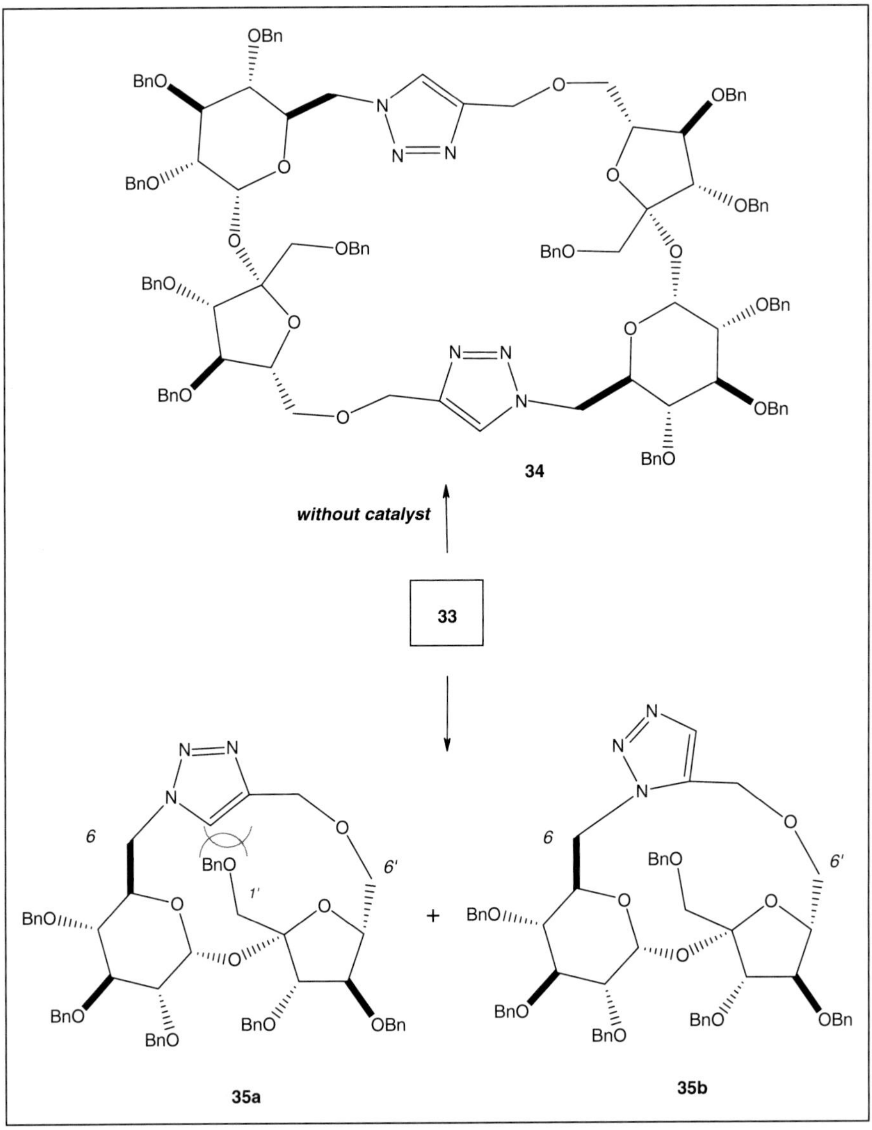

FIGURE 9.13 The click approach to sucrose macrocycles from azidoacetylene **33**.

The 1,3-dipolar cycloaddition reaction performed in toluene and without catalyst provided ***only*** the dimer **34** (in 45% yield). When this process was catalyzed with Cu(I) species in acetonitrile, monomer **35** was formed (ca. 50%) with only traces of the dimer. Surprisingly the monomeric product consisted of two regioisomeric triazoles: **35a** and **35b** (Fig. 9.13) [49].

The unexpected formation of the 1,5-product **35b** might be eventually explained by steric factors. In the expected (from mechanistic point of view) 1,4-isomer **35a,**

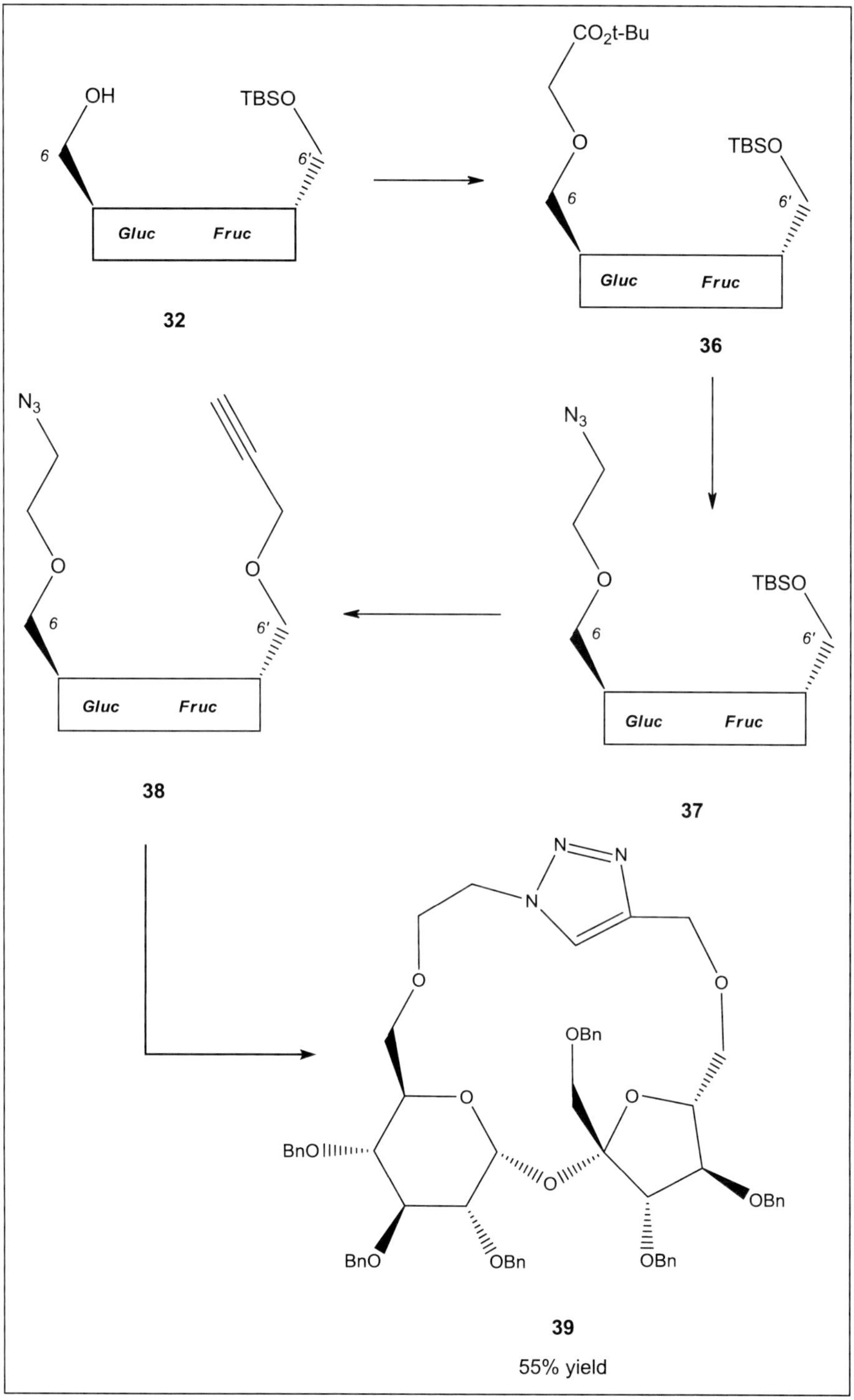

FIGURE 9.14 Synthesis of monomeric sucrose macrocycle by click approach from elongated azidoacetylene **38**.

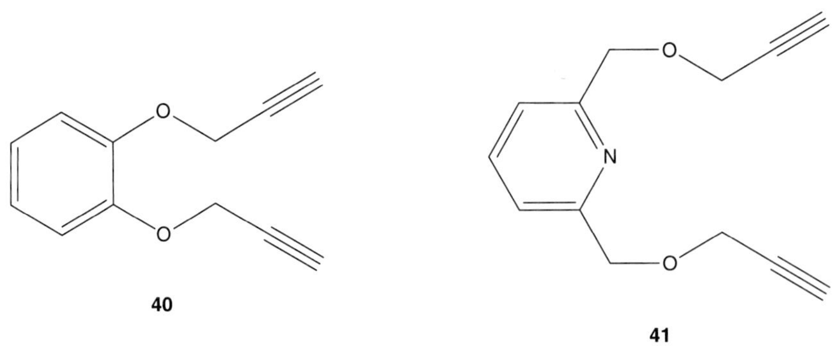

FIGURE 9.15 Linkers used in the synthesis of dimeric sucroses by click approach.

the cavity is rather small, so strong repulsion between the triazole proton and the benzyloxy group at the C1′-position makes this structure not very favored.

If this assumption is correct, enlargement of the cavity should result in the formation of the mechanistically favored 1,4-isomer exclusively. Indeed, the "click" cyclization of the elongated azidoacetylene **38** catalyzed with Cu(I) species led to the expected monomer **39** in good yield. The elongated azidoacetylene **38** was prepared by a rather classical approach taking advantage from the high affinity of the terminal fructose hydroxyl group at the C6′ toward silylating agents and further standard transformations (Fig. 9.14) [50].

To prepare more complex symmetrical sucrose macrocycles, we decided to connect two sucrose molecules via dialkyne linkers **40** and **41** (Fig. 9.15) obtained from catechol or lutidine by standard methodology [51].

Each linker might react efficiently with two molecules of sucrose that were functionalized with an azide at one of the terminal positions. Coupling of sucrose azide **42** (readily available from the mono-silylated derivative **32**) with both linkers under the click reaction conditions (Cu(I)-cat/DIPEA/acetonitrile for **40**; $CuSO_4$, sodium ascorbate, t-BuOH/H_2O for **41**) afforded the C2-symmetrical precursors of macrocycles: **43** and **44** in good yields (Fig. 9.16) [51].

Further steps leading to macrocycles included activation of both terminal hydroxyl groups as mesylates and subsequent reaction with ethylene diamine. The last step was very demanding [51].

The dimesylate derived from **43** did not react with ethylene diamine under the standard conditions (Fig. 9.17). The product **45** could be obtained only in the presence of the template, (*R*)-(-)-2-phenylglycine methyl ester hydrochloride. This template was also effective in the cyclization of the dimesylate derived from **44**; without the template, the yield of the corresponding macrocycle was very low, while in the presence of this amino acid derivative 25% of the product **46** was obtained.

9.2 CONCLUSION

A few examples of the preparation of sucrose macrocycles by "click" methodology shown here point out the usefulness of this approach in the synthesis of such complex

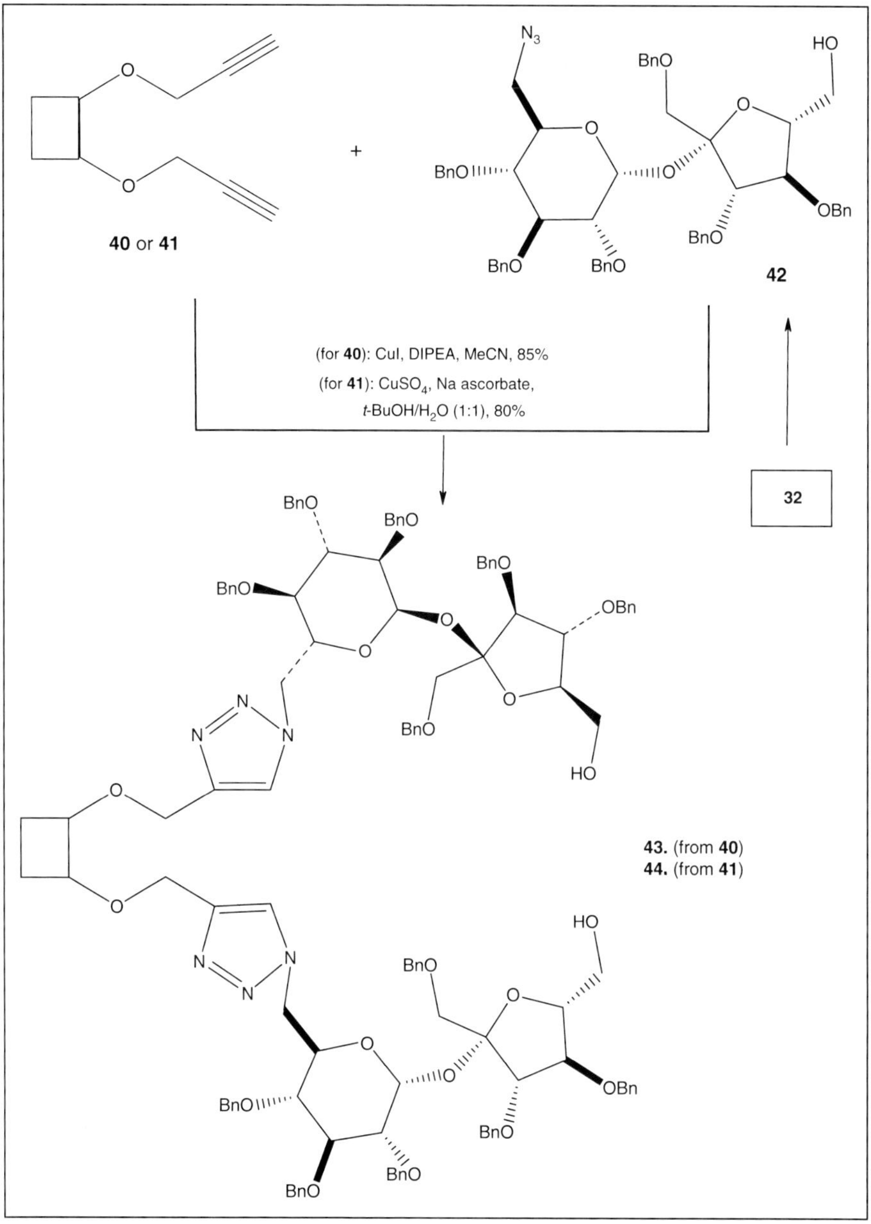

FIGURE 9.16 Synthesis of dimeric linear C2-symmetrical sucroses by click approach.

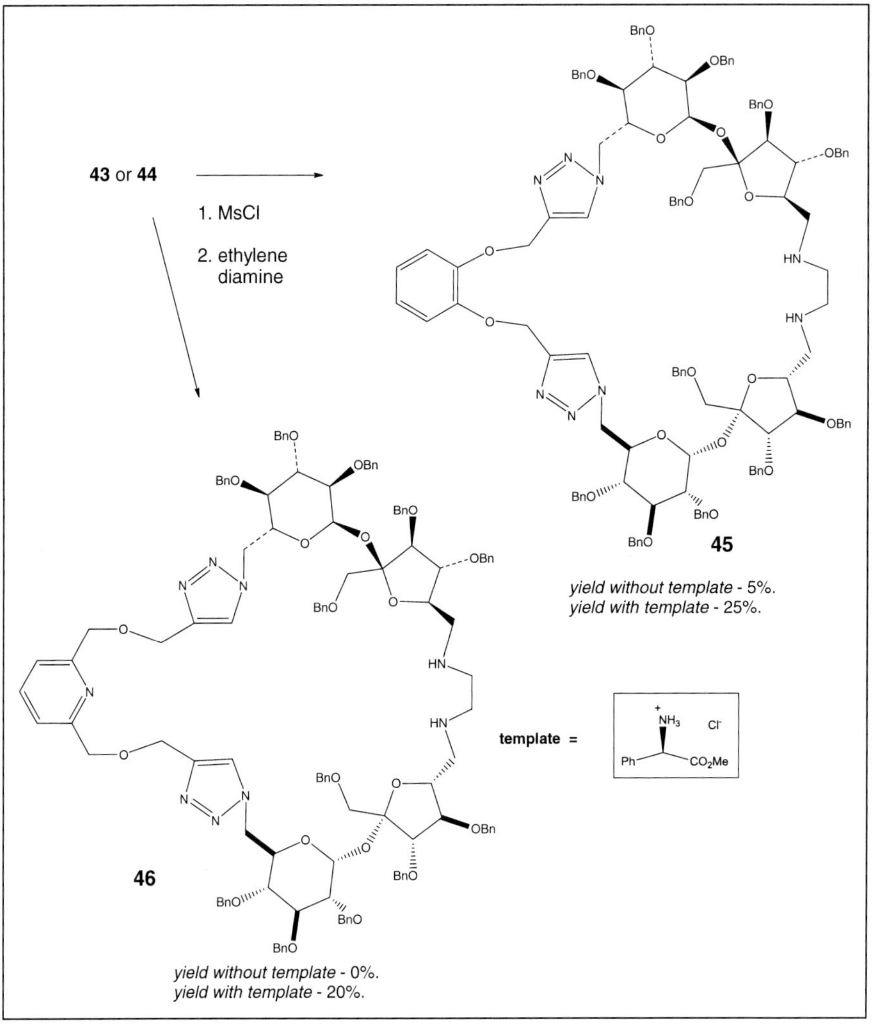

FIGURE 9.17 Cyclization of the linear symmetrical sucroses.

receptors. The functional groups (azide, acetylene) engaged in the preparation of macrocycles can be placed at the desired position of sucrose molecule; such derivatives are easily available from the parent disaccharide. The availability of differently activated sucrose derivatives may open a route to much complex, sucrose-derived products by their "click" coupling with other interesting molecules such as other (biologically important) sugars, amino acids, or peptides.

Although we have not yet prepared such "click-derived" macrocycles in a free form, we believe they are available. We have already reported that fully benzylated sucrose macrocycles (e.g., **27** in Fig. 9.10) can be easily deprotected [42]. This result opens a possibility to study the complexation of the sucrose “click” receptors in a “more biological” (i.e., aqueous) medium.

ACKNOWLEDGMENT

The support from Grant: POIG.01.01.02-14-102/09 (part-financed by the European Union within the European Regional Development Fund) is acknowledged.

REFERENCES

1. Kolb, H. C.; Finn, M. G.; Sharpless, K. B. *Angew. Chem. Int. Ed.* **2001**, *40*, 2004–2021.
2. Huisgen, R. *1,3-Dipolar Cycloaddition Chemistry*, A. Padwa, Ed., Wiley: New York, **1984**; pp 1–176.
3. Padwa, A. *Comprehensive Organic Synthesis*, B. M. Trost, Ed., Pergamon: Oxford, Vol. 4, **1991**; pp 1069–1109.
4. Rostovtsev, V. V.; Green, L. G.; Fokin, V. V.; Sharpless, K. B. *Angew. Chem. Int. Ed.* **2002**, *41*, 2596–2599.
5. T ornøe, C. M.; Christensen, C.; Meldal, M. *J. Org. Chem.* *67*, 3057–3064.
6. Miller, N.; Williams, G. M.; Brimble, M. A. *Org. Lett.* **2009**, *11*, 2409–2412.
7. Wu, P.; Feldman, A. K.; Nugent, A. K.; Hawker, C. J.; Scheel, A.; Voit, B.; Pyun, J.; Fréchet, J. M. J.; Sharpless, K. B.; Fokin, V. V. *Angew. Chem. Int. Ed.* **2004**, *43*, 3928–3932.
8. Chassaing, S.; Kumarraja, M.; Sido, A. S. S.; Pale, P.; Sommer, J. *Org. Lett.* **2007**, *9*, 883–886.
9. Alix, A.; Chassaing, S.; Pale, P.; Sommer, J. *Tetrahedron* **2008**, *64*, 8922–8929.
10. Fournier, D.; Hoogenboom, R.; Schubert, U. S. *Chem. Soc. Rev.* **2007**, *36*, 1369–1380.
11. Moses, J. E.; Moorhouse, A. D. *Chem. Soc. Rev.* **2007**, *36*, 1249–1263.
12. Alvarez, R.; Velazquez, S.; San-Felix, A.; Aquaro, S.; De Clercq, E.; Perno, C.-F.; Karlsson, A.; Balzarini, J.; Camarasa, M. J. *J. Med. Chem.* **1994**, *37*, 4185–4194.
13. Genin, M. J.; Allwine, D. A.; Anderson, D. J.; Barbachyn, M. R.; Emmert, D. E.; Garmon, S. A.; Graber, D. R.; Grega, K. C.; Hester, J. B.; Hutchinson, D. K.; Morris, J.; Reischer, R. J.; Ford, C. W.; Zurenko, G. E.; Hamel, J. C.; Schaadt, R. D.; Stapert, D.; Yagi, B. H. *J. Med. Chem.* **2000**, *43*, 953–970.
14. Kuijpers, B. H. M.; Groothuys, S.; Keereweer, A. R.; Quaedflieg, P. J. L. M.; Blaauw, R. H.; van Delft, F. L.; Rutjes, F. P. J. T. *Org. Lett.* **2004**, *6*, 3123–3126.
15. Srinivas, Ch.; Fang, X.; Wang, Q. *Tetrahedron Lett.* **2005**, *46*, 2331–2336.
16. Hotha, S.; Kashyap, S. J. *J. Org. Chem.* **2006**, *71*, 364–367.
17. Collman, J. P.; Devaraj, N. K.; Chidsey, C. E. D. *Langmuir* **2004**, *20*, 1051–1053.
18. Bodine, K. D.; Gin, D. Y.; Gin, M. S. *J. Am. Chem. Soc.* **2004**, *126*, 1638–1639.
19. Dorner, S.; Westermann, B. *Chem. Commun.* **2005**, *22*, 2852–2854.
20. Billing, J. F.; Nilsson, U. J. *J. Org. Chem.* **2005**, *70*, 4847–4850.
21. Laurent, B. A.; Grayson, S. M. *J. Am. Chem. Soc.* **2006**, *128*, 4238–4239.
22. Perspective: Sumerlin, B. S.; Vogt, A. P. *Macromolecules* **2010**, *43*, 1–13.

23. Bera, S.; Lindhardt, R. J. *J. Org. Chem.* **2011**, *76*, 3181–3193.
24. Marra, A.; Vecchi, A.; Chiappe, C.; Melai, B.; Dondoni, A. *J. Org. Chem.* **2008**, *73*, 2458–2461.
25. Weïwer, M.; Chen, C.-C.; Kemp, M. M.; Linhardt, R. J. *Eur. J. Org. Chem.* **2009**, 2611–2620.
26. Wilkinson, B. L.; Bornaghi, L. F.; Houston, T. A.; Innocenti, A.; Supuran, C. T.; Poulsen, S.-A. *J. Med. Chem.* **2006**, *49*, 6539–6548.
27. Branderhorst, H. M.; Kooij, R.; Salminen, A.; Jongeneel, L. H.; Arnusch, C. J.; Liskamp, R. M. J.; Finne, J.; Pieters, R. J. *Org. Biomol. Chem.* **2008**, *6*, 1425–1434.
28. Moni, L.; Ciogli, A.; Acquarica, I. D.; Dondoni, A.; Gasparrini, F.; Marra, A. *Chem. Eur. J.* **2010**, *16*, 5712–5722.
29. Dörner, S.; Westermann, B. *Chem. Commun.* **2005**, 2852–2854.
30. http://www.fas.usda.gov/htp/sugar/2011/SugarMay2011.pdf.
31. Khan, R. *Pure Appl. Chem.* **1984**, *56*, 833–844 and references therein.
32. Chan, J. Y. C.; Hough, L.; Richardson, A. C. *J. Chem. Soc., Perkin Trans. 1*, **1985**, 1457–1462.
33. de Raadt, A.; Stütz, A. E. *Tetrahedron Lett.* **1992**, *33*, 189–192.
34. Garndnig, G.; Legler, G.; Stütz, A. E. *Carbohydr. Res.* **1996**, *287*, 49–57.
35. Lichtenthaler, F. W. *Acc. Chem. Res.* **2002**, *35*, 728–737.
36. Queneau, Y.; Jarosz, S.; Lewandowski, B.; Fitremann, J. *Adv. Carbohydr. Chem. Biochem.* **2007**, *61*, 221–300.
37. Lichtenthaler, F. W. *Carbohydr. Res.* **1998**, *313*, 69–90.
38. *Carbohydrates as Organic Raw Materials* VCH Verlagsgesellschaft mbH, Weinheim, Lichtenthaler, F. W. Ed. **1991**.
39. Lichtenthaler, F.W.; Mondel, S. *Pure Appl. Chem.* **1997**, *69*, 1853–1866.
40. Descotes, G.; Gagnaire, J.; Bouchu, A.; Thevenet, S.; Giry-Panaud, N.; Salanski, P.; Belniak, S.; Wernicke, A.; Porwanski, S.; Queneau, Y. *Polish J. Chem.* **1999**, *73*, 1069–1077.
41. Mach, M.; Jarosz, S.; Listkowski, A. *J. Carbohydr. Chem.* **2001**, *20*, 485–493.
42. Jarosz, S.; Listkowski, A.; Lewandowski, B.; Ciunik, Z.; Brzuszkiewicz, A. *Tetrahedron* **2005**, *61*, 8485–8492.
43. Lewandowski, B.; Jarosz, S. *Chem. Commun.* **2008**, 6399–6401.
44. Ziegler, K.; Eberle, H.; Ohlinger, H. *Liebigs Ann Chim.* **1933**, *504*, 94–130.
45. Pietraszkiewicz, M.; Salański, P.; Jurczak, J. *J. Chem. Soc. Chem. Commun.* **1983**, 1184–1185.
46. *Template Synthesis of Macrocyclic Compounds* Eds. Gerbeleu, N.V.; Arion, V.B.; Burgess, J., Wiley-VCH: Weinheim, **1999**.
47. Busch, D. H.; Vance, A. L.; Kolchinski, G. *Comprehensive Supramolecular Chemistry*, Lehn, J-M. Ed., Elsevier Science: New York, **1996**; Vol. *9*, pp 1–42.
48. Jarosz, S.; Listkowski, A. *Can. J. Chem.* **2006**, *84*, 492–496.
49. Jarosz, S.; Lewandowski, B.; Listkowski, A. *Synthesis* **2008**, *6*, 913–916.
50. Lewandowski, B.; Jarosz, S. *Synth. Comm.* **2011**, *41*, 1–8.
51. Lewandowski, B.; Jarosz, S. *Org. Lett.* **2010**, *12*, 2532–2535.

PART IV

CARBOHYDRATE CLICK CHEMISTRY IN BIOMEDICAL SCIENCES

10

NEOGLYCOPROTEIN SYNTHESIS USING THE COPPER-CATALYZED AZIDE–ALKYNE CLICK REACTION AND NATIVE CHEMICAL LIGATION

JOANNA M. WOJNAR, DONG JUN LEE, CLIVE W. EVANS, KALYANESWAR MANDAL, STEPHEN B. H. KENT, AND MARGARET A. BRIMBLE

10.1 INTRODUCTION

Glycosylation is the most common and complex post-translational protein modification, introducing a tremendous structural diversity to glycoproteins [1]. The glycan (complex oligosaccharide) can modulate the physicochemical properties of the protein to which it is attached, affecting solubility, viscosity, charge, conformation and dynamics [2]. It can also influence peptide secondary and tertiary structures, increase stability by bestowing protease protection, modulate biological activity, and provide recognition motifs for lectins (carbohydrate binding proteins) [3]. Glycoproteins are involved in numerous varied and essential processes such as immune defense, inflammation, cell adhesion and migration, signalling, endocytosis, and cell growth and differentiation, making them ubiquitous and essential components of the cell [2]. About 50% of all human proteins are thought to be glycosylated [4], with the majority of membrane and extracellular proteins bearing oligosaccharide groups [3a].

Glycosylation in eukaryotes is performed by a plethora of enzymes via eleven biochemical pathways that transfer complex, preformed carbohydrate structures, or a variety of extension or branching building blocks, onto newly synthesized peptides

Click Chemistry in Glycoscience: New Developments and Strategies, First Edition.
Edited by Zbigniew J. Witczak and Roman Bielski.

and proteins [1, 3a]. Imperfections in this process (i.e., aberrant forms of glycosylation) are associated with a variety of diseases, including rheumatoid arthritis, an assortment of congenital disorders, certain muscular dystrophies, and cancer [5]. Glycan changes can mark malignancy, tumor cell invasiveness and metastasis potential [5d, 6]. Glycopeptides are therefore attractive targets for the development of therapeutics, but unfortunately progress has been limited by the complexity and heterogeneity of the oligosaccharides [7].

Almost all glycoproteins exhibit microheterogeneity, that is, a polymorphism associated with their glycan [8]. These various glycoforms—where the peptide backbone is identical but the site and/or nature of the glycosylation can differ—can have a range of different physical, biochemical, and biological properties [9]. While initially they were thought to be of a random nature resulting from a lack of fidelity during their synthesis, today it is apparent that the populations of glycoforms are highly regulated, with the glycans being used by the organism as a means to fine-tune the biological activity of the protein [8, 9]. Unfortunately, these variable patterns of glycosylation present a range of problems to the researcher as the exact structure–activity relationship is difficult to determine, as well as causing regulatory difficulties [9b].

Clearly, a need exists for readily obtainable sources of fully characterized, homogeneous glycoforms. Regrettably, oligosaccharides—unlike other biopolymers—are produced by a template-independent pathway, and so direct genetic methods for manipulating glycosylation patterns and generating a single, homogenous glycoform do not exist [7]. Recombinant expression techniques may not be suitable, as glycosylation is organism specific; artificial cultivation conditions may also lead to aberrant glycosylation patterns [1]. Furthermore, the heterogeneity of the glycoprotein mixture makes isolation of pure glycoforms from natural sources very difficult, requiring abundant source material and many rounds of repetitive chromatography [9b].

One solution to this problem is chemical synthesis which can provide access to homogenous glycopeptides, enabling a quantitative structure activity assessment [1]. Such homogeneous glycopeptides can be used along with synthetic peptide segments in the chemical construction of glycoprotein molecules of defined structure. De novo chemical synthesis of proteins offers many advantages: it provides exquisite atom-level control, also allowing incorporation of unnatural amino acids, labels and markers which may help in unravelling the mechanism of function [10].

While many methods and techniques exist for the rapid and efficient chemical synthesis of peptides (such as Nobel-Prize winning solid-phase peptide synthesis (SPPS) [11]) and proteins (native chemical ligation (NCL) [12]), the limiting factor for the synthesis of glycopeptides is often the construction and attachment of the oligosaccharide. Two main synthetic strategies are usually pursued (Scheme 10.1) [2, 13]. One is to construct the putative glycan–protein link early, in the form of a building block which is then incorporated into SPPS (linear synthesis). Alternatively, the glycan–protein link can be formed late in the synthesis, when the peptide scaffold is complete (convergent synthesis).

Unfortunately, both routes are not without their drawbacks. Construction of the glycosylated amino acid building block is often complex and time consuming, and requires the protection of the carbohydrate hydroxyl groups. Subsequent removal of

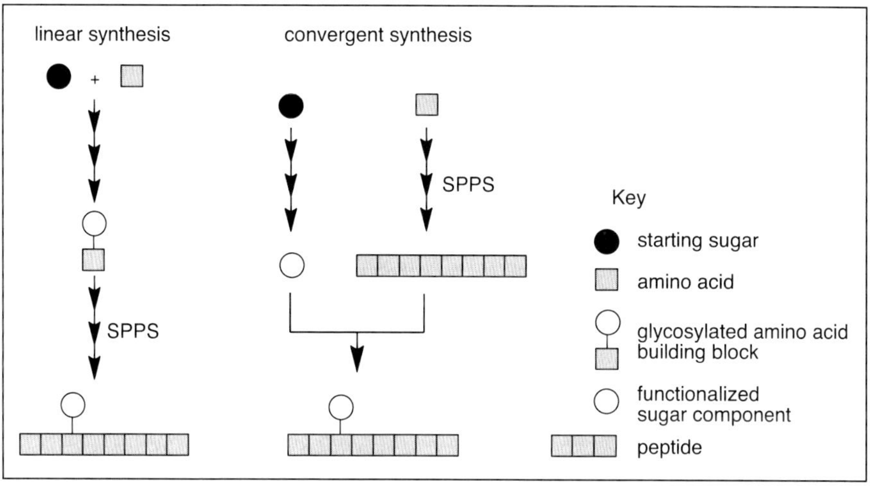

SCHEME 10.1 Linear and convergent synthetic strategies employed in the synthesis of glycopeptides.

protecting groups such as acetates can be an issue as some deprotection methods can affect the epimerization of the α-position of certain amino acids in the protein [4]. In addition, the acid sensitivity of the carbohydrate linkage means it is only compatible with the milder fluorenylmethyloxycarbonyl chloride (Fmoc)–SPPS chemistry [1, 4], precluding the use of Boc-chemistry SPPS for thioester synthesis and subsequent NCL. Site-selective glycosylation post peptide synthesis, on the other hand, requires extensive protection of the peptide as the delicate glycosylation reaction requires strictly anhydrous conditions incompatible with free amino acid side chains [7], and often suffers from low yield [14]. Synthesis of glycoproteins therefore remains a significant challenge, performed in only a few highly specialized laboratories in the world [2].

An alternate approach involves the synthesis of glycopeptide and glycoprotein mimetics that involve unnatural glycan–peptide linkages. These "neoglycopeptides" and "neoglycoproteins" are of great interest as they can simplify a complex, multi-step synthesis. Naturally, the mimetics should retain as much as possible the geometry and electronics of the glycosidic bond, and therefore hopefully maintain native biological activity. An unnatural but isosteric linkage may in fact lead to analogs that are more stable to chemical and metabolic (enzymatic) degradation, leading to longer half-lives and more potent biological activity [2].

10.2 STRATEGIES FOR THE SYNTHESIS OF NEOGLYCOPEPTIDES AS GLYCOPEPTIDE MIMETICS

With the above ideas in mind, we began our investigations in this area, combining the two core strengths of our two groups (Brimble and Kent): synthetic organic

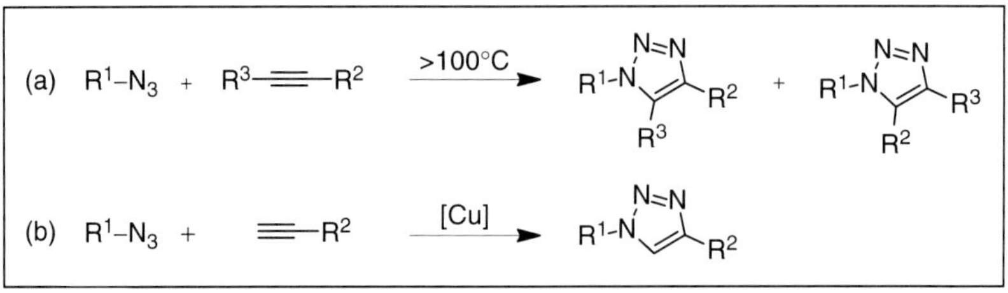

SCHEME 10.2 (a) The thermal 1,3-dipolar cycloaddition usually results in a mixture of both regioisomers after prolonged heating. (b) The CuAAC produces the single 1,4-disubstituted triazole at room temperature.

chemistry, and peptide and protein syntheses. A variety of non-natural glycosidic bonds between the glycan and amino acid have been explored, and are reviewed elsewhere [2, 15]. Given our recent forays into "click" chemistry [16], we focused on the use of the 1,3-dipolar cycloaddition of an organic azide and a terminal alkyne to form a triazole-linked glycoamino acid.

The classical 1,3-dipolar cycloaddition between an azide and an alkyne (Scheme 10.2a) was studied for many years by Huisgen [17], but found limited application due to its low reaction rate and lack of regioselectivity. More recently in 2002, Tornøe and Meldal [18] and simultaneously the Sharpless laboratories [19] described the use of Cu(I) catalysis to significantly accelerate the reaction by seven orders of magnitude. In contrast to the classical, thermal reaction, the Cu(I)-catalyzed azide–alkyne 1,3-dipolar cycloaddition (CuAAC, Scheme 10.2b) shows high regioselectivity, yielding the 1,4-regioisomer nearly exclusively [20]. In addition, the reaction proceeds in many different protic and aprotic solvents, is not significantly affected by the steric and electronic properties of the attached groups, occurs selectively even in the presence of other, unprotected organic functional groups, and is often quantitative, obviating the need for purification steps [21]. Unsurprisingly, the CuAAC reaction has become the most extensively utilised [21b] of the so-called "click" reactions as described by Sharpless [22].

The 1,2,3-triazole linkage itself is quite stable to a range of hydrolytic, oxidizing, and reducing conditions, even at high temperatures [21a]. It is also relatively planar, has a strong dipole moment, and has hydrogen bond accepting ability, characteristics that make it a useful isosteric amide-bond replacement in peptidomimetic chemistry [20, 21b, 23]. As the triazole linkage is non-symmetrical, it can be obtained both through a reaction of a glycosyl azide with an alkynyl amino acid (Scheme 10.3a), and a glycosyl alkyne and an azido amino acid (Scheme 10.3b), resulting in a subtly different orientation of the triazole, as depicted in Scheme 10.3.

The Rutjes lab was the first to synthesize triazole-linked glycoamino acid building blocks in both orientations, from both azido-functionalized glycosides and acetylenic amino acids [24], as well as the inverse, from acetylenic sugars and azido-containing amino acids [24a] (which were independently also reported by Dondoni et al. [25]). Since then, several research groups have been focusing on triazole-linked neoglycopeptides [26]; a comprehensive review is beyond the scope of this article. We

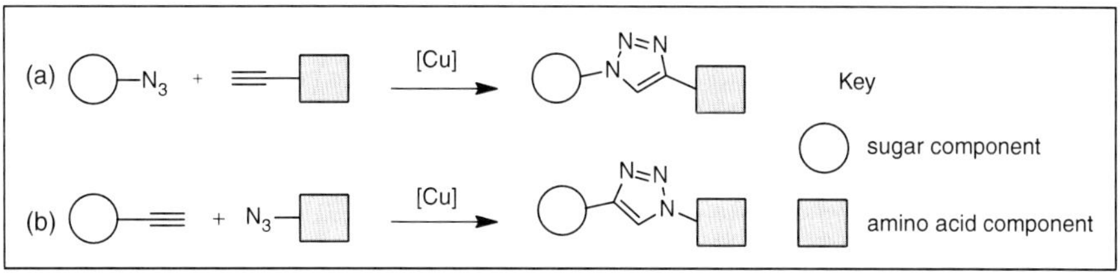

SCHEME 10.3 The two possible triazole orientations dependant on whether a sugar-azide amino acid alkyne (a) or a sugar alkyne and amino acid azide is used (b).

herein report a summary of our recent work on several neoglycopeptides synthesized through the use of the copper-catalyzed azide–alkyne click reaction.

10.3 SYNTHESIS OF NEOGLYCOPEPTIDE MIMICS OF FISH ANTIFREEZE GLYCOPEPTIDES

10.3.1 Combining Azido Peptides with Alkynyl Sugars

We first focused on the fish antifreeze glycoproteins (AFGP), which mainly consist of a repetitive three amino acid unit (Ala-Ala-Thr*), with the threonine hydroxyl bearing the disaccharide β-D-galactose-(1→3)-α-D-*N*-acetyl-galactosamine. Several groups have tackled the synthesis of the natural disaccharide-containing peptide, but it remains a complex and lengthy process [27]. Fortunately, previous studies have shown that the structurally simpler α-*N*-acetylgalactosamine (GalNAc) monoglycoside-containing AFGPs retained comparable antifreeze activity to the more complex naturally occurring antifreeze glycopeptides [28]. We therefore targeted a simpler, triazole-linked monoglycoside neoglycopeptide.

Our first approach utilized the neoglycosyl amino acid building blocks **1** and **2**, synthesized from the commercially available *N*-acetyl-D-galactosamine or 1,3,4,6-tetra-*O*-acety-2-azido-2-deoxy-D-galactose [29], respectively (Scheme 10.4). We found that the rate of conversion during the CuAAC reaction could be dramatically accelerated using microwave irradiation: less than 5 minutes for full conversion using 5 mol% $CuSO_4{\cdot}5H_2O$ or 5 mol% $Cu(OAc)_2$, and 15 mol% sodium ascorbate [30]. Building blocks **1** and **2** were then successfully incorporated into the SPPS of fish antifreeze neoglycopeptides **3** and **4**.

Attempting a convergent route to the same neoglycopeptide targets, azidoalanine (Aza) was incorporated into the peptide backbone under standard SPPS conditions. Subsequent CuAAC reaction with propargyl sugar **5** or **6** under microwave conditions, followed by acetate deprotection yielded the neoglycopeptides **7** and **8** (Scheme 10.5) [30].

During this and related unpublished work we discovered, however, that the azide moiety is not always stable to the SPPS reaction conditions, although this seems to be very sequence dependent. For example, we noticed that the azide can be partially reduced to the amine during the cleavage of the peptide from the resin, an observation

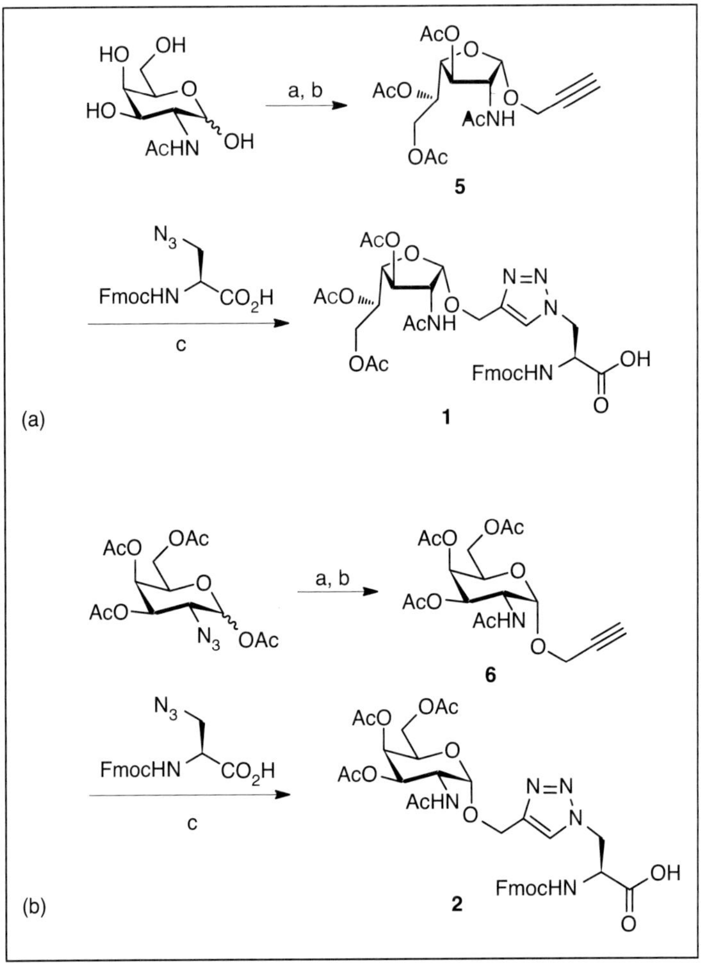

SCHEME 10.4 (a) Synthesis of pyranose building block **1** [30a]. Reaction conditions: a) NaH, propargyl bromide, dry DMF, RT, 2.5 hours; b) NEt_3/Ac_2O, DMAP, RT, overnight; c) 5 mol% $Cu(OAc)_2$, 15 mol% Na ascorbate, 1:1 H_2O/*t*BuOH, MW 80°C, <5 minutes. (b) Synthesis of hexose building block **2** [30b]. Reaction conditions: a) propargyl alcohol, $BF_3{\cdot}Et_2O$, CH_2Cl_2, 35°C, 7 hours; b) Zn, AcOH then Ac_2O/pyridine; c) 5 mol% $Cu(OAc)_2$, 15 mol% Na ascorbate, 1:1 H_2O/*t*BuOH, MW 80°C, <5 minutes.

corroborated by other reports [32]. We have also observed a sequence–dependent elimination of the azide during repetitive Fmoc deprotection cycles under microwave conditions (data not shown). Other reports also suggest that the azide moiety is not compatible with Boc chemistry, specifically the HF cleavage [33]. Our subsequent investigations have therefore made use of propargylglycine (Pra) for SPPS, as the alkyne functionality is fully stable under both Fmoc and Boc chemistry reaction conditions.

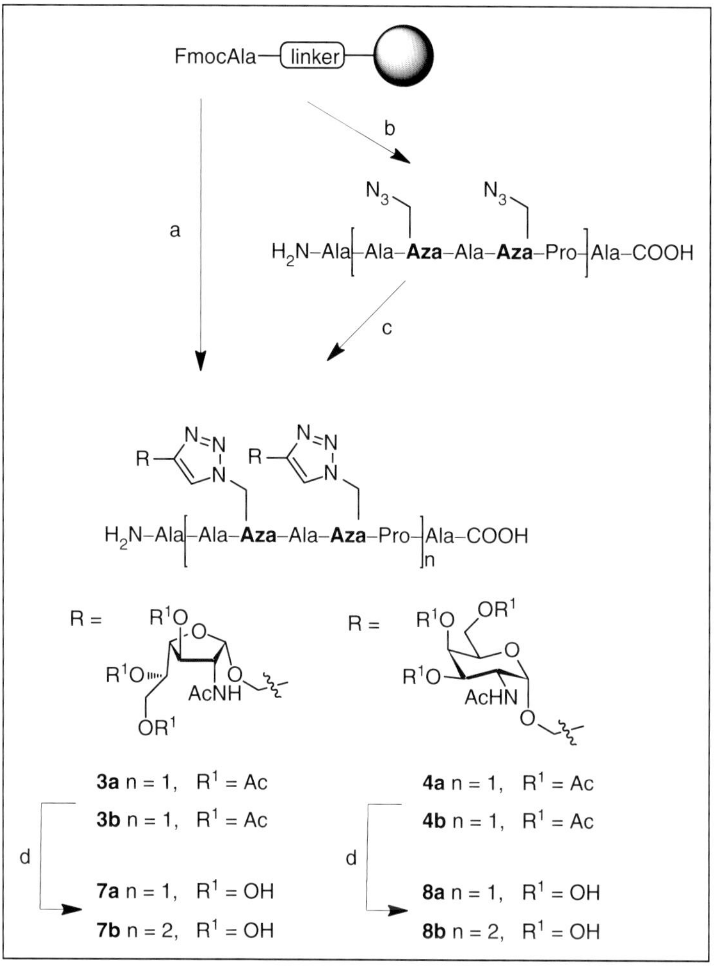

SCHEME 10.5 Two routes to the fish antifreeze neoglycopeptides **7** and **8** [30a, 31]. Reaction conditions: a) MW Fmoc SPPS with **1** or **2**; b) Fmoc SPPS with Aza; c) **5** or **6**, $CuSO_4{\cdot}5H_2O$, Na ascorbate, 1:1 H_2O/*t*BuOH, MW 80°C; d) 1M NaOMe/MeOH or NaOH.

10.4 SYNTHESIS OF NEOGLYCOPEPTIDE MIMICS OF THE MUCINS

10.4.1 Combining Alkynyl Peptides and Azido Sugars

We embarked upon the synthesis of a focused library of neoglycopeptides of the 20 amino acid membrane-bound tumor-associated MUC1 domain [34]. The MUC1 glycoprotein contains an extracellular domain with a variable number of tandem repeats, which possess five potential *O*-glycosylation sites (Fig. 10.1a). A series of propargylated peptides (**9–14**, Fig. 10.1b) were synthesized using standard Fmoc SPPS, incorporating Pra at selected Thr/Ser positions. These peptides were then

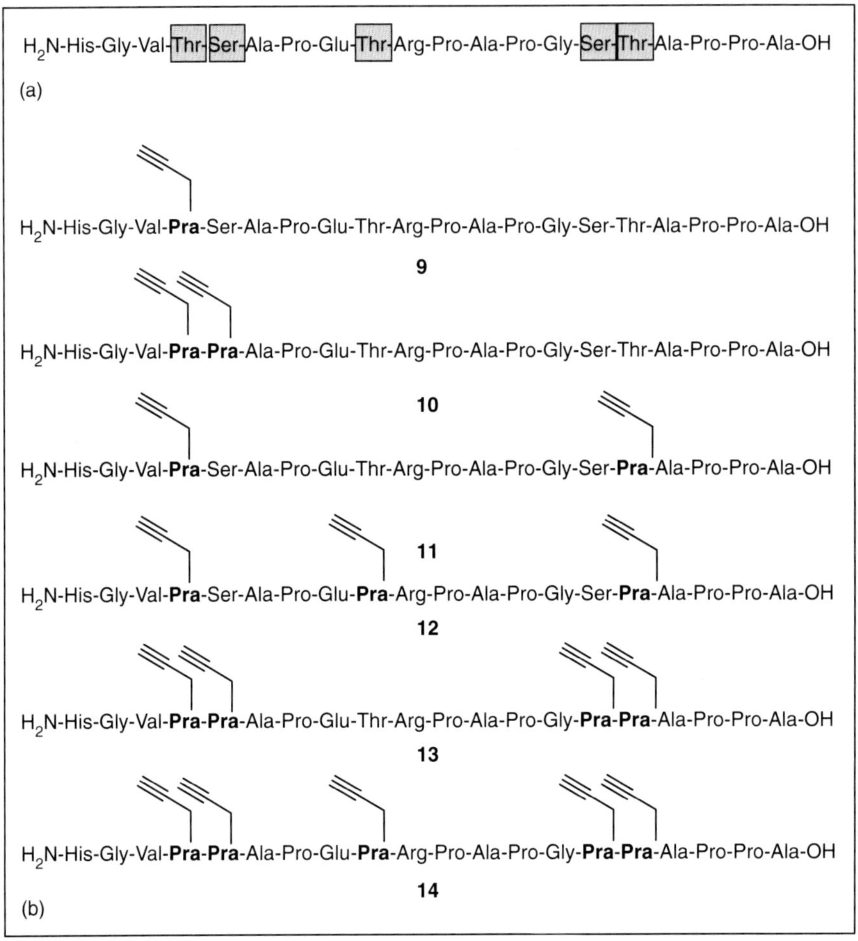

FIGURE 10.1 (a) The MUC1 repeat sequence, with the five possible glycosylation sites highlighted. (b) The propargylated peptides **9–14** synthesized [34].

subsequently subjected to CuAAC reactions with alkynyl sugars using standard conditions, but unfortunately large amounts of copper catalyst and sodium ascorbate reducing agent were required to enable the reaction to proceed. Under these reaction conditions, significant amounts of side products were observed and that complicated the subsequent purification.

The use of large quantities of copper catalyst seems to be a common phenomenon, with equimolar or even excess copper used during the on-resin global click glycoconjugation of alkynyl-substituted peptoids [35], the synthesis of triazole-linked peptide–oligonucleotide conjugates [36], and the ligation of a peptide–azide and a peptide–alkyne to form dimers [37]. This is in stark contrast to the usual catalytic 0.2 eq. [24], 0.1 eq. [25, 26c], and even down to 0.05 [30a] copper equivalents used for the synthesis of the triazole building blocks. The reason for this disparity is not currently understood.

Unfortunately, such large quantities of copper catalyst and ascorbate reducing agent are undesirable due to copper-mediated generation of reactive oxygen species and possible covalent modification of peptide functional groups by ascorbate byproducts, as well as the presence of residual metal after purification [38]. We therefore needed to reevaluate the click conditions. Firstly, we turned to the much milder and biologically friendly reducing agent *tris*(carboxyethyl)phosphine (TCEP). The water soluble TCEP effectively reduces Cu(II) to Cu(I), without significant concomitant reduction of aliphatic azides. It also significantly stabilizes the catalytic Cu(I) oxidation state [39]. Secondly, we explored several different solvent systems, finally settling on 6 M guanidine hydrochloride (GuHCl)/0.2 M Na_2HPO_4 [40]. GuHCl is often used to minimize aggregation and facilitate chemical modification, as it disrupts protein secondary and tertiary structure.

Pleasingly, under these conditions the reaction between the unprotected sugar azide **15** and the propargylated peptides (**9–14**) proceeded to completion both at room temperature and 50°C. As seen previously, the use of microwave irradiation also accelerated the reaction times, up to fourfold. Using the microwave conditions, multiple and even contiguous click reactions were successfully executed, as exemplified by the facile synthesis of the penta-clicked neoglycopeptide **16** (Scheme 10.6).

The GuHCl was critical to the success of this reaction, both because of its solubilizing activity as a chaotrope and also because of its denaturing function. Peptides very

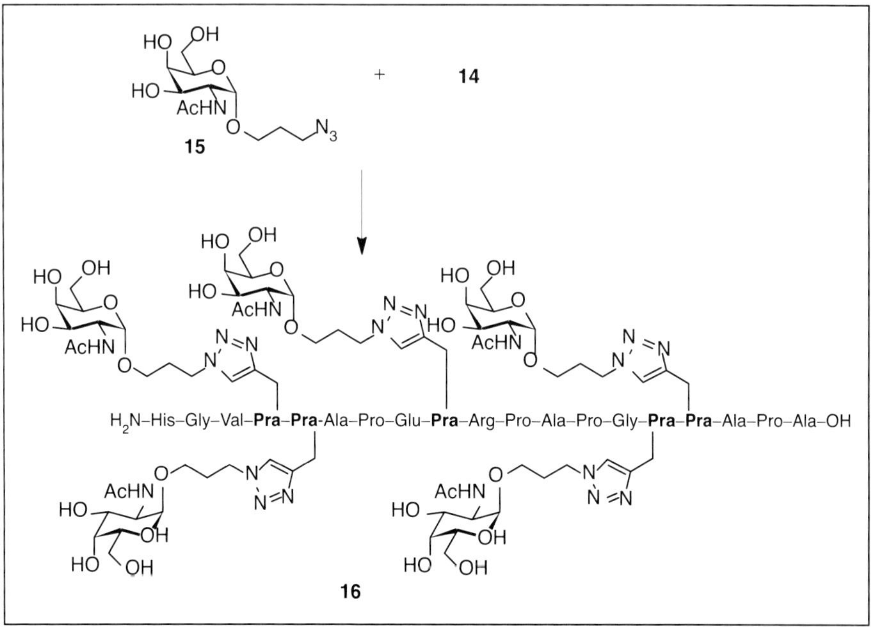

SCHEME 10.6 Synthesis of penta-clicked neoglycopeptide **16** [34]. Reaction conditions: $CuSO_4{\cdot}5H_2O$, TCEP, 6 M GuHCl/0.2 M Na_2HPO_4 buffer (pH 7), MW 50°C, 4 hours.

easily form hydrogen–bonded secondary structures such as α-helices or β-sheets. Even sequences as short as ten amino acids can fold to adopt a well-defined three-dimensional structure [41]. This phenomenon inevitably means that some residues may be internal and away from the surface, making them less accessible than others, and hence drastically reducing their reactivity [42]. In 2007, van Kasteren et al. [43] reported the incorporation, through site-directed mutagenesis, of the unnatural amino acids azidohomoalanine (Aha) and homopropargylglycine (Hpg) into two large proteins, both over 50 kDa in size. The researchers were then able to utilize the CuAAC reaction to quantitatively attach several azido- and propargyl-containing sugars to the proteins, with no loss of biological activity of the neoglycoprotein. However, they observed that reactivity at specific sites was correlated to the accessibility of the residue in question, to the extent that they were able to effect complete transformation at one residue, with completely no reaction observed under the same conditions at a different position. We suspect that longer polypeptides may encounter similar problems and hence the denaturing action of the GuHCl may be critical to the success of the click reaction.

Interestingly, the GuHCl appeared to have additional effects beyond just a denaturing function: the reduction of $CuSO_4$ by TCEP was accelerated, and the Cu(I) species formed were less susceptible to air oxidation. Cu(I) species are generally thermodynamically unstable, easily oxidizing to Cu(II) and/or disproportionating to Cu(0). To maintain sufficient Cu(I) ion concentrations usually requires either inert atmospheres and anhydrous solvents or large quantities of reducing agent [44], which (as mentioned above) may be detrimental to the reaction; although more recently the use of specific tris(heterocyclemethyl)amine ligands to stabilize Cu(I) in aerobic, aqueous conditions has also been described [38, 44, 45]. Therefore the Cu(I) stabilizing property of the 6 M GuHCl solution is a great advantage in addition to its peptide-denaturing ability, making it an attractive choice of solvent for the CuAAC reaction on longer peptides and proteins.

10.5 APPLICATION OF NCL TO THE SYNTHESIS OF NEOGLYCOPEPTIDES

10.5.1 A One-Pot Approach

The abovementioned reports of neoglycopeptide synthesis still face the inherent limitation of SPPS: the practically achievable peptide size is usually no more than 30–40 amino acids. Peptide chains larger than that are difficult to obtain due to the accumulation of byproducts and poor yield [46]. To access chemically synthesized protein constructs for study, different techniques must be applied, such as NCL developed by Dawson and Kent [12]. NCL involves the chemoselective, thiol-catalysed reaction of a *C*-terminal thioester with an *N*-terminal cysteine residue to afford a native amide linkage through an *S* to *N* acyl shift, as depicted in Scheme 10.7. This reaction has dramatically increased the size of proteins accessible by chemical synthesis, the largest reported currently being the 304 amino acid K48-linked tetraubiquitin protein [47].

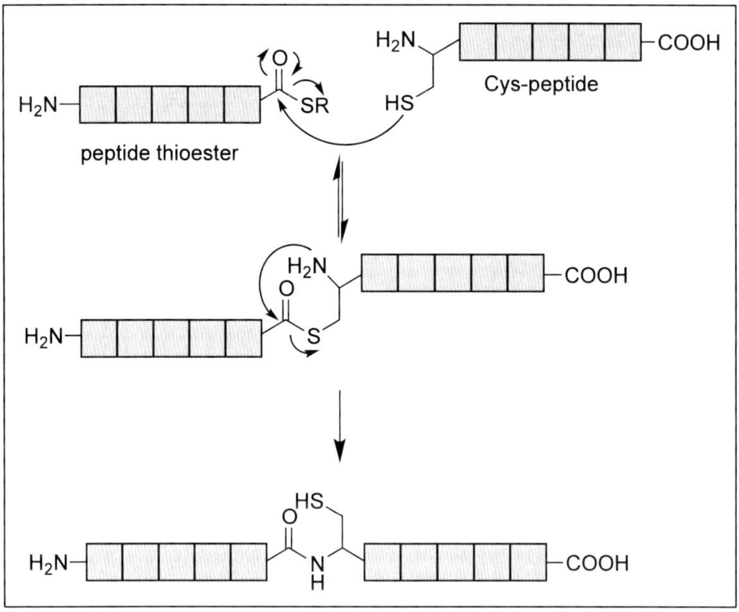

SCHEME 10.7 Mechanism of the native chemical ligation reaction.

Previous work by Macmillan and Blanc [26c] has shown that the triazole linkage is stable to NCL conditions. Other groups had also used a combination of NCL and CuAAC in the synthesis of lipopeptides [48] and *N*-terminally-linked protein dimers [37], but no comprehensive investigation of click conditions between sugar–azides and alkyne–peptides containing all 20 amino acids found in proteins, has been carried out.

We therefore began this investigation by designing and synthesizing three model peptides (**17–19**, Fig. 10.2) containing all 20 genetically encoded amino acids, as well as the thiazolidine (Thz), thioester, and acetamidomethyl (Acm) moieties used in the total chemical synthesis of proteins through NCL. The unprotected peptides **17–19** were then subjected to click conditions with GalNAc-N_3 (**20**) using a series of conditions, with the best results obtained in 6 M GuHCl/0.2 M Na_2HPO_4 (pH 7) and using TCEP as the reducing agent [40]. The click reaction was compatible with all unprotected amino acid side-chain functionalities, the peptide thioester and the Acm protecting group. The Thz moiety (used as a masking group for cysteine during the synthesis of large proteins by NCL) was not stable under these conditions—clean conversion to the cysteine was observed in under 1 hour. Therefore, the click reaction cannot be used directly to make a Thz-neoglycopeptide from a Thz-propargyl peptide. However, a Cys to Thz conversion occurs readily in the presence of formaldehyde [49], and hence may be applied to convert a Cys-neoglycopeptide to its corresponding Thz-neoglycopeptide.

As the best solvent conditions for the click reaction were the same as those for NCL, we postulated that consecutive native chemical ligation and click reactions

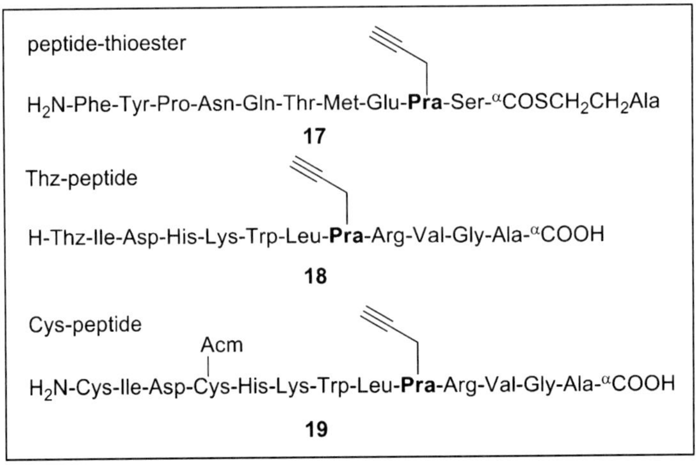

FIGURE 10.2 Three model peptides synthesized by Lee et al. [40].

could be performed in a one-pot fashion, removing the need for intervening work-up and chromatography. Peptide-thioester **17** and Cys-peptide **19** were subjected to NCL conditions overnight. When ligation was complete, $CuSO_4$ was added to the reaction, followed by GalNAc-N_3 (**20**) (Scheme 10.8). Pleasingly, after 2 hours at room temperature, the formation of the clicked triazole peptides was observed with the two different mono-triazole products, and the desired di-triazole peptide **21** distinguished by LC-MS. After 6 hours, all mono-triazole products had undergone complete conversion to the double click neoglycopeptide **21**, which was readily purified by high-performance liquid chromatography (HPLC).

This study demonstrated that the two ligation chemistries—NCL and CuAAC—are fully compatible and can be carried out sequentially in a one-pot fashion, reducing the number of isolation or purification steps. Subsequently, Steinhagen et al. have shown that these chemistries can be also performed simultaneously [50]. Unprotected peptides and azido-sugars can be successfully combined through native chemical ligation and click chemistry in an efficient, one-pot synthesis of a neoglycopeptide. This opens up the whole native chemical ligation toolkit (including kinetically controlled ligation [51], auxiliary-mediated native chemical ligation, and native chemical ligation–desulfurization [52]) for the synthesis of larger neoglycopeptides and complete neoglycoproteins.

10.6 LARGE COMPLEX NEOGLYCOPROTEINS

10.6.1 Combining Click Chemistry with NCL for the de novo Synthesis of Large Neoglycoprotein Constructs

To date, most mimetics of whole glycoproteins are accessed through biomolecular expression systems [53]. This is often achieved by carrying out chemoselective conjugations between unnatural functionalities (such as an alkyne) within the peptide

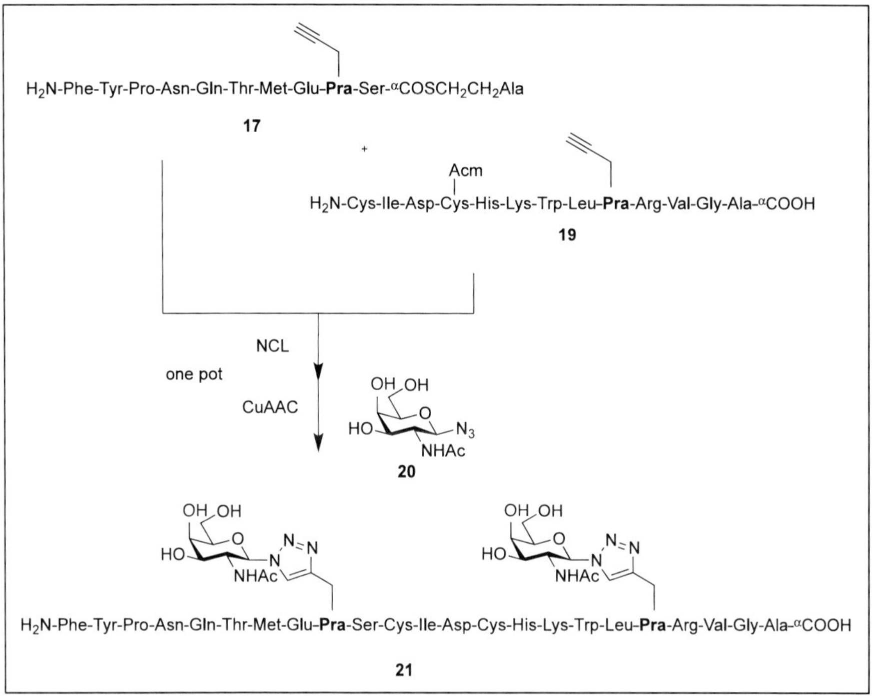

SCHEME 10.8 Ligation of peptide thioester and Cys-peptide, followed by one-pot CuAAC reaction with GalNAc-N_3 [40].

sequence, which have been incorporated through site-specific genetic mutations, with glycosyl partners (such as glycosyl azides). However, global, simultaneous conjugation can lack control, with different residues having varying levels of reactivity that ultimately leads to a complex mixture of fully and partially clicked neoglycoproteins which may or may not be separable. Most importantly, this method is limited to a single type of sugar to be conjugated, and thus does not permit a diversity of glycans to be installed onto the protein of interest.

In contrast to this work, total chemical synthesis was used to make a "synthetic erythropoiesis protein" with full biological activity, in which the complex glycan was replaced by a branched olig(ethyleneoxide-amide) moiety of defined chemical structure [54]. Our goal is to undertake the total chemical synthesis of neoglycoproteins. The main advantage of total chemical synthesis is that the type and extent of glycoconjugation can be controlled in a site-specific manner. We wished to take advantage of this and utilize the methodology developed within our research groups in order to synthesize a large and complex neoglycoprotein in a fully convergent manner. Our initial target was the synthesis of a click neoglycoprotein analog of erythropoietin (EPO).

Erythropoietin is the principal hormone that regulates mammalian erythrocyte differentiation [55]. The EPO polypeptide chain contains 165 amino acids, and has four glycosylation sites—three N-linked, and an O-linked; this glycosylation is important for its *in vivo* activity [56], although non-glycosylated EPO retains full receptor binding activity *in vitro*. Glycosylated human EPO is only produced recombinantly in Chinese hamster ovary cells [57], and currently a number of commercial preparations of recombinant human EPO are available for medical use. Several research groups [54, 58] have attempted the synthesis of EPO analogs that may display *in vivo* activities; however, the only significant mimetic is that reported by Kochendoerfer et al. [54] discussed above.

We devised an efficient, fully convergent route to allow four different types of sugars to be conjugated onto the peptide backbone to afford a diversely glycosylated neoglycoprotein analog of EPO. The synthetic strategy is outlined in Scheme 10.9.

We propose to utilize a five peptide segment strategy, based on the work of Liu [59] whereby four of the natural glycosylation sites are substituted with propargylglycine, and take advantage of the different native chemical ligation strategies for the click

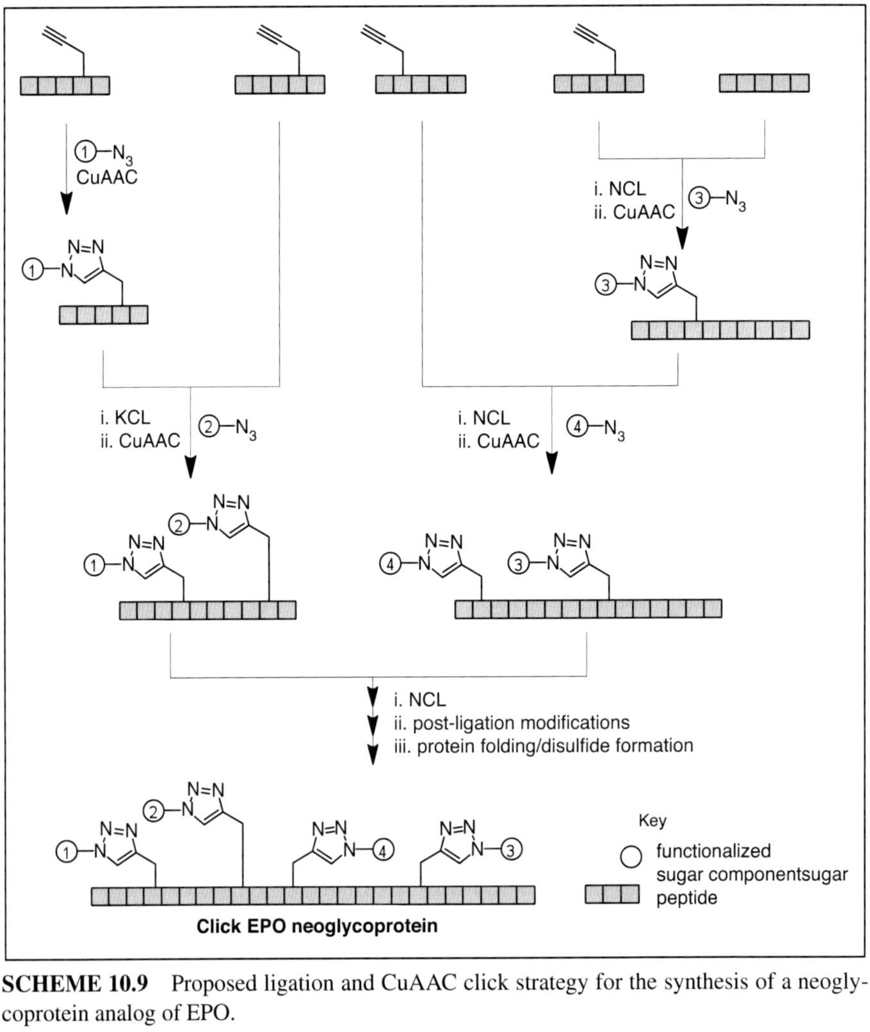

SCHEME 10.9 Proposed ligation and CuAAC click strategy for the synthesis of a neoglycoprotein analog of EPO.

conjugation of different glycosyl azides. Using NCL and kinetically controlled ligation, together with click chemistry, the successful synthesis of a full-length click EPO polypeptide containing four different types of sugars will be carried out. The success of this study will demonstrate the viability of our devised methodology [34, 40] as a versatile chemical toolkit for the fully convergent synthesis of complex neoglycoproteins using click chemistry, native chemical ligation, and kinetically controlled ligation. The synthesis of a neoglycoEPO will also provide a route to the systematic exploration of the effect of glycan structure on the *in vivo* biological activity of erythropoietin. This work is currently underway in our laboratories, and will be reported in due course.

10.7 CONCLUSION

The synthesis of glycoprotein mimics, or neoglycoproteins, is emerging as an exciting area of research as it allows the rapid and efficient production of glycosylated protein constructs. By replacing the sensitive and synthetically difficult peptide N-glycosidic linkage with an isosteric moiety, many challenges of current glycopeptide synthesis can be overcome. In particular, the robust and reliable copper-catalyzed azide–alkyne "click" reaction is an attractive choice to replace the glycosylation reaction. The triazole linkage thus generated may in fact impart extra metabolic and chemical stability to the neoglycopeptide.

By combining our synthetic organic chemistry, and peptide and protein synthesis experience, our research groups embarked on a series of projects to develop tools and methodologies for the synthesis of neoglycopeptides and neoglycoproteins. Both the use of triazole-linked amino acid building blocks in SPPS, as well as performing click chemistry on assembled peptides, were explored [30, 31]. The advantages of using the denaturing 6 M GuHCl/0.2 M Na_2HPO_4 buffer for the CuAAC reaction on larger peptides [34], and the powerful combination of a one-pot native chemical ligation and CuAAC click reaction were demonstrated [40]. These tools are currently being applied to the synthesis of a large click neoglycoprotein mimic of EPO.

The click chemistry/native chemical ligation combination developed herein brings chemists one step closer to being able to conjugate synthetic proteins, assembled using chemical synthesis, with glycosides in a site-specific manner. The only difference between the synthetic neoglycoprotein and the native glycoprotein is the nature of the glycoside–peptide linkage. Such neoglycopeptides/neoglycoproteins provide a valuable focused library of molecular probes of defined structure that provide useful tools for biological evaluation.

REFERENCES

1. Haase, C.; Seitz, O. *Top. Curr. Chem.* **2007**, *267*, 1–36.
2. Specker, D.; Wittmann, V. *Top. Curr. Chem.* **2007**, *267*, 65–107.
3. (a) Lauc, G.; Rudan, I.; Campbell, H.; Rudd, P. M. *Mol. BioSyst.* **2010**, *6*, 329–335. (b) Wittmann, V. In *Glycoscience;* Coté, G., Flitsch, S., Ito, Y., Kondo, H., Nishimura, S.I., Yu, B., Eds.; Springer: New York, 2008; pp 1771–1793.

4. Kajihara, Y.; Yamamoto, N.; Okamoto, R.; Hirano, K.; Murase, T. *Chem. Rec.* **2010**, *10*, 80–100.
5. (a) Arnold, J. N.; Wormald, M. R.; Sim, R. B.; Rudd, P. M.; Dwek, R. A. *Annu. Rev. Immunol.* **2007**, *25*, 21–50. (b) Dube, D. H.; Bertozzi, C. R. *Nat. Rev. Drug Discovery* **2005**, *4*, 477–488. (c) Ohtsubo, K.; Marth, J. D. *Cell* **2006**, *126*, 855–867. (d) Thayer, D. A.; Wong, C.H. *Top. Curr. Chem.* **2007**, *267*, 37–63.
6. Reis, C. A.; Osorio, H.; Silva, L.; Gomes, C.; David, L. *J. Clin. Pathol.* **2010**, *63*, 322–329.
7. Pratt, M. R.; Bertozzi, C. R. *Chem. Soc. Rev.* **2005**, *34*, 58–68.
8. Wittmann, V. In *Glycoscience*; Coté, G., Flitsch, S., Ito, Y., Kondo, H., Nishimura, S.I., Yu, B., Eds.; Springer: New York, **2008**; pp 1735–1770.
9. (a) Bill, R. M.; Flitsch, S. L. *Chem. Biol.* **1996**, *3*, 145–149. (b) Davis, B. G. *Chem. Rev. (Washington, DC, U. S.).***2002**, *102*, 579–601.
10. (a) Kent, S. B. H. *Chem. Soc. Rev.* **2009**, *38*, 338–351. (b) Kent, S. *Curr. Opin. Biotechnol.* **2004**, *15*, 607–614.
11. Merrifield, R. B. *J. Am. Chem. Soc.* **1963**, *85*, 2149–2154.
12. Dawson, P. E.; Muir, T. W.; Clark-Lewis, I.; Kent, S. B. H. *Science* **1994**, *266*, 776–779.
13. Nicotra, F.; Cipolla, L.; Peri, F.; La Ferla, B.; Redaelli, C. *Adv. Carbohydr. Chem. Biochem.* **2008**, *61*, 353–398.
14. Gamblin, D. P.; Scanlan, E. M.; Davis, B. G. *Chem. Rev.* **2009**, *109*, 131–163.
15. Lee, Y. C.; Begley, T. P. In *Wiley Encyclopedia of Chemical Biology*, Begley, T. P., Ed.; John Wiley & Sons, Inc.: Hoboken, NJ, **2007**, pp 1–10. DOI: 10.1002/9780470048672.wecb379.
16. (a) Rathwell, K.; Sperry, J.; Brimble, M. A. *Tetrahedron* **2010**, *66*, 4002–4009. (b) Choi, K. W.; Brimble, M. A. *Org. Biomol. Chem.* **2008**, *6*, 3518–3526.
17. (a) Huisgen, R. *Proc. Chem. Soc., London.* **1961**, 357–369. (b) Huisgen, R.; Szeimies, G.; Möbius, L. *Chem. Ber.* **1967**, *100*, 2494–2507.
18. Tornøe, C. W.; Christensen, C.; Meldal, M. *J. Org. Chem.* **2002**, *67*, 3057–3064.
19. Rostovtsev, V. V.; Green, L. G.; Fokin, V. V.; Sharpless, K. B. *Angew. Chem., Int. Ed.* **2002**, *41*, 2596–2599.
20. Holub, J. M.; Kirshenbaum, K. *Chem. Soc. Rev.* **2010**, *39*, 1325–1337.
21. (a) Hein, J. E.; Fokin, V. V. *Chem. Soc. Rev.* **2010**, *39*, 1302–1315. (b) Aragão-Leoneti, V.; Campo, V. L.; Gomes, A. S.; Field, R. A.; Carvalho, I. *Tetrahedron* **2010**, *66*, 9475–9492. (c) Meldal, M.; Tornøe, C. W. *Chem. Rev.* **2008**, *108*, 2952–3015.
22. Kolb, H. C.; Finn, M. G.; Sharpless, K. B. *Angew. Chem., Int. Ed.* **2001**, *40*, 2004–2021.
23. (a) Pedersen, D. S.; Abell, A. *Eur. J. Org. Chem.* **2011**, 2399–2411. (b) Angell, Y. L.; Burgess, K. *Chem. Soc. Rev.* **2007**, *36*, 1674–1689.
24. (a) Kuijpers, B. H. M.; Groothuys, S.; Keereweer, A. R.; Quaedflieg, P. J. L. M.; Blaauw, R. H.; van Delft, F. L.; Rutjes, F. P. J. T. *Org. Lett.* **2004**, *6*, 3123–3126. (b) Groothuys, S.; Kuijpers, B. H. M.; Quaedflieg, P. J. L. M.; Roelen, H. C. P. F.; Wiertz, R. W.; Blaauw, R. H.; van Delft, F. L.; Rutjes, F. P. J. T. *Synthesis-Stuttgart* **2006**, *18,* 3146–3152. (c) Kuijpers, B. H. M.; Groothuys, S.; Hawner, C.; ten Dam, J.; Quaedflieg, P. J. L. M.; Schoemaker, H. E.; van Delft, F. L.; Rutjes, F. P. J. T. *Org. Process Res. Dev.* **2008**, *12*, 503–511.
25. Dondoni, A.; Giovannini, P. P.; Massi, A. *Org. Lett.* **2004**, *6*, 2929–2932.

26. (a) Lin, H. N.; Walsh, C. T. *J. Am. Chem. Soc.* **2004**, *126*, 13998–14003. (b) Miller, N.; Williams, G. M.; Brimble, M. A. *Int. J. Pept. Res. Ther.* **2010**, *16*, 125–132. (c) Macmillan, D.; Blanc, J. *Org. Biomol. Chem.* **2006**, *4*, 2847–2850.
27. (a) Peltier, R.; Brimble, M. A.; Wojnar, J. M.; Williams, D. E.; Evans, C. W.; DeVries, A. L. *Chem. Sci.* **2010**, *1*, 538–551. (b) Wojnar, J. M.; Evans, C. W.; DeVries, A. L.; Brimble, M. A. *Aust. J. Chem.* **2011**, *64*, 723–731.
28. Tachibana, Y.; Fletcher, G. L.; Fujitani, N.; Tsuda, S.; Monde, K.; Nishimura, S. I. *Angew. Chem., Int. Ed.* **2004**, *43*, 856–862.
29. Goddard-Borger, E. D.; Stick, R. V. *Org. Lett.* **2007**, *9*, 3797–3800.
30. (a) Miller, N.; Williams, G. M.; Brimble, M. A. *Org. Lett.* **2009**, *11*, 2409–2412. (b) Miller, N.; Williams, G. M.; Brimble, M. A. *Org. Lett.* **2010**, *12*, 1375–1376.
31. Miller, N.; Brimble, M. A. Unpublished work. **2010**.
32. (a) Schneggenburger, P. E.; Worbs, B.; Diederichsen, U. *J. Pept. Sci.* **2010**, *16*, 10–14. (b) Katayama, H.; Hojo, H.; Ohira, T.; Nakahara, Y. *Tetrahedron Lett.* **2008**, *49*, 5492–5494.
33. Punna, S.; Kuzelka, J.; Wang, Q.; Finn, M. G. *Angew. Chem., Int. Ed.* **2005**, *44*, 2215–2220.
34. Lee, D. J.; Harris, P. W. R.; Brimble, M. A. *Org. Biomol. Chem.* **2011**, *9*, 1621–1626.
35. Norgren, A. S.; Budke, C.; Majer, Z.; Heggemann, C.; Koop, T.; Sewald, N. *Synthesis-Stuttgart.* **2009**, 488–494.
36. Gogoi, K.; Mane, M. V.; Kunte, S. S.; Kumar, V. A. *Nucleic Acids Res.* **2007**, *35*, e139.
37. Xiao, J.; Tolbert, T. J. *Org. Lett.* **2009**, *11*, 4144–4147.
38. Hong, V.; Presolski, S. I.; Ma, C.; Finn, M. G. *Angew. Chem., Int. Ed.* **2009**, *48*, 9879–9883.
39. Wang, Q.; Chan, T. R.; Hilgraf, R.; Fokin, V. V.; Sharpless, K. B.; Finn, M. G. *J. Am. Chem. Soc.* **2003**, *125*, 3192–3193.
40. Lee, D. J.; Mandal, K.; Harris, P. W. R.; Brimble, M. A.; Kent, S. B. H. *Org. Lett.* **2009**, *11*, 5270–5273.
41. Honda, S.; Akiba, T.; Kato, Y. S.; Sawada, Y.; Sekijima, M.; Ishimura, M.; Ooishi, A.; Watanabe, H.; Odahara, T.; Harata, K. *J. Am. Chem. Soc.* **2008**, *130*, 15327–15331.
42. Mendoza, V. L.; Vachet, R. W. *Mass Spectrom. Rev.* **2009**, *28*, 785–815.
43. van Kasteren, S. I.; Kramer, H. B.; Jensen, H. H.; Campbell, S. J.; Kirkpatrick, J.; Oldham, N. J.; Anthony, D. C.; Davis, B. G. *Nature* **2007**, *446*, 1105–1109.
44. Chan, T. R.; Hilgraf, R.; Sharpless, K. B.; Fokin, V. V. *Org. Lett.* **2004**, *6*, 2853–2855.
45. Presolski, S. I.; Hong, V.; Cho, S.-H.; Finn, M. G. *J. Am. Chem. Soc.* **2010**, *132*, 14570–14576.
46. Kent, S. B. *Annu. Rev. Biochem.* **1988**, *57*, 957–989.
47. Kumar, K. S. A.; Bavikar, S. N.; Spasser, L.; Moyal, T.; Ohayon, S.; Brik, A. *Angew. Chem., Int. Ed.* **2011**, *50*, 6137–6141.
48. Musiol, H. J.; Dong, S.; Kaiser, M.; Bausinger, R.; Zumbusch, A.; Bertsch, U.; Moroder, L. *ChemBioChem.* **2005**, *6*, 625–628.
49. (a) Villain, M.; Vizzavona, J.; Rose, K. *Chem. Biol.* **2001**, *8*, 673–679. (b) Gesquière, J. C.; Diesis, E.; Tartar, A. *J. Chem. Soc., Chem. Commun.* **1990**, 1402–1403.
50. Steinhagen, M.; Holland-Nell, K.; Meldal, M.; Beck-Sickinger, A. *Biopolymers* **2011**, *96*, 454–454.
51. Bang, D.; Pentelute, B. L.; Kent, S. B. H. *Angew. Chem., Int. Ed.* **2006**, *45*, 3985–3988.
52. Rohde, H.; Seitz, O. *Pept. Sci.* **2010**, *94*, 551–559.

53. (a) Chalker, J. M.; Bernardes, G. J. L.; Davis, B. G. *Acc. Chem. Res.* **2011**, *44*, 730–741. (b) Gamblin, D. P.; van Kasteren, S. I.; Chalker, J. M.; Davis, B. G. *FEBS J.* **2008**, *275*, 1949–1959. (c) Magliery, T. J. *Med. Chem. Rev. Online.* **2005**, *2*, 303–323. (d) Matsubara, N.; Oiwa, K.; Hohsaka, T.; Sadamoto, R.; Niikura, K.; Fukuhara, N.; Takimoto, A.; Kondo, H.; Nishimura, S.I. *Chem.Eur. J.* **2005**, *11*, 6974–6981. (e) van Kasteren, S. I.; Kramer, H. B.; Gamblin, D. P.; Davis, B. G. *Nat. Protoc.* **2007**, *2*, 3185–3194.

54. Kochendoerfer, G. G.; Chen, S.Y.; Mao, F.; Cressman, S.; Traviglia, S.; Shao, H.; Hunter, C. L.; Low, D. W.; Cagle, E. N.; Carnevali, M.; Gueriguian, V.; Keogh, P. J.; Porter, H.; Stratton, S. M.; Wiedeke, M. C.; Wilken, J.; Tang, J.; Levy, J. J.; Miranda, L. P.; Crnogorac, M. M.; Kalbag, S.; Botti, P.; Schindler-Horvat, J.; Savatski, L.; Adamson, J. W.; Kung, A.; Kent, S. B. H.; Bradburne, J. A. *Science* **2003**, *299*, 884–887.

55. (a) Miyake, T.; Kung, C. K. H.; Goldwasser, E. *J. Biol. Chem.* **1977**, *252*, 5558–5564. (b) Graber, S. E.; Kranz, S. B. *Annu. Rev. Med.* **1978**, *29*, 51–66.

56. (a) Lai, P. H.; Everett, R.; Wang, F. F.; Arakawa, T.; Goldwasser, E. *J. Biol. Chem.* **1986**, *261*, 3116–3121. (b) Sasaki, H.; Bothner, B.; Dell, A.; Fukuda, M. *J. Biol. Chem.* **1987**, *262*, 12059–12076. (c) Takeuchi, M.; Takasaki, S.; Miyazaki, H.; Kato, T.; Hoshi, S.; Kochibe, N.; Kobata, A. *J. Biol. Chem.* **1988**, *263*, 3657–3663. (d) Tsuda, E.; Goto, M.; Murakami, A.; Akai, K.; Ueda, M.; Kawanishi, G.; Takahashi, N.; Sasaki, R.; Chiba, H.; Ishihara, H.; Mori, M.; Tejima, S.; Endo, S.; Arata, Y. *Biochemistry* **1988**, *27*, 5646–5654.

57. Lin, F. K.; Suggs, S.; Lin, C.H.; Browne, J. K.; Smalling, R.; Egrie, J. C.; Chen, K. K.; Fox, G. M.; Martin, F.; Stabinsky, Z.; Badrawi, S. M.; Lai, P.-H.; Goldwasser, E. *Proc. Natl. Acad. Sci. U. S. A.* **1985**, *82*, 7580–7584.

58. (a) Hirano, K.; Macmillan, D.; Tezuka, K.; Tsuji, T.; Kajihara, Y. *Angew. Chem., Int. Ed.* **2009**, *48*, 9557–9560. (b) Kan, C.; Trzupek, J. D.; Wu, B.; Wan, Q.; Chen, G.; Tan, Z.; Yuan, Y.; Danishefsky, S. J. *J. Am. Chem. Soc.* **2009**, *131*, 5438–5443. (c) Tan, Z.; Shang, S.; Halkina, T.; Yuan, Y.; Danishefsky, S. J. *J. Am. Chem. Soc.* **2009**, *131*, 5424–5431. (d) Yuan, Y.; Chen, J.; Wan, Q.; Tan, Z.; Chen, G.; Kan, C.; Danishefsky, S. J. *J. Am. Chem. Soc.* **2009**, *131*, 5432–5437. (e) Robertson, N.; Ramage, R. *J. Chem. Soc., Perkin Trans. 1* **1999**, 1015–1021. (f) Richardson, J. P.; Macmillan, D. *Org. Biomol. Chem.* **2008**, *6*, 3977–3982.

59. Liu, S.; Pentelute, B. L.; Kent, S. B. H. *Angew. Chem. Int. Ed.* **2012**, *51*(4), 993–999.

11

BIOMEDICAL APPLICATIONS OF "CLICK"-MODIFIED CYCLODEXTRINS

ZHENSHAN JIA, RAKESH K. SINGH, AND DONG WANG

11.1 INTRODUCTION

Cyclodextrins (CDs) are water-soluble cyclic oligosaccharides derived from starch, which contain six α-CDs, seven β-CDs, eight γ-CDs, (Scheme 11.1) or more α-D-glucopyranoside units linked one to four [1]. Extensive crystallographic x-ray studies of CDs and their molecular complexes have revealed their homochiral doughnut-like annular structure with the wide and narrow hydrophilic rims decorated by the secondary and primary hydroxyl groups respectively, and the hydrophobic annular interior lined with skeletal carbons and ethereal oxygens. Because of this very unique structural amphiphilicity distribution pattern, CDs, especially the most commonly used α-, β-, and γ-CDs have been employed widely to form inclusion complexes with many lipophilic compounds. Used as solubilizers, CDs have already found many applications in the pharmaceutical industry [2]. A total of 35 different pharmaceutical products have been marketed as solid or solution-based CD complex formulations with many more in different phases of clinical development. Biologically, the small CDs are generally considered inert and have been approved by regulatory agencies to be used as formulation excipients in food, cosmetics, and pharmaceutical products. Many chemically modified CDs have stereoselectivity, which leads to the development of chiral-specific molecular sensors and chromatography media for highly sensitive detection and separation [3]. Concurrent with the surge of nanotechnology

Click Chemistry in Glycoscience: New Developments and Strategies, First Edition.
Edited by Zbigniew J. Witczak and Roman Bielski.

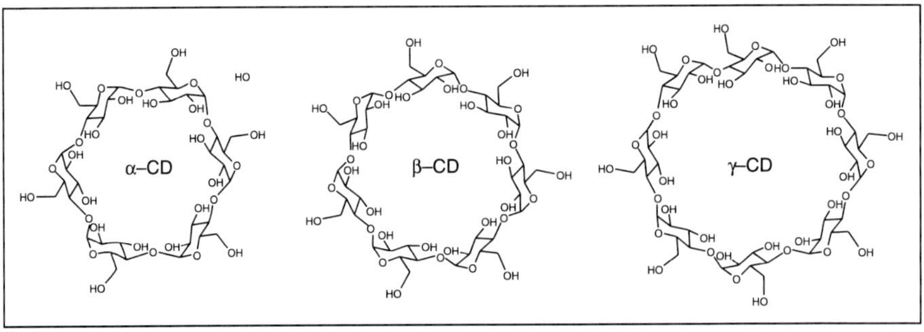

SCHEME 11.1 Representative structures of commonly used cyclodextrins.

development in recent years, many nanomedicine and drug delivery designs incorporated CDs as major structural elements, capitalizing on their unique dendrite-like functionalities and the hosting capability of the annular cavity [4]. To explore its vast spectrum of applications, it is critically important to recognize that CDs have to be modified chemically to achieve their desirable functions. In this chapter, we will discuss how "click" reactions [5], a new chemistry tool, would potentiate the biomedical applications of the CDs.

11.2 CHEMICAL MODIFICATION OF CYCLODEXTRINS

Most biomedical applications of CDs involve the chemical conjugations of CD structures to other functional structure elements. Due to the availability of the multiple hydroxyl groups on the rims of CDs, most of the chemical modifications of these functional groups are therefore the first step in derivatization of CD. According to their different *pKa* values and nucleophilicity, primary or secondary hydroxyl groups can be selectively modified. Under neutral, acidic, and weak basic conditions, the primary hydroxyl groups have relatively higher nucleophilic substitution potential than secondary hydroxyl groups. While under strong basic conditions, nucleophilic substitution mainly takes place on secondary hydroxyl groups. No selectivity can be achieved among the primary hydroxyl groups on the narrow rim or among secondary hydroxyl groups on the wide rim in the process of chemical modification because of their structural similarity. However, monosubstituted CDs can be obtained by controlling the conversion ratio and applying different purification methodologies. For example, treatment of β-CD with p-toluenesulfonic anhydride under basic aqueous conditions would lead to monotosylation of the primary hydroxyl group on β-CD [6]. Derived from these intermediates, other functionalities such as halide and azide can be introduced. Interestingly, it is possible to obtain disubstituted CDs by using bifunctional reagents with geometry matching the distance between two primary hydroxyl groups on the narrow rim. Similarly, selective multisubstitution of CDs can also be achieved. Therefore, in theory, by controlling the pH value of the reaction media, the amount and geometry of the reagents, a variety of substituted CDs with defined structure can be prepared [7].

Most CD functionalizations, however, are not single-step substitutions, but involve multiple steps of derivatization. To get the objective products, strict reaction conditions have to be met. Temperature, solvent, water, oxygen, reaction time, and reagent ratios are the most common concerns. Therefore, CD modifications often give poor yield and selectivity. Furthermore, the competing reaction sites and spatial hindrance usually would lead to structure uncertainty and decrease in reaction efficiency. Due to these limitations of the conventional methodologies, the field has been constantly in need of more efficient tools for CDs' derivatization with better functional group tolerance and mild reaction condition requirements [7].

Originally conceptualized by Sharpless et al. [5], "click" chemistry encompasses a group of reactions that are modular, high yielding, stereospecific, and generate inoffensive byproducts. The main characteristics of "click" reactions include simple reaction conditions, readily available starting materials and reagents, the use of no solvent or a solvent that is benign or easily removed, and simple product isolation [5]. Among the identified "click" reactions, Cu(I)-catalyzed azide–alkyne cycloaddition, namely the Huisgen 1,3-dipolar cycloaddition (HDC) reaction, is the most widely reported. With azides and terminal alkynes as starting materials, this reaction exclusively forms 1,4-substituted products. The reaction can be carried out within a wide temperature range (0°C –160°C), in a variety of solvent systems (including water), and over a wide range of pH values (5–12). It proceeds as much as 10^7 times faster than the uncatalyzed version. Furthermore, it is unaffected by steric factors [8]. Different primary, secondary, tertiary, and aromatic azides readily participate in this reaction. It also has excellent tolerance for various acetylene reactants [9].

Clearly, these unique features of the Cu(I)-catalyzed HDC reaction make it a powerful tool in organic coupling reactions. For the complex CD derivatization/conjugation, which have often been plagued by rigorous reaction conditions, multiple reaction sites and spatial hindrance, these exceptional features of HDC reactions were especially valuable.

11.3 BIOMEDICAL APPLICATIONS

Since "click" reactions were first introduced to the modification of CDs, a large number of CD derivatives have been rationally designed and prepared. In the following sections, some recent representative developments using novel "click"-modified CDs are discussed in detail, focusing on their biomedical applications in the area of chiral separation, functional biomaterials, biosensors, and drug delivery.

11.3.1 Chiral Separation

Chirality of active pharmaceutical ingredients (APIs) is essential for their biological activities. Different enantiomers or diastereomers may have significantly different, or even completely reversed activities [10]. Therefore, effective detection and separation of chiral isomers are of great importance for the pharmaceutical industry. Chromatography techniques, especially high-performance liquid chromatography

(HPLC), are widely used in enantiomeric purity analysis and chiral isomers' separation. Various chiral stationary phases (CSPs) were developed and can be divided into Pirkle-type, protein-based, polysaccharide-based macrocyclic CSPs, crown ethers, imprinted polymers, chiral ligand-exchange CSPs, and CD-based CSPs.

Due to their homochiral doughnut-like annular nature, CDs are widely used in the preparation of chiral stationary phase. To achieve separation, the inner hydrophobic surface of the homochiral CDs would complex with chiral molecules to form transient diastereomers. The primary and secondary hydroxyl groups could also afford hydrogen bonds to assist in the chiral separation process. Many immobilization methods have been developed. While the CD-based CSPs demonstrate superior functionalities, one issue that often plagues the development is the immobilization chemistry. For example, the linkages (urethane or amide) between the CDs and the stationary phase are not always stable enough in aqueous mobile phase; due to the absence of reaction selectivity among CDs' hydroxyl groups, uncontrolled CDs' immobilization could lead to multiple linkages to the stationary phase with poor reproducibility [11].

Due to their high efficacy, simplicity, and selectivity, "click" reactions were explored to address these problems in CD immobilization. In their recent publication [12], Zhang et al. prepared β-CD-based chiral separation materials with two reverse modification strategies, that is, alkyne-modified β-CD on the azide-modified silica and azide-modified β-CD on alkyne-modified silica (Scheme 11.2). These modified β-CDs were directly derived from monotosylated β-CD. Catalyzed by copper sulfate and sodium ascorbate, the modified β-CDs were successfully immobilized on the silica supports. The β-CD loading on the silica supports was calculated to be 0.13 $\mu mol/m^2$. The analyses of the chiral recognition behaviors on flavanone compounds show that the "click" β-CD-based CSP exhibited good ability in the enantioresolution.

To enhance the linker stability and improve enantioseparation, Wang et al. developed three "click" β-CD-based chiral CSPs (i.e. CCN-CSP, CCM-CSP, CCP-CSP) (Scheme 11.3) by HDC coupling of alkynyl-modified silica with mono-6-azido-β-CD, mono-6-azido-permethylated-β-CD, mono-6-azido-perphenyl-carbamated-β-CD, respectively [13, 14]. It is important to note that azidation precedes perfunctionalization of the hydroxyl groups in the synthesis route, which demonstrates the chemical tolerance of the alkyne/azide handles. Another interesting approach the group took was their use of soluble organic catalyst CuI(PPh_3), which

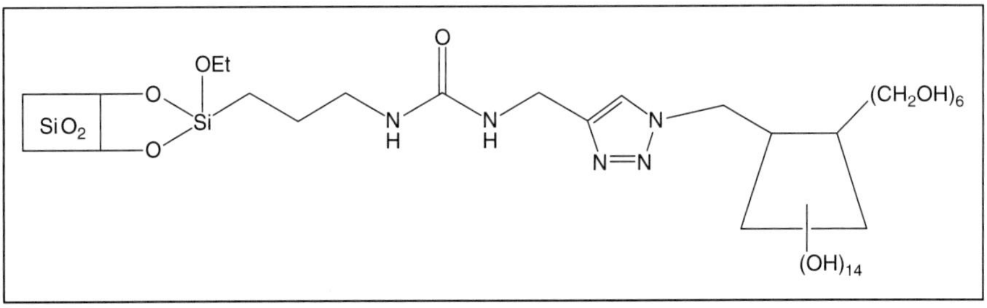

SCHEME 11.2 "Click"-immobilized β-CD on silica support. (Adapted from ref. 12.)

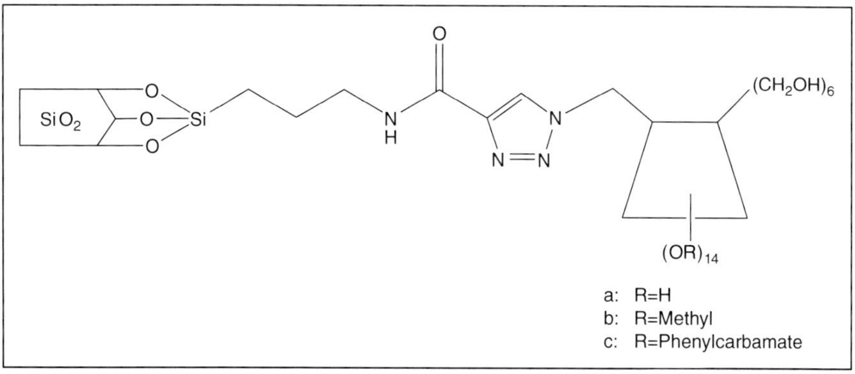

SCHEME 11.3 "Click"-immobilized perfunctionalized β-CD on silica support. (Adapted from ref. 14.)

significantly increased the loading of β-CD on the silica supports to 0.51 μmol/m^2. When testing the new CSPs in capillary electrochromatography to separate racemic 1-(4-chlorophenyl)-ethanols, impressive selectivity and resolution were obtained. They further investigate the application of the new CSPs in separation of dansyl amino acids, flavonoids, and other pharmaceutical products using HPLC. The results showed fine chiral recognition abilities, which might be attributed to the inclusion complexation, and the different functional groups on the CD rims and the hydrogen bonds of nitrogen in triazole linkage with the guest molecules.

Interestingly, the three CSPs exhibited quite different chiral recognition properties. CCN-CSP showed good resolution of various aromatic acids particularly dansyl amino acids while CCM-CSP and CCP-CSP did not, probably due to the hydrogen bonds of hydroxyl groups of the CD rims with carboxylate and amine moieties. For aromatic alcohol and some β-blockers, CCP-CSP showed much better resolution, probably due to π–π interactions. The alicyclic ionone derivatives could only be separated on the methylated CCM-CSP. All the three CSPs could resolve the flavonoid analytes. The researchers believe these differential enantiorecognition capabilities might be due to the different functional groups on the CD rim. For example, hydroxyl groups could form hydrogen bonds, methoxy groups could afford hydrophobic interactions, and phenyl carbamate groups could introduce π–π interactions and dipole–dipole interactions.

In order to further improve the stability of the linkage between β-CD and CSP, the same team prepared β-CD-based CSP with multiple triazole linkage through "click" coupling of alkynyl-modified silica with heptakis (6-deoxy-6-azido)-β-CD (CSP1) and heptakis (6-deoxy-6-azido-phenylcarbamoylated)-β-CD (CSP2) (Scheme 11.4). The chiral separation experiment suggested that CSP1 gave good separation results for hydroxyl, carboxylic and dansyl amino acids, indoprofen, and ketoprofen, while CSP2 gave better resolution for flavonoids.

Direct analysis of a biological sample with HPLC has been always a challenge for analytical chemists. Designing and preparing restricted-access material (RAM) used as the stationary phase of HPLC may be a feasible and promising solution,

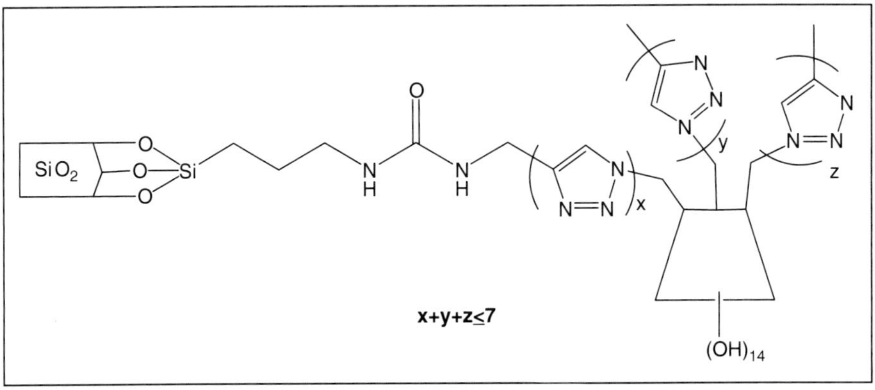

SCHEME 11.4 "Click"-immobilization of β-CD on silica support with multiple linkages. (Adapted from ref. 13.)

which has drawn extensive interests. RAM is usually composed of two layers, the inner layer is accessible to small molecules and immobilized with chiral functional groups for chiral separation; the external layer is biocompatible but non-adsorptive to biomolecules. In this way, the material can be stable to biological samples, adsorb and separate small molecules, and exclude large biomolecules such as proteins. Due to its wonderful chiral discrimination capability, β-CD has always been selected to be immobilized on the inner layer of the RAM.

Using "click" HDC reaction, Wang et al. developed a novel RAM and applied it in chiral separation [15]. At first, they grafted two layers of poly(glycidyl methacrylate) (pGMA) on the surface of porous silica gel. β-CDs were then immobilized as the inner layers by a reaction between the oxirane rings of GMA and the hydroxyl groups of CDs to give a CD-sil-2 material. As can be expected, the multiple hydroxyl reaction sites led to the uncertainty of the linker position of β-CD [9]. To address this issue, they turned to "click" chemistry. After immobilization of 3-(2-bromoisobutyryl)propyltriethoxysilane on silica gel, atom transfer radical polymerization (ATRP) was carried out to introduce the alkyne-functionalized inner surface. The "click" conjugation with mono-(6-azido-6-deoxy)-β-CD and the subsequent ATRP gave the CD-click-RAM, which showed better enantioseparation efficiency than CD-sil-2 (Scheme 11.5). The authors attributed the better separation results to the more homogeneous linkage of CD with inner support and the stronger hydrogen bonds of the analytes with triazole.

Guerrouache et al. developed a β-CD "click"-grafted macroporous organic polymer monolith used as stationary phases in capillary electrochromatography and nanoliquid chromatography [16]. The photo-initiated copolymerization of N-acryloxysuccinimide and ethylene dimethacrylate in toluene using azobisisobutyronitrile (AIBN) as initiator produced a macroporous organic polymer monolith, which was alkyne functionalized by nucleophilic substitution of succinimide moieties with propargylamine. Through the "click" HDC reaction with the mono-(6-azido-6-deoxy)-β-CD, the polymer monolith was successfully grafted with β-CD,

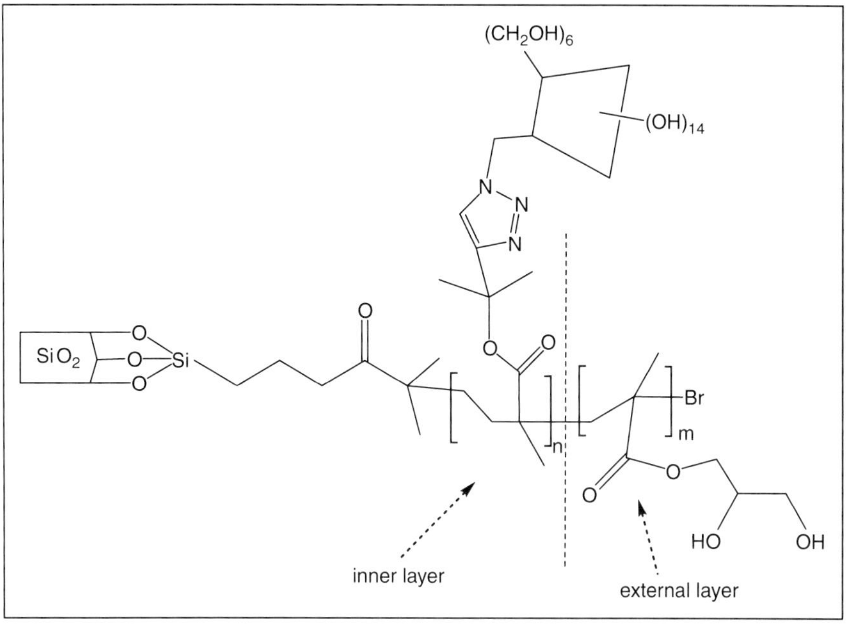

SCHEME 11.5 The structure of "click" RAM (restricted-access material). (Adapted from ref. 15.)

the chiral selector. The chiral separation of flavanone enantiomers was achieved in both capillary electrochromatography and nano-liquid chromatography with good resolution.

11.3.2 Functional Biomaterials

Due to the presence of multiple reactive hydroxyl groups and their inclusion ability, CDs have also been used extensively in the development of various biomaterials, such as polyrotaxane (PR), hydrogel, and functionalized polymers. As a novel chemical tool, "click" reactions were used to replace the conventional methodologies in the design and development of these biomaterials, motivated by their high efficiency, functional tolerance, and mild reaction conditions.

11.3.2.1 Polyrotaxanes Polyrotaxanes (PRs) are supramolecular assemblies in which cyclic molecules (CD, calixarene, cucurbituril, or crown ether) are threaded onto a linear polymer chain capped with bulky end groups. Among all PRs investigated, the PR consisting of CDs and poly(ethylene glycol) (PEG) were the most widely investigated. Since first discovered by Harada et al. [17, 18], their applications in biomaterials and drug delivery have been extensively explored.

The spontaneous supramolecular assembly of CDs onto PEG chain is an enthalpy-driven process. Such threading is not a thermodynamically stable construct, which is subject to entropy-driven dethreading upon dilution. Due to this instability, such assembly was termed as pseudopolyrotaxane. To stabilize this supramolecular assembly and prevent dethreading, bulky chemical structures are often conjugated to cap the PEG chain termini. As the pseudopolyrotaxanes are often formed in aqueous media, the chain stopper conjugation has to be water tolerable and highly efficient to prevent the dethreading of CDs from happening. Apparently, "click" HDC reaction meets these criteria.

Using "click" HDC reaction, Wu et al. prepared a PR with rhodamine as the stopper. They first capped the PEG-diol with a carboxylic azide to form azidoester at the chain termini, threaded CD, and then "click" conjugated with alkyne-functionalized rhodamine B (RhB-alk) [19]. Because of the biodegradability of the esters, the threaded CDs can be released after the hydrolysis of ester bond to serve as drug carriers. Clearly, the use of RhB-alk as the stopper also has the additional benefit of visualizing the PR. As a further improvement, the same team later proposed to cap their PRs with β-CD as stopper in a one-pot synthesis approach in water to produce a terminally recognizable "full-CD" PR [20]. Similar to rhodamine B, the β-CD stopper was also "clicked" on due to its high efficiency and tolerance of water. Not only does the terminal β-CDs effectively prevented α-CDs from dethreading, the terminal β-CDs cavity could complex with specific molecules or functionalities to form unique supramolecular structures as well. For example, adamantane fits the β-CD annulus very well. Therefore, by conjugation of adamantane to the chain termini of PEG-diol and mixing the resulting product with β-CD-terminated PEG, a very interesting supramolecular block copolymer assembly of alternate rod (PR) and coil (diadamantyl-terminated PEG) can be formed.

Molecular brushes are graft copolymers with densely distributed side chains. These unique structures may lead to many exciting biomedical applications. For most of the molecular brushes, the side chains are grafted on a fixed backbone through covalent bonds. For PR, however, the threaded CD rings can slide and be chemically modified. Therefore, a PR-based molecular brush may offer very unique properties. Recently, Wu et al. prepared a novel supramolecular polymer brush consisting of a PR backbone, hydrophilic sliding side chains, and hydrophobic sliding C16 side chains. In order to introduce these functionalities, all hydroxyl groups of CD were converted to azide. The side functionalities were then "clicked" onto the modified CD ring with extremely high efficiency ($\sim$100% grafting density). Such synthetic strategy would allow the amphiphilicity of the molecular brush to be easily tuned by the feed-in ratio of hydrophilic and hydrophobic side chains.

To investigate the multivalent interactions in a ligand-receptor binding system, Hyun et al. prepared azidated PRs comprising PEG and mono-, di-, and triazidated α-CDs in a water/DMSO solution in a one-pot synthesis [21]. The pre-azidated α-CDs were threaded onto a diamino PEG to give a pseudopolyrotaxane, which was then end-capped by two bulky groups. Because of the precise control of the sites and numbers of hydroxyl group azidation on each α-CD, the ligand density of polyrotaxane was also controlled precisely. By "click" HDC reaction, the propargyl-modified mannose

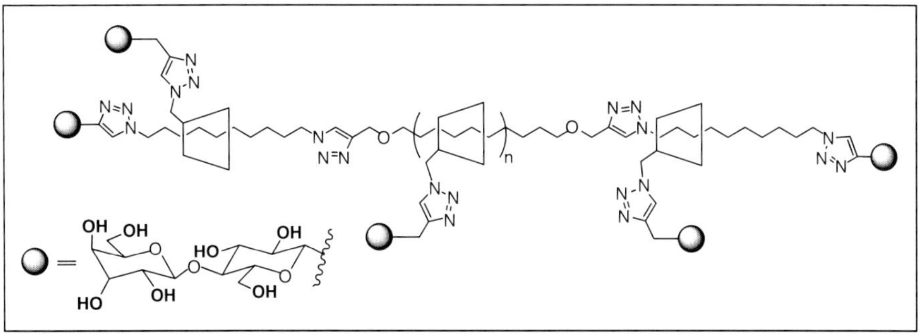

SCHEME 11.6 General structure of lactosyl groups functionalized oligorotaxane. (Adapted from ref. 22.)

was effectively conjugated with polyrotaxane. The resulting ligand-density-controlled PRs were then evaluated for a variety of biological applications.

Similar to this approach, Chwalek et al. developed low-molecular-weight oligorotaxanes with adjustable ligand densities as potential multivalent lectin inhibitors [22]. The oligorotaxane has five α-CDs and seven lactosyl groups "clicked" on as pendent functionalities. In testing its binding ability to Arachis hypogaea peanut lectin, the oligorotaxane exhibited a threefold increase per lactose unit (Scheme 11.6).

11.3.2.2 Hydrogels Hydrogels are highly hydrated cross-linked polymer networks that have been used extensively in biomedical applications [23]. As a critical element for the network formation, cross-linking agents with multiple functionalities would join the linear polymers together to form the network. As a biocompatible multifunctional molecule, CDs are natural selection for the cross-linking agents. As an example for this design, Xu et al. recently developed a porous "click" hydrogel using β-CD as the cross-linking structure [24]. They first prepared 2-(2-bromoisobutyl-oxy) ethyl methacrylate (BIEMA) to copolymerize with N-isopropylacrylamide (NIPAAm) at different ratios; after azidation, it produced a copolymer with pendent azide groups distributed along the backbones. The seven primary hydroxy groups of β-CD reacted with alkynecarboxylic acid catalyzed by N,N′-dicyclohexylcarbodiimide (DCC) to introduce seven alkynyl groups for cross-linking with azide groups. After mixing the copolymer, the functionalized β-CD and the catalyst, the "click" hydrogel formation was completed within several minutes (Scheme 11.7). It is notable that PNIPAAm is a thermosensitive polymer, which can be employed as on–off switches for drug release purpose. Below the lower critical solution temperature (LCST), the hydrogel exhibits a swollen state, while above LCST, the phase separation takes place. To explore the therapeutic potential of the "click" CD hydrogel, fish DNA was selected as the prototypic drug in the drug release study and the DNA-loaded hydrogel demonstrated good thermosensitivity and drug releasing profile.

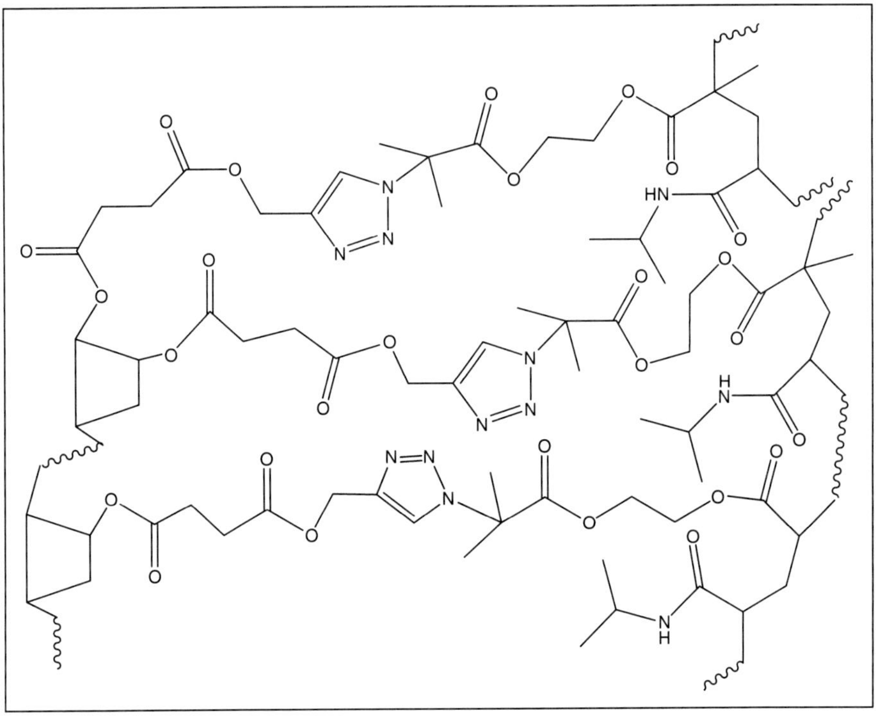

SCHEME 11.7 Representative structure of the "click" CD hydrogel. (Adapted from ref. 24.)

11.3.2.3 Functional Biomedical Polymers Poly(vinylidene fluoride) (PVDF) is an important biopolymer due to its unique piezoelectric, ferroelectric, and dielectric properties, good biocompatibility, and excellent membrane-forming capability [25, 26]. For many of its potential biomedical applications, it is critical to be able to manipulate and control the surface properties of PVDF. Recently, Cai et al. prepared PVDF-g-PPMA copolymers bearing pendent propargyl functionalities by thermally induced graft copolymerization of propargyl methacrylate (PMA) from the ozone pre-activated PVDF backbones [27]. By phase inversion in aqueous media, microporous membranes with pendent propargyl functionalities were then fabricated from the functionalized PVDF, which would serve as a "clickable" platform for further tailoring of the surface functionalities. Catalyzed by CuBr, azido-β-CD reacted with the grafted propargyl groups to give the PVDF-g-P[PMA-"click"-β-CD] membrane. After complexation with the adamantyl group of diadamantyl-poly(ethylene oxide) (DA-PEO) guest polymer, the modified membrane exhibited good resistance to protein adsorption and fouling under continuous flow conditions.

The unique hydrophobic annulus structure of CDs affords them with the capability to complex with a variety of different low-molecular-weight hydrophobic molecules. Such capacity is very important in food industry in terms of protecting lipophilic food components, stabilize fragrances, flavors, and especially suppress unpleasant odor or bitter taste. It is well known that flavonoids (such as naringin and neohesperidine)

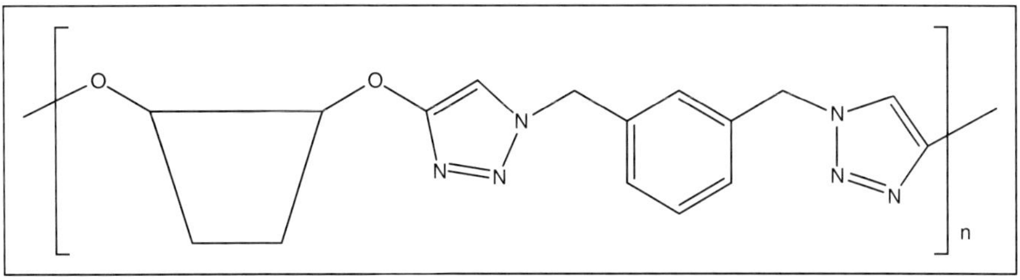

SCHEME 11.8 Representative structure of debittering "click" CD polymer. (Adapted from ref. 28.)

and limonoids (limonin) are responsible for the bitterness in citrus juice. To suppress the bitterness in citrus juice, Binello et al. prepared a β-CD copolymer via "click" HDC reaction, which can efficiently sequester naringin from aqueous solution [28]. They first converted two secondary hydroxyl groups of β-CD into alkynyl groups, then copolymerize them with 1,3-bis(azidomethyl)benzene to give a reticulated water insoluble polymer (Scheme 11.8). The debittering performance is better than that of commercial porous polymer of Amberlite XAD-16, and can be used for at least three entrapment/desorption cycles without appreciable loss of sequestering power.

11.3.3 Biological Sensors

Due to the inclusion ability and the multiple hydroxyl-decorated rims of CDs, another exciting application area for these cyclic molecules is to develop biological sensors to probe the existence and concentration of a particular substance with important biological functions. These substances may include metal ions, organic pesticides, and fatty acids.

Some metal ions play important roles in the normal functions of biological systems, with others being extremely toxic. Therefore, the development of effective and efficient metal ion detecting methods becomes critically important for clinical diagnosis of trace element deficiencies or heavy metal intoxication. Using β-CD, Zhang et al. developed two effective fluorescent sensors for detection of cadmium ion, which can cause acute and chronic toxicity in humans [29]. As the first step of the sensor design, mono(6-deoxy-6-azido)-β-CD and 8-propargyloxyquinoline were prepared (Scheme 11.9). They were then "clicked" together to yield the sensor molecule, which has weak fluorescence. When exposed to Cd^{2+}, the sensor will be able to emit significantly enhanced fluorescence signal (1.7-fold increase). The researchers believed that the observed enhancement was attributed to the cooperative coordination of Cd^{2+}, 8-hydroxyquinoline, and triazole, which blocked the intramolecular photo-induced electron transfer. The sensor also showed good selectivity with no other metal ion being able to generate such fluorescence enhancement. Because of the high affinity of the β-CD annulus to adamantyl group, the sensor can easily sequester 1-admantanecarboxylic acid sodium salt (ADCA). In the presence of ADCA, the authors observed a sevenfold fluorescence enhancement than the sensor without ADCA. In these studies, it is very important to note that the HDC

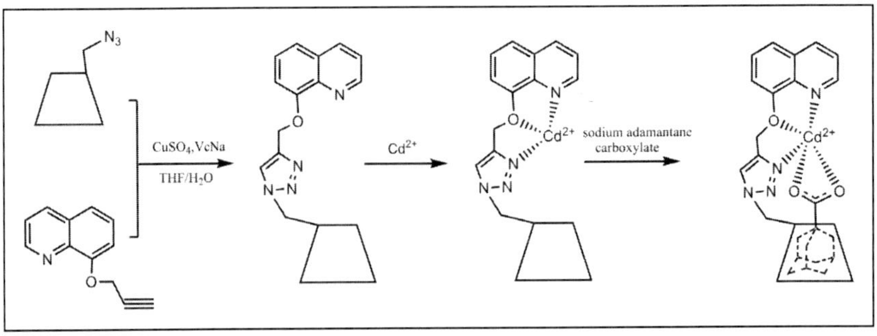

SCHEME 11.9 The design and synthesis of "click" CD-containing Cd^{2+} sensor molecules. (Adapted from ref. 29.)

"click" reaction not only provides a stable linker between the CD and the chelation functionality, but also contributed the triazole moiety as additional chelation structure for Cd^{2+}. The high inclusion potential of β-CD to adamantyl group allows the introduction of a carboxylic group, which contributed to the formation of an even stronger binding model. Due to its satisfactory solubility and high sensing ability towards Cd^{2+}, these two fluorescence sensor molecules may have wide application in clinical diagnosis and environmental monitoring.

The presence of pesticide residues in farm products has been linked to the etiology of some very devastating diseases, such as cancer and Parkinson's disease [30], which make the sensitive detection of pesticide residues a critical issue for food safety. To address this issue, Mallard-Favier et al. developed a fluorescent tripod detection system for pesticides such as α-cypermethrin, pendimethalin, thiram, and imidacloprid. Through microwave-assisted "click" chemistry, they prepared a peracetylated CD trimer bearing three 1,2,3-triazole linkers, which exhibited two absorption maxima at 277 nm and 326 nm. They found fluorescence characteristic emissions at 315 nm and 329 nm at an excitation at 227 nm. When complexed with pesticide such as pendimethalin at increasing concentrations, the fluorescence intensity decreased until extinction [31]. The absence of wavelength shift excluded the correlations between the aromatic core of the trimer and the pesticide due to π–π stacking interactions. The authors ascribe the quenching of fluorescence to the inclusion of the pesticide molecules into the CD annulus.

Fujimoto et al. recently developed a group of fluorescent sensors for the detection of unsaturated fatty acids (Scheme 11.10) [32]. Inspired by the structures of immunoglobulin antibodies, the sensors are designed to be a pair of complementary-sequenced oligonucleotides (ODN) possessing the pyrene and guest-binding moieties. Because fatty acids can form 1:2 complexes with CD, one fatty acid can promote the hybridization of ODNs resulting in the pyrene-emission switching from monomer to excimer. To implement their design, they first examined the fluorescent transformation caused by bisadamantyl guest molecule, which can complex with two β-CDs due to the well-known high affinity between adamantyl groups with β-CD. The pyrene-emission switching took place from 379 nm (monomer emission) to 500 nm (excimer emission) and an isoemissive point at 446 nm. Similar emission

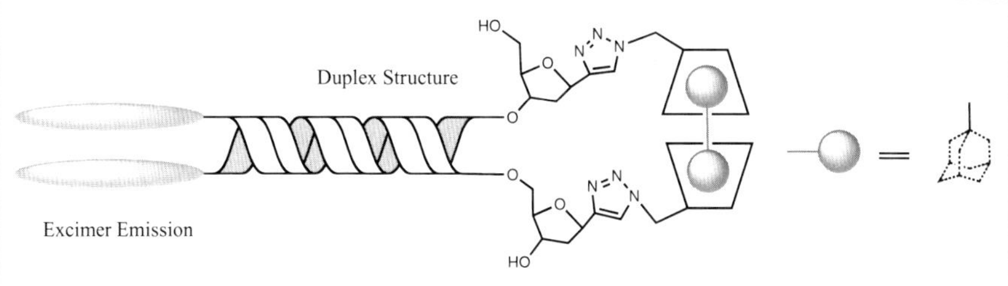

SCHEME 11.10 Working mechanism of proposed "click" CD-containing fatty acid sensor. (Adapted from ref. 32.)

switchings were also observed using unsaturated fatty acids including oleic, elaidic, and arachidonic acid as guest molecules, which testified their original design.

11.3.4 Drug Delivery

Among all the areas of biomedical applications of "click"-modified CDs, drug delivery contributed the largest number of publications. This may be partially attributed to the extensive use of CDs by the pharmaceutical industry in conventional formulations and novel nanomedicine developments. "Click" reaction, a powerful modification tool, has played important roles in preparing CD-containing drug delivery systems. In the following examples, we will present some of the most interesting developments of using "click"-modified CD for drug delivery systems.

11.3.4.1 Gene Delivery Carriers Gene therapy is a very exciting research area in the field of drug delivery. Instead of the traditional payload of low molecular organic compounds, genetic materials are packaged for specific intracellular delivery in order to correct a mutated gene or to introduce a new protein expression for the healing of a particular pathological condition. The extremely sophisticated task of subcellular gene insertion or replacement and the numerous anatomical barriers the gene delivery must overcome necessitate the development of powerful delivery strategies. Though presented with outstanding transfection efficiency, virus-based gene delivery has been plagued with various safety concerns, which lead to extensive research effort into the development of polymeric material-based nonviral gene delivery, which has much better safety profile [33–35].

For nonviral delivery of genetic materials, most of the effort has been focused on carrier materials that would protect the payload and facilitate gene transfection. In the development of polymeric gene carriers, one major challenge has been the structural polydispersity. From this aspect, dendrimers, which have well-defined architectures, precise molecular weight, and multivalent functionalization sites, are very good candidates as gene delivery vehicles.

Inspired by the clinic potential of dendrimers and the distinctive structure and properties of β-CD, Srinivasachari et al. developed a novel family of polycationic β-CD "click" clusters with discrete, monodisperse, and symmetric geometries for

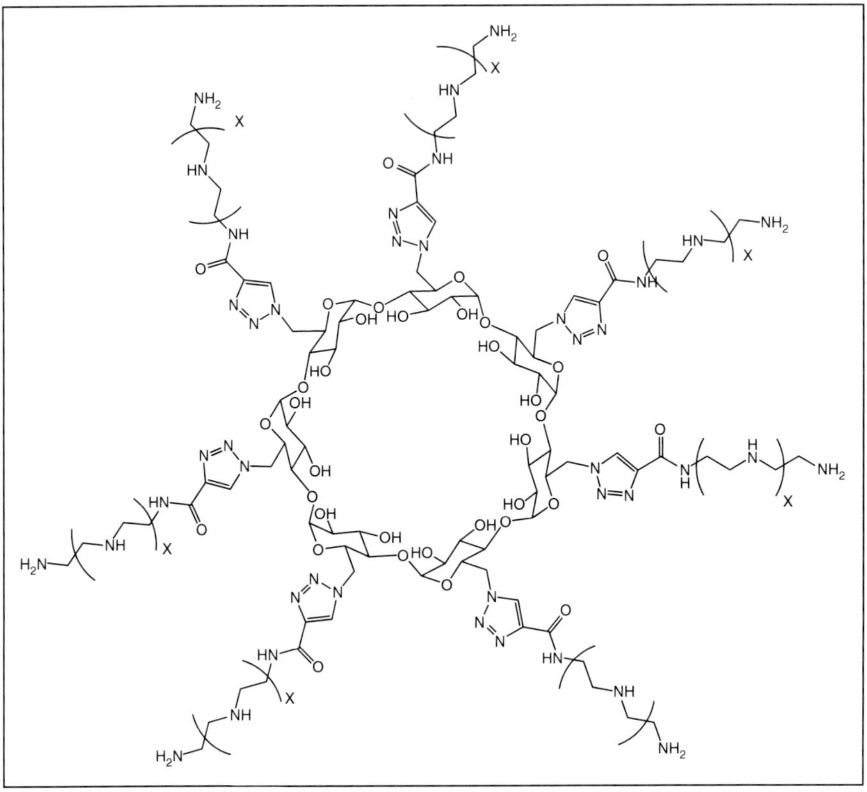

SCHEME 11.11 The structure of "click" CD-based dendrimer for gene delivery. (Adapted from ref. 36.)

gene delivery [37]. They found that the novel polymer design could effectively complex, condense, and deliver plasmid DNA (pDNA) with exceptional biocompatibility and efficacy. The dendrimers were synthesized by direct "click" coupling of acetylated-per-azido-β-CD and a series of alkyne-functionalized cationic dendrons (Scheme 11.11). Different analyses revealed that the cluster bound and complexed with pDNA into the spherical nanoparticles in the size range of 80–130 nm. The stable polycationic nanoparticles deliver pDNA into HeLa and H9c2(2-1) cells effectively and protect pDNA degradation from nuclease.

To develop linear CD-containing cationic polymers for efficient gene delivery, the same research group copolymerized acetylated-diazido-β-CD and dialkyne-oligoethyleneamines via "click" HDC reaction with very high efficiency (Scheme 11.12) [36]. By changing the repeating amine number and polymerization degree, two series of polymers with different amine density or different molecular weight were successfully synthesized. The influence of the oligoethylene amine density and the polymer length were also evaluated. Their results showed that the polymer complex could effectively deliver pDNA into HeLa cells and the complex also protects the payload from nuclease degradation.

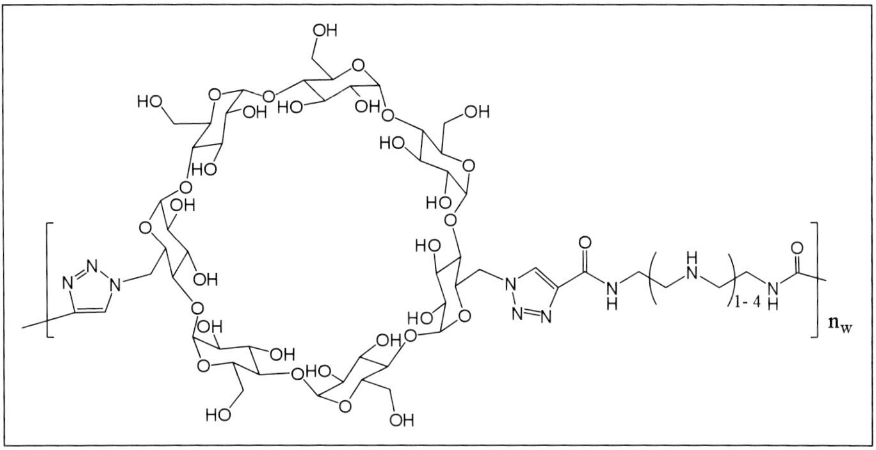

SCHEME 11.12 The structure of "click" copolymer of β-CD and dialkyne-oligo-ethyleneamines. (Adapted from ref. 37.)

As another interesting approach, Mendez-Ardoy et al. developed a family of discrete and well-characterized amphiphilic polycationic β-CD "click" clusters as gene delivery vehicles (Scheme 11.13). The cationic structure for DNA complexation was first installed on the primary hydroxyl group of β-CD by "click" HDC reaction. Hydrophobic tails were then conjugated to the secondary hydroxyl group. Such CD-centered amphiphilic compounds could self-assemble and complex with pDNA. By tuning the density of positive charges, their distance to the CDs' core and the length

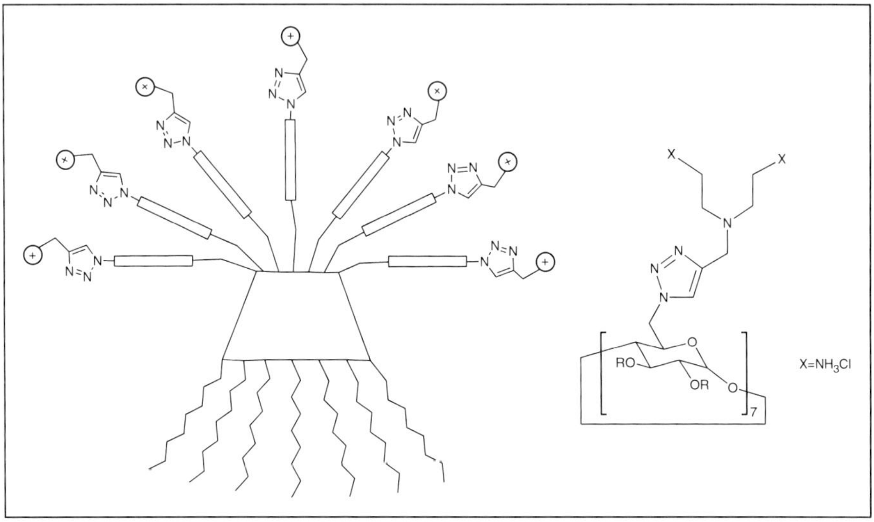

SCHEME 11.13 Representative structure of β-CD-based "click" cluster for gene delivery. (Adapted from ref. 38.)

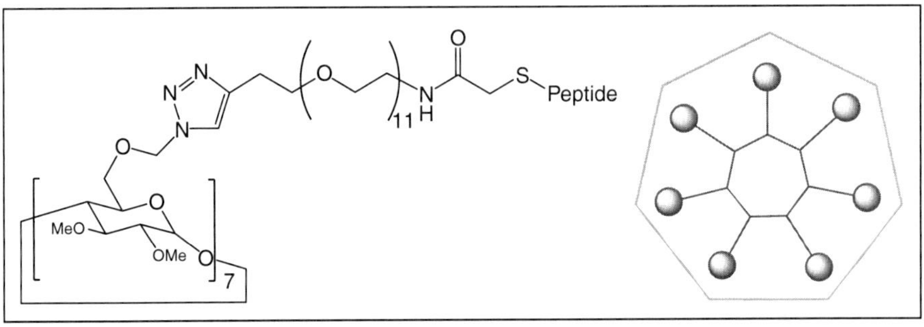

SCHEME 11.14 The design of "click" CD-based anthrax toxin antagonist. (Adapted from ref. 39.)

of the lipophilic tails, the "click" clusters' self-assembling capability in complexing with pDNA could be fine-tuned to facilitate the membrane-crossing and transfection capabilities of the CD–pDNA complex [38].

11.3.4.2 β-CD-based Anthrax Toxin Antagonist Anthrax is an acute infectious disease caused by the bacterium *Bacillus anthracis*. Anthrax toxin, a three-protein exotoxin secreted by virulent strains of *Bacillus anthracis*, plays a vital role in the lethality of anthrax infection. Anthrax toxin is composed of two enzymes: lethal factor (LF), edema factor (EF), and protective antigen (PA). PA is proteolytically processed into fragments PA_{63} that oligomerize to heptamers $[PA_{63}]_7$. The heptamers complex with LF and EF, and then translocated into the cytosol after endocytosis. As a very smart approach, Joshi et al. used β-CD as a scaffold and "click"-grafted seven PA inhibitory peptides to the β-CD primary hydroxyl groups via PEG linker, which produce an anthrax toxin antagonist with a nice spatial match to the $[PA_{63}]_7$ complex [39]. In the structural design, the selection of symmetric β-CD as the core scaffolding and the highly efficient "click" conjugation strategy are the key to construct the molecule skeleton and achieve the inhibition effect (Scheme 11.14). Due to the rational design, the "click" CD-based antagonist demonstrated a more than 100,000-fold increase in inhibitory activity than the monovalent peptide inhibitor.

11.3.4.3 Development of Delivery Carriers Using the Guest/Host Capability of CDs Aliphatic polyesters, with fine biocompatibility and biodegradability, are widely used in the bioengineering field (e.g., implant materials, prosthetics, surgical sutures, bone screw) [40]. Aliphatic polyesters can be obtained via organic synthetic method or fermentation. One limitation of the aliphatic polyesters, however, is their lack of functionality, which makes it difficult for encapsulation and delivery of water-soluble drugs and biologics. To address this issue, Jazkewitsch et al. introduce pendent β-CD to functionalized bacterial polyesters via "click" HDC reaction [41]. Two isomeric alkyne-functionalized ε-caprolactones, prepared by Baeyer–Villiger oxidation, were copolymerized with ε-caprolactone using $Sn(Oct)_2$ and adipic acid as co-initiators to give a pendent alkyne-functionalized poly(ε-caprolactone) (PCL) with carboxylic acid at the chain termini, which was capped with 2,2-diphenylethanamine

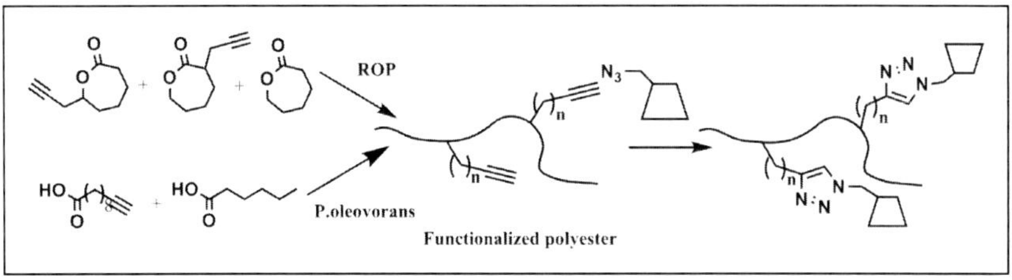

SCHEME 11.15 The design and synthesis of "click" CD-decorated PCL. (Adapted from ref. 41.)

via amidation. Catalyzed by Cu(I), mono-(6-azido-6-deoxy)-β-CD (β-CD-N_3) was "clicked" onto the PCL as pendent functionalities with high efficiency (Scheme 11.15). As a proof of principle for the "click" CD-functionalized PCL to be used as a delivery vehicle, potassium adamantyl carboxylate was complexed with the β-CD-grafted PCL. Dynamic light scattering (DLS) measurement suggests that the hydrodynamic diameter of the polyesters was increased substantially as a result of electrostatic repulsion of negatively charged carboxylate groups. Similar to this approach, "click" CD-decorated dextran and glycodendrimer were also developed [27,28].

Utilizing the guest/host capability of CDs, noncovalently connected micelles (NCCMs) were developed as a novel drug delivery system. For regular micelle formulations, the inner core of micelles is hydrophobic and serves as a nano-depot for the payload and the outer shells are hydrophilic, which maintains the micelle stability in water. Typically, micelles were formulated via the self-assembly of amphiphilic molecules. For NCCMs, the most unique feature is that the linker between the hydrophilic and hydrophobic segments of the amphiphilic molecules is of noncovalent nature. For example, Quan et al. developed an NCCM for cancer chemotherapy. In their design, the hydrophobic segment is PCL terminated with an adamantyl group; and the hydrophilic segment is a copolymer of NIPAAm and N-acryloxysuccinimide (NAS), which is terminated with a smaller phenyl group. An α/β-CD dimmer was "click" synthesized to join the hydrophilic and the hydrophobic segment together. Due to different cavity sizes, α-CD prefers to complex with phenyl group while β-CD favors the complexation with adamantyl group (Scheme 11.16). By this host/guest

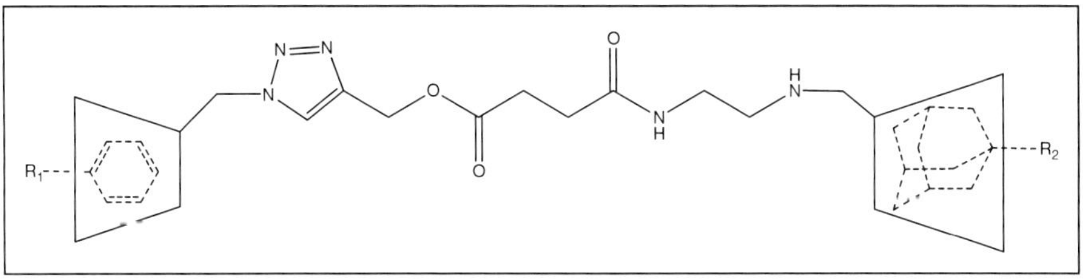

SCHEME 11.16 The general design of "click" α/β-CD dimmer that noncovalently assembles the hydrophilic/hydrophobic segments of the micelle construct. (Adapted from ref. 42.)

interaction, the hydrophilic and hydrophobic segments were joined together noncovalently to allow the self-assembly and micellization of chemotherapeutic agents in aqueous solution [42].

11.3.4.4 Development of Osteotropic CDs for Musculoskeletal Disease The skeletal system is the only tissue *in vivo* that contains calcium phosphate in the form of hydroxyapatite (HA). The nanosized HA needle crystals embeds in an organic matrix, mainly type-I collagen to form a hard and tough composite material, the bone. The skeletal system's first function is to provide a mechanical support and protection for our body. It also provides a rich source of inorganic ions, which helps to maintain the homeostasis of calcium and phosphate. Together with tendons, ligaments, and muscles, the skeletal system provides mobility, which is critical for the life of mammals. Three types of cells populate the skeletal system: osteoclasts, osteoblasts, and osteocytes. The abnormity of their biological functions leads to a wide array of musculoskeletal diseases [43].

One major issue of drug development for musculoskeletal diseases is the fact that most of the bone drugs are not osteotropic, which often lead to limited efficacy and severe safety concerns due to the drugs' ubiquitous distribution in the soft tissue. To address this issue, several groups have focused on the development of bone-targeted prodrugs [44]. As an example, Liu et al. proposed the development of the osteotropic CDs (Scheme 11.17). The general idea is to synthesize CD derivatives that have strong bone/tooth-binding capability. As part of the critical element of the derivatives, the CD annuli will interact with hydrophobic drugs to form molecular complexes. Upon exposure of the osteotropic CD/drug complex to the hard tissues, they will be immediately immobilized on the hard tissue surface. The complexed drug will be gradually released from the complex via extracellular fluid exchange/dilution.

One issue surfaced during the synthesis of the osteotropic CDs is the difficulty of conjugating osteotropic moieties (e.g., bisphosphonates, oligo glutamic acid

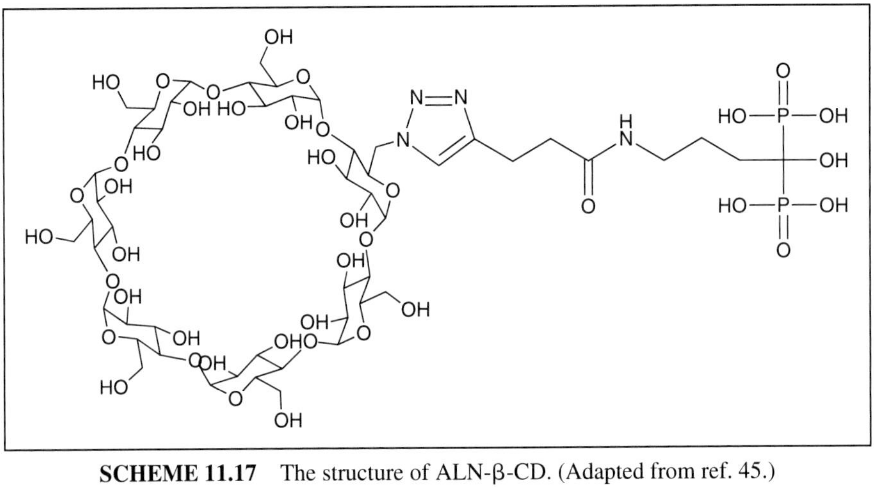

SCHEME 11.17 The structure of ALN-β-CD. (Adapted from ref. 45.)

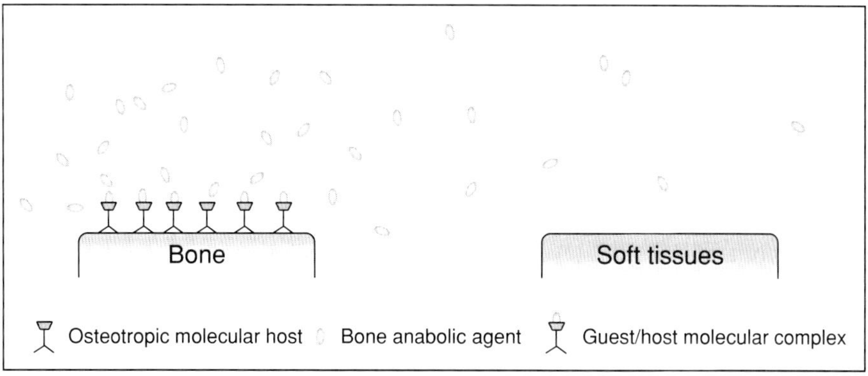

FIGURE 11.1 The proposed bone anabolic mechanism of ALN-β-CD as a molecular sequestrant. (See color plate section.)

peptides) to CD rings. Because these targeting moieties are only soluble in water, the final conjugation step has to be done in water, which restricts the use of many reactions. "Click" HDC reaction was eventually selected by the team to conjugate alendronate (ALN) to β-CD, due to its high efficiency and most importantly its tolerance of water. When tested in a simulated oral environment, ALN-β-CD/dexamethasone (Dex) complex was shown to be able to bind to model tooth surface immediately upon exposure and gradually release the complexed Dex. It was proposed that such formulation could be used clinically as a mouthwash for the treatment of oral mucositis and vesiculo-erosive oral mucosal disease (V-EOMD) [45]. As another potential application, the team proposed that ALN-β-CD might act as a molecular sequestrant. Once anchored on bone, the β-CD annulus would sequester/enrich endogenous bone anabolic agents such as prostaglandin E_2, which would lead to a local bone anabolic reaction (Figure 11.1). Indeed, when tested in a rat mandibular bone model, the local injection of ALN-β-CD to the mandibular bone caused substantial localized bone formation. Such therapy is highly favored by periodontists for periodontal bone repair via simple injection compared to the painful open gum surgery repair.

Inspired by the successful synthesis and applications of ALN-β-CD and the extensive development of polyrotaxane as a drug delivery carrier, the same team then synthesized ALN-α-CD and used it to develop a linear multifunctional "click" PEG-based osteotropic polyrotaxane for diagnosis and treatment evaluation of cancer bone metastasis. This system is designed to address a critical challenge in developing osteotropic macromolecular prodrugs. It has been found in the past that the increase of bone-targeting moiety in polymeric prodrug will enhance their osteotropicity [46]. Due to the fact that these targeting moieties are only soluble in water, their incorporation into the prodrug has always been difficult and nonconsistent from batch to batch. By threading ALN-α-CD onto PEG and followed by "click" copolymerization, one could freely adjust the content of ALN-pendent functionality of the polyrotaxane [47]. Indeed, by increasing the ALN-α-CD feed-in ratio, the team could gradually

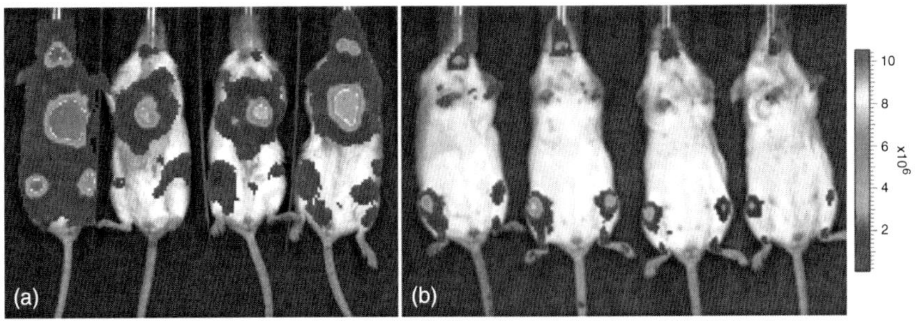

FIGURE 11.2 The early identification of breast cancer bone metastasis lesion by NIR dye-labeled osteotropic polyrotaxane. (a) Luciferase activity associated with breast cancer cells was mainly found at knee joints (bone metastatic sites) and heart (due to intracardial injection of Luciferase-expressing breast cancer cells); (b) High signal intensity of NIR-dye-labeled osteotropic polyrotaxane was found at knee joints but not the heart, which identifies these positions as cancer bone metastasis. (See color plate section.)

increase the osteotropicity of the polymer conjugates. When tested in a spontaneous breast cancer bone metastasis model, the team observed that the near-infrared (NIR) dye-labeled osteotropic polyrotaxane could discover micrometastasis sites, which are invisible to μ-CT scan (Figure 11.2). This was confirmed by the presence of GFP^+ breast cancer cells at the metastasis sites. This observation may be attributed to the leaky vasculature associated with micrometastatic lesion, which allows the polymeric conjugates' site-specific extravasation and the freshly eroded local bone surface that favors the binding to the lesion. An osteotropic docetaxol polyrotaxane conjugate has been synthesized and it is currently under evaluation for the improved treatment of breast cancer bone metastasis.

11.4 SUMMARY

In this chapter, we have presented some of the most exciting recent developments of CDs and their various biomedical applications. Often, these transforming developments are natural extensions from the previous work, which was not possible in the past. The introduction of "click" reactions revolutionized the development of CD derivatives. Most notably, the high reaction efficiency, stereospecificity, simple reaction conditions, and tolerance of various solvent systems make "click" reactions excellent tools for CDs' modification. The field is still at its early stage of development. In the next few years, we anticipate that many more exciting CDs' derivatives will be developed and used in various biomedical applications. Due to the concern of Cu catalyst residue in the final product, we believe Cu-free "click" HDC reaction will be used more often, especially for products to be used *in vivo*. For the same reason, other types of "click" reactions (e.g., Michael addition) may find use in CDs' modifications.

ACKNOWLEDGMENT

This work is supported in part by Susan G. Komen for the Cure Grants Program KG090860 and NIH R01 grant AR053325.

REFERENCES

1. Easton, C. J.; Lincoln, S. F. *Modified Cyclodextrins – Scaffolds and Templates for Supermolecular Chemistry*; Imperial College Press: London, **1999**; pp 1–32.
2. Brewster, M. E.; Loftsson, F. *Adv. Drug. Deliv. Rev.* **2007**, *59*(7), 645–666.
3. Easton, C. J.; Lincoln, S. F. *Modified Cyclodextrins – Scaffolds and Templates for Supermolecular Chemistry*; Imperial College Press: London, **1999**; pp 71–96.
4. Kanwar, J. R.; Long, B. M.; Kanwar, R. K. *Curr. Med. Chem.* **2011**, *18*(14), 2079–2085.
5. Kolb, H. C.; Finn, M. G.; Sharpless, K. B. *Angew. Chem Int. Ed. Engl.* **2001**, *40*(11), 2004–2021.
6. Takahashi, K.; Hattori, K.; Toda, F. *Tetrahedron Lett.* **1984**, *25*(31), 3331–3334.
7. Easton, C. J.; Lincoln, S. F. *Modified Cyclodextrins – Scaffolds and Templates for Supermolecular Chemistry*; Imperial College Press: London, **1999**; pp 43–64.
8. Hein, C. D.; Liu, X. M.; Wang, D. *Pharm. Res.* **2008**, *25*(10), 2216–2230.
9. Rostovtsev, V. V.; Green, L. G.; Fokin, V. V.; Sharpless, K. B. *Angew. Chem. Int. Ed. Engl.* **2002**, *41*(14), 2596–2599.
10. Mohan, S. J.; Mohan, E. C.; Yamsani, M. R. *In.l J. Pharm. Sci. Nanotech.* **2009**, *1*(4), 309–316.
11. Muderawan, I. W.; Ong, T. T.; Ng, S. C. *J. Sep. Sci.* **2006**, *29*(12), 1849–1871.
12. Zhang, Y.; Guo, Z.; Ye, J.; Xu, Q.; Liang, X.; Lei, A. *J. Chromatogr. A.* **2008**, *1191*(1–2), 188–192.
13. Wang, Y.; Xiao, Y.; Tan, T. T. Y.; Ng, S.-C. *Tetrahedron Lett.* **2008**, *49*(35), 5190–5191.
14. Wang, Y.; Young, D. J.; Tan, T. T. Y.; Ng, S. C. *J. Chromatogr. A.* **2010**, *1217*(31), 5103–5108.
15. Wang, H.; Jiang, P.; Zhang, M.; Dong, X. *J. Chromatogr. A.* **2011**, *1218*(9), 1310–1313.
16. Guerrouache, M.; Millot, M. C.; Carbonnier, B. *Macromol. Rapid Commun.* **2009**, *30*(2), 109–113.
17. Harada, A.; Kamachi, M. *Macromolecules* **1990**, *23*(10), 2821–2823.
18. Harada, A.; Kamachi, M. *Nature* **1992**, *356*, 325–327.
19. Wu, J. Y.; Gao, C. *Macromol. Chem. Phy.* **2009**, *210*(20), 1697–1708.
20. Wu, J. Y.; He, H.; Gao, C. *Macromolecules* **2010**, *43*(5), 2252–2260.
21. Hyun, H.; Yui, N. *Macromol. Rapid Commun.* **2011**, *32*(3), 326–331.
22. Chwalek, M.; Auzely, R.; Fort, S. *Org. Biomol. Chem.* **2009**, *7*(8), 1680–1688.
23. Bae, Y. H., Kim, S. W. *Advanced Drug Deliv. Rev.* **1993**, *11*(1–2), 109–135.
24. Xu, X. D.; Chen, C. S.; Lu, B.; Wang, Z. C.; Cheng, S. X.; Zhang, X. Z.; Zhuo, R. X. *Macromol. Rapid Commun.* **2009**, *30*(3), 157–164.
25. Li, K. *Chem. Eng. Technol.* **2002**, *25*, 203–206.

26. Souzy, R.; Ameduri, B.; Boutevin, B. *Prog. Polym. Sci.* **2004**, *29*(2), 75–106.
27. Cai, T.; Neoh, K. G.; Kang, E. T. *Macromolecules* **2011**, *44*(11), 4258–4268.
28. Binello, A.; Robaldo, B.; Barge, A.; Cavalli, R.; Cravotto, G. *J. Appl. Polym. Sci.* **2008**, *107*(4), 2549–2557.
29. Zhang, Y. M.; Chen, Y.; Li, Z.Q.; Li, N.; Liu, Y. *Bioorg. Med. Chem.* **2010**, *18*(4), 1415–1420.
30. Ascherio, A.; Chen, H.; Weisskopf, M. G.; O'Reilly, E.; McCullough, M. L.; Calle, E. E.; Schwarzschild, M. A.; Thun, M. J. *Ann. Neurol.* **2006**, *60*(2), 197–203.
31. Mallard-Favier, I.; Blach, P.; Cazier, F.; Delattre, F. *Carbohydr. Res.* **2009**, *344*(2), 161–166.
32. Fujimoto, K.; Yamada, S.; Inouye, M. *Chem. Commun. (Camb).* **2009**, (46), 7164–7166.
33. Braun, C. S.; Vetro, J.A.; Tomalia, D. A.; Koe, G. S.; Koe, J. G.; Middaugh, C. R. *J. Pharm. Sci.* **2005**, *94*(2), 423–436.
34. Luo, D.; Haverstick, K.; Belcheva, N.; Han, E.; Saltzman, W. M. *Macromolecules* **2002**, *35*(9), 3456–3462.
35. Ohsaki, M.; Okuda T.; Wada, A.; Hirayama, T.; Niidome, T.; Aoyagi, H. *Bioconjug. Chem.* **2002**, *13*(3), 510–517.
36. Srinivasachari, S.; Fichter, K. M.; Reineke, T. M. *J. Am. Chem. Soc.* **2008**, *130*(14), 4618–4627.
37. Srinivasachari, S.; Reineke, T. M. *Biomaterials* **2009**, *30*(5), 928–938.
38. Mendez-Ardoy, A.; Gomez-Garcia, M.; Mellet, C. O.; Sevillano, N.; Girón, M.D.; Salto, R.; Santoyo-González, F.; García Fernández, J.M. *Org. Biomol. Chem.* **2009**, *7*(13), 2681–2684.
39. Joshi, A.; Kate, S.; Poon, V.; Mondal, D.; Boggara, M. B.; Saraph, A.; Martin, J. T.; McAlpine, R.; Day, R.; Garcia, A. E.; Mogridge, J.; Kane, R. S. *Biomacromolecules* **2011**, *12*(3), 791–796.
40. Serrano, M. C.; Chung, E. J.; Ameer, G. A. *Adv. Funct. Mater.* **2010**, *20*(2), 192–208.
41. Jazkewitsch, O.; Mondrzyk, A.; Staffel, A.; Ritter, H. *Macromolecules* **2011**, *44*(6), 1365–1371.
42. Quan, C.; Chen, J.; Wang, H.; Li, C.; Zhang, X. *ACS Nano* **2010**, *4*(7), 4211–4219.
43. Rodan, G. A.; Martin, T. J. *Science* **2000**, *289*, 1508–1504.
44. Wang, D.; Miller, S. C.; Kopeček, J. *Adv. Drug Deliv. Rev.* **2005**, *57*(7), 935–937.
45. Liu, X. M.; Lee, H.T.; Reinhardt, R.A.; Marky, L.A.; Wang, D. *J. Control Release* **2007**, *122*(1), 54–62.
46. Wang, D.; Sima, M.; Mosley, R. L.; Davda, J. P.; Tietze, N.; Miller, S. C.; Gwilt, P. R.; Kopecková, P.; Kopecek, J. *Mol. Pharm.* **2006**, *3*(6), 717–725.
47. Hein, C. D.; Liu, X. M.; Chen, F.; Cullen, D. M.; Wang, D. *Macromol Biosci.* **2010**, *10*(12), 1544–1556.

12

TRIAZOLYL GLYCOCONJUGATES IN MEDICINAL CHEMISTRY

Rama Pati Tripathi, Pratibha Dwivedi, Anindra Sharma, Divya Kushwaha, and Vinod Kumar Tiwari

12.1 INTRODUCTION

The group of biomolecules called carbohydrates play critical roles in many physiological as well as pathological cell functions by virtue of their structural complexity and ubiquitous nature [1]. They are crucial to cellular recognition events, including signal transduction, inflammation, immune response, apoptosis, tumor metastasis, and viral and bacterial infections [2, 3]. They are important adhesion receptors in cell–cell and cell–matrix interactions, and as such are absolutely necessary for leukocyte extravasation during inflammation and lymphocyte recirculation. Many microbes, including viruses, bacteria and their toxins, have evolved to bind cell surface carbohydrates, a binding that is a prerequisite for infection to occur [4]. Carbohydrates and their derivatives include many other important biomolecules that play key roles in the fertilization, blood clotting, hormone trafficking, cellular differentiation, and development. Carbohydrates are significant determinants of self/non-self, such that anti-carbohydrate antibodies may initiate graft rejection following transplantation between individuals of different ABO blood groups or between species. Since the phenotypic changes due to carbohydrate units play a prominent role in cell differentiation and tumor progression, it offers a novel approach to the immunotherapeutic strategies in the treatment of malignant diseases. Several chiral hydroxyl groups in carbohydrates, upon appropriate modifications and utilization, would stimulate

Click Chemistry in Glycoscience: New Developments and Strategies, First Edition.
Edited by Zbigniew J. Witczak and Roman Bielski.

chemical diversities, that is glycoconjugates/glycohybrids are of great significance in drug design and development [5, 6]. The multivalent nature of these molecules is frequently used to increase the affinities to the targets in different biological processes, such as the binding of bacteria, bacterial toxins [7], galectins [8], and other lectins [9].

Triazole ring embodied in molecules are not just passive connecting units; in fact, they may interplay with suitable biological targets owing to their high dipole moment, which allows strong hydrogen-bonding interaction with targets. Moreover, this heterocycle is stable toward metabolic degradation and can improve the solubility of molecules, thereby establishing triazoles as a potential pharmacophore [10].

Triazole unit is an important pharmacophore in several anti-HIV, antiviral [11], antibacterial [12], antifungal [13], antiparasitic [14], anti-inflammatory, and antitubercular agents [15], specially, 1,2,3-triazoles do not occur in nature, but they are endowed with numerous biological activities. Therefore, synthesis of triazolyl glycoconjugates possessing a variety of biological activities and their development as a pharmacophore, is a valuable area of research in the field of medicinal chemistry. The present review is aimed at recent developments in carbohydrate-based triazolyl scaffolds in medicinal chemistry.

12.2 Cu(I)-CATALYZED ALKYNE–AZIDE CYCLOADDITION (CuAAC) REACTION

The Cu(I)-catalyzed azide–alkyne cycloaddition (CuAAC) reaction, discovered by two independent Research Groups of Sharpless [16] and Meldal [17] in 2002, proceeds under mild conditions to give exclusively 1,4-disubstituted 1,2,3-triazoles, contrary to its classical thermally induced Huisgen-type reaction, which gives a mixture of 1,4 and 1,5 disubstituted products. The protocol is useful because it utilizes biologically inert starting materials alkyne/azide to deliver chemically stable 1,2,3-triazoles at 10^7 times faster rate than uncatalyzed reaction. Also, compatibility of reaction condition with aqueous medium and well tolerance to a variety of functional groups and pH enhances the biological significance of the approach. Similar to CuAAC, Ruthenium-catalyzed reaction of alkyne with azide has been reported, which affords exclusively 1,5-disubstituted triazoles regioselectively opposite to Cu-assisted click reaction [18].

Especially in the case of carbohydrate chemistry, compatibility of Cu(I)-catalyzed coupling of carbohydrate-based azides and alkynes to preappended protecting groups, as well as relative inertness of triazole ring toward the stringent protection/deprotection and glycosylation steps establish this approach as a patent tool in accessing diverse carbohydrate mimetics with improved affinity. Therefore, click chemistry has emerged as a powerful conjugation strategy for the synthesis of 1,2,3-triazole-containing glycoconjugates of biomedical interest [19–29].

Carbohydrate chemistry, through provision of "click chemistry" offers a potentially more reliable route to homogeneous glycoconjugates leading to either exact copies of naturally occurring glycoconjugates or, alternatively, incorporating

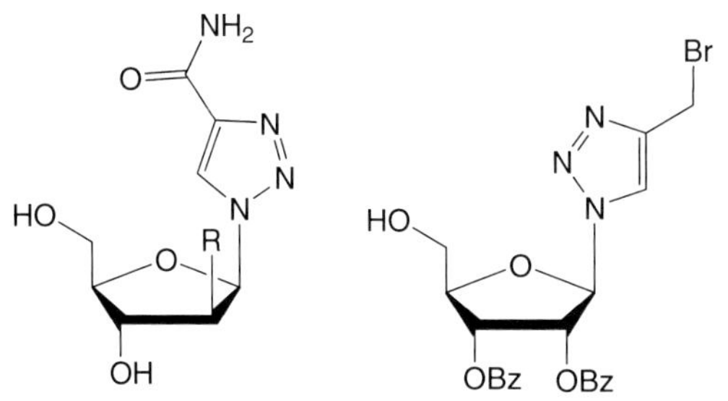

1a **1b**

FIGURE 12.1 Sugar-based triazoles known before invention of CuAAC reaction.

unnatural glycosidic linkages. An appreciable benefit of the latter approach is that the "artificial" glycoconjugate may retain the geometric and spatial characteristics of the native glycoform with sufficient hydrolytic stability toward glycosidases and glycosyltrasferases and also inhibit these enzymes to some extent, so increasing the potential for a wide array of *in vivo* applications [30]. In medicinal chemistry, the 1,4-disubstituted 1-*H*-1,2,3-triazoles have been considered significant for diverse range of *in vivo* applications, and indeed before the discovery of CuAAC reaction, more than 7000 1,4-disubstituted triazoles have been known (Fig. 12.1).

More recently, Cu(I)-catalyzed click reaction has emerged as powerful conjugation strategy for the discovery and optimization of structurally diverse glycoconjugates of biomedical interest [29].

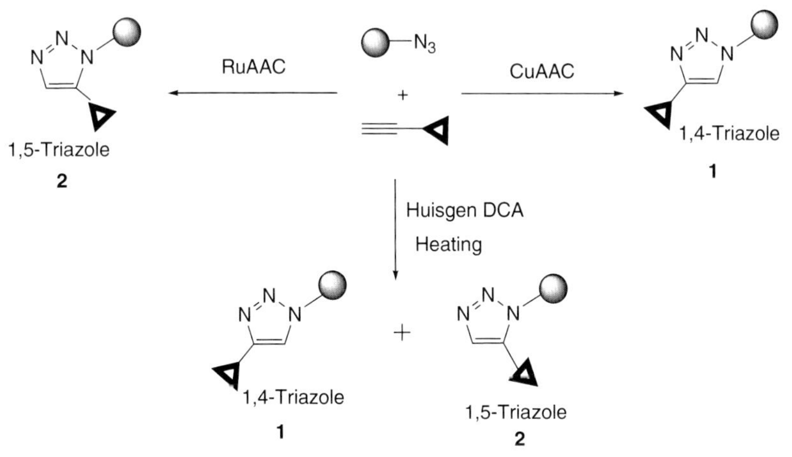

SCHEME 12.1 Regioselectivity in alkyne-azide 1,3-dipolar cycloaddition reaction.

12.3 MECHANISM OF Cu-CATALYZED CLICK REACTION

The cycloadditions generally proceed through a concerted mechanism. However, experimental kinetic data and molecular modeling of the Huisgen cycloaddition have suggested a stepwise mechanism to these reactions [31, 32]. Based on the experimental evidences available, it has been proposed that mechanism of Cu(I)-catalyzed 1,3 dipolar cycloaddition reaction involves bimolecular Cu species. The first step of catalytic cycle involves π-complexation of the Cu(I) dimer to the alkyne **II**, which leads to deprotonation of terminal hydrogen to deliver Cu-acetylide **III**. In the following step, *N*(1) of azide complexes with Cu(I) metal center displacing one of the ligands from another Cu resulting in Cu-acetylide complex **IV**, which in turn activates *N*(3) of azide for nucleophilic attack at C(5) to give metallacycle **V** [33]. In the successive step, **V** undergo ring contraction when lone pair of electrons of *N*(1) attacks C(5) to give the respective intermediate **VI,** which has been isolated [34]. Finally, protonation of **VI** with solvent or base eventually releases 1,4 disubstituted 1,2,3 triazole **VII** product and facilitates dissociation of labile copper complex, which could undergo second catalytic cycle (Scheme 12.2).

12.4 TRIAZOLE RING AS BIOISOSTERES OF AMIDE

Amide bond represents the main chemical bond that links amino acid building blocks together to constitute proteins, where this functionality plays a major role in the elaboration and composition of biological systems [35]. Triazoles exhibit chemical and electronic resemblance to amide bond, which make this heterocycle moiety biologically more relevant, and eventually addresses CuAAC as a more valuable approach because it introduces the triazole skeleton to diverse building blocks with apparent ease (Fig. 12.2) [36].

12.5 TRIAZOLYL GLYCOCONJUGATES IN MEDICINAL CHEMISTRY

Involvement in crucial life-governing metabolic processes such as cell–cell interaction, cell migration, pathogen defense, etc. has established the carbohydrates as promising scaffolds in drug discovery and development [37]. Despite the complexity in synthesis, especially concerning anomeric stereochemistry, poor pharmacological properties, and modest affinity for various receptors, the compatibility of Cu(I)-catalyzed coupling of carbohydrate-based azides and alkynes to pre-appended protecting groups, as well as relative inertness of triazole ring toward the string of protection/deprotection and glycosylation steps, has established click chemistry as a potent tool for accessing diverse carbohydrate mimetics with improved affinity. A large number of biologically active glycosylated 1,2,3-triazole moieties have been designed to show good pharmacokinetic properties.

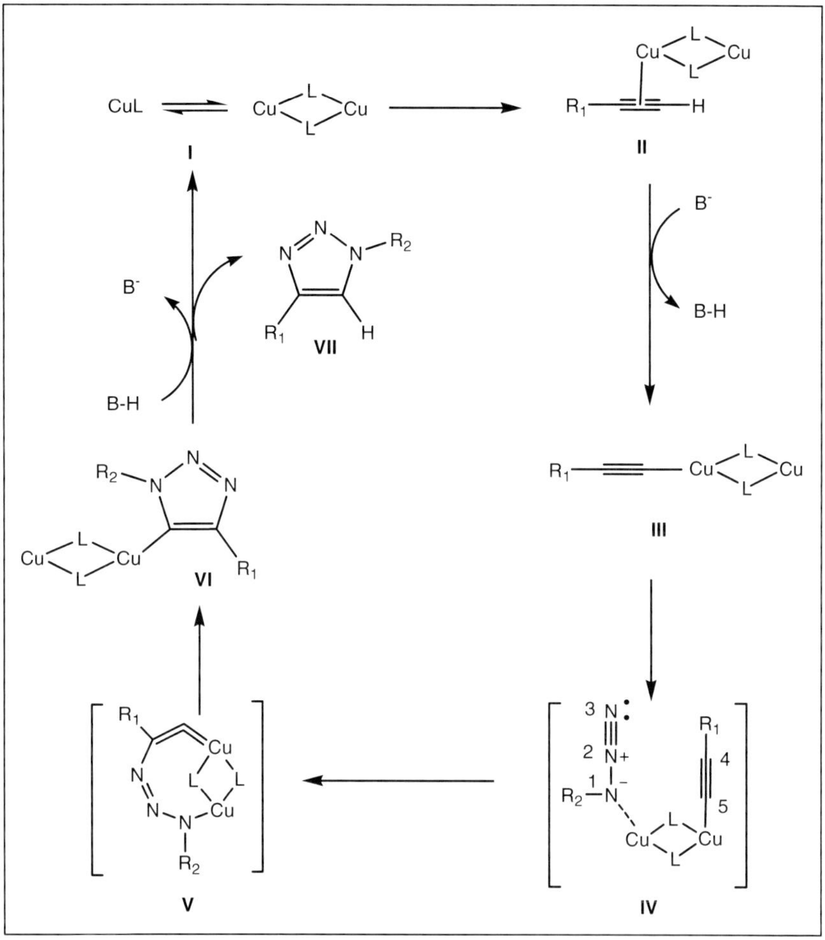

SCHEME 12.2 Well-accepted mechanism for CuAAC reaction.

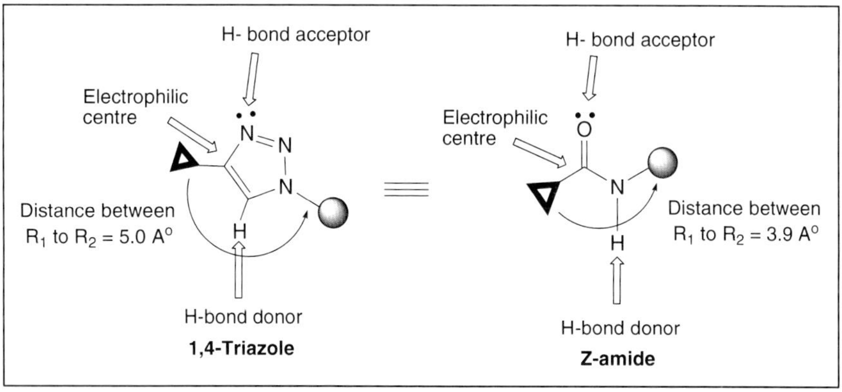

FIGURE 12.2 Hypothesis on structural similarity between 1,4-triazole and Z-amide.

12.5.1 Anticancer and Anti-Inflammatory Activities

Cancer is a major public health burden worldwide. A large number of anticancer agents including taxol, vinblastine, vincristine, camptothecin derivatives, opotecan and irinotecan, and etoposide derived from epipodophyllotoxin are in clinical use all over the world. While several promising agents such as flavopiridol, roscovitine, combretastatin A-4, betulinic acid, and silvestrol are in clinical or preclinical development. But increasing occurrence of cancer cases has triggered the search for efficient anticancer agents for the prevention and cure of the disease.

Cell-surface of cancerous cells is abundant in oligosaccharide antigens, which could act as markers for active or passive cancer. It has been observed that glycosyl mimetics of glycosylated sugar residing over cancerous cells can serve as an effective antigen to evoke antitumor immune response in body, when conjugated with some large biomolecules. Therefore, Danishefsky group synthesized the potent anticancer vaccine in which antigenic oligosaccharides were linked to keyhole limpet hemocyanin (KLH) polypeptide carriers by means of triazole moiety (**3**) (Fig. 12.3) [38].

Carvalho et al. prepared a library of 1,4-disubstituted 1,2,3-triazole galactose derivatives using Cu(I)-catalyzed azide–alkyne 1,3-dipolar cycloaddition reaction. The compounds demonstrated significant inhibition of *Trypanosoma cruzi trans*-sialidase (TcTS) enzyme that plays a key role in recognition and invasion of host cells and enables the parasite to escape from human immune response. In TcTS assay, the compound (**4**) demonstrated maximum inhibition (37%), while other synthesized sugar-triazoles displayed moderate to weak activity ranging between 0.5 mM and 1 mM. The *in vitro* assay against trypomastigotes of *T. cruzi* Y strain using benzimidazole as a reference drug, the compound (**5**) demonstrated most powerful inhibition with optimum activity in 100s of μm range [14].

The α-GalCer is the most potent agonistic antigen of the T cell receptor of natural killer T cells. Lee and coworkers have synthesized a series of 1,2,3-triazole-containing α-GalCer analogs (**6**) by varying the length of lipid chain attached. It has been observed that replacement of amide group with isosteric triazole enhances the IL-4 versus IFN-γ bias of released cytokines. Also, the analogs carrying long-chained triazoles exhibited stimulatory effect on cytokine production comparable to α-GalCer and showed a stronger Th2 cytokine response [39]. Likewise, Slamova et al. have reported the synthesis β-2-acetamido-2-deoxygluco GlcNAc-triazole (**7**) by click reaction under microwave condition, which was found to be hydrolyzed by *Talaromyces flavus* CCF 2686 β-*N*-acetylhexosaminidase and was furthermore established to act as a strong ligand of rat and human natural killer cell activating receptors (Fig. 12.4) [40].

In another effort, 1,2,3-triazole moiety was linked with biologically active 1,2,4-oxadiazole heterocycles at one end and glucopyranosyl function at another end. Using Cu(I)-catalyzed click reaction between 2,3,4,6-tetra-O-acetyl-β-D-glucopyranosyl azide and propynyl 3-[3-aryl)-1,2,4-oxadiazol-5-yl] propionates, dos Anjos et al. prepared a library of glycosyl-triazole linked 1,2,4-oxadiazole (**8**). The whole library showed weak cytotoxic activity and demonstrated 22–25% cell-growth

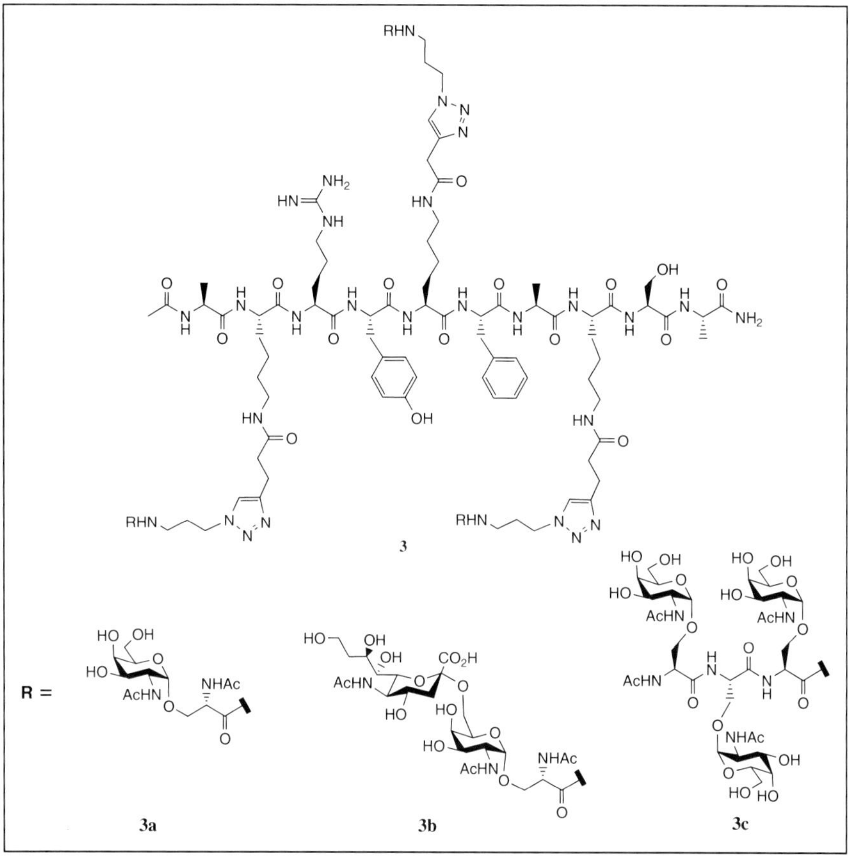

FIGURE 12.3 Glycosylated carrier protein (KLH) to derive synthetic anticancer vaccines.

inhibition against cell strains NCI-H_{292} (lung carcinoma) and HEp-2 (larynx carcinoma) [41].

The tubulin/microtubule system is involved in the key step of mitosis, therefore cell division can be prevented and apoptosis can be induced by disturbing their dynamic equilibrium. To date, paclitaxel (taxol), discodermolide, or epothilones are the most common microtubule-stabilizing antitubulin agents. Manach et al. synthesized a library of paclitaxel (taxol) mimics (**9**) by stereoselective β-glycosylation of L-glucurono-γ-lactone followed by click reaction with aromatic azides. Synthesized molecules were anticipated to serve as potential inhibitors of tubulin depolymerization and showed cytotoxic properties (Fig. 12.5) [42].

Chen et al. using Cu(I)-catalyzed 1,3-dipolar cycloaddition reaction of azide and alkyne synthesized a *C*-3 symmetric (1-6)-*N*-acetyl-β-D-glucosamine octadecasaccharide (**10**), which demonstrated significant antitumor activity against H22 in the preliminary mice test (Fig. 12.6) [43].

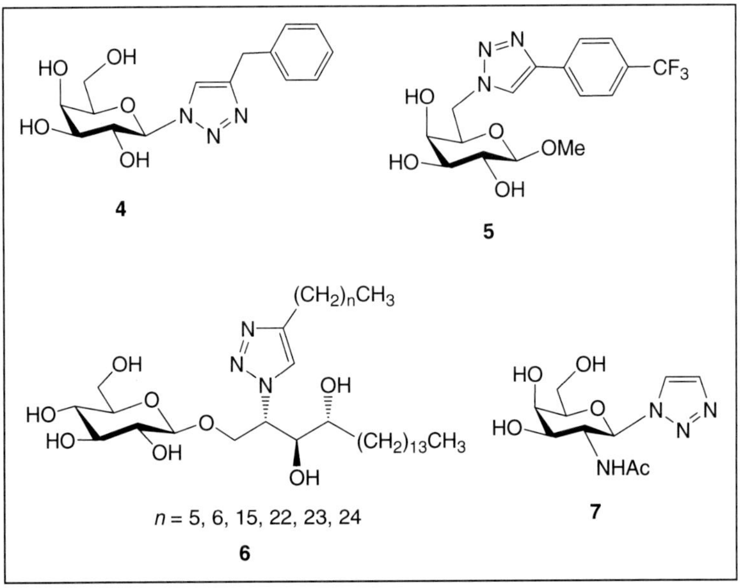

FIGURE 12.4 Biologically active Glycosylated Triazoles.

Campo et al. explored CuAAC-assisted cyclo-oligomerization of isomeric azido-alkyne derivatives of β-D-galactopyranose and obtained dimeric to pentameric cyclic pseudo-oligosaccharide macrocycles, which were subsequently examined for their compatibility with enzymatic sialylation by *Trypanosoma cruzi trans-sialidase*, in search of potent inhibitors against the enzyme. The bifunctional galactose precursors

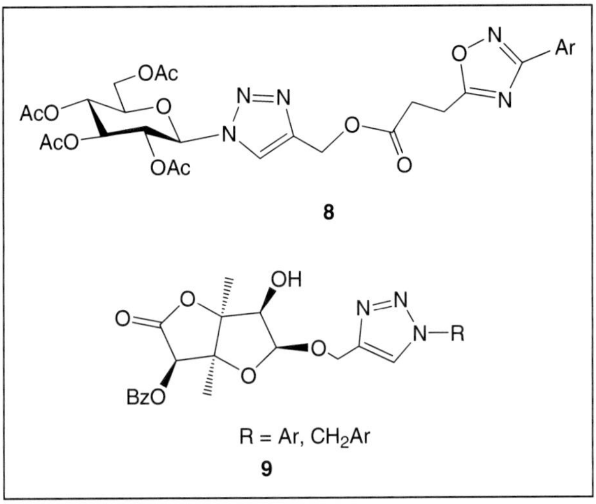

FIGURE 12.5 Biologically active Triazoles.

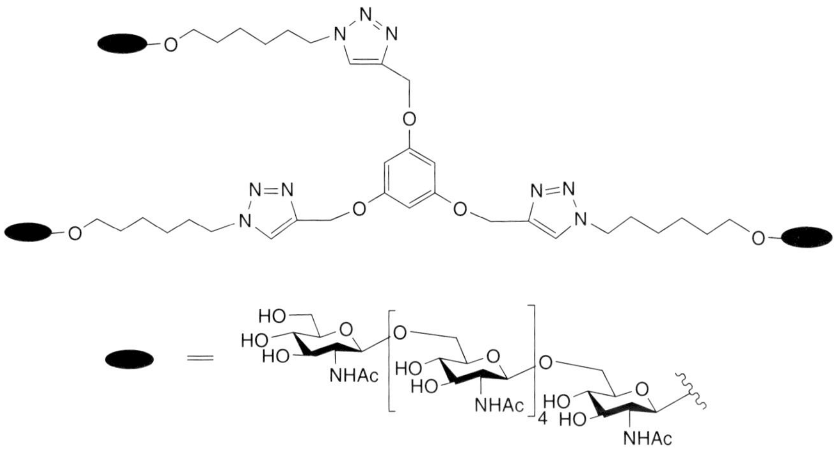

10

FIGURE 12.6 Triazolyl glycocluster having antitumor property.

(**11**), (**12**) bearing azido groups either at anomeric position or at *C*-6 and an alkyne functionality in the form of either a 6-*O*- or 1-*O*-propargyl on cyclo-oligomerization via click reaction using $CuSO_4$/Cu turnings in DMF afforded corresponding macrocycles. These macrocycles served as a template for enzymatic sialylation with TcTS in presence of MUNANA as sialic acid donor delivering (**13**), (**14**), (**15**), and (**16**), respectively (Scheme 12.3) [44].

Galectins are a family of cytosolic β-D-galactoside binding proteins, which strongly modulate their expression during development, differentiation stages, and under different physiological or pathological conditions. Studies demonstrated that Gal-3 is involved in colon cancer metastasis, brain tumor progression, inhibition of metastasis-associated cancer cell adhesion, and may play a key role in innate immunity. Other reports suggest that Gal-3 and Gal-1 can regulate apoptosis processes, and Gal-1 acts as an insoluble host factor that promotes HIV-1 infectivity through stabilization of virus attachment to host cells [45].

Carbohydrate ligands of natural occurrence exhibit weak affinity for galectins, therefore galactosides and lactosides-based monovalent and multivalent triazole clusters were prepared by Giguere et al. with the anticipation of better binding capacity and selective inhibition of Gal-1 over the other galectins. Evaluation of monovalent compounds showed maximum inhibitory activity for triazolyl galactoside (**17**) with inhibitory properties of 1250 μM for both inhibitors whereas trivalent lactoside (**18**) registered inhibitory properties of 20 μM for galectins [45].

Tejler et al. prepared the acetylene derivatives of various natural and nonnatural amino acids and clustered them with complex 2-azidoethyl β-D-galactopyranosyl-(1→4)-β-D-glucopyranoside via regioselective copper(I)-mediated 1,3-dipolar

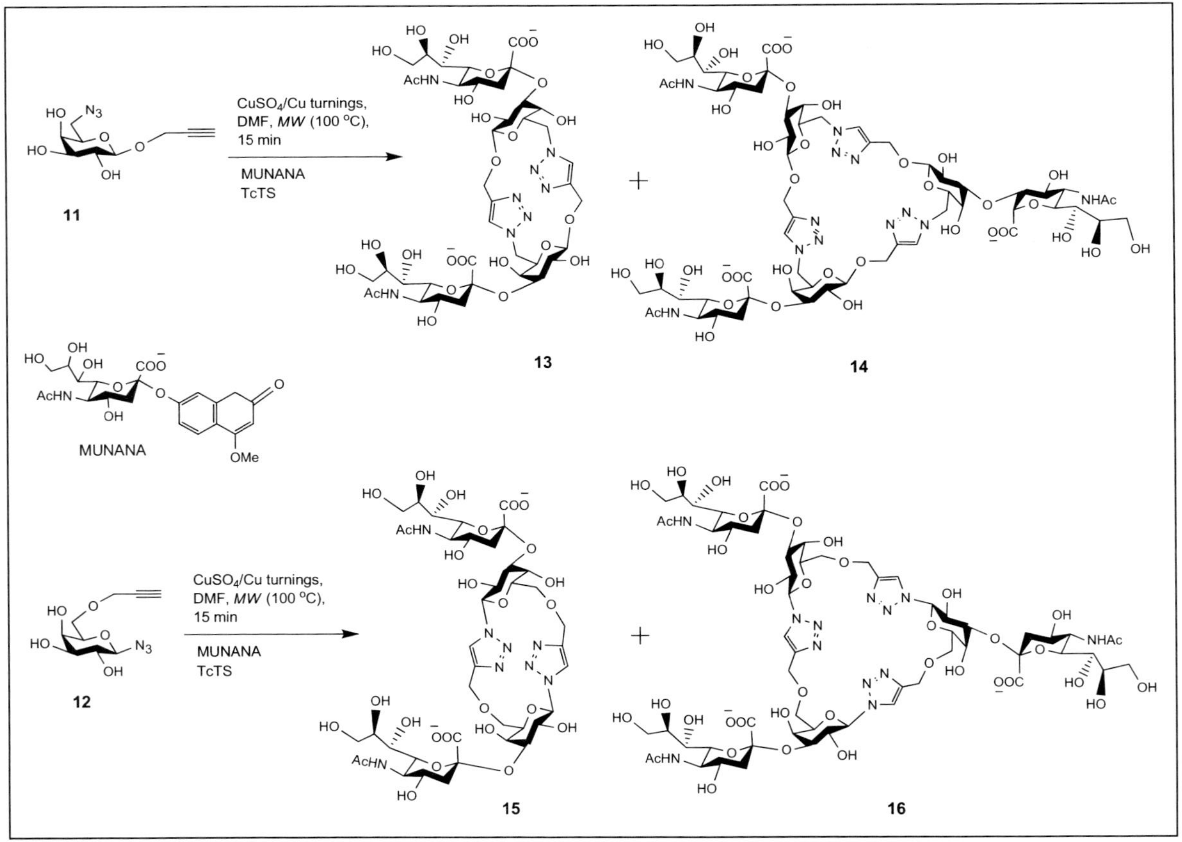

SCHEME 12.3 Cyclooligomerization of precursors (**11**), (**12**) via click reaction and concurrent sialylation to obtain cyclic pseudo-oligosaccharides (**13–14**), (**15–16**).

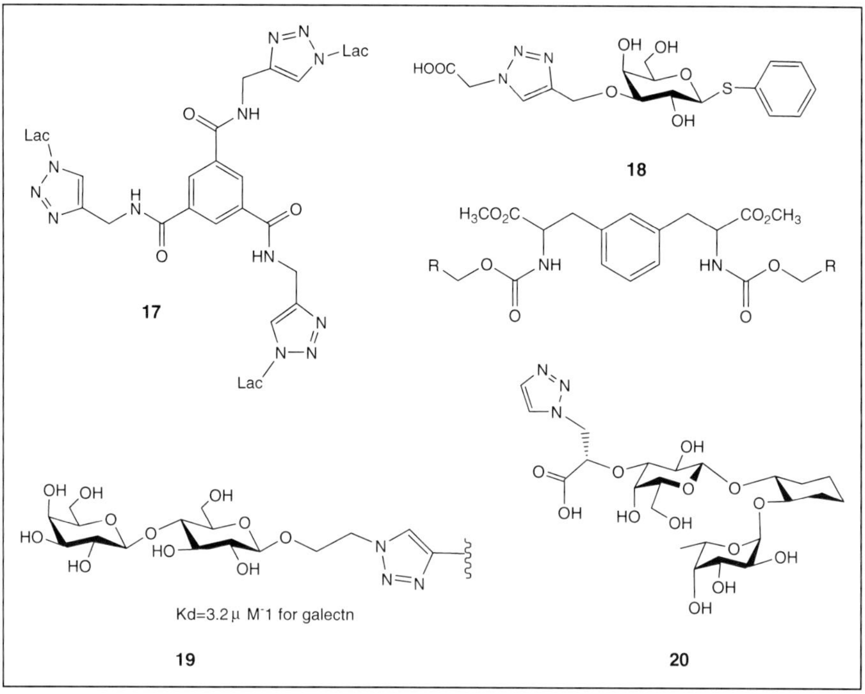

FIGURE 12.7 Biologically active triazolyl glycoconjugates.

cycloaddition reaction to afford a panel of mono-, di- and trivalent lactoside derivatives. Among all, the divalent cluster **19** with K_d value as low as 3.2 μM was identified as a promising galectin-1 inhibitor (Fig. 12.7) having a relative potency of 30 per lactose unit as compared to the natural disaccharide ligand lactose [46].

Ernst et al. utilized click chemistry in preparation of a library of selectin antagonists based on sialyl Lewis^x (sLe^x) structure where Neu5Ac was replaced by (*S*)-isoserine. The sLe^x residing on physiological selectin ligands are responsible for inflammation, a protective attempt of body against pathogens. A series of 7 different alkynes were inserted over sLe^x based triazole but only the unsubstituted triazole and the azido compound (**20**) displayed slightly improved affinities for E-selectin, while substituted compounds showed affinity similar to sLe^x (Fig. 12.7) [47].

Sialyl Lewis^x (sLe^x) and sialyl Lewis^a (sLe^a), the tetrasaccharides usually attached to *O*-glycans on the surface of cells, are crucial for the cell–cell recognition processes such as fertilization, embryogenesis, lymphocyte trafficking, immune responses, and cancer metastasis. The enzyme fucosyltransferase catalyzes the last step of glycosylation of sLe^x and sLe^a by facilitating transfer of L-fucose moiety from guanosine diphosphate β-L-fucose (GDP-fucose) to a specific hydroxyl group of sialyl *N*-acetyllactosamine. Wong et al. using Cu(I)-catalyzed 1,3-dipolar cycloaddition prepared a library of 85 GDP-triazoles by keeping GDP as common scaffold, and varying hydrophobic group and linker chain attached to it. The compound (**21**) with an

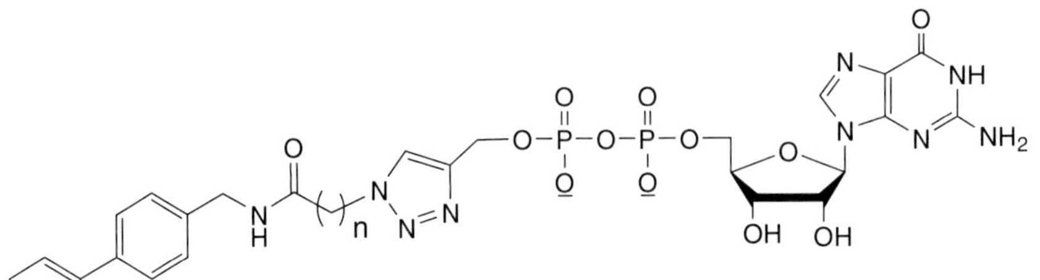

FIGURE 12.8 Triazole synthesis in microtiter plate for *in situ* screening of Fucosyltransferase VI inhibitor.

IC_{50} value of 0.15 μM demonstrated specific optimum inhibition for human R-1,3-fucosyltransferase VI (Fuc-T VI) enzyme (Fig. 12.8) [48].

Presence of more triazole branching units and multivalent glycopyranoside units generally enhance the anti-inflammatory properties of dendrimers, and indeed the higher generation dendrimers exhibit better anti-inflammatory activity than the lower generation. Rajakumar and Anandhan successfully utilized alkyne–azide click reaction in the development of glycodendrimers (**22**) (Fig. 12.9) containing 12 and 18 α-D-glycopyranosides at the periphery, and triazole as branching unit. In anti-inflammatory assay, the glycodendrimer (**22**) demonstrated significant activity [49].

12.5.2 Antituberculosis Agent

Tuberculosis (TB), caused by *Mycobacterium tuberculosis*, is one of the major causes of death and suffering worldwide. Directly observed therapy short-course (DOTS), a World Health Organization (WHO)-approved treatment for TB, involves a three- or four-drug regimen that comprises isoniazid, rifampin, pyrazinamide, and/or ethambutol for a minimum of 6 months. However, the increasing microbial resistance, toxicity, and side effects of currently used anti-tuberculosis drugs also emphasize the urgent need for the development of more effective anti-tuberculosis agents [50].

The mycobacterial α-(1,6)-mannosyltransferase enzyme plays a key role in biosynthesis of the cell envelope of *Mycobacterium tuberculosis*, therefore development of inhibitors to this enzyme would effectively combat tuberculosis. Dondoni and coworkers synthesized a new class of (1,6)-oligomannose analogs having triazole bridges using click chemistry. Ethynyl α-*C*-mannoside upon 1,3 dipolar coupling with alkyl 6-azido-α-*C*-mannoside derivatives furnished linear oligomers (triazolo-pentamannose derivative) in good yield [51]. Moreover, a number of triazole-linked 1,6-oligomannosides up to a hexadecamer (**23**) were also prepared by the Cu(I)-catalyzed azide–alkyne cycloaddition. Among all, the hexamannoside and octamannoside with IC_{50} values ranging between 0.14 mM and 0.22 mM

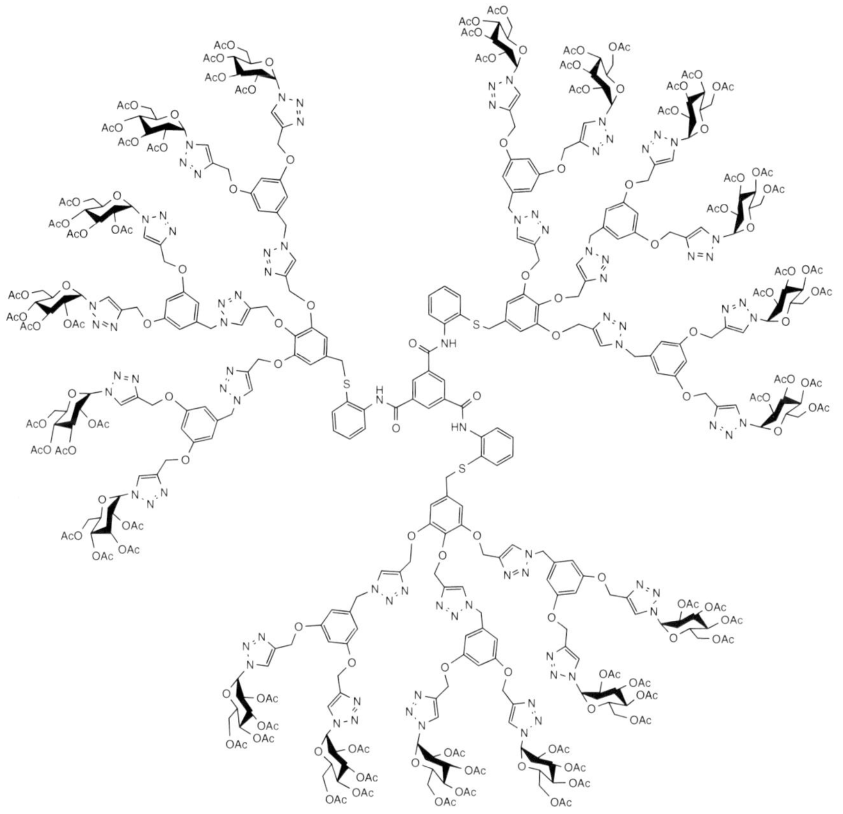

22

FIGURE 12.9 Anti-inflammatory glyco-dendrimers with benzene 1,3,5 carboxamide core.

demonstrated maximum inhibition against mycobacterial α-(1,6)-mannosyltransferases (Fig. 12.10) [52].

Our group has synthesized 5-Azido-5-deoxyxylo-, ribo-, and arabinofuranoses, which upon 1,3 dipolar cycloaddition with different alkynes resulted in a series of 1,2,3-triazolyl glycoconjugates. All the compounds were evaluated for antitubercular activity against *Mycobacterium tuberculosis* $H_{37}Rv$, where compound (**24**) showed mild antitubercular activity *in vitro* with MIC 12.5 μg/mL [15].

12.5.3 Antifungal and Antibacterial

The incidence of life-threatening fungal infections has tremendously increased in the last two decades due to greater use of immunosuppressive drugs, prolonged use of broad spectrum antibiotics, widespread use of indwelling catheters, and also in

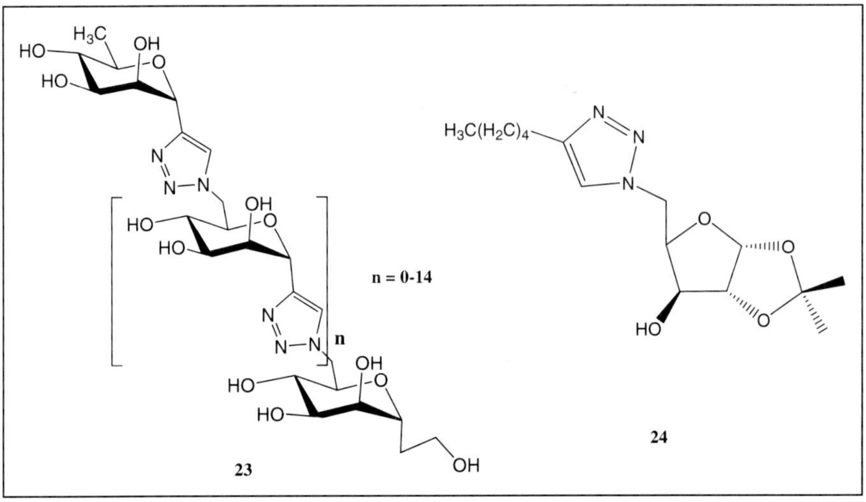

FIGURE 12.10 1,6-oligomannose analogs having triazole bridges.

cancer and AIDS. The presently marketed antifungal and antibacterial drugs are either highly toxic or becoming ineffective due to the appearance of resistant strains. This necessitates continued research into new classes of antimicrobial agents. 1,2,3-Triazole-containing molecules is one of these classes [53a].

A series of sugar modules carrying alkyne/azide functionality were synthesized by glycosidase-mediated glycosylation of alkynyl alcohols and azide-containing alcohols with unprotected sugars in one step, and their subsequent click with a variety of alkyne/azide counterparts resulted in triazole-containing glycosides (**25**) and their subsequent submission to bioassays showed that this procedure is a feasible approach to the development of antifungal drugs [53b].

Zhang et al. developed a facile synthetic strategy for the preparation of numerous neomycin B derivatives (**26**) by modifications at the 5″ position. The structure-activity relationship (SAR) against aminoglycoside resistant bacteria equipped with various aminoglycoside-modifying enzymes (AMEs) resulted into the identification of numerous derivatives with an increased antibacterial activity compared to neomycin. Prominent antibacterial activity against methicillin-resistant *Staphylococcus aureus* (MRSA) and vancomycin resistant *enterococci* (VRE) has been observed in some compounds [54].

Reddy et al. have reported numerous glycal-derived tetrahydrofuran 1,2,3-triazoles and screened them for antimicrobial activities. Among all, the compound (**28**) with a MIC value of 25 μg/mL exhibited potent antimicrobial activity whereas compound (**27**) displayed significant antibacterial and antifungal activity with MIC value of 12.5 μg/mL (Fig. 12.11) [13].

Tyrocidine (Tyc), a class of cyclic decapeptides possess antibacterial property due to its ability to penetrate deep inside the lipid bilayer of cell membranes, however, toxicity toward human blood and reproductive cells have anchored tremendous efforts

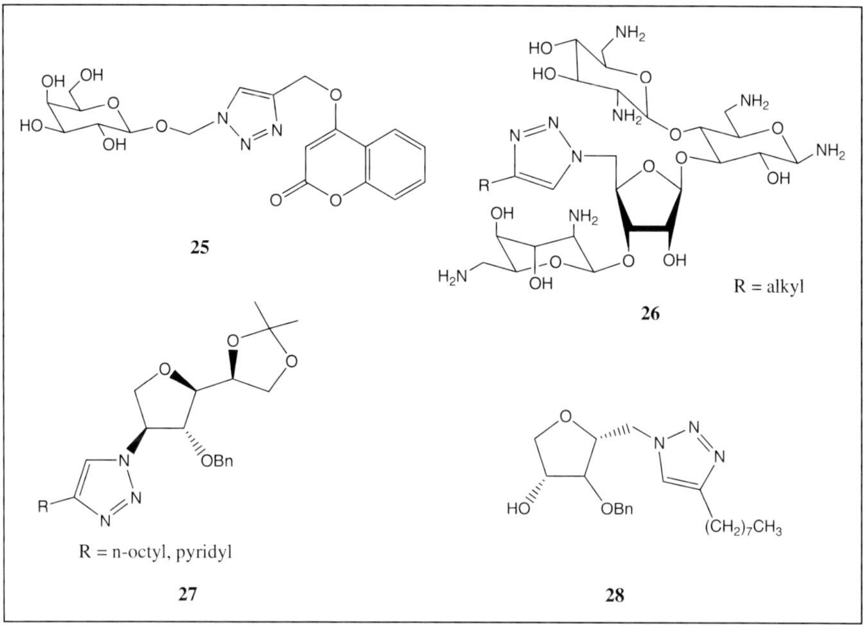

FIGURE 12.11 Antifungal and Antibacterial triazolyl glycoconjugates.

for the synthesis of numerous glycosylated analogs with the anticipation of better therapeutic index. Walsh and Lin derived Tyc analogs containing mono-, di-, and tri-propargyl glycines using a chemoenzymatic peptide coupling approach. A subsequent click reaction with a library of 21 azido sugars afforded 247 carbohydrate-modified tyrocidine analogs. Screening of all the compounds for minimal inhibitory concentration (MIC) and minimal hemolytic concentration (MHC) index demonstrated the 6-times better therapeutic index of triazole-grafted glycopeptides (**29, 30**) than natural tyrocidine (Tyc) (Fig. 12.12) [55].

Vancomycin, a glycosylated natural product from *Amycolatopsis orientalis,* is a last defense against infections caused by methicillin-resistant Gram-positive bacteria, such as *Staphylococcus aureus* or *Enterococcus faecium*. It inhibits cell wall synthesis in Gram-positive bacteria by binding to D-Ala-D-Ala. Thorson et al. applied chemoenzymatic and click approach for the synthesis of a library of diverse monoglycosylated Vancomycin variants. All the derivatives were assessed, and derivative (**32**) registered twice enhanced antibacterial activity against both *S. aureus* and *E. faecium* while being only one-half as potent as vancomycin against *B. Subtilis* (Fig. 12.13) [56].

12.5.4 Antiviral Activities

Monovalent and multivalent sialic acid containing scaffolds modified at *C*-4 or *C*-7 positions serve as highly potent biologically active molecules [57]. Jiang et al.

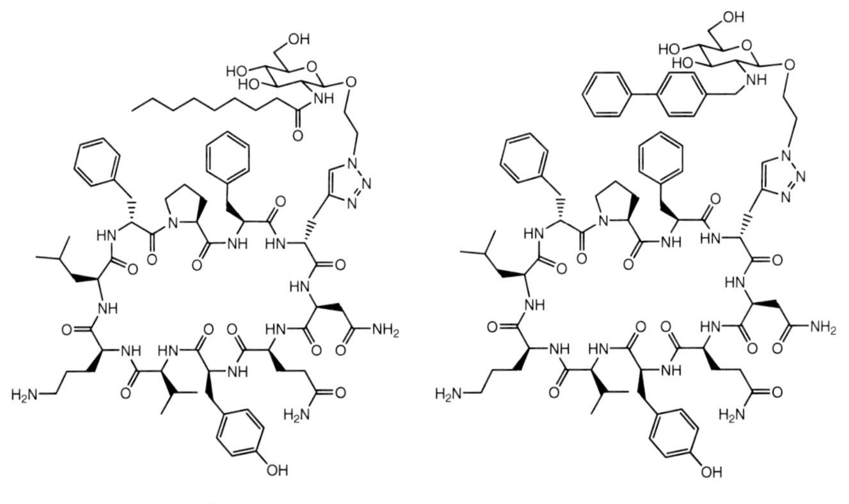

29

30

FIGURE 12.12 Tyrocidine based potent glycopeptide antibiotics Tyc4PG-14 (73) and Tyc4PG-15.

synthesized analogs of drug zanamivir, a neuraminidase inhibitor used in the treatment and prophylaxis of avian influenza (H5N1) caused by influenza A viruses [56]. The neuraminidase enzyme stimulates cleavage of sialic acid from glycoproteins and glycolipids of the host cells, which is a key step for replication of influenza virus. A library of 16 *C*-4 triazole substituted zanamivir analogs were synthesized using

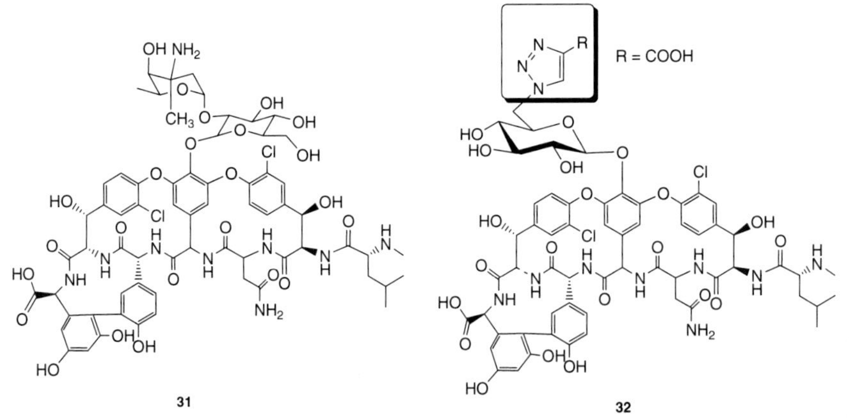

31

Vancomycin

32

Twice as potent as Vancomycin

FIGURE 12.13 Vancomycin analouge.

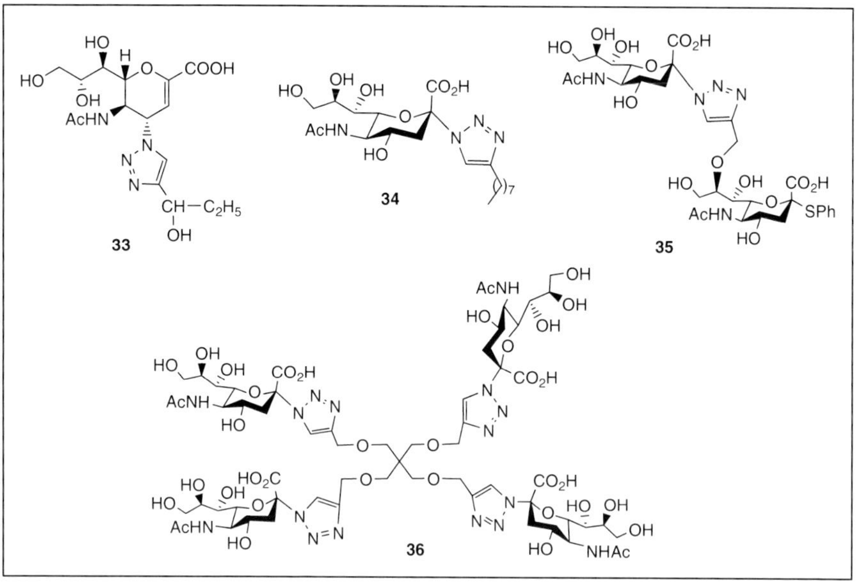

FIGURE 12.14 Sialic acid based potent neuraminidase inhibitors.

click chemistry. Among the synthesized neuraminidase inhibitors, the compound (**33**) with an IC_{50} value of 6.4 μM exhibited anti-AIV activity as compared to zanamivir (IC_{50} = 2.8 μM), the reference drug [58].

Linhardt et al. prepared a library of 1,2,3-triazoles of sialic acid using CuAAC reaction of methyl-[5-acetamido-3,5-dideoxy-2-azido-D-glycero-α-D-galacto-non-2-ulopyranoside]onate with terminal alkyne functionalized molecules and screened them for neuraminidase inhibitory activity (Fig. 12.14). The click reaction of compound sialyl azide with dec-1-yne afforded 5-acetamido-3,5-dideoxy-2-(4-octyl-1*H*-1,2,3-triazol-1-yl)-D-glycero-α-D-galacto-non-2-ulopyranosidic acid (**34**), a potential lead having neuraminidase inhibitory activity (IC_{50} value of 28 μM) comparable to the known sialidase inhibitor Neu5Ac2en. Similarly, sialic acid disaccharide mimic (**35**) (IC_{50} = 17 μM) and dendrimer (**36**) (IC_{50} = 20 μM) demonstrated significant neuraminidase inhibitory activity (Fig. 12.14) [59].

12.5.5 Miscellaneous Activities

A diverse range of physiological and pathological processes that involve rupture or the construction of glycosidic linkages can be regulated by identifying inhibitors of glycosidases and glycosyl transferases that catalyze these processes. Therefore, they may serve as highly promising therapeutic agents for the treatment of diseases such as diabetes [60,61], viral infections [62,63] or cancer metastasis [64]. The 1,2,3-triazoles are poorly basic in nature and as a result remain non-protonated under physiological pH. Studies have revealed that the nonprotonated sp^2-hybridized nitrogen atoms of

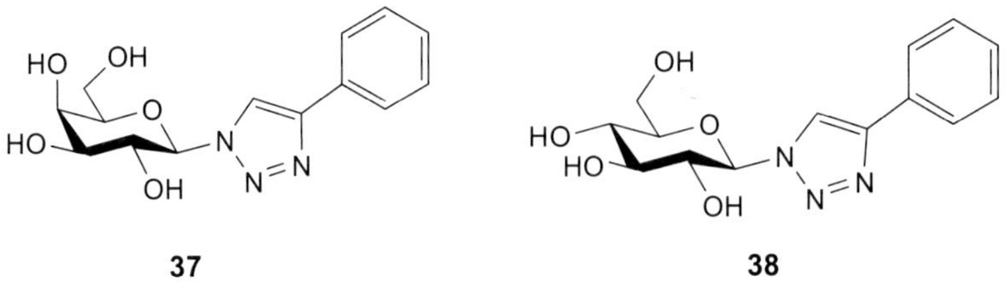

FIGURE 12.15 1-Glycosyl-4-phenyl triazoles potential inhibitors of glycosidases.

1,2,3-triazoles may better mimic the partial positive charge at the anomeric carbon in transition state of the glucosidase-catalyzed reaction, thereby may actually behave as more specific glycosidase inhibitor.

Basu and Rossi synthesized 1-β-D-glucosyl- and 1-β-D-galactosyl-4-phenyl triazole using Cu-catalyzed [3 + 2] cycloaddition reaction (Fig. 12.15). The inhibitory effect of compounds against β-glucosidase enzyme was evaluated in percentage assay against three β-glycosidase enzymes, sweet almond glucosidase (SAG), *Escherichia coli* galactosidase (ECG), and bovine liver galactosidase (BLG). The galactose derivative (**37**) demonstrated moderate inhibition for ECG, whereas glucose triazole (**38**) exhibited better inhibition of BLG than the galactose derivative (**37**) and none of them displayed good inhibition for SAG, when compared to glycosidase inhibitors deoxygalactonojirimycin and deoxynojirimycin (Fig. 12.14) [65].

In search of more potent inhibitors for angiogenesis than 1-deoxynojiromycin (**39**) and aryl-1,2,3-triazole (**40**), which target two different pathways for angiogenesis therapy, Zhou and coworkers prepared a hybrid of them (Fig. 12.15). Thus, *N*-alkylation of 1-DNJ followed by 1,3 dipolar cycloaddition resulted in novel bifunctional compounds (**41–43**), out of which (**43**) with an IC_{50} value of 1.15 μM exhibited a better α-glycosidase activity than DNJ ($IC_{50} = 1.67$ μM). Moreover, the compound (**43**) with an IC_{50} value of 0.105 mM demonstrated improved inhibition of bovine aortic endothelial cells proliferation (BAEC) than compound (**40**) ($IC_{50} = 0.347$ mM) (Fig. 12.16) [66].

A series of saccharidyl-triazoles based on anti-diabetic drug acarbose structure were synthesized by Perion and group using click chemistry. The bioassay the synthesized triazoles significantly inhibited the α-glycosidase enzyme [67]. Field et al. synthesized a library of glucosidase inhibitors by the reaction of α- and β-glucopyranosyl azides with diverse alkynes carrying various functional groups, which were supposed to interact differently with the active site of enzyme [68].

Diot et al. synthesized multivalent imino sugars derived from the glycosidase inhibitor 1-DNJ by using CuAAC reaction (Fig. 12.17) and studied the effect of multivalency on selectivity and affinity of different glycosidases. The 1,3-dipolar coupling of azido functionalized DNJ derivatives with alkynes containing oligoethylene furnished mono-, di- and tri-valent imino sugars. The results of bioactivity evaluation

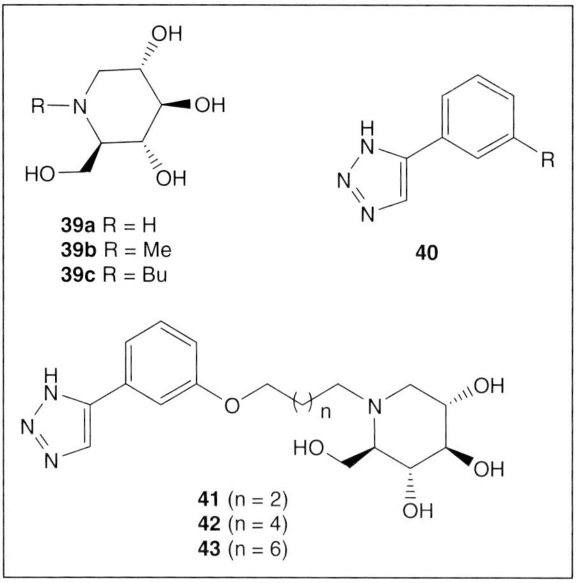

FIGURE 12.16 Synthesis of angiogenesis antagonists, 1-Deoxynojiromycin analogs.

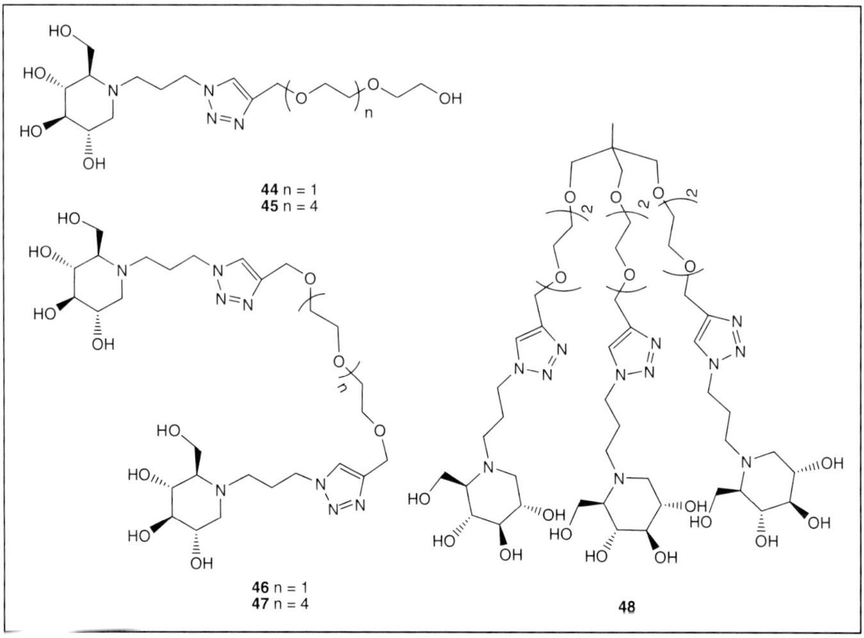

FIGURE 12.17 Multivalent imino sugars.

TABLE 12.1 Glycosidase Inhibitory Activities (Inhibition Constant *K*i, μM) for 1-DNJ and Its Multivalent Azasugar Derivatives (44–48)

Enzyme	1-DNJ	**44**	**45**	**46**	**47**	**48**
β- Galactosidase	42	950 ± 20	n.i.	n.i.	760 ± 10	790 ± 10
α- Galactosidase	–	120 ± 5	95 ± 5	215 ± 5	105 ± 5	225 ± 5
β- Glucosidase	47	47 ± 2	100 ± 5	70 ± 2	70 ± 2	400 ± 10
α- Glucosidase	25	n.i.	n.i.	n.i.	n.i.	n.i.
β- Mannosidase	–	n.i.	n.i.	n.i.	772 ± 10	n.i.
α- Mannosidase	270	230 ± 10	265 ± 10	120 ± 5	90 ± 3	35 ± 2
Amyloglucosidase	2.1	28 ± 1	30 ± 2	17 ± 1	20 ± 2	11 ± 1

n.i. = no inhibition detected for 2 mM.

against different glycosidase enzymes have been summarized in Table 12.1 that manifested the absence of any regular variation in activity with valency for all glycosidases [69].

Our group at CDRI investigated a series of 1,2,3-1*H*-triazolyl glycohybrids containing two or more than two sugar units or a chromenone moiety via CuAAC reaction of glycosyl azides with 2,3-unsaturated alkynyl glycosides or propargyloxy coumarins (Fig. 12.18). The synthesized triazoles were screened for α-glucosidase, glycogen phosphorylase, and glucose-6-phosphatase inhibitory activities, out of which triazole glycohybrids (**49–52**) demonstrated promising inhibitory activities (Table 12.2) [10].

Focusing on development of potent β-galactosidase inhibitors of *E. coli*, Kovensky et al. investigated selective azidation in preparing polyazide derived platform for the installation of various alkyne substituted 1-thio-β-D-galactosides moieties via CuAAC reaction. The stability of *S*-linked glycosides toward hydrolytic conditions and specific affinity of the selected enzyme for galactosides established the alkyne

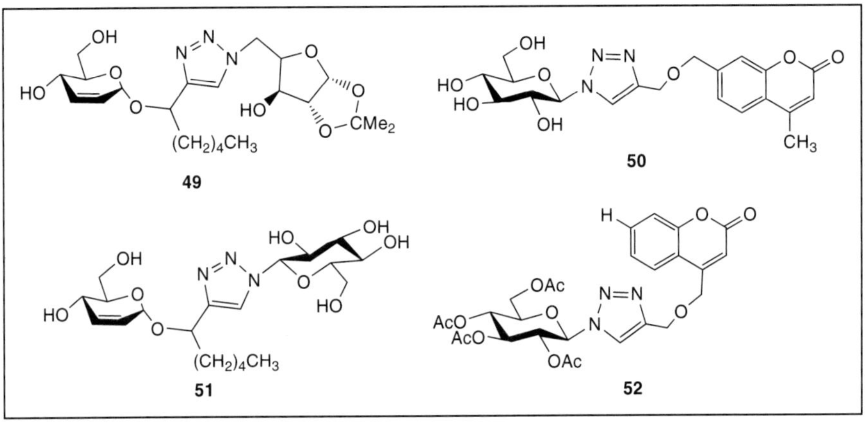

FIGURE 12.18 1,2,3-1*H*-triazolyl glycohybrids, inhibitors of various glycosidase enzymes.

TABLE 12.2 Comparison of α-Glucosidase, Glycogen Phosphorylase, and Glucose-6-Phosphatase Inhibitory Activities of 1,2,3-1*H*-Triazolyl Glycohybrids (49–52) at 100 μmol Conc w.r.t. Acarbose and Sodium-Ortho-Vandate

1*H*-Triazolyl Glycohybrids	α-Glucosidase Inhibitory Activities	Glycogen Phosphorylase Inhibitory Activities	Glucose-6-Phosphatase Inhibitory Activities
Acarbose	60%	–	–
Sodium-ortho-vandate	–	–	36%
49	–	47.7%	≈ 25%
50	–	34.1%	–
51	29.6%	20%	≈ 25%
52	31.4%	–	–

armed-oligoethyl linked thiogalactosides as appropriate choice for click reaction precursor furnishing clusters (**53**) and (**54**) (Fig. 12.19) [70].

Gouin et al. have synthesized carbohydrate-centered multi-mannoside clusters through an efficient four step strategy including direct regioselective azidation of unprotected D-glucose, maltose, and maltotriose cores using triphenylphosphine-carbon tetrabromide-sodium azide and their successive acetylation to prepare template over which acetyl protected mannosyl alkynes were grafted. Click reaction followed by deprotection of clustered sugars afforded multivalent mannoside-based clusters (**55–56**) exhibiting remarkable cluster effects toward concanavalin A (Con A) lectin by an enzyme-linked lectin assay (ELLA) (Fig. 12.20) [71].

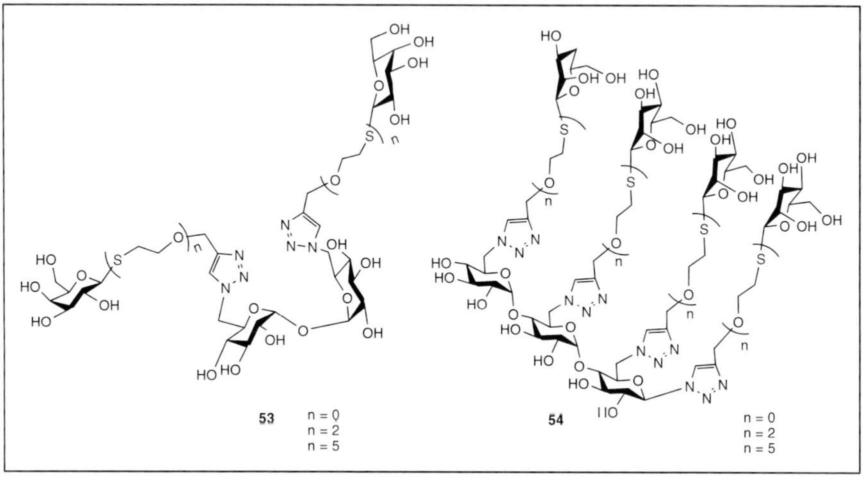

FIGURE 12.19 1-Thio-β-D-galactose derived glyco-clusters exhibiting β-galactosidase inhibitor.

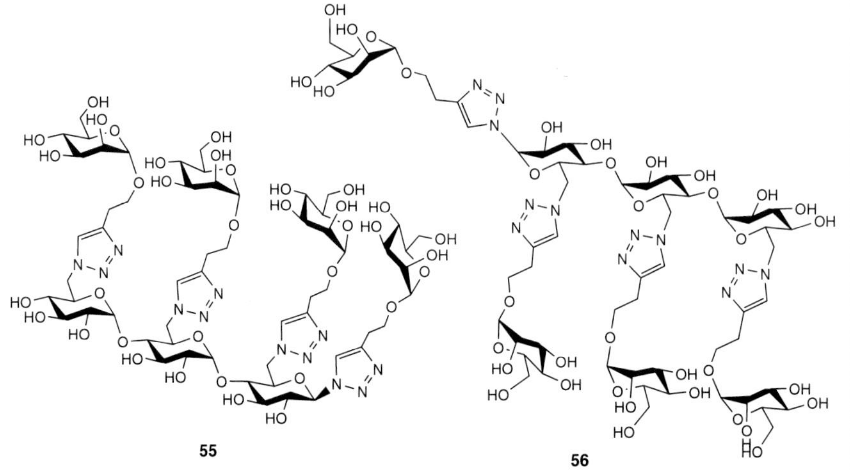

55

56

FIGURE 12.20 Glyco-clusters exhibiting lectin inhibition.

Wong and group prepared very first carbohydrate microarrays via CuAAC reaction using long aliphatic chains bearing terminal alkyne for the derivatization of polystyrene microtiter plate through non-covalent glycolipid interactions. A series of 11 diverse azido-sugars were prepared for grafting onto alkyne-functionalized plate via click reaction followed by evaluation of the extent of fucosylation and/or sialylation against potent lectin inhibitors (Scheme 12.4) [72].

Reineke et al. synthesized the diazo trehalose monomer (**70**) that underwent polymerization with various dialkyne-oligoamine monomers containing 1, 2, and 3-Boc-protected secondary amines to afford three water soluble biocompatible polycations (**73a–c**) identified as effective nucleic acid carriers (Scheme 12.5) [73].

In another approach, an efficient one-pot three enzyme based chemoenzymatic strategy has been developed by Chen's Research Group at UC Davis for an easy access of diverse sialic acid containing molecules [57]. These highly potential molecules have been successfully utilized to achieve structurally defined sialic acid containing macrocyclic oligosaccharides of different sizes. In this context, several oligosaccharides of varying lengths have been prepared using chemoenzymatic method, which contain galactose at the non-reducing end and propargyl group at the reducing end. Further, their chemoenzymatic sialylation led to well-defined sialosides, equipped with azide and alkyne functionalities at the two ends, for example, (**74**) on successive cyclization reaction under click condition affords sialic acid-containing complex macrocycle (**75**) in good yield (Scheme 12.6) [74].

The knowledge of various biochemical and physiological processes at the molecular level, for example, molecular components of living cells, their organization, interaction, movement, formation and elimination etc. can be traced out with the help

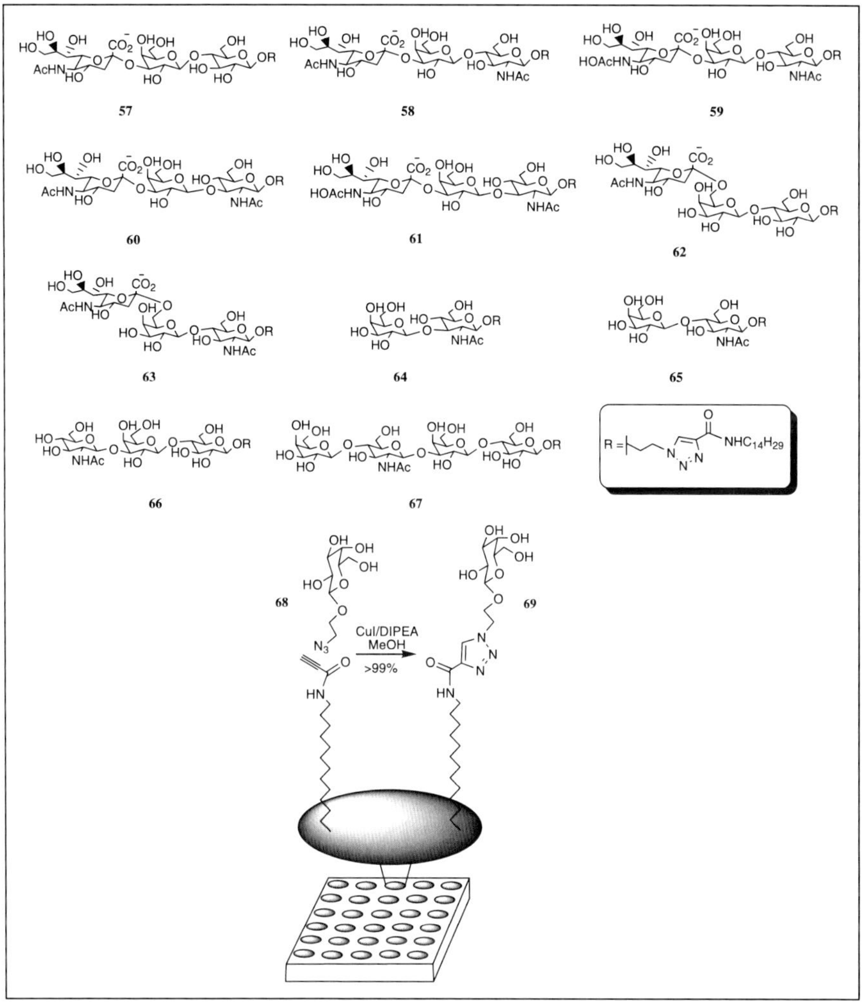

SCHEME 12.4 Sugar arrays synthesized *in situ* in the microtiter plate.

of radioisotopes capable of imaging these specific events *in vivo*. In recent years, the discovery of new biological targets has imposed the development of innovative and efficient methodologies that can ensure a firm attachment of molecules of biomedical interest to readily available radionuclides having suitable decay characteristics. Bryson et al. synthesized a novel contrast agent, β-cyclodextrin click cluster (**76**) for macromolecular Magnetic Resonance Imaging (MRI). The cluster consisting of seven paramagnetic arms with two water exchange sites each (Fig. 12.21) can exert an exceptionally high relaxivity profile (43.4 mM^{-1} s^{-1} at 9.4 T) at high magnetic field and enhances contrast on a human MRI scanner. Moreover, the useful inclusion properties exhibited by β-cyclodextrin also makes this an excellent host scaffold to

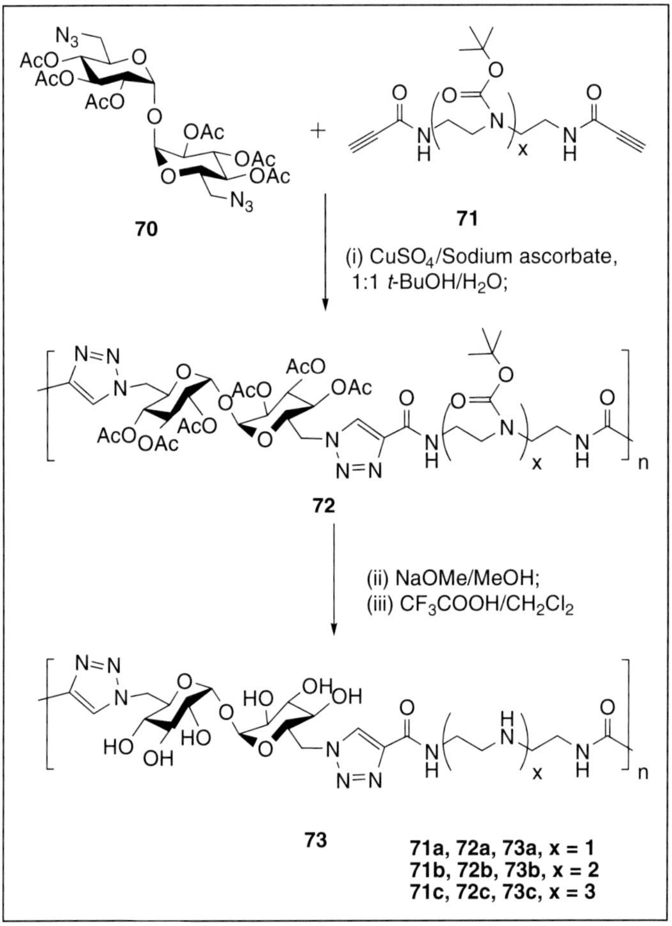

SCHEME 12.5 Click mediated polymerization.

functionalize via non-covalent assembly with biological receptor-specific targeting moieties [75].

Wu et al. synthesized unsymmetrical dendrimers (**77**) (Fig. 12.22) containing mannose units and coumarin fluorescent units via click chemistry [76]. The synthesized dendrimers were identified as highly efficient, dual-purpose recognition/detection agents for the inhibition of hemagglutination, a specific form of agglutination that involves red blood cells (RBCs). The dendrimer (**75**) under hemagglutination assay using the mannose binding protein concanvalin A and rabbit red blood cells, exhibited 240-fold greater potency than monomeric mannose, corresponding to a relative activity of 15 per sugar moiety when compared to mannose (activity = 1) [76].

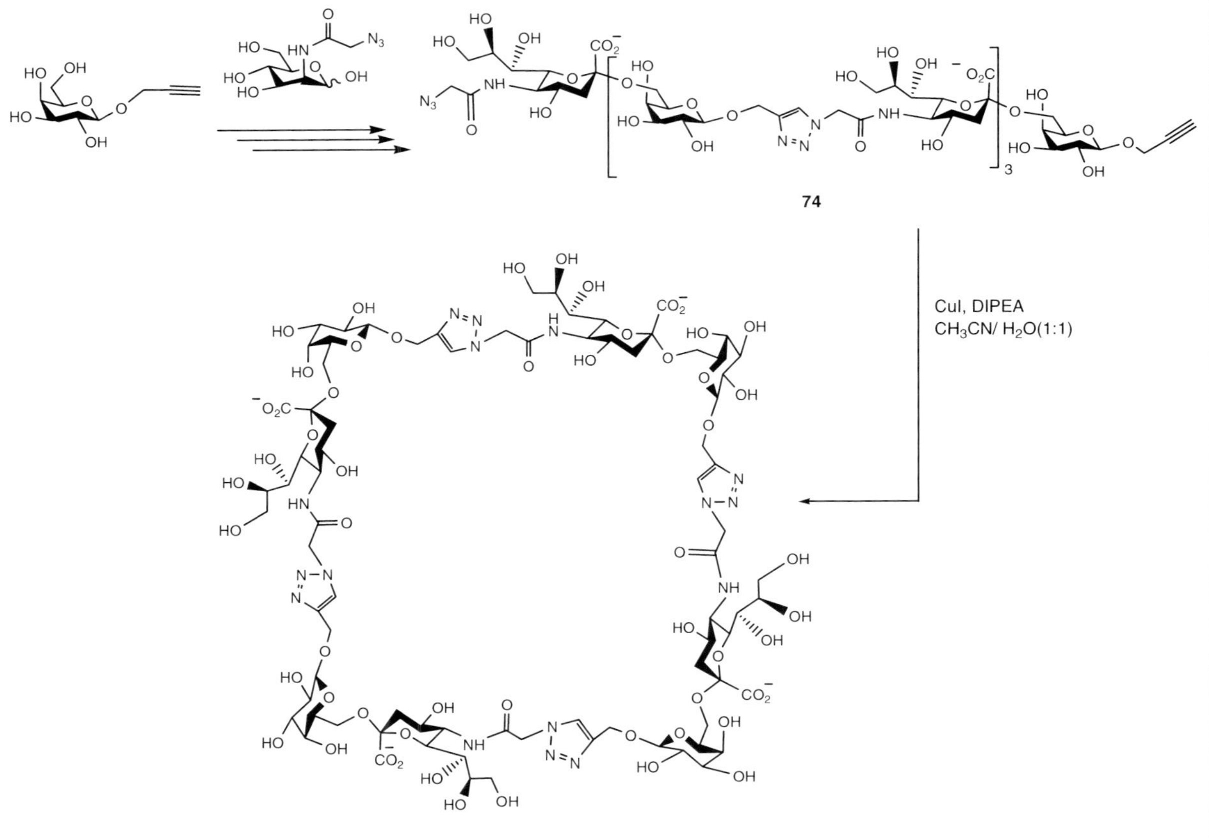

SCHEME 12.6 Chemoenzymatic approach for sialosides and their application in macrocylization.

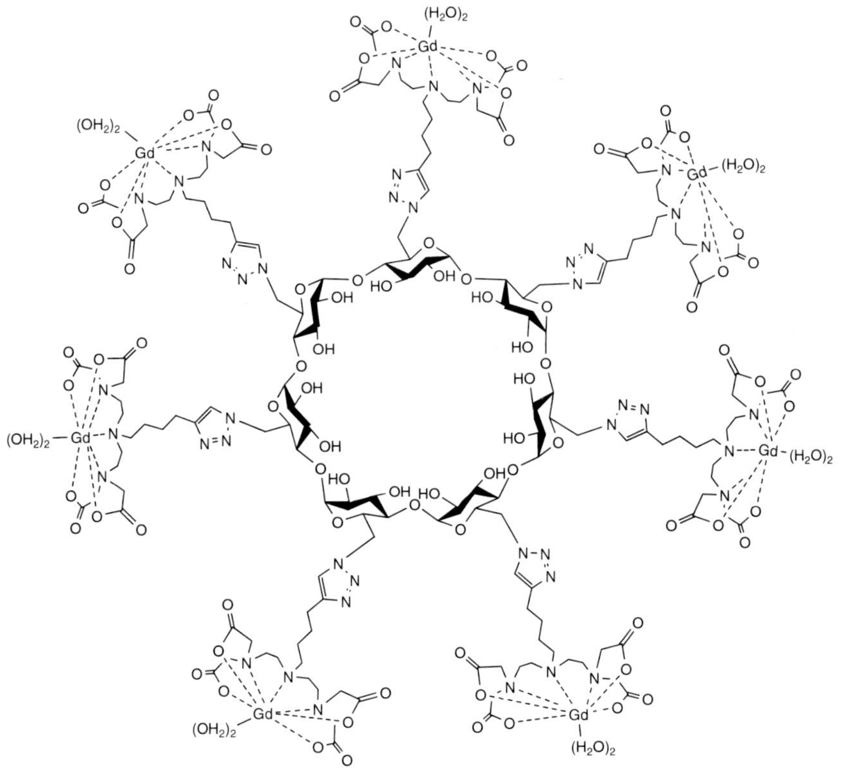

76

FIGURE 12.21 The paramagnetic β-cyclodextrin click cluster.

12.6 CONCLUSION AND FUTURE PERSPECTIVE

Carbohydrates constitute a major class biomolecules lying inside or on the surface of cell and govern several pathologically and physiologically vital biological events such as cellular recognition, adhesion, migration, invasion, communication, bacterial/viral infection, tumor metastasis, post-translational modifications of protein etc. Therefore, it is challenging to craft glycoconjugates, which can fulfil the objective of understanding role of carbohydrates in these pivotal interactions and also regulate these processes majorly by acting as inhibitors of those enzymatic transformations. Several 1,2,3 triazole core containing drug like molecules with good pharmacokinetic properties have been synthesized, however, there is lot more to explore about this vast pool of triazolyl glycoconjugates. The triazole forming click reaction promisingly facilitates creation lead and its optimization due to its modular nature, reliance on extremely short sequences of near-perfect reactions. Therefore, it's time to recognize

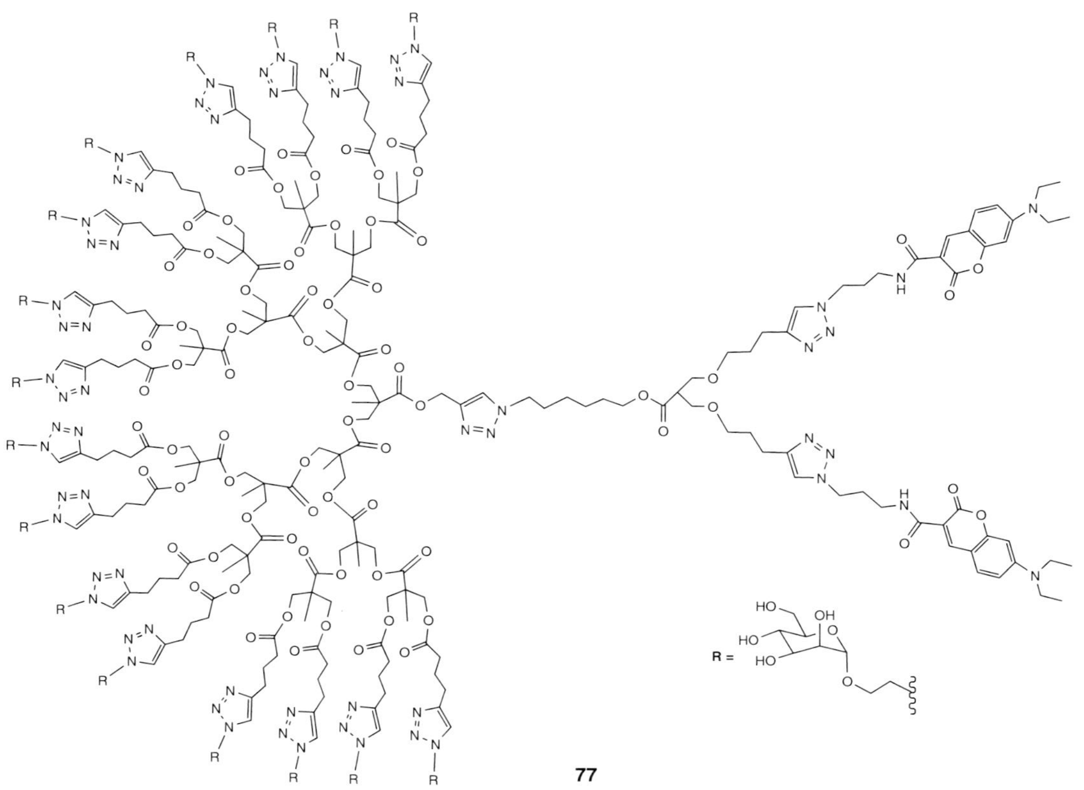

77

FIGURE 12.22 Unsymmetrical dendrimers containing mannose coumarin units.

the real potential of versatile click reaction and design highly active, selective, pharmacologically important triazolyl scaffolds, which can contribute to major advances in pharmaceutical sciences.

ACKNOWLEDGMENT

Authors thank CSIR and DRDO, India for grant-in aid.

REFERENCES

1. (a) Tiwari, V. K.; Mishra, R. C.; Sharma, A.; Tripathi, R. P. *Mini-Rev. Med. Chem.* **2012**, *12*, 1497–1519. (b) Cao, H.; Jeol, H.; Chen, X. *Opportunity, Challenges and Scope of Natural Products in Medicinal Chemistry*, V. K. Tiwari and B. B. Mishra, Eds., Research Signpost Publications, 2011; p 411.
2. Oliveira, R. N.; Sinou, D.; Srivastava, R. M. J. *Carbohydr. Chem.* **2006**, *25*, 407–425.
3. Bertozzi, C. R.; Kiessling, L. L. *Science* **2001**, *291*, 2357–2364.
4. Holgersson, J.; Gustafsson, A.; Breimer, M. E. *Immunology Cell Biology* **2005**, *83*, 694–708.
5. Lindhorst, T. K.; Dubber, M.; Krallmann-Wenzel, U.; Ehlers, S. *Eur. J. Org. Chem.* **2000**, (11), 2027–2034.
6. Joosten, J. A. F.; Loimaranta, V.; Appeldoorn, C. C. M.; Haataja, S.; Ait El Maate, F.; Liskamp, R. M. J.; Finne, J.; Pieters, R. J. *J. Med. Chem.* **2004**, *47*, 6499–6508.
7. Fan, E.; Zhang, Z.; Minke, W. E.; Hou, Z.; Verlinde, C. L. M. J.; Hol, W. G. J. *J. Am. Chem. Soc.* **2000**, *122*, 2663–2664.
8. Vrasidas, I.; Andre, S.; Valentini, P.; Bock, C.; Lensch, M.; Kaltner, H.; Liskamp, R. M. J.; Gabius, H. J.; Pieters, R. J. *Org. Biomol. Chem.* **2003**, *1*, 803–810.
9. Wittmann, V.; Seeberger, S. *Angew. Chem. Int. Ed.* **2004**, *43*, 900–903.
10. Anand, N.; Jaiswal, N.; Pandey, S. K.; Srivastava, A. K.; Tripathi, R. P. *Carbohydr. Res.* **2011**, *346* 16–25.
11. Velazquez, S.; Alvarez, R.; Perez, C.; Gago, F.; De, C.; Balzarini, J.; Camarasa, M. J. *Antiviral Chem. Chemother.* **1998**, *9*, 481–489.
12. Wilkinson, B. L.; Long, H.; Sim, E.; Fairbanks, A. J. *Bioorg. Med. Chem. Lett.* **2008**, *18*, 6265–6267.
13. Reddy, L. V. R.; Mishra, N. N.; Shukla, P. K.; Yadav, G.; Srivastava, R.; Shaw, A. K. *Carbohydr. Res.* **2010**, *345*, 1515–1521.
14. Carvalho, I.; Andrade, P.; Campo, V. L.; Guedes, P. M.; Sesti-Costa, R; Silva, J. S.; Schenkman, S.; Dedola, S.; Hill, L.; Rejzek, M.; Nepogodiev, S. A.; Field, R. A. *Bioorg. Med. Chem.* **2010**, *18*, 2412–2427.
15. Singh, B. K.; Yadav, A. K.; Kumar, B.; Gaikwad, A.; Sinha, S. K.; Chaturvedi, V.; Tripathi, R. P. *Carbohydr. Res.* **2008**, *343*, 1153–1162.
16. Kolb, H. C.; Finn, M. G.; Sharpless, K. B. *Angew. Chem. Int. Ed.* **2001**, *40*, 2004–2021.
17. Meldal, C. W.; Tornoe, C.; Meldal, M. *J. Org. Chem.* **2002**, *67*, 3057–3062.

18. Zhang, L.; Chen, X.; Xue, P.; Sun, H. H. Y.; Williams, I. D.; Sharpless, K. B.; Fokin, V. V.; Jia, G. *J. Am. Chem. Soc.* **2005**, *127*, 15998–15999.
19. Rasmussen, L. K.; Boren, B. C.; Fokin, V. V. *Org. Lett.* **2007**, *9*, 5337–5339.
20. Dondoni, A. *Chem. Asian J.* **2007**, *2*, 700–708.
21. Huisgen, R. *1,3-Dipolar Cycloaddition Chemistry*, Wiley-VCH: New York, **1984**.
22. Hein, J. E.; Fokin, V. V. *Chem. Soc. Rev.* **2010**, *39*, 1302–1315.
23. Meldal, M.; Tornoe, C. W. *Chem. Rev.* **2008**, *108*, 2952–3015.
24. Dondoni, A.; Marra, A. *Chem. Rev.* **2010**, *110*, 4949–4977.
25. Holub, J. M.; Kirshenbaum, K. *Chem. Soc. Rev.* **2010**, *39*, 1325–1337.
26. Tron, G. C.; Pirali, T.; Billington, R. A.; Canonico, P. L.; Sorba, G.; Genazzani, A. *Med. Res. Rev.* **2008**, *28*, 278–308.
27. Kappe, C. O.; Eycken, E. V. D. *Chem. Soc. Rev.* **2010**, *39*, 1280–1290.
28. Aragao-Leoneti, V.; Campo, V. L.; Gomes, A. S.; Field, R. A.; Carvalho, I. *Tetrahedron* **2010**, *66*, 9475–9492.
29. Kushwaha, D.; Dwivedi, P.; Kuanar, S. K.; Tiwari, V. K. *Curr. Org. Synth.* **2012**, *9*(2), 1–45.
30. Wilkinson, B. L.; Bornaghi, L. F.; Poulsen, S. A.; Houston, T. A. *Tetrahedron* **2006**, *62*, 8115–8125.
31. Rodionov, V. O.; Fokin, V. V.; Finn, M. G. *Angew. Chem. Int. Ed.* **2005**, *44*, 2210–2215.
32. Bock, V. D.; Hiemstra, H.; Maarseveen, J. H. V. *Eur. J. Org. Chem.* **2006**, (1), 51–68.
33. (a) Straub, B. F. *Chem. Commun.* **2007**, 3868–3870. (b) Himo, F.; Lovell, T.; Hilgraf, R.; Rostovtsev, V. V.; Noodleman, L.; Sharpless, K. B.; Fokin, V. V. *J. Am. Chem. Soc.* **2005**, *127*, 210–216. (c) Ahlquist, M.; Fokin, V. V. *Organometallics* **2007**, *26*, 4389–4439. (d) Zhou, Y.; Lecourt, T.; Micouin, L. *Angew. Chem. Int. Ed.* **2010**, *49*, 2607–2610.
34. Nolte, C.; Mayer, P.; Straub, B. F. *Angew. Chem. Int. Ed.* **2007**, *46*, 2101–2103.
35. Kale, R. R.; Prasad, V.; Tiwari, V. K. *Lett. Org. Chem.* **2010**, *7*, 136–143.
36. Brik, A.; Alexandratos, J.; Lin, Y. C.; Elder, J. H.; Olson, A. J.; Wlodawer, A.; Goodsell, D. S.; Wong, C. H. *Chem. Bio. Chem.* **2005**, *6*, 1167–1169.
37. (a) Wong, C. H. *Carbohydrate-based Drug Discovery*, Wiley-VCH: New York, **2003**. (b) Witczak, Z. J. *Carbohydr. Chem.* **2010**, *36*, 176–193.
38. Wan, Q.; Chen, J.; Chen, G.; Danishefsky, S. J. *J. Org. Chem.* **2006**, *71*, 8244–8249.
39. Lee, T.; Cho, M.; Ko, S.-Y.; Youn, H.-J.; Baek, D. J.; Cho, W.-J.; Kang, C.-Y.; Kim, S. *J. Med. Chem.* **2007**, *50*, 585–589.
40. Slamova, K.; Marhol, P.; Bezouska, K.; Lindkvist, L.; Hansen, S. G.; Kren, V.; Jensen, H. H. *Bioorg. Med. Chem. Lett.* **2010**, *20*, 4263–4265.
41. dos Anjos, J. V.; Neves, R. A. W.; do Nascimento, S. C.; Srivastava, R. M.; de Melo, S. J.; Sinou, D. *Eur. J. Med. Chem.* **2009**, *4*, 3571–3576.
42. Manach, C. L.; Baron, A.; Guillot, R.; Vauzeilles, B.; Beau, J. M. *Tetrahedron Lett.* **2011**, *52*, 1462–1465.
43. Chen, Q.; Yang, F.; Du, Y. G. *Carbohydr. Res.* **2005**, *340*, 2476–2482.
44. Campo, V. L.; Carvalho, I.; Silva, C. H. T. P.; Schenkman, S.; Hill, L.; Nepogodiev, S.; Field, R. A. *Chem. Sci.* **2010**, *1*, 507–514.
45. Giguere, D.; Patnam, R.; Bellefleur, M. A.; St-Pierre, C.; Sato, S.; Roy, R. *Chem. Commun.* **2006**, *22*, 2379–2381.

46. Tejler, J.; Tullberg, E.; Frejd, T.; Leffler, H.; Nilsson, U. J. *Carbohydr. Res.* **2006**, *341*, 1353–1362.
47. Titz, A.; Patton, J.; Smiesko, M.; Radic, Z.; Schwardt, O.; Magnani, J. L.; Ernst, B. *Bioorg. Med. Chem.* **2010**, *18*, 19–27.
48. Fazio, F.; Bryan, M. C.; Blixt, O.; Paulson, J. C.; Wong, C. H. *J. Am. Chem. Soc.* **2002**, *124*, 14397–14402.
49. Rajakumar, P.; Anandhan, R. *Eur. J. Med. Chem.* **2011**, *46*, 4687–4695.
50. Tripathi, R. P.; Tewari, N.; Dwivedi, N.; Tiwari, V. K. *Med. Chem. Rev.* **2005**, *25*, 91–125.
51. Cheshev, P.; Marra, A.; Dondoni, A. *Org. Biomol. Chem.* **2006**, *4*, 3225–3227.
52. Conte, M. L.; Marra, A.; Chambery, A.; Gurcha, S. S.; Besra, G. S.; Dondoni, A. *J. Org. Chem.* **2010**, *75*, 6326–6336.
53. (a) Lu, W-Y.; Sun, X. W.; Zhu, C.; Xu, J-H.; Lin, G-Q. *Tetrahedron* **2010**, *66*, 750–757. (b) Agalave, S. G.; Maujan, S. R.; Pore, V. S. *Chem. Asian J.* **2011**, *6*, 2696-2718.
54. Zhang, J.; Chiang, F. I.; Wu, L.; Czyryca, P. G.; Li, D.; Chang, C. W. T. *J. Med. Chem.* **2008**, *51*, 7563–7573.
55. Lin, H.; Walsh, C. T. *J. Am. Chem. Soc.* **2004**, *126*, 13998–14003.
56. Fu, X.; Albermann, C.; Jiang, J.; Liao, J.; Zhang, C.; Thorson, J. S. *Nat. Biotechnol.* **2003**, *21*, 1467–1469.
57. Yu, H.; Cheng, J.; Ding, L.; Khedri, Z.; Chen, Y.; Chin, S.; Lau, K.; Tiwari, V. K.; Chen, X. *J. Am. Chem. Soc.* **2009**, *131*, 18467–18477.
58. Li, J.; Zheng, M.; Tang, W.; He, P. L.; Zhu, W.; Li, T.; Zuo, J. P.; Liu, H.; Jiang, H. *Bioorg. Med. Chem. Lett.* **2006**, *16*, 5009–5013.
59. Weiwer, M.; Chen, C. C.; Kemp, M. M.; Linhardt, R. J. *Eur. J. Org. Chem.* **2009**, (16), 2611–2620.
60. Butters, T. D.; Dwek, R. A.; Platt, F. M. *Curr. Top. Med. Chem.* **2003**, *3*, 561–574.
61. Platt, F. M.; Neises, G. R.; Reinkensmeier, G.; Townsend, M. J.; Perry, V. H.; Proia, R. L.; Winchester, B. G.; Dwek, R. A.; Butters, T. D. *Science* **1997**, *276*, 428–431.
62. Greimel, P.; Spreitz, J.; Stutz, A. E.; Wrodnigg, T. M. *Curr. Top. Med. Chem.* **2003**, *3*, 513–523.
63. Zitzmann, N.; Mehta, A. S.; Carrouee, S.; Butters, T. D.; Platt, F. M.; McCauley, J.; Blumberg, B. S.; Dwek, R. A.; Block, T. M. *Proc. Natl. Acad. Sci. USA* **1999**, *96*, 11878–11882.
64. Gross, P. E.; Baker, M. A.; Carver, J. P.; Dennis, J. W. *Clin. Cancer Res.* **1995**, *1*, 935–944.
65. Rossi, L. L.; Basu, A. *Bioorg. Med. Chem. Lett.* **2005**, *15*, 3596–3599.
66. Zhou, Y.; Zhao, Y.; O'Boyle, K. M.; Murphy, P. V. *Bioorg. Med. Chem. Lett.* **2008**, *18*, 954–958.
67. Perion, R.; Ferrieres, V.; Garcia-Moreno, M. I.; Mellet, C. O.; Duval, R.; Fernandez, J. M. G.; Plusquellec, D. *Tetrahedron* **2005**, *61*, 9118–9128.
68. Dedola, S.; Hughes, D. L.; Nepogodiev, S. A.; Rejzek, M.; Field, R. A. *Carbohydr. Res.* **2010**, *345*, 1123–1134.
69. Diot, J.; Garcia-Moreno, M. I.; Gouin, S. G.; Mellet, C. O.; Haupt, K.; Kovensky, J. *Org. Biomol. Chem.* **2009**, *7*, 357–363.
70. Cagnoni, A. J.; Varela, O.; Gouin, S. G.; Kovensky, J.; Uhrig, M. L. *J. Org. Chem.* **2011**, *76*, 3064–3077.

71. Gouin, S. G.; Vanquelef, E.; Fernandez, J. M. G.; Mellet, C. O.; Dupradeau, F. Y.; Kovensky, J. *J. Org. Chem.* **2007**, *72*, 9032–9045.

72. Fazio, F.; Bryan, M. C.; Blixt, O.; Paulson, J. C.; Wong, C. H. *J. Am. Chem. Soc.* **2002**, *124*, 14397–14402.

73. Srinivasachari, S.; Liu, Y.; Zhang, G.; Prevette, L.; Reineke, T. M. *J. Am. Chem. Soc.* **2006**, *128*, 8176–8184.

74. Muthana, S.; Yu, H.; Cao, H.; Cheng, J.; Chen, X. *J. Org. Chem.* **2009**, *74*, 2928–2936.

75. Bryson, J. M.; Chu, W. J.; Lee, J. H.; Reineke, T. M. *Bioconjugate Chem.* **2008**, *19*, 1505–1509.

76. Wu, P.; Malkoch, M.; Hunt, J. N.; Kaltgrad, R. V. E.; Finn, M. G.; Fokin, V. V.; Sharpless, K. B.; Hawker, C. J. *Chem. Commun.* **2005**, (46), 5775–5777.

13

CLICK CHEMISTRY APPLIED TO CARBOHYDRATE-BASED DRUG DISCOVERY

Vanessa Leiria Campo and Ivone Carvalho

13.1 INTRODUCTION

The discovery of new bioactive molecules based on screening of analog libraries and structure–activity relationship has been accomplished by a range of chemical procedures, in which the cycloadditions, particularly the Huisgen 1,3-dipolar cycloaddition catalyzed by Cu(I), designated as "copper-catalyzed azide/alkyne cycloaddition" (CuAAC), are the most extensively utilized "click chemistry" [1–3]. These reactions are easily performed, fast, highly selective, insensitive to oxygen and water, do not require excessive chromatographic purifications, and can lead to great structural diversity in high yields. Moreover, the physicochemical properties of the formed triazole group are particularly favorable, since it acts as a rigid link, displaying pharmacophore groups in well-defined spatial orientation, and neither hydrolyzable, oxidizable, or reducible [3, 4]. Concerning carbohydrate chemistry, CuAAC reactions have been increasingly applied for synthesis of carbohydrates and neoglycoconjugates directed to the understanding of biological pathways mechanisms and drug discovery, including the generation of potential substrates/inhibitors for a range of different enzymes, glycopolymers able to establish multivalent interactions with defined target proteins and biological probes for selective labeling of biomolecules either in cells or within living organisms [5, 6].

Click Chemistry in Glycoscience: New Developments and Strategies, First Edition.
Edited by Zbigniew J. Witczak and Roman Bielski.

Therefore, the purpose of this chapter is to provide a general overview of the potential application of CuAAC reactions in the synthesis of bioactive carbohydrate-based ligands for lectins (galectins and selectins), antitumor vaccines and substrates/inhibitors for fucosyltransferases, manosyltransferases, glucosidases, neuraminidase and *trans*-sialidase, all of them built up from a common 1,4-disubstituted 1,2,3-triazole scaffold.

13.2 CLICK CHEMISTRY AND CARBOHYDRATE-BASED LIGANDS FOR LECTINS (GALECTINS AND SELECTINS)

Lectins are carbohydrate-specific proteins that act as recognition molecules in various biological systems. These proteins bind to carbohydrates reversibly and with high specificity to mono/oligosaccharides, being also known as glycan-binding proteins. They are expressed by animals, plants, fungi, bacteria, and viruses and are involved in several biological and pathological processes [7, 8]. The carbohydrate recognition capability inherent to lectins is often associated to a distinct polypeptide region in these proteins, denominated Carbohydrate Recognition Domain (CRD) [9–11]. Based on the amino acid sequence and structural homology of their CRDs, animal lectins are classified into different families (types) identified as S-, C-, I-, M-, L-, P-, and R-types [11]. In this chapter, we will focus on the development of specific ligands to galectins (S-type) and selectins (C-type) by means of different strategies utilizing Cu(I)-assisted 1,3-dipolar azide–alkyne cycloaddition (CuAAC) reactions.

13.2.1 Galectins

Galectins, existing in the form of 15 different subtypes in mammals, share primary structural homology in their CRDs, which are constituted by approximately 135 amino acid residues. Through specific interactions with only few CRD residues, galectins are able to bind to β-galactose molecules, especially those found in *N*-acetyl-lactosamine (LacNAc) sequences. Thus, terminal β-galactose residues and the size of the poly-lactosamine chain seem to represent critical requirements for the carbohydrate recognition capacity of these lectins [12].

Galectins are expressed by a variety of cell types and can participate in cell–cell and cell–extracellular matrix interactions, besides modulating several cellular functions [13, 14]. Therefore, these proteins may play an essential role in the development of inflammatory response, autoimmune diseases, atherosclerosis, infectious processes, and cancer [13, 15]. Galectins, mainly subtypes -1 and -3, are overexpressed in several types of tumor cells and are then involved in important functions related to cancer, such as tumorigenesis, neoplastic transformation, tumor cell survival, angiogenesis, tumor metastasis, and regulation of apoptosis. Moreover, they can help tumors to escape from immune surveillance through modulation of immune and inflammatory responses [16, 17].

Taking into account the low affinities of naturally occurring carbohydrate ligands for galectins, besides their low physiological stabilities due to acid sensitive glycosidic

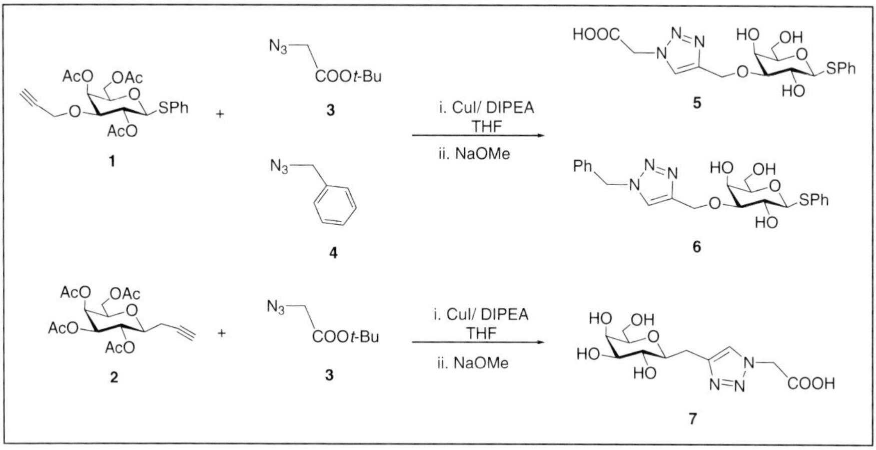

SCHEME 13.1 Synthesis of triazoles **5–7** by click reaction.

bonds, and their high polarity, synthesis of anomeric and O-3 triazole analogs of galactosides has been developed to obtain higher affinity inhibitors of galectins 1- and -3. Thus, click 1,3-dipolar cycloaddition reactions between galactosyl alkynes **1** and **2** with azides **3** or **4**, using CuI/DIPEA as catalytic system in THF, provided the triazole–galactosyl derivatives **5–7** (Scheme 13.1) [18]. The best inhibitory activities in the qualitative hemagglutination inhibition assays were verified for compounds **1** and **5**, which showed to be 40 times more active than galactose at 1.25 mM concentration.

The CRD of galectins is composed by four subsites (A–D) which, along their 135 aa, form a groove able to bind up to a tetrasaccharide. Galactose binds to the most conserved subsite C whereas the second more important subsite D is occupied by another pyranoside represented by (1→4)Glc/GlcNAc or (1→3) GlcNAc/GalNAc linked to galactose (subsite C) [19]. Therefore, the affinity enhancement for the different subsites of galectins by combining structural fragments to galactoside or lactoside derivatives may represent a significant approach. In this perspective, two galactosyl oximes having C3-triazole fragments were prepared by CuAAC reactions between the 3-azido-Gal-indol-carbaldoxime **8** and the corresponding alkynes methyl propiolate **9** and phenyl acetylene **10**, in the presence of Cu wire in propanol (Scheme 13.2). The obtained compounds **11** and **12** showed promising affinities for galectin-3, with Kd values of 11 μM and 17 μM, respectively, in Fluorescence polarization assays. The improved affinity and selectivity of **11** and **12** for galectin-3 was further justified through their modeling into the binding sites of different galectins. It was then observed that, apart from interacting with subsites B and C present in all galectins, these compounds were able to establish an extra strong interaction, through their indole aldoxime fragments, with the Arg144 residue present in a distinct cavity existent only in galectin-3 [19].

Following the strategy of getting ligands able to interact with the different subsites (A–D) of galectins, a set of 3-triazol-galactosides, 3′-triazol-N-acetyllactosamine (LacNAc), and di-triazol-thiodigalactoside derivatives were synthesized, under

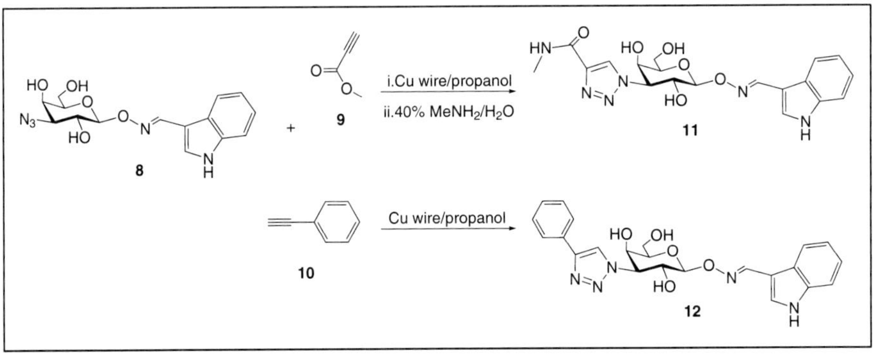

SCHEME 13.2 Synthesis of galactosyl oximes **11** and **12** with subsite B-binding 1,2,3-triazoles at C3.

Cu(I)-catalyzed reactions, as potential galectin inhibitors [20]. Initially, a series of triazoles was obtained at the C3 position of galactose by CuAAC reaction (CuI/DIPEA in toluene, rt-40°C) of 3-azide-GalSMe **13** with different monosubstituted-aryl alkynes, as exemplified for triazole **14** (Scheme 13.3). Subsequently, considering that LacNAc are far superior to galactose as natural ligands for galectins, several LacNAc derivatives carrying 4-carbamoyl and 4-aryl triazoles at the galactose C3′ were obtained by click reactions of LacNAc C3′-azido **15** and corresponding alkynes, as shown for compound **16** (Scheme 13.3). Lastly, a panel of thiodigalactoside triazole amides was synthesized in steps, starting with the preparation of the thiodigalactoside triazole ester **17** by cycloaddition reaction between the galacto-azide **18** and methyl propiolate **9**. The formed triazole was then treated with HBr and to the resulting bromide **19** was added dried sodium sulfide, which led to its dimerization. As the last step, reactions of **17** with a series of primary amines in methanol afforded the respective amides, as illustrated for the thiodigalactoside triazole butyl-amide **20** (Scheme 13.3). In general, the obtained compounds proved to be better inhibitors of galectin-3 if compared to other galectins subtypes (7, 8N and 9N), in competitive Fluorescence polarization assays. The highest inhibitory activities toward galectin-3 were verified for compounds **16** (Kd 0.66 μM) and **20** (Kd 0.029 μM), related to LacNAc- and thiodigalactoside–triazoles series, respectively [20].

Since lectin–carbohydrate interactions are generally weak, special efforts are often required to achieve tighter binding of carbohydrate-based ligands to lectins. One of the established strategies to improve such interactions is based on the concept of the "glycoside cluster effect," which is related to the number and the geometry of carbohydrate residues and also depends on their steric bulk, density, and relative distance, as well as on their three-dimensional arrangement [21]. According to this concept, the affinity enhancements observed for multimeric ligands over monomeric species is often greater than predicted from the sum of the constitutive interactions. Within this context, multimeric lactosides based on carbohydrate scaffolds with valences ranging from 1 to 4 and different linker lengths were synthesized by CuAAC reactions as inhibitors of galectins-1 and -3 (Scheme 13.4) [22]. To achieve

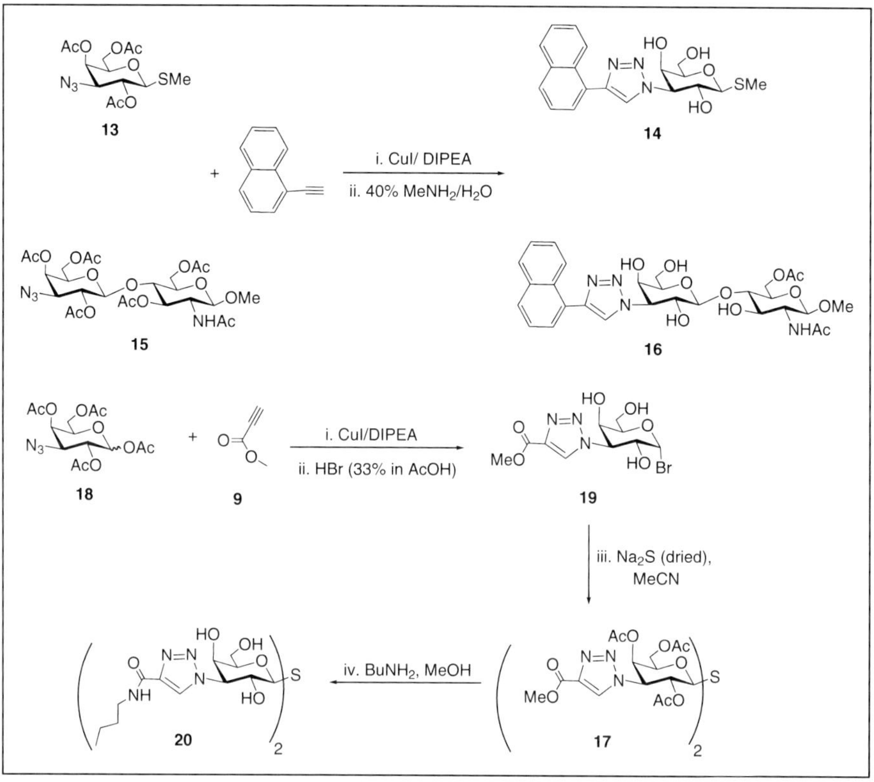

SCHEME 13.3 Synthesis of naphtyl-3-triazol-galactoside **14**, naphtyl-3′-triazol-N-acetyllactosamine **16**, and butylamine-di-triazol-thiodigalactoside **20** by Cu(I)-catalyzed reactions.

this purpose, commercially available glucose, maltose, and maltotriose were used as core structures to generate multivalency, by means of the preparation of their corresponding azido derivatives, in parallel to the synthesis of lactoside moieties tethered by alkyne-containing oligo(ethyleneglycol) (EG) linkers. Subsequently, click reactions between the obtained azido-sugar cores and alkyne lactosides, in the presence of $CuSO_4$/sodium ascorbate under microwave irradiation, led to a series of mono- and multivalent lactoside triazoles such as compounds **21**–**26** (Scheme 13.4) [22]. According to competitive Fluorescent polarization assays, the highest affinities for galectins-1 and -3 were observed for the tetravalent derivatives **26** (Kd 42 μM) and **25** (Kd 16 μM), respectively, confirming their intrinsic "glycoside cluster effect."

13.2.2 Selectins

Selectins are C-type lectins expressed in vascular endothelium and in circulating leukocytes, playing a key role in the early inflammatory response through the control of lymphocyte homing and of leukocyte trafficking to sites of inflammation. These

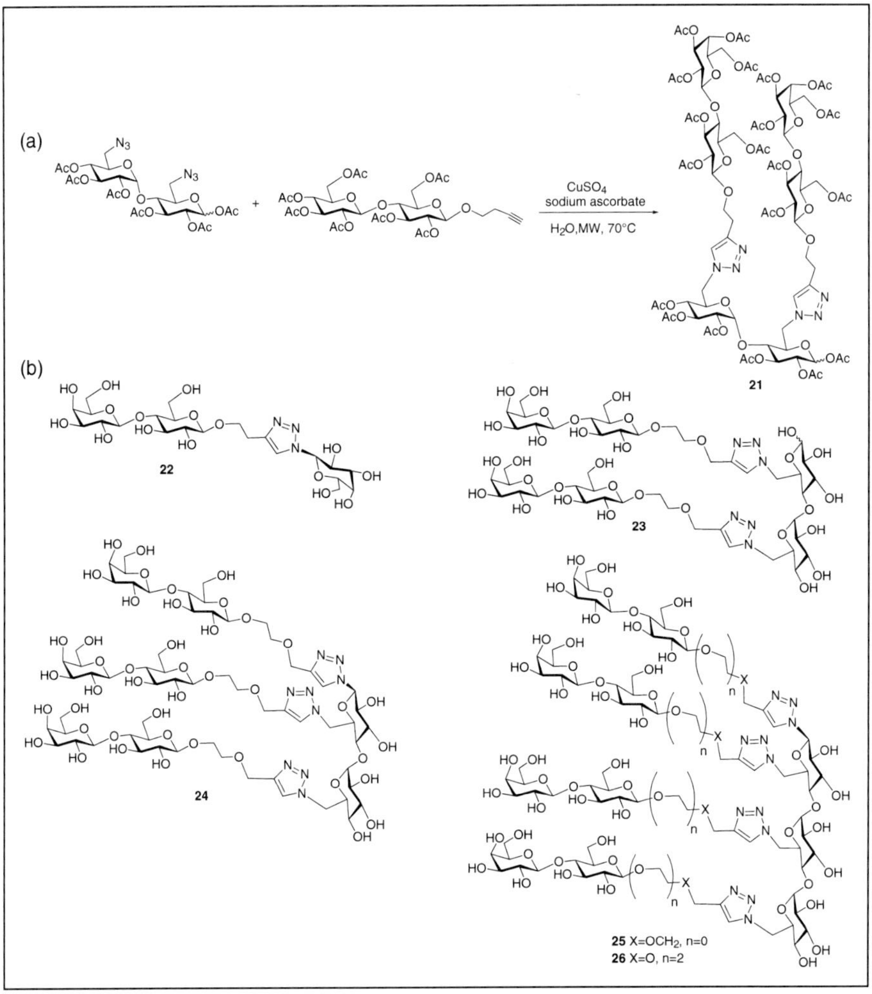

SCHEME 13.4 (a) General procedure of Cu(I)-catalyzed reactions between azido-sugar cores and alkyne lactosides, exemplified for **21**. (b) Structures of mono- **22** and multivalent lactoside triazoles **23–26**.

selective cell adhesion molecules interact with leukocytes, leading, at first, to their characteristic tethering and rolling on the vascular endothelium. Subsequently, adhesion and migration of leukocytes through the endothelial monolayer, whereby they interact transiently with selectins, lead to their extravasation, originally intended as a defensive strategy [23]. However, when occurring in excess, extravasation of leukocytes to inflammation sites can cause severe acute or chronic reactions, such as stroke, psoriasis, and rheumatoid arthritis. Thus, the inhibition of selectins can be considered as an important approach for the treatment of inflammatory diseases. To date, only three selectins have been identified: L-selectin, expressed on all leukocytes; E-selectin, expressed by cytokine-activated endothelial cells; and P-selectin, expressed

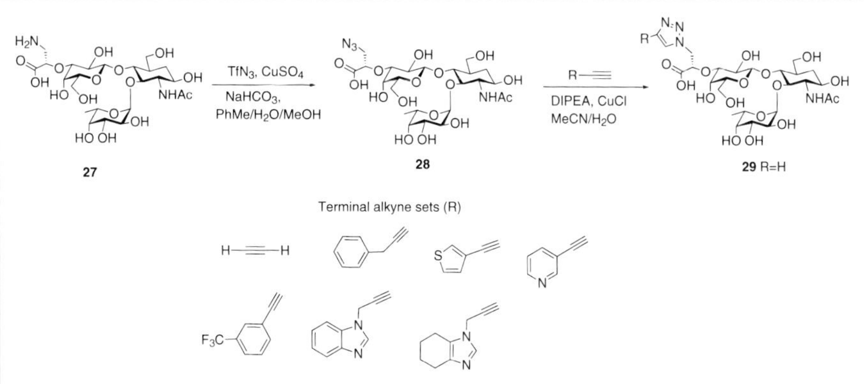

SCHEME 13.5 Synthesis of E-selectin antagonists derived from 1,3-dipolar cycloadditions.

on activated platelets and endothelial cells. All of these selectins have in common the modest specificity and affinity for the terminal tetrasaccharide epitope sialyl Lewis X (sLex), constituent of natural selectin ligands [24], which can, therefore, act as a prototype for development of selectin antagonists.

Accordingly, in the search for effective selectin inhibitors, a series of E-selectin antagonists derived from 1,3-dipolar cycloadditions was developed based on the structure of sLex, composed by NeuAcα2–3Galβ1–4(Fucα1–3)GlcNAcβ-1-R (Scheme 13.5) [24,25]. So, a sLex derivative **27**, containing a residue of (S)-isoserine in place of the Neu5Ac sugar, was first prepared, followed by conversion of amine group of **27** into azide **28**, which was then reacted with different commercially available alkynes by click reactions using CuCl/DIPEA in MeOH/H_2O (Scheme 13.5). Unexpectedly, azide **28** and the unsubstituted triazole **29** showed the best affinities for E-selectin among the obtained triazoles, with IC_{50s} of 0.29 and 0.48 mM, respectively [25].

In continuous efforts to get sLex mimetics as selectin ligands, the "shape-modulating linker" concept, which involves the design of partially flexible hybrids by connecting their rigid(ified) components with rotatable covalent bonds, has been applied for the synthesis of carbohydrate-β-lactam hybrids containing the 1,2,3-triazole moiety. These compounds, containing the triazole as a shape-modulating linker, are expected to exhibit rigidified backbone and simultaneously enhanced resistance to chemical and enzymatic hydrolysis by proteases [26]. Therefore, click cycloaddition reactions of propargyl-β-lactam **30**, obtained from the α-propargylserinate **31** and aminoester **32**, with O-protected 2-azidosugars derived from D-mannose and L-fucose led to β-lactam glycoconjugates such as **33–35** (Scheme 13.6). Next, compounds **33–35** were submitted to *N*-denosylation and *in situ* *N*-acylation, followed by *O*-deprotection to get the desired ligands **36–38** for lectin binding tests (Scheme 13.6) [26]. The evaluation of the capacity of these molecules to act as true glycomimetics with modulable shape has been performed by a combined NMR/docking approach employing a model fucose-binding lectin, *Ulex Europaeus Lectin I* (UEL-I), as receptor. Initial experiments involving the L-β-Fucose/β-lactam

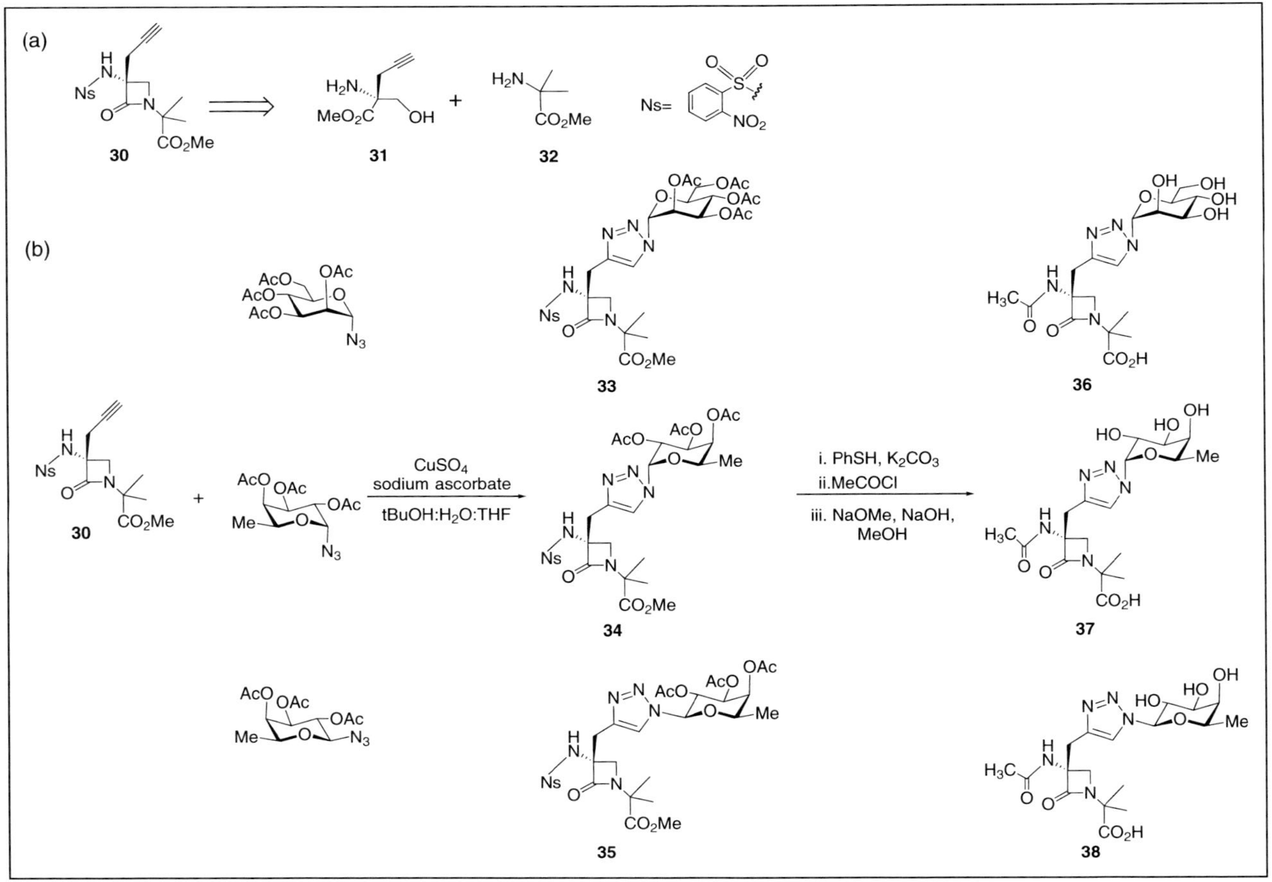

SCHEME 13.6 (a) Retrosynthesis of propargyl-β-lactam **30**. (b) Synthesis of compounds **36–38** by cycloaddition reactions.

hybrid **38** showed that it is properly recognized by *Ulex Europaeus Lectin-I* when it adopts an extended conformation.

Taking into account the advantages of "glycoside cluster effect" for improvement of carbohydrate–lectins interactions, as previously described in section 13.2.1, a new class of functional, multivalent glycoarchitectures based on hyperbranched polyglycerols (HPGs) conjugated to galactose sugar moieties were developed by click chemistry with the aim of evaluating their multivalent effect on selectin binding [23]. In fact, HPGs represent ideal scaffolds for the multivalent presentation of saccharides due to their biocompatible and carbohydrate-like properties. Hence, alkyne groups were introduced into the HPG **39** by reacting with propargyl bromide in the presence of NaH whereas galactose was reacted with a propylene spacer containing a terminal azido group. In sequence, click reactions between alkyne-polymer **40** with protected, unprotected and sulfated azido-galactose derivatives **41a–c**, using $CuSO_4$/sodium ascorbate, in THF/H_2O, afforded glycopolymers **42a–c** (Scheme 13.7). Then, selectin

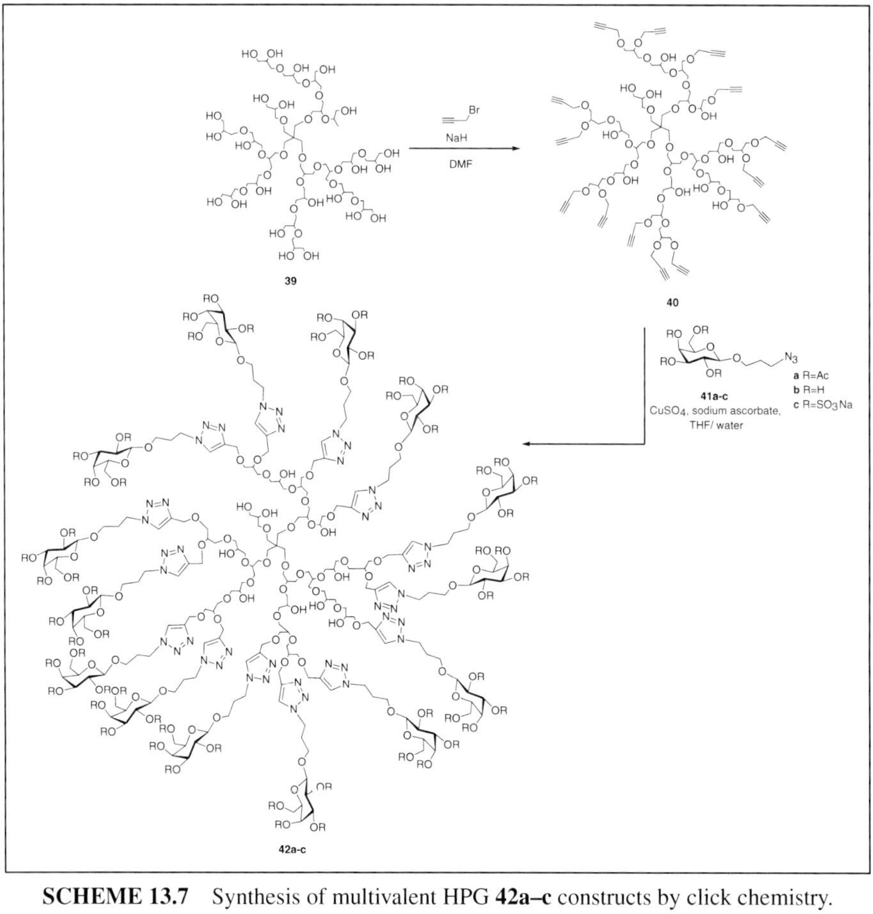

SCHEME 13.7 Synthesis of multivalent HPG **42a–c** constructs by click chemistry.

recognition activities were investigated using surface plasmon resonance (SPR) binding assays [27]. The sulfated-galactose multivalent construct **42c** showed the highest binding affinities to L- and to P-selectin, with IC_{50} values of 1 and 7 nM, respectively. This result corroborates prior findings that polyanionic carbohydrate ligands display marked binding enhancement to selectins [28].

13.3 CLICK CHEMISTRY AND GLYCOCONJUGATE ANTITUMOR VACCINES

Reverse vaccinology comprises an innovative approach involving the recognition and characterization of key epitopes on selected antigens, which are bound by neutralizing or inhibitory antibodies, later on used for guiding effective vaccine design. Once the target is established, small and synthetically accessible planned structures may be built mimicking only important epitopic regions, which allows focusing on the immune reaction [29]. Hence, the role played by chemical synthesis in the rational design and optimization of therapeutic agents able to interfere in the immune response processes is of great relevance. [29].

Regarding tumoral antigens, carbohydrates are considered the most important targets for inducing active immunity, considering that they are found in several *O*-glycosylated mucin glycoproteins and glycolipids expressed on a variety of tumor types, acting as broad-spectrum markers associated with malignancy [30]. Therefore, several tumor-associated glycoprotein and glycolipid antigens, such as the mucin-*O*-linked truncated glycans Tn, sialyl-Tn (STn), and Thomsen-Friedenreich (TF), as well as the lipid tail-containing Globo-H, gangliosides (GM2, GD2, GD3), and Lewis (X, Y), have been adopted as targets for the development of therapeutic vaccines against cancer [31].

Antigenic carbohydrates alone trigger a T cell-independent immune response through direct activation of B lymphocytes, resulting in the production of low affinity IgM antibodies, solely. Thus, application of carbohydrates as vaccines is conditioned to their coupling to immunogenic carrier proteins or peptides, such as bovine serum albumin (BSA) and keyhole limpet hemocyanin (KLH). Such approach ensures the stimulation of T lymphocytes, which do not recognize intact antigens but are sensitized by antigenic peptide fragments processed and presented via Major Histocompatibility Complex [32]. Another strategy to improve immunogenicity of carbohydrate-based vaccines involves the concept of multivalency to mimic more properly the dense display of carbohydrates on cell surfaces. Tumor-associated carbohydrate antigens grouped in dendrimers or clusters are then far more efficient in eliciting a protective response than the isolate analog [33].

Cu(I)-assisted 1,3-dipolar azide–alkyne cycloaddition (CuAAC) reactions have been largely applied for the development of neoglycopeptides containing the triazole moiety as synthetic antitumor vaccines. Taking into account the well-known difficulties found in traditional solid phase synthesis (SPS) procedures for joining oligosaccharides to polypeptides, the use of click reactions has proved to be advantageous for synthesis of complex glycopeptides via coupling of azide-containing

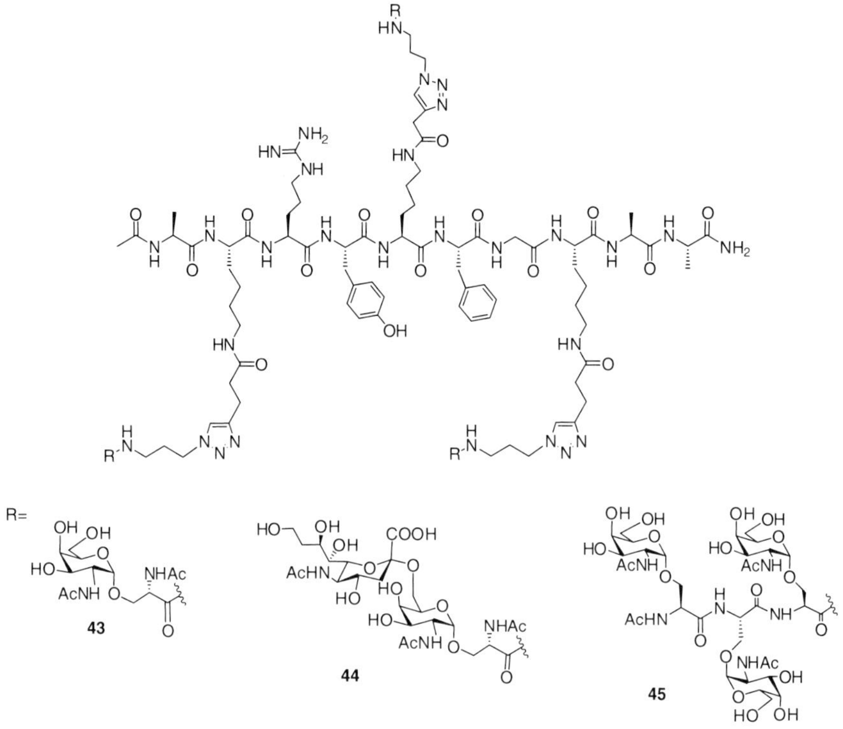

FIGURE 13.1 Triazole-linked glycopeptide vaccines derived from click chemistry reactions between alkyne-based peptide and azide-containing glycosyl amino acids.

carbohydrates or glycosyl amino acids (building blocks) to peptide chains possessing acetylene units, or vice versa. In this regard, glycopeptides **43**, **44,** and **45** containing monomeric or trimeric Tn and STn antigens were synthesized as potential anticancer vaccines by CuAAC reactions. The Tn (monomer and trimer) or STn (monomer) azide-glycosyl amino acids were coupled to pentynoic acid amides of peptide-based lysine side chains, utilizing Cu(0) nanosize powder as catalyst, in PBS at room temperature (Fig. 13.1) [34]. It is noteworthy to mention that the azide derivatives of Tn and STn antigens were previously prepared by conjugation of 3-azido propanolamine to free carboxylic acid of the corresponding serine amino acids, being the Tn cluster obtained by successive coupling reactions of its monomeric unit. Concerning the alkyne-based peptide, which is rich in lysine in order to simulate the environment of the carrier protein, it was prepared by conjugation of NH_2-Lys to *N*-succinimidyl-4-pentynoate [34].

Among the different types of heavily *O*-glycosylated mucin glycoproteins, MUC1, which contains numerous tandem repeats of the sequence HGVTSAPDTR-PAPGSTAPPA with five potential *O*-glycosylation sites, are one of the most explored targets for development of antitumor vaccines, as they are expressed on almost all

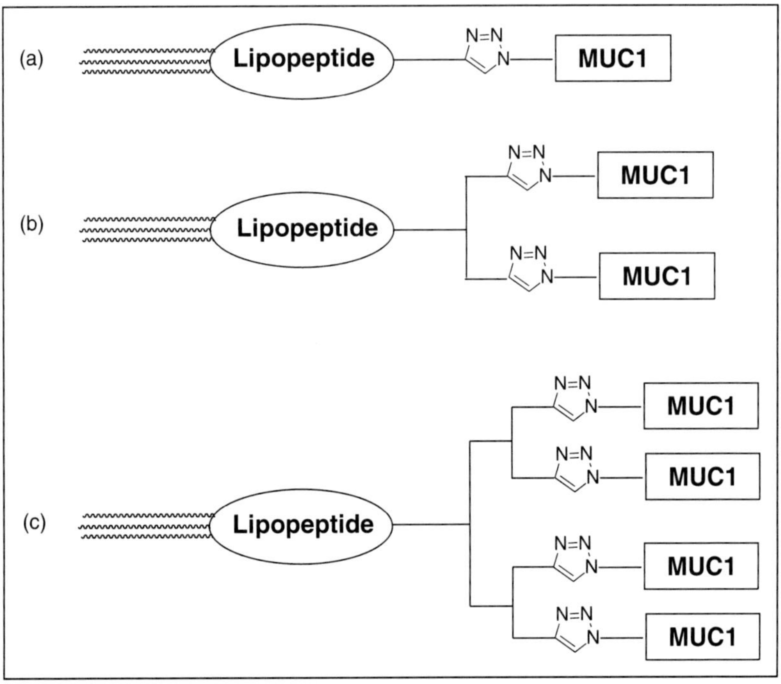

FIGURE 13.2 Structure of the mono-, di-, and tetravalent MUC1 glycopeptide conjugates with the TLR2 lipopeptide ligand. (a) monovalent; (b) divalent; (c) tetravalent.

epithelial tissues and are highly overexpressed on tumor cells [32]. On the other hand, tripalmitoyl-(S)-glycerylcysteine lipopeptides, such as $Pam_3CysSer\text{-}(Lys)_4$ (Pam_3CSK_4) derived from bacterial lipopeptides, are toll-like receptor 2 (TLR2) ligands and have shown capacity to act as strong B lymphocyte and macrophage activators [35]. In this way, they can be used as immune stimulants in synthetic vaccines, besides the fact of having much fewer side effects if compared with protein conjugates. In this context, a different approach involving the conjugation of MUC1-derived glycopeptides to $Pam_3CysSer\text{-}(Lys)_4$ (Pam_3CSK_4) lipopeptides by 1,3-dipolar cycloaddition reaction was applied with the aim to get potential antitumor vaccines able to elicit a strong humoral immune response (Fig. 13.2). For this purpose, four glycopeptides (**46–49**) differing in the tumor-carbohydrate antigens (Tn or T) attached to the MUC1 peptide sequence were firstly synthesized under established SPS conditions, followed by coupling of their N-terminus to a spacer functionalized with an azido group [36] (Scheme 13.8). Regarding Pam_3CSK_4 lipopeptides, distinct mono-, di-, and tetra-alkyne functionalized TLR ligand lipopeptides were prepared by means of suitably designed multilysine cores assembly with subsequent coupling of the alkyne-functionalized spacer. Finally, copper(I)-mediated click reactions for conjugation of glycopeptides with the TLR ligand lipopeptides were carried out in the presence of $Cu(OAc)_2$ and sodium ascorbate as catalytic system, in water at

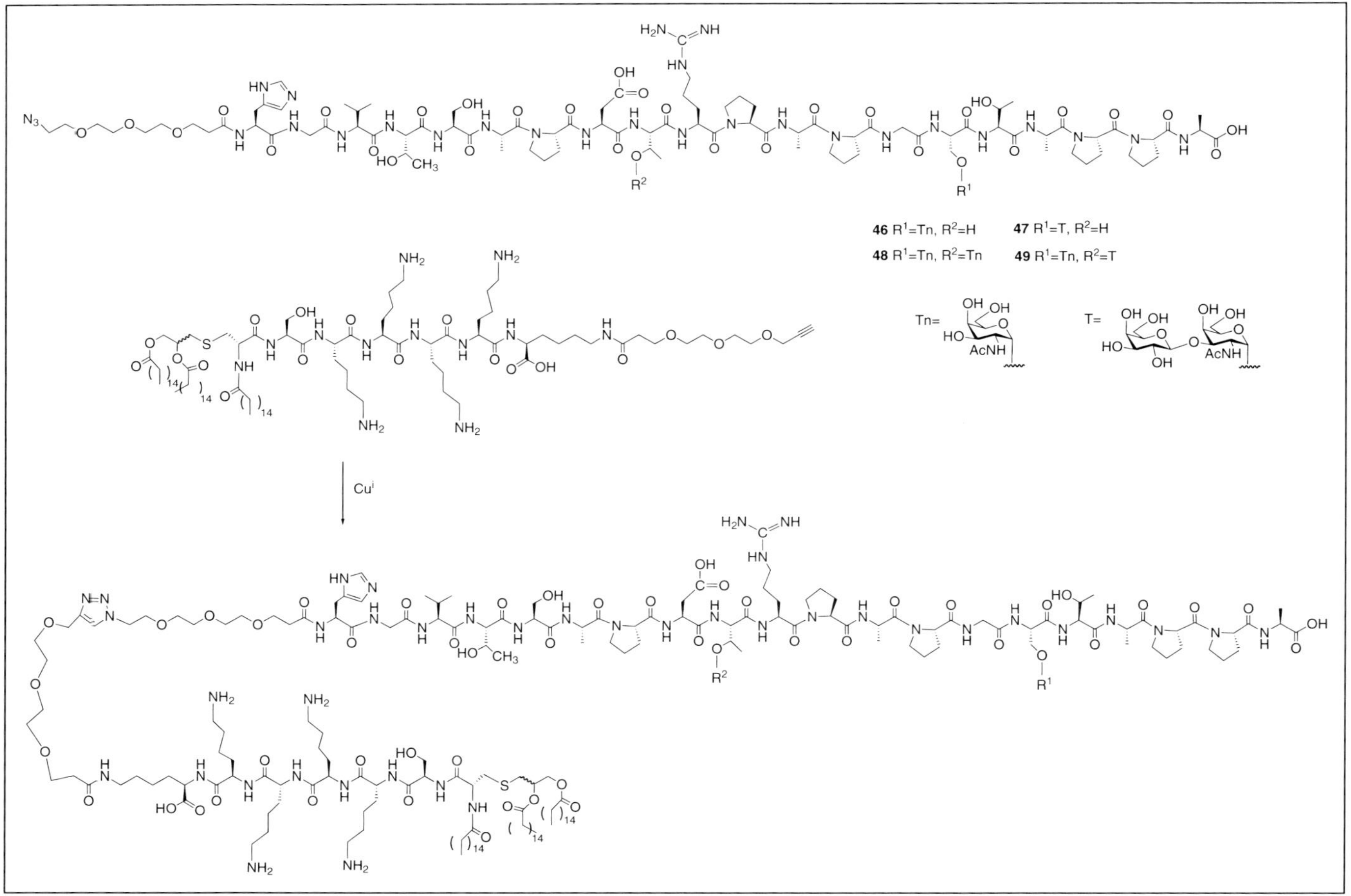

SCHEME 13.8 Copper(I)-mediated click reactions for conjugation of one glycopeptide with a lipopeptide.

40°C, affording the corresponding MUC1 glycopeptides-TLR lipopeptides conjugates (Fig. 13.2, Scheme 13.8) [36].

The aberrant expression of MUC1 glycoproteins on tumor cells was also recently explored for development of MUC1 neoglycopeptides, containing a variable number of GalNAcα1-*O*-Thr/Ser (Tn) mimics, by 1,3-dipolar cycloaddition reactions of unprotected sugar-azides with peptides containing one or more propargyl groups (Scheme 13.9) [37]. The azido-neoglycosyl amino acid mimic (Tn) **50** was obtained by glycosylation reaction of GalNAc with 3-tosil-propanol, followed by azide nucleophilic displacement whereas the propargylated MUC1-derived-peptides were prepared by incorporation of Fmoc-L-propargylglycine residues to the pepide chain, replacing Thr/Ser at selected positions in the MUC1 sequence. Subsequently, the click reactions were performed utilizing the catalytic system $CuSO_4$/tris(carboxyethyl)phosphine (TCEP) in 6 M guanidine hydrochloride (GnHCl)/0.2 M Na_2HPO_4 buffer, under microwave irradiation (25 W, 60°C, 4h), affording a set of mono to penta-clicked MUC-1 derived neoglycopeptides, such as **51** (Scheme 13.9) [37].

Lipid A, which is a fragment of lipopolysaccharides (LPS) present on the cell surface of Gram-negative bacteria, is known for its inherent capacity to bind to Toll-like receptor 4 (TLR4), activating a cascade of immunological responses, including the production of a number of cytokines and chemokines (TNF-α, TNF-β, IL-6, and IFN-β) [38]. For that reason, the design and synthesis of lipid A derivatives such as monophosphoryl lipid A (MPLA) have been carried out in order to investigate their potential immunostimulatory or adjuvant activity. Once they have been coupled with peptides and carbohydrates for the assembly of multicomponent glycoconjugates, they can pave the way for development of functional conjugate vaccines. Following the continuous use of click chemistry to generate neoglycoconjugates as antitumor vaccines, the synthesis of MPLA containing alkyne functionality **52** was then developed in several steps with a view to be further conjugated to the azide-derivative of N-phenyl-acetyl GM3 (GM3NPhAc) **53**, which is an unnatural analog of GM3 ganglioside, overexpressed in tumor cells. Click reaction between **52** and **53** was then performed utilizing CuI/DIPEA in MeOH/THF at room temperature, furnishing the neoglycoconjugate **54** (Scheme 13.10) [39].

13.4 CLICK CHEMISTRY AND CARBOHYDRATE-BASED ENZYME INHIBITORS

13.4.1 Fucosyltransferases (Fuc-T)

Fucosyltransferases (Fuc-T) catalyze the transfer of the L-fucose from guanosinediphosphate-β-L-fucose (GDP-fucose) to the oligosaccharide portion of cell surface glycoconjugates, which are involved in many physiological and pathological processes such as fertilization, embryogenesis, lymphocyte trafficking, immune responses, and cancer metastasis. By exploring the specificity of the substrate GDP-fucose for the target enzyme, a chemical diversity of 85 GDP-triazoles containing

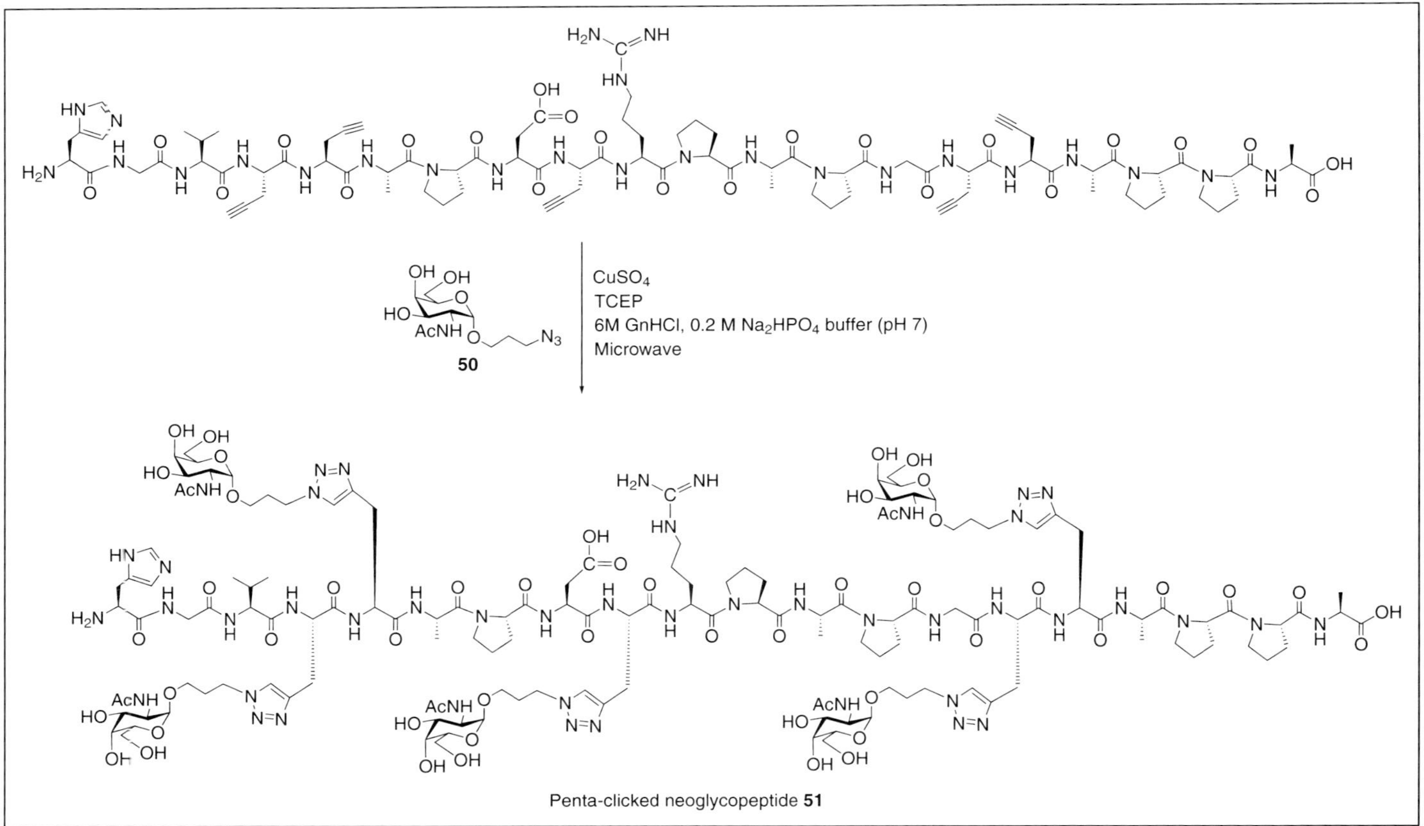

SCHEME 13.9 Synthesis of neoglycopeptide **51** by click chemistry.

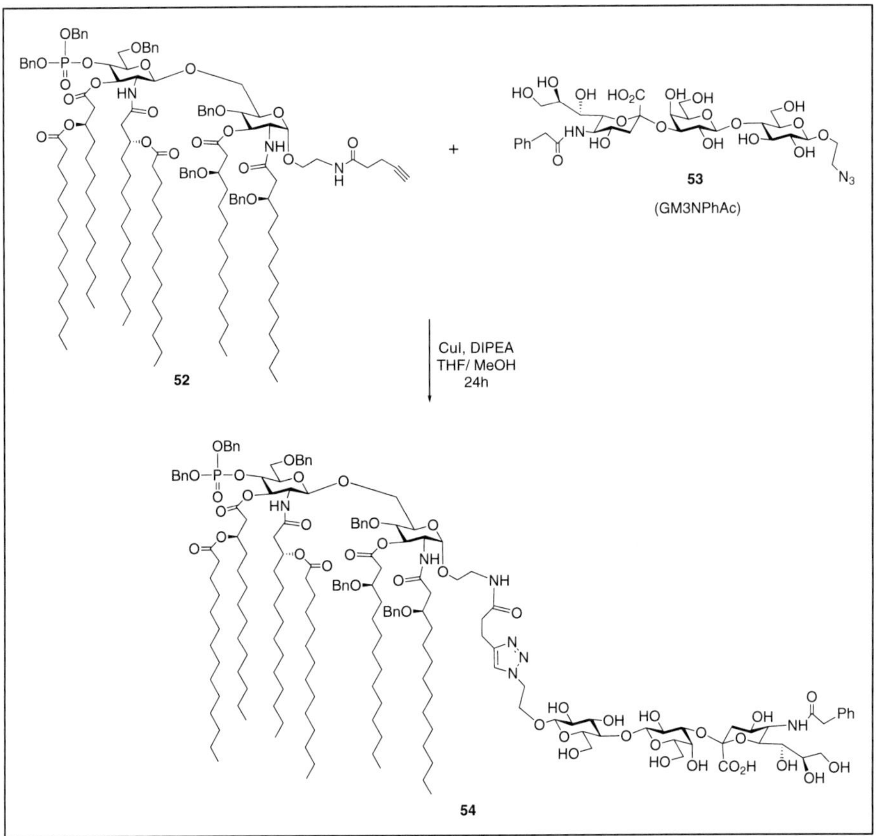

SCHEME 13.10 Coupling of alkyne-containing MPLA **52** with the azide-derived GM3NPhAc **53**, through click reaction, to afford the neoglycoconjugate **54**.

a variety of linker length was synthesized to mimic the GDP moiety and interact in the hydrophobic pocket adjacent to the binding site of the acceptor molecule using a click chemistry strategy (Scheme 13.11) [40]. The inhibition of the human α-1,3-fucosyltransferase VI (Fuc-T VI) was evaluated, and showed that molecule **55** had the highest activity, with *K*i (comp) 62 nM, being a competitive inhibitor against GDP-fucose and the first nanomolar and most potent inhibitor of Fuc-Ts already described. On the other hand, a noncompetitive mechanism was observed when the enzymatic assays were performed against the acceptor molecule, *N*-acetyllactosamine (LacNAc), with a *K*i (noncomp) of 221 ± 8 nM [40].

13.4.2 Mannosyltransferases

Mannosyltransferase(s) may play a role for *Leishmania* amastigotes survival in macrophages, since it is involved in the biosynthesis of β-1,2-mannan and/or

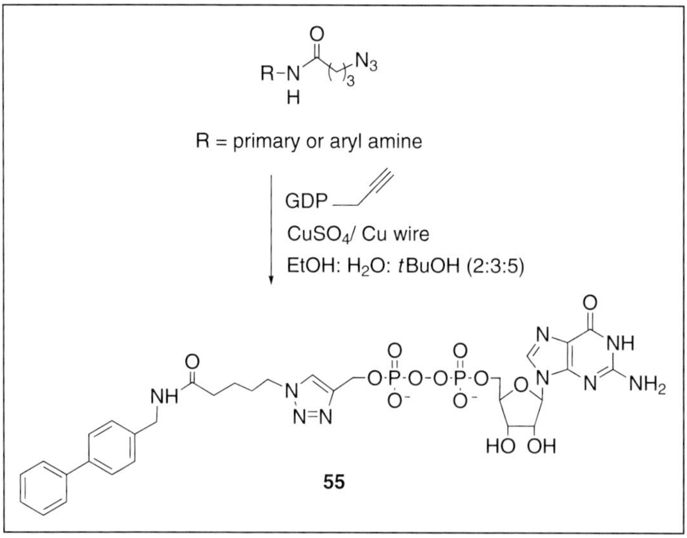

SCHEME 13.11 Synthesis of α-1,3-fucosyltransferase VI (Fuc-T VI) inhibitor **55** by Cu(I) conditions.

mannose-containing surface glycoconjugates by elongating short mannose oligosaccharides, with or without the cyclic phosphate, through the addition of β-1,2-linked mannosyl residues. In order to determine the substrate tolerance and, consequently, the active-site architecture of the *Leishmania* β-1,2-mannosyltransferases, the click chemistry approach was used to construct a library comprising 28 mannose derivatives from the coupling of either an azide or an alkyne-functionalized mannose, such as the 6-azido mannoside **56** and propargyl mannoside **57**, with an alkyne- or azide-based diversity set, respectively, and screen them directly from reaction mixtures (Scheme 13.12) [41]. Thus, screening of the series against *L. mexicana* β-1,2-mannosyl transferases was performed using radiolabeled GDP-[^{3}H]Man as donor. The propargyl

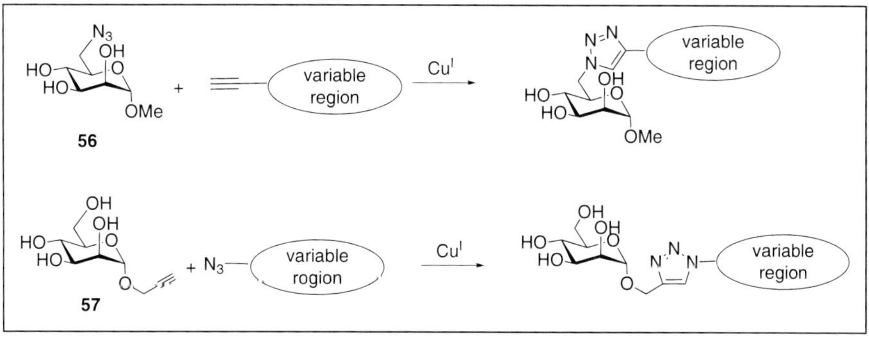

SCHEME 13.12 Synthesis of mannose derivatives by click chemistry approach using azide **56** or alkyne **57**.

mannoside 57 and its 1,2,3-triazole derivatives were effective substrates, since the partition of the reaction mixture between water and butan-1-ol, and analysis by HPTLC detected by fluorography, showed the presence of disaccharides and higher oligomers connected by β-1,2-linkage. In contrast, no substrate activity was observed for methyl 6-azido-α-mannopyranoside 56 nor for any of its 1,2,3-triazole derivatives when tested against β-1,2-mannosyltransferases from *L. Mexicana* [41].

13.4.3 Glucosidases

Glucosidases are involved in several biochemical processes such as degradation of diet polysaccharides to furnish monosaccharide units, which are then able to be metabolically absorbed and used by the organism, lysosomal glycoconjugate catabolism and glycoprotein processing, and biosynthesis of oligosaccharide units in glycoproteins or glycolipids. Glucosidase inhibitors are currently of interest owing to their promising therapeutic potential in the treatment of disorders such as diabetes, human immunodeficiency virus (HIV) infection, metastatic cancer and lysosomal storage diseases. In this regard, many attempts have been made to utilize click chemistry for the synthesis of potential glucosidase inhibitors [42].

The search for favorable binding interactions by mimicking glycosyl amines functionalities, which can be protonated to interact with the carboxylate moieties on glucosidase active site, led to the synthesis of glucosyl triazoles via a copper-catalyzed [3 + 2] cycloaddition between the appropriate acetyl-protected β-azido glycosides and phenylacetylene (Scheme 13.13) [43]. The biological assays carried out against three β-glycosidases, sweet almond glucosidase (SAG), *Escherichia coli* galactosidase (ECG), and bovine liver galactosidase (BLG) showed that only weak inhibition was observed for compounds **58** (30%) and **59** (48%) against ECG and BLG glycosidases, respectively, when tested 0.24 mM (Table 13.1) [43].

Glucosylceramide synthase (GCS) is an essential glucosyltransferase involved in the biosynthesis of glucosylceramide, which is subsequently hydrolyzed into ceramide and glucose by the action of the β-glucocerebrosidase (β-GCase). Defects in the lysosomal degradation of glycosphingolipids due to the deficient activity of β-GCase led to glucosylceramide accumulation in macrophages and the development of the Gaucher disease.

The iminosugar *N*-butyl-1-deoxynojirimycin (*N*-Bu-DNJ) is a potent inhibitor of GCS, which reduces the glucosylceramide biosynthesis and minimizes the deficient

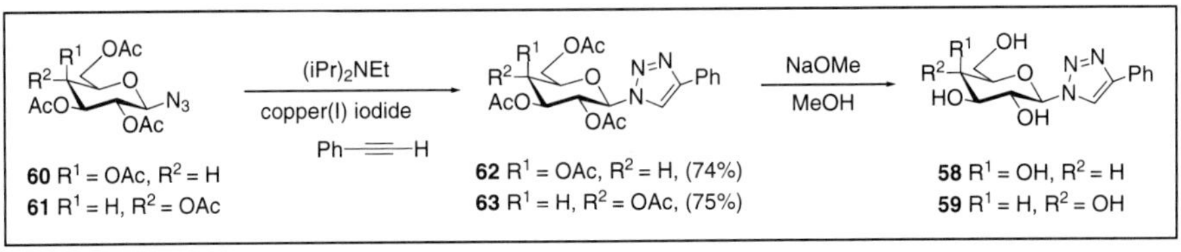

SCHEME 13.13 Synthesis of glycosyl triazoles **58** and **59**, via copper-catalysed 3 + 2 cycloaddition, as glucosidase inhibitors.

TABLE 13.1 Inhibition of Glycosidases by Triazoles

	Enzyme		
Compound[a]	SAG (%)	ECG (%)	BLG (%)
58	<20	30 ± 11	<20
59	<20	<20	48 ± 11
DNJGal	<20	91 ± 1	<20
DNJGluco	50 ± 8	<20	<20

[a] All compounds were screened at the concentration of 0.24 mM.

activity of β-GCase. Additionally, iminosugars containing hydrophobic groups may act as chaperones to stabilize N370S β-GCase, the most common mutation causing Gaucher disease. Based on these findings, 14 new hydrophobic iminosugars were synthesized by coupling alkynyl-armed substituents to an azido-functionalized *N*-propyl-DNJ core, using commercially available hydrophobic alkynes, in microwave CuAAC-assisted reactions catalyzed by $CuSO_4$/sodium ascorbate in a mixture of dioxane-water (4:1) (Fig. 13.3) [44]. The triazole-iminosugars **64–77** were then tested against a set of commercial glucosidase and β-GCase, besides mutant Gaucher lymphoblasts with the most common N370S mutation to assess their chaperone activity. Most of the compounds showed inhibitory activity toward commercial β-glucosidase (9 to 669 μM from bovine liver and 0.59 to 177 μM from almond), amyloglucosidase (1 to >2 mM), and baker's yeast α-glucosidase (20 μM to >2 mM). As observed for β-glucosidases, the highest inhibitory potency toward N370S β-GCase was displayed by derivatives bearing hydrophobic *N*-substituents, such as compounds **66**, **67**, **71** and **72**, which were able to inhibit the enzyme in a range of 25–30% at 5 μM. Regarding the chaperone activity in mutant Gaucher cells, even though compound **72** was the most active of the series, promoting a twofold increase in enzyme activity at 20 μM, no direct correlation was obtained for β-GCase and cellular chaperone activities [44].

The regio- and stereocontrolled synthesis of novel non-fused 1,2,3-triazolo-pseudooligosaccharides was carried out with the aim to improve acarbose activity, a potent α-glucosidase inhibitor currently used in the treatment of diabetes. Using 2-butyn-1,4-diol as the dipolarophile reagent and 2-azidoethanol, as a model, and then azido-deoxyglycosides in a copper-catalyzed azide–alkyne cycloaddition (CuAAC), a series of 1,4,5-trisubstituted 1,2,3-triazole derivatives were obtained in high yield (Fig. 13.4) [45]. As a result, 6-deoxyglucosyl triazole **78** displayed good competitive inhibition toward α-glucosidase (Ki 73 mM), being fourfold more potent as compared to unglycosylated triazole **79** and twice more active than acarbose for this particular enzyme. Nevertheless, the tests performed with a panel of glycosidases revealed that pseudooligosaccharides **80**, **81**, **82** and **83**, comprising the trisaccharide core present in acarbose, were not active. Therefore, these results suggested that the bis(hydroxymethyl)triazole unit is not a functional surrogate of the natural aminocyclitol moiety (valienamine) present in acarbose [45].

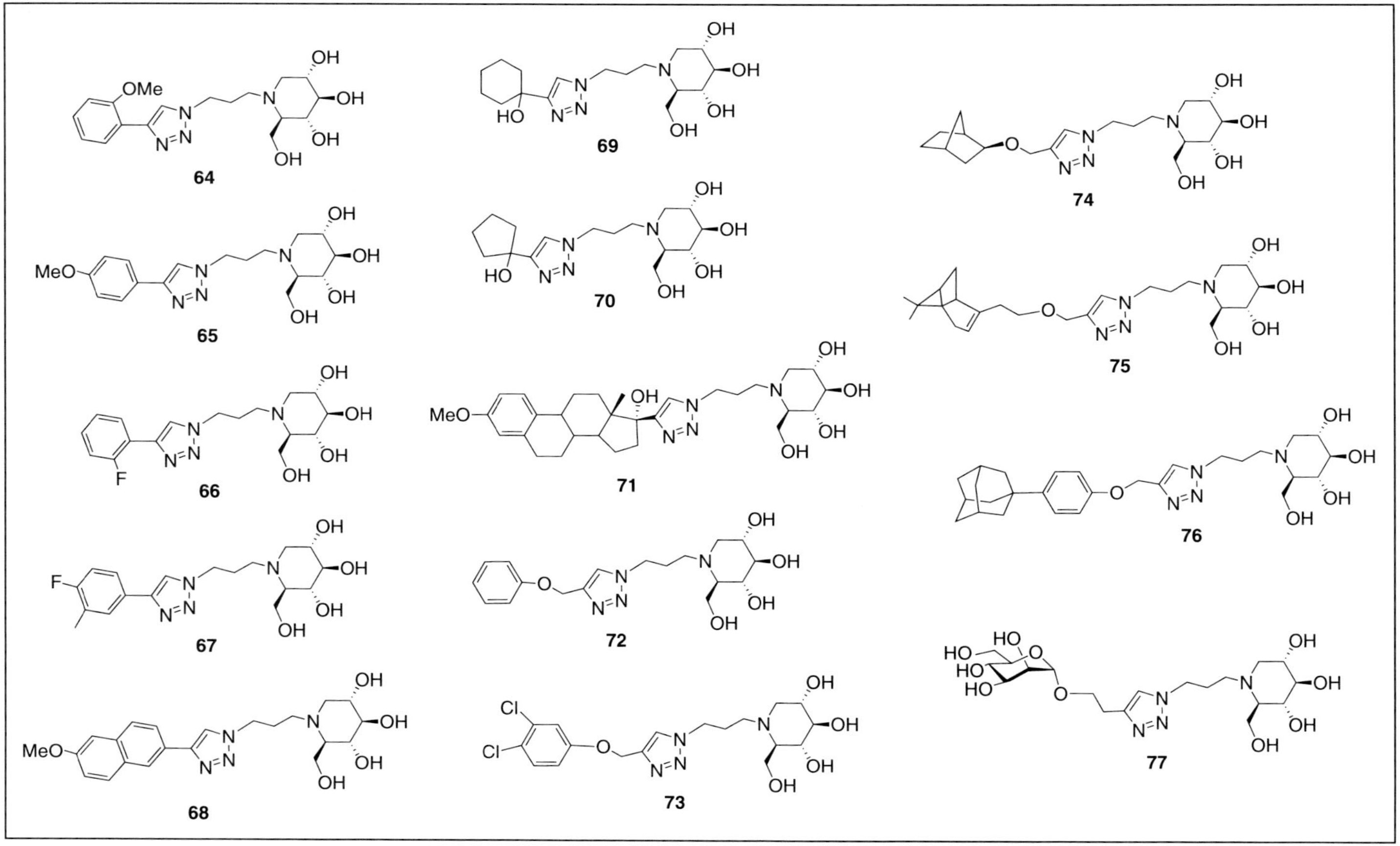

FIGURE 13.3 Structures of triazole-iminosugars **64–77** obtained by CuAAC reactions.

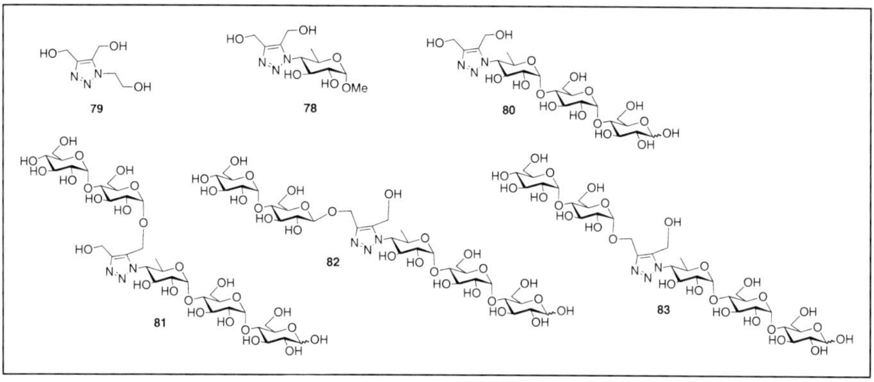

FIGURE 13.4 Structures of 1,2,3-triazolo-pseudooligosaccharides triazole mimetics **80–83**, designed from acarbose.

A small library of 21 α-D- and β-D-glucopyranosyl triazoles having variable aglycones was synthesized, via CuAAC reactions, as potential glycosidase inhibitors [39]. Hence, the coupling between β-D-glucopyranosyl azide **84** or the corresponding α isomer **85** with selected alkynes bearing a wide variety of functional groups, under reaction conditions involving the use of the catalytic system $CuSO_4$/sodium ascorbate, afforded compounds **86a–92a/86b–92b** (Fig. 13.5). Alternatively, the application of the catalytic system $CuSO_4$/Cu turnings using β-D-glucopyranosyl azide **84** and other alkynes, gave compounds **93–99** (Fig. 13.5) [46]. The impact of glycosyl azide anomeric configuration on the rate of CuAAC reactions was investigated to distinguish α- and β-glucopyranosyl azides reactivities. While reactions with β-glucopyranosyl azide were typically complete in 10–45 minutes, similar reaction conditions performed with α-glucopyranosyl azide often required 45–120 minutes. This difference was further investigated using competition reactions and rationalized on the basis of X-ray crystallographic data, which revealed significant differences in bond lengths within the azido groups of α- and β-anomers [47].

Iminosugars, such as DNJ, have great potency and specificity to inhibit glucosidases due to their ability to mimic the transition state of natural glucosidase substrates [42]. Based on the single binding center and on the conformational and electrostatic influences of DNJ in a panel of glucosidases, the synthesis of multivalent iminosugars was carried out with the aim to modulate affinity and selectivity for glycosidases [48]. Thus, mono-(**100**, **101**), di-(**102**, **103**), and trivalent derivatives **104** were obtained by a click chemistry approach using the 1,3-dipolar cycloaddition between azide-functionalized 1-DNJ derivatives and alkynyl-armed oligo(ethyleneglycol) (EG) scaffolds in the presence of $CuSO_4$/sodium ascorbate in a mixture of dioxane and water [48] (Fig. 13.6). CuAAC was highly regioselective and afforded only 1,4-disubstituted triazole structures, which bear different number of active DNJ moieties and length between the functional epitopes. The inhibition assays were performed against several glycosidases using 1-DNJ for comparative purpose. In general, the incorporation of EG spacers, containing the triazole unit at the nitrogen ring, had a deleterious

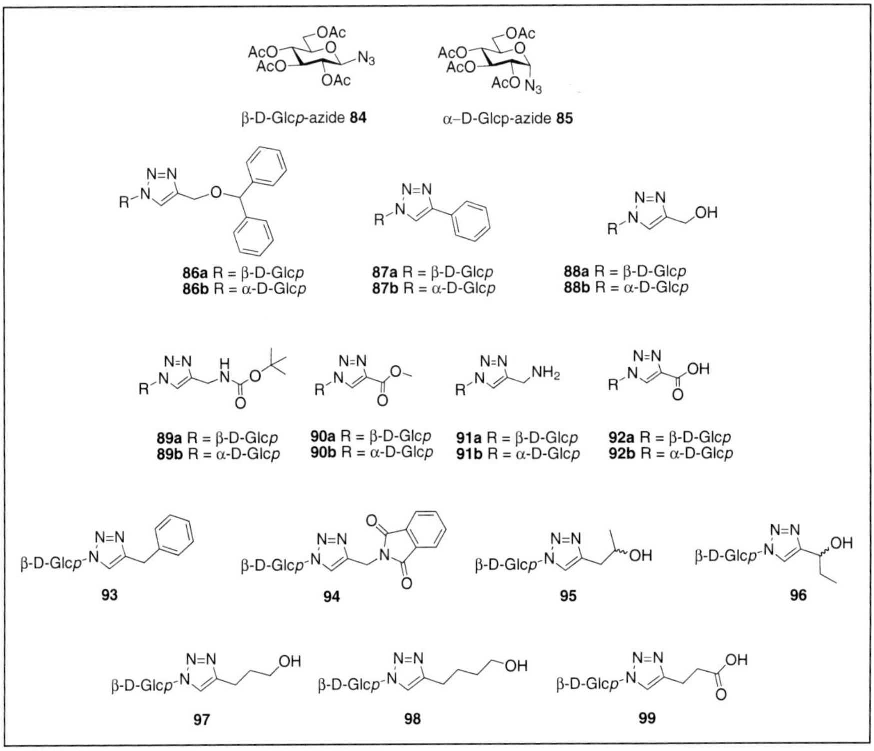

FIGURE 13.5 Structures of glucopyranosyl triazoles **86a–92b**, **93–99**, as potential glycosidase inhibitors.

effect in the inhibitory potency of most glycosidases and the goal to enhance affinity of multivalent iminosugars has not been achieved, despite the selectivity and activity obtained for β-glucosidase, α-mannosidase, and amyloglucosidase (Table 13.2) [48].

The search for hybrid inhibitors able to target two different enzymes of the angiogenesis pathway has been pursued by combining 1-deoxynojirimycin (DNJ) and 5-aryl-1,2,3-triazole in a single hybrid molecule connected by different linkers length (Fig. 13.7) [49]. In this regard, DNJ core was used as α-glucosidase inhibitor to disturb the oligosaccharide biosynthesis on endothelial cell surface and 5-aryl-1,2,3-triazole unit as potential methionine aminopeptidase II (MetAP2) inhibitor, whose activity is assessed by BAEC proliferation [49].

The resulting bifunctional derivatives were evaluated for their α-glucosidase (from *Bacillus stearothermophilus*) inhibitory activity, being compound **107** (IC_{50} 1.15 μM) more potent than DNJ (IC_{50}1.67μM), and compounds **105** (IC_{50} 6.07 μM) and **106** (IC_{50} 2.41 μM) being more efficient to alter the biosynthesis of oligosaccharides on endothelial cell surfaces when compared with DNJ. Therefore, the seven methylene linker that holds the two active units in compound **107** may contribute for favorable interactions in the active site, since the precursor aryl-1,2,3-triazole did not exhibit

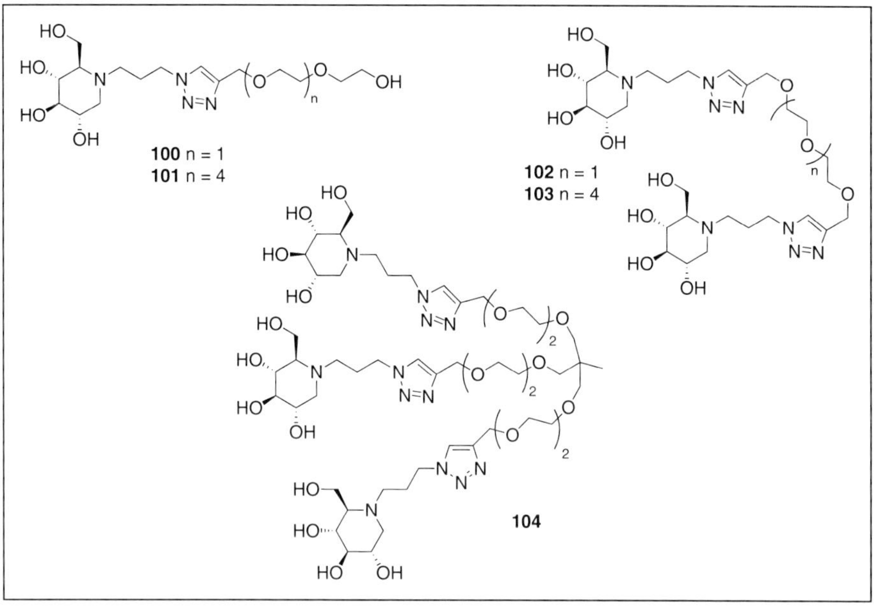

FIGURE 13.6 Structures of synthetic iminosugars **100–104** obtained by CuAAC reactions.

TABLE 13.2 Glycosidase Inhibitory Activities (K_i, μM) for DNJ and Synthetic Iminosugars 100–104 Obtained by "Click Chemistry"

Enzyme	DNJ	100	101	102	103	104
β-galactosidase (bovine liver)	42	950 ± 20	n.i.	n.i.	760 ± 10	790 ± 10
α-galactosidase (green coffee beans)	–	120 ± 5	95 ± 5	215 ± 5	105 ± 5	225 ± 5
β-glucosidase (almonds, pH 7.3)	47	47 ± 2	100 ± 5	70 ± 2	70 ± 2	400 ± 10
α-glucosidase (bakers' yeast)	25	n.i.	n.i.	n.i.	n.i.	n.i.
β-mannosidase (*Helix pomatia*)	–	n.i.	n.i.	n.i.	772 ± 10	n.i.
α-mannosidase (Jack bean)	270	230 ± 10	265 ± 10	120 ± 5	90 ± 3	35 ± 2
isomaltase	11	1050 ± 20	480 ± 15	n.i.	770 ± 15	1440 ± 20
naringinase (*Penicillium decumbens*)	–	80 ± 3	95 ± 5	60 ± 3	55 ± 2	75 ± 3
amyloglucosidase (*Aspergillus niger*)	2.1	28 ± 1	30 ± 2	17 ± 1	20 ± 2	11 ± 1

n.i. = no inhibition at 2 mM.

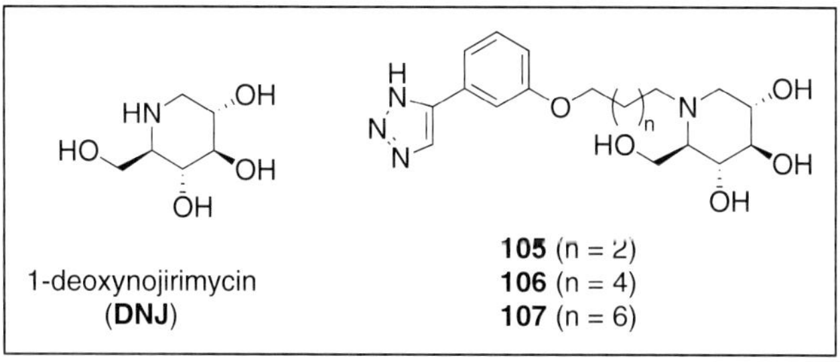

FIGURE 13.7 Structures of DNJ and hybrids **105**, **106,** and **107**.

any α-glucosidase activity. Additionally, the same compound proved to be a potent inhibitor of the BAEC cells proliferation (IC_{50} 105 μM) and capillary tube formation (at 0.2 mM), as observed in capillary-like blood vessels that are formed *in vivo* during angiogenesis [49].

13.4.4 Neuraminidase

Neuraminidase plays an essential role in the influenza infection (AIV) owing to its ability to promote the cleavage of sialic acid units from glycoproteins and glycolipids found in the viral envelope, which disrupt the protective mucous layer of the upper respiratory tract, exposing the epithelial cells for viral adsorption. After viral hemagglutinin binding to cell surface and invasion, neuraminidases also play a role in the escape of new virions from infected cells. Thus, neuraminidases (or sialidases) have been targeted to stop the viral infection by preventing the release of viruses from infected cells [50].

Neuraminidase inhibitors are selective antiviral agents since the viral active site is different from comparable mammalian enzymes. For instance, the inhibitors zanamivir and oseltamivir were successfully designed on a mechanism-based transition state of neuraminic acid (NANA), in which the distorted and charged pyranose ring has a planar trigonal center to fit the active site. Nevertheless, genetic mutations have major influence on the variety of influenza viruses, which lead to the development of resistance [51].

The synthesis of potential neuraminidase inhibitors that mimic the transition state have been described and several attempts have been made to utilize click chemistry to produce more effective and longlasting therapeutic drugs for anti-AIV infection. In this regard, a library of 16 novel 4-triazole-modified zanamivir analogs were synthesized by the Huisgen 1,3-dipolar cycloaddition of structural diverse alkynes and 2,3-didehydro-*N*-acetyl-neuraminic acid (DANA) derivatives bearing azide group at position C-4, which mimic the zanamivir–guanidyl function (Scheme 13.14) [52].

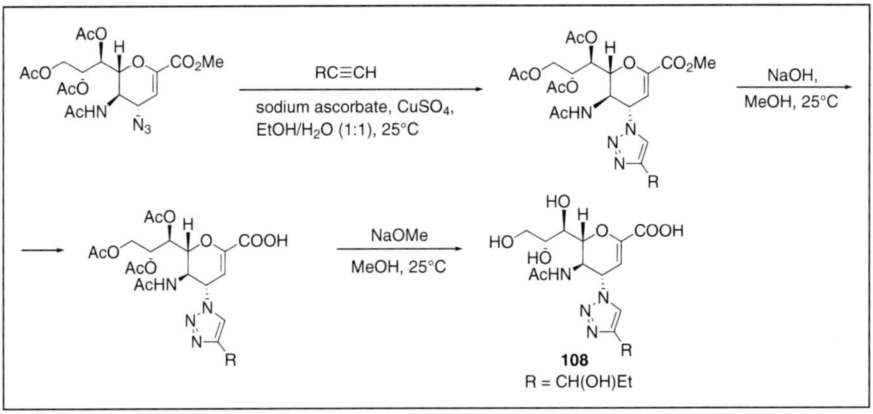

SCHEME 13.14 Synthesis of zanamivir analog **108** by click chemistry.

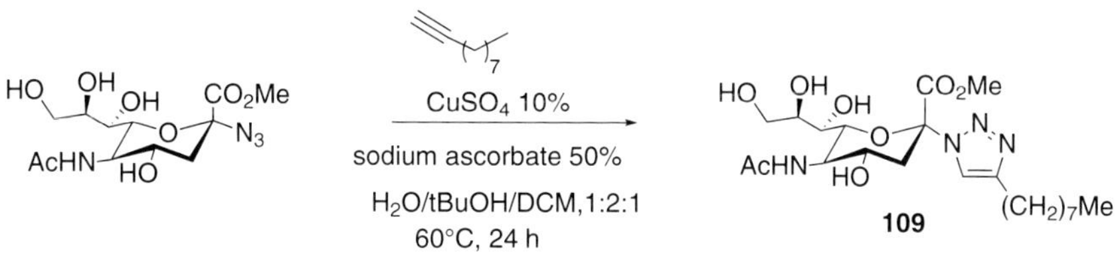

SCHEME 13.15 Synthesis of the 1,2,3-triazole-linked sialic acid **109**.

Antiviral efficacy of the library was assessed against infected cell (MDCK cell) at 50 μM and expressed as percent protective rate against virus infection, being verified either no activity or percent values varying from 9% to >61.4%. The highest antiviral activity was displayed by 4-hydroxypropyl triazole derivative **108** (IC_{50} 6.4 μM), which had comparable activity to zanamivir (IC_{50} 2.8 μM) (Scheme 13.14) [52].

In an alternative approach, with the intention to overcome the liability of natural O-glycosides, non-natural *N*-glycosides of sialic acid resistant to neuraminidase-catalyzed hydrolysis were produced. Thus, from the copper-catalyzed azide–alkyne Huisgen cycloaddition, a small library of 1,2,3-triazole-linked sialic acid derivatives was synthesized using a key intermediate of azide NANA linked at position C-2, as exemplified for compound **109** (Scheme 13.15) [53]. In spite of the fact that hindrance of the C-2 position, stereoelectronic effects from electron-withdrawing acetyl-protecting groups or inappropriate conformation of azido sialic acid may account for initial reaction failure, high yield and complete regioselectivity was achieved when the acetyl protecting group was removed before the click reaction. Next, in order to enhance the affinity of these inhibitors for neuraminidase by multivalent 1,2,3-triazole-linked sialic acid approach, a disaccharide **110** and a dendrimer **111** mimics were also synthesized using the click reaction (Fig. 13.8) [53]. In the fluorimetric evaluation of neuraminidase, compound **109**, containing a longer and more flexible side chain, showed to be the strongest inhibitor of the monomeric triazole series, with IC_{50} 28 μM whereas sialic acid disaccharide **110** and dendrimer **111** showed slightly better inhibitions profile, with IC_{50} 17 and 20 μM, respectively.

Mammalian exo-sialidases NEU1, NEU2, NEU3, and NEU4 are involved in a large number of natural processes, including neurite outgrowth, immune cell activation, cellular communication, signaling, adhesion, as well as the apoptosis and metastasis of malignant cells [54–59]. Despite the nanomolar efficacy of zanamivir and oseltamivir against viral neuraminidase, they have limited potency against all of the human sialidases and no tight-binding (<1 μM), so that selective inhibitors for individual isozymes have been described. Based on the plasma membrane-associated NEU3 homology model using site-directed mutagenesis, and molecular docking of a well-known sialidase inhibitor, 2,3-didehydro-N-acetyl-neuraminic acid (DANA), a small library of DANA analogs using CuI-catalyzed azide–alkyne 1,3-dipolar cycloaddition (CuAAC) was assembled to test their substrate and/or inhibition activities [60]. Hence, condensation between N5-azidoacetyl- **112** or 9-azido-9-deoxy **113** DANA

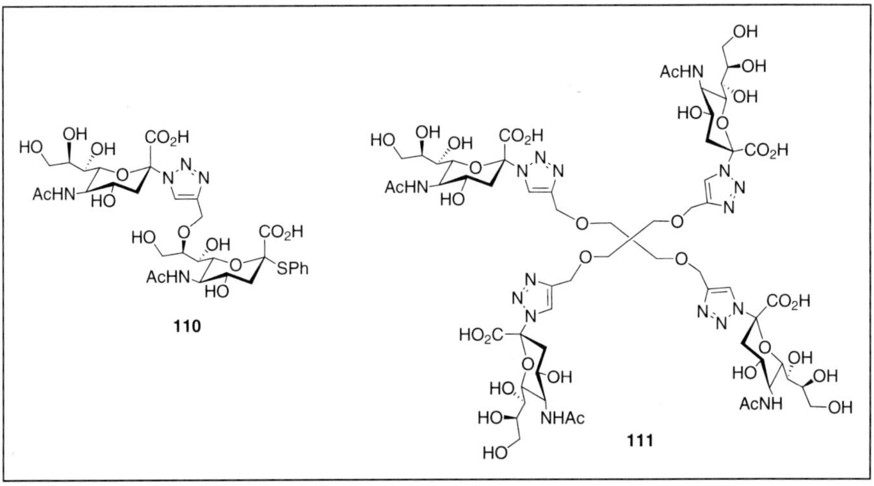

FIGURE 13.8 Sialic acid disaccharide mimic **110** and dendrimer of sialic acid **111** as neuraminidase inhibitors.

derivatives and a panel of terminal acetylenic alcohols, ether, phenyl, and hexyl derivatives afforded 18 1,4-disubstitute triazoles (**114a–i/115a–i**) (Scheme 13.16), which were assayed against NEU3 using a fluorescence assay based on the hydrolysis of 4-methylumbelliferyl α-D-N-acetylneuraminic acid (MUNANA). As a result, the modification of DANA at C-9 position containing a triazole with hydrophobic

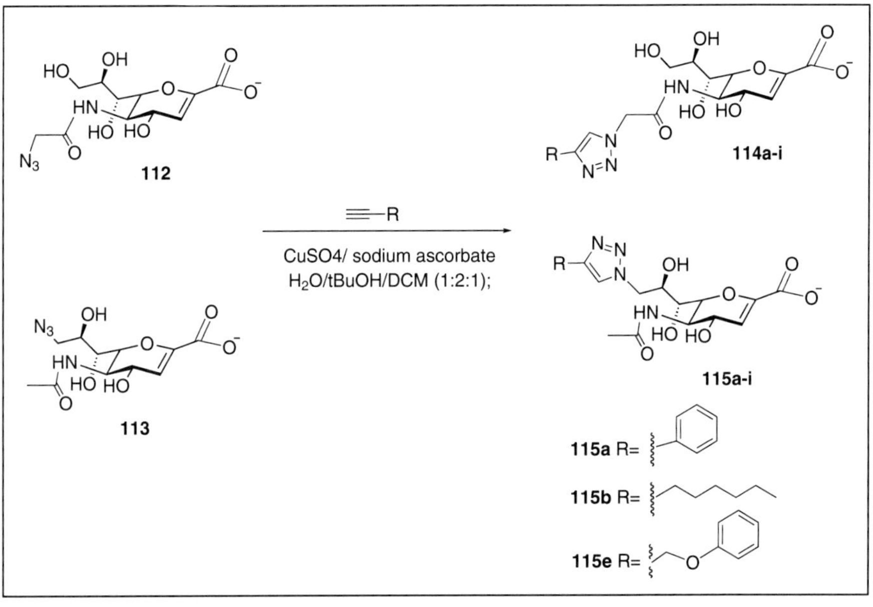

SCHEME 13.16 Synthesis of N5Ac (**114a–i**) and C9 (**115a–i**) modified analogs of DANA by CuAAC.

groups lacking a hydrogen bond donor gave the most active compounds of the series, such as phenyl **115a** (20 ± 10 μM), hexyl **115b** (23 ± 4 μM), and phenoxymethyl **115e** (45 ± 3 μM) derivatives (Scheme 13.16), comparable to DANA (approximately 40 μM). On the other hand, the presence of substituted DANA-triazoles at N5Ac **114a–i** showed reduced potency (>500 μM) probably due to the sterical hindrance at NEU3 binding site [60].

13.4.5 *Trypanosoma cruzi Trans*-Sialidase

The parasite *Trypanosoma cruzi* sheds a developmentally regulated enzyme called *trans*-sialidase, which is involved in the interactions between host cells and parasite that are crucial for cellular invasion processes and infection. This fundamental role played by *Trypanosoma cruzi trans*-sialidase (TcTS) besides the fact that it is not present in host cells validates this enzyme as a potential therapeutic target for Chagas' disease [61]. Thus far, success in the development of inhibitors of this enzyme has been extremely limited.

Therefore, a small library of non-hydrolysable 1,4-disubstituted 1,2,3-triazole galactosides was synthesized by means of CuAAC reactions of galactose derivatives containing an azide group at the C-6 **116** or C-1 **117** positions with a panel of 23 structurally diverse terminal alkynes (**118–140**), generating 46 disubstituted triazoles represented by **141c/d**-**163c/d** in good-to-excellent yield and with complete regioselectivity (Scheme 13.17) [62]. The obtained triazole galactosides were then submitted to inhibition assays of TcTS and to *in vitro* trypanocidal activity against trypomastigote forms of *T. cruzi* Y strain. The strongest TcTS inhibition (37%), at 1 mM concentration, was verified for compound **144d**, while compound **153c** showed the best trypanocidal activity at four different concentrations (0.1 to 1 mM), with an IC_{50} of 0.18 mM (Fig. 13.9) [62].

The existence of multivalent display of TcTS on the parasite surface and its ability to bind galactoside substrates represents a great opportunity for exploiting multivalent interactions for the design of effective ligands for this enzyme. In this regard, cyclooligomerization reactions of bifunctional galactose monomers bearing azido and propargyl groups, using Cu(I)-assisted azido–alkyne 1,3–dipolar cycloaddition, were recently investigated with the aim to obtain new cyclic *pseudo*-oligosaccharides that can be used for the study of TcTS activity and in the search for potential inhibitors of this enzyme [63]. CuAAC reactions from building blocks **164** and **165** in DMF, utilizing $CuSO_4$/Cu turnings as a catalytic system, afforded a series of macrocycles with two to five repeating units (Scheme 13.18). The obtained glycomacrocycles were subsequently submitted to enzymatic assays with TcTS, in the presence of MUNANA as a sialic acid-donor substrate, to give the corresponding sialylated products, as illustrated for compounds **166** and **167** (Scheme 13.18) [63].

More recently, a series of sialylmimetic neoglycoconjugates were designed and synthesized as mimetics of the terminal sugars α-D-Neu5Ac(2→3)-β-D-Gal of *T. cruzi* mucins with the aim to get inhibitors able to establish strong and multivalent interactions with TcTS active site, besides exerting trypanocidal activity [64]. This series of compounds, represented by 1,4-disubstituted 1,2,3-triazole-sialic acid

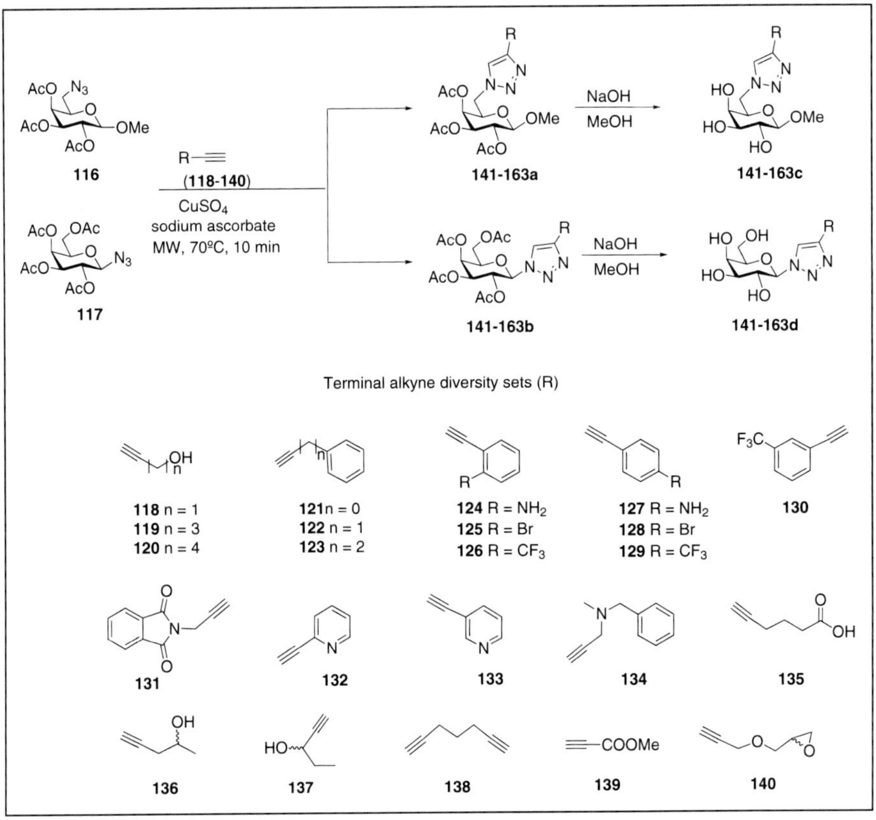

SCHEME 13.17 Synthesis of a library of 1,2,3-triazole-galactosides **141–163**, obtained from 1,3-dipolar cycloaddition reactions as potential TcTS inhibitors.

derivatives containing galactose modified at either C-1 **168** or C-6 **169** positions, glucose **170** or gulose **171** at C-3 positions, and by the amino acid derivative 1,2,3-triazole fused threonine-3-*O*-galactose **172**, was obtained by Cu(I)-catalysed cycloaddition reaction between the azido-functionalized sugars 1-N_3-Gal **173**, 6-N_3-Gal **174**, 3-N_3-Glc **175** and 3-N_3-Gul **176** with the corresponding alkyne-based 2-propynyl-sialic

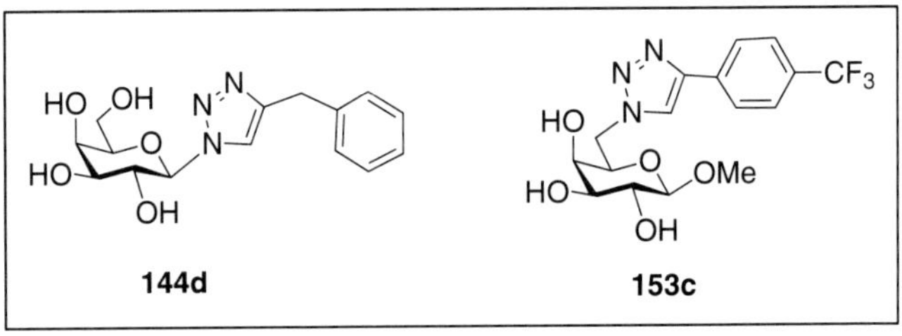

FIGURE 13.9 Galactose derivatives **144d** and **153c** that showed highest TcTS inhibition and trypanocidal activity.

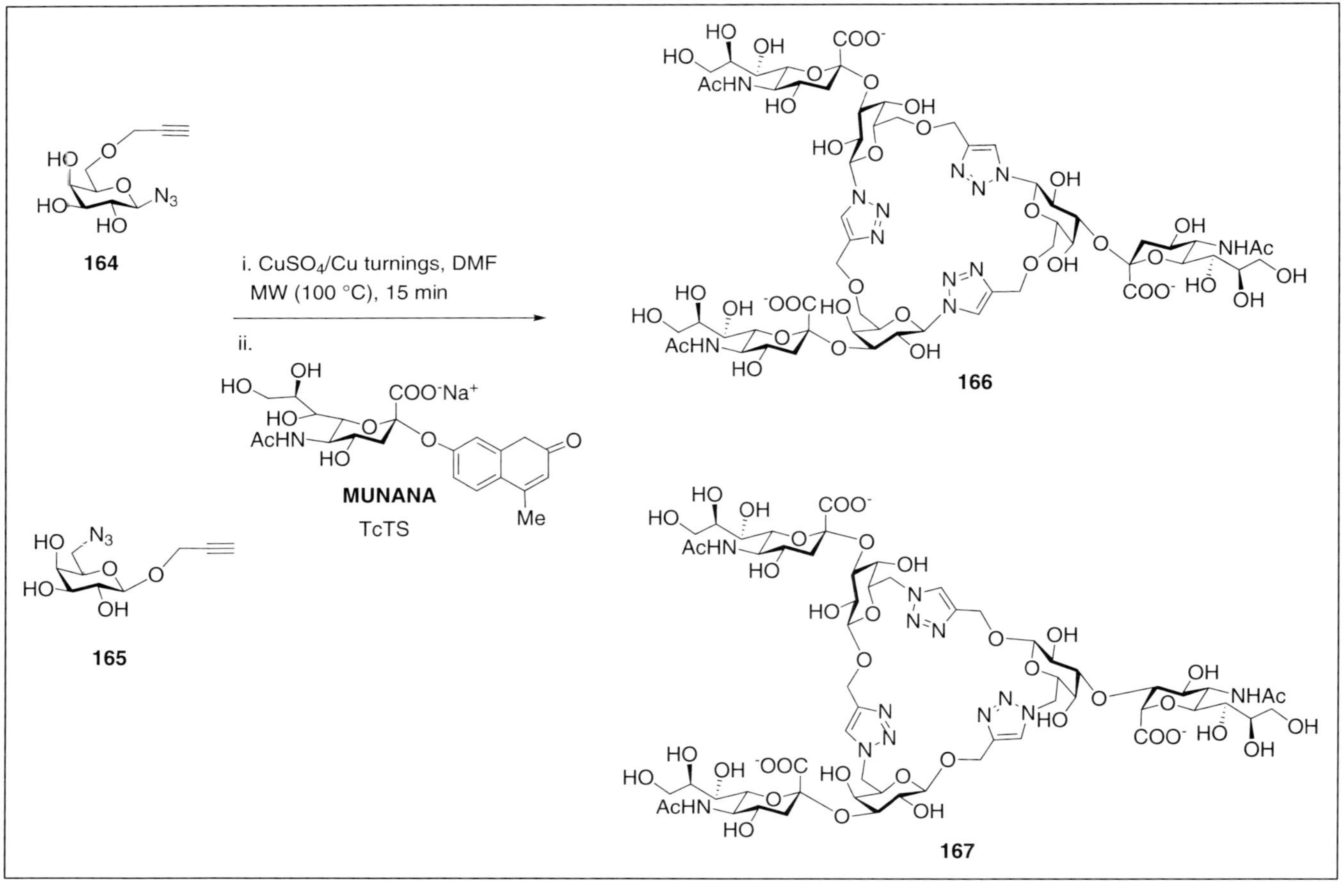

SCHEME 13.18 Synthesis of glycomacrocycles **166** and **167** by click chemistry from building blocks **164** and **165**.

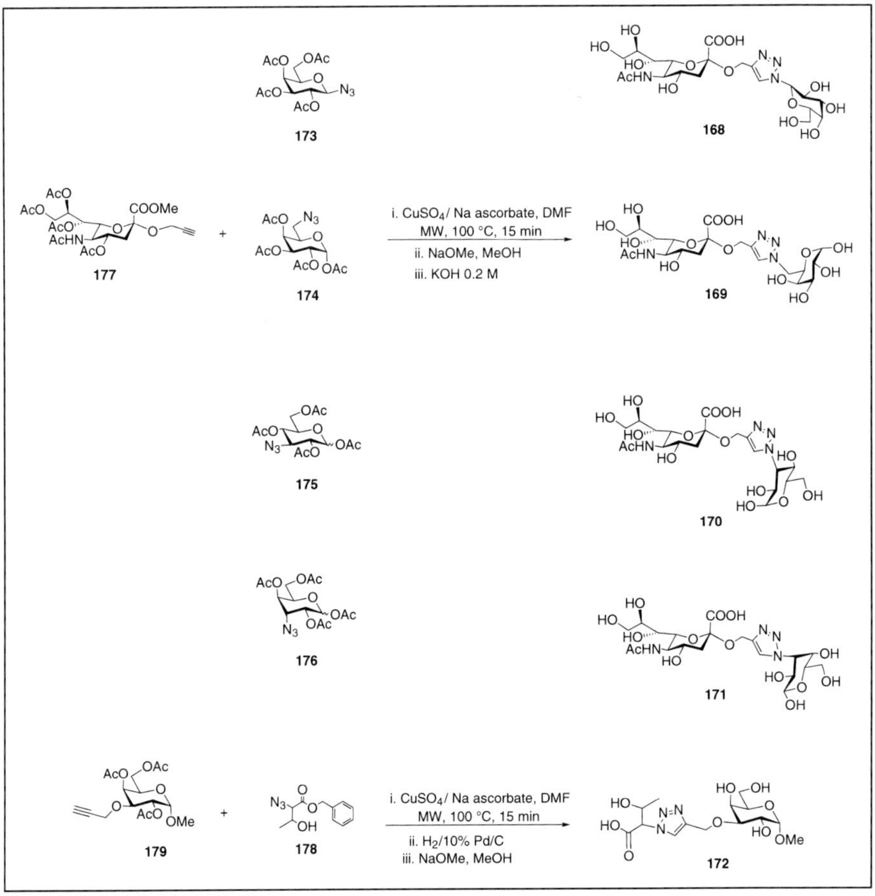

SCHEME 13.19 Synthesis of 1,2,3-triazole-linked sialic acid and amino acid glycosides **168–172** by click chemistry.

acid **177**, as well as by click chemistry reaction between the amino acid N_3-ThrOBn **178** with 3-O-propynyl-GalOMe **179** (Scheme 13.19) [64]. The 1,2,3-triazole linked sialic acid-6-O-galactose **169** and the sialic acid-galactopyranoside **168** showed high TcTS inhibitory activity at 1 mM (approx. 90%), whilst only the former displayed relevant trypanocidal activity (IC_{50} 260 μM). These results highlight the 1,2,3-triazole linked sialic acid-6-O-galactose **169** as a prototype for further design of new neoglycoconjugates against Chagas' disease.

REFERENCES

1. Kolb, H. C.; Finn, M. G.; Sharpless, K. B. *Angew. Chem., Int. Ed.* **2001**, *40*, 2004–2021.
2. Rostovtsev, V. V.; Green, L. G.; Fokin, V. V.; Sharpless, K. B. *Angew. Chem., Int. Ed.* **2002**, *41*, 2596–2599.

3. Aragão-Leoneti, V.; Campo, V. L.; Gomes, A. S.; Field, R. A.; Carvalho, I. *Tetrahedron* **2010**, *66*, 9475–9492.
4. Lee, B. Y.; Park, S. R.; Jeon, H. B.; Kim, K. S. *Tetrahedron Lett.* **2006**, *47*, 5105–5109.
5. Lutz, J. F. *Angew. Chem., Int. Ed.* **2007**, *46*, 1018–1025.
6. Baskin, J. M.; Bertozzi, C. R. *QSAR Comb. Sci.* **2007**, *26*, 1211–1219.
7. Kilpatrick, D. C. *Biochimica et Biophysica Acta* **2002**, *1572*, 187–197.
8. Sharon, N.; Lis, H. *Glycobiology* **2004**, *4*, 53R–62R.
9. Dodd, R. B.; Drickamer, K. *Glycobiology* **2001**, *11*, 71R–79R.
10. Sharon, N. *Trends. Biochem. Sci.* **1993**, *18*, 221–226.
11. Varki, A.; Cummings, R. D.; Esko, J.; Freeze, H.; Hart, G.; Marth, J. *Essentials of Glycobiology*, 1st ed.; Cold Spring Harbor Laboratory Press, Inc.: Boston, **1999**.
12. Stowell, S. R.; Karmakar, S.; Stowell, C. J.; Dias-Baruffi, M.; Mcever, R. P.; Cummings, R. D. *Blood* **2007**, *1*, 219–227.
13. Cummings, R. D.; Liu, F. T. Galectins, chapter 33. In: A. Varki, R. D. Cummings, J. D. Esko, H. H. Freeze, P. Stanley, C. R. Bertozzi, G. W. Hart, M. E. Etzler. Eds. *Essentials of Glycobiology*. 2nd ed.; Cold Spring Harbor Laboratory Press, Inc.: New York, **2009**.
14. Camby, I.; Le Mercier, M.; Lefranc, F.; Kiss, R. *Glycobiology* **2006**, *16*, 137–157.
15. Rabinovich, G. A.; Baum, L. G.; Paganelli, R.; Natoli, C.; Liu, F. T.; Lacobelli, S. *Trends in Immunology* **2002**, *23*, 313–320.
16. Ingrassia, L.; Camby, I.; Lefranc, F.; Mathieu, V.; Nshimyumukiza, P.; Darro, F.; kiss, R. *Curr. Med. Chem.* **2006**, *13*, 3513–3527.
17. Lahm, H.; Andre, S.; Hoeflich, A.; Kaltner, H.; Siebert, H. C.; Sordat, B.; Von der lieth, C. W.; Wolf, E.; Gabius, H. J. *Glycoconjugate J.* **2004**, *20*, 227–238.
18. Gigue're, D.; Patnam, R.; Bellefleur, M-A.; St-Pierre, C.; Satob, S.; Roy, R. *Chem. Commun.* **2006**, 2379–2381.
19. Tejler, J.; Salameh, B.; Lefflerc, H.; Nilsson, U. J. *Org. Biomol. Chem.* **2009**, *7*, 3982–3990.
20. Salameh, B. A.; Cumpstey, I.; Sundin, A.; Leffler, H.; Nilsson, U. J. *Bioorg. Med. Chem.* **2010**, *18*, 5367–5378.
21. Galante, E.; Geraci, C.; Sciuto, S.; Campo, V. L.; Carvalho, I.; Sesti-Costa, R.; Guedes, P. M. M.; Silva, J. S.; Hill, L.; Nepogodiev, S. A.; Field, R. A. *Tetrahedron* **2011**, *67*, 5902–5912.
22. Gouin, S. G.; Fernandez, J. M. G.; Vanquelef, E.; Dupradeau, F.-Y.; Salomonsson, E.; Leffler, H.; Ortega-Muñoz, M.; Nilsson, U. J.; Kovensky, J. *ChemBioChem.* **2010**, *11*, 1430–1442.
23. Cummings, R. D.; Lowe, J. B.; Selectins, chapter 26. In: A. Varki, R. D. Cummings, J. D. Esko, H. H. Freeze, P. Stanley, C. R. Bertozzi, G. W. Hart, M. E. Etzler. Eds. *Essentials of Glycobiology*. 2nd ed.; Cold Spring Harbor Laboratory Press, Inc.: New York, **2009**.
24. Phillips, M. L.; Nudelman, E.; Gaeta, F. C.; Perez, M.; Singhal, A. K.; Hakomori, S.; Paulson, J. C. *Science* **1990**, *250*, 1130–1132.
25. Titz, A.; Patton, J.; Smiesko, M.; Radic, Z.; Schwardt, O.; Magnani, J. L.; Ernst, B. *Bioorg. Med. Chem.* **2010**, *18*, 19–27.
26. Palomo, C.; Aizpurua, J. M.; Balentova, E.; Azcune, I.; Santos, J. I.; Barbero, J. J.; Canãda, J.; Miranda, J. I. *Org. Lett.* **2008**, *10*, 2227–2230.
27. Papp, I.; Dernedde, J.; Enders, S.; Haag, R. *Chem. Commun.* **2008**, 5851–5853.

28. Rele, S. M.; Cui, W.; Wang, L.; Hou, S.; Barr-Zarse, G.; Tatton, D.; Gnanou, Y.; Esko, J. D.; Chaikof, E. L. *J. Am. Chem. Soc.* **2005**, *127*, 10132–10133.
29. Robinson, J. A. *Chimia*, **2007**, *61*, 84–92.
30. Cunto-Amesty, G.; Monzavi-Karbassi, B.; Luo, P.; Jousheghany, F.; Kieber-Emmons, T. *Int. J. Parasitol.*, **2003**, *33*, 597–613.
31. Vichier-Guerre, S.; Lo-Man, R.; Huteau, V.; Deriaud, E.; Leclerc, C.; Bay, S. *Bioorg. Med. Chem.* **2004**, *14*, 3567–3570.
32. Brocke, C.; Kunz, H. *Bioorg. Med. Chem.* **2002**, *10*, 3085–3112.
33. Slovin, S. F.; Keding, S. J.; Ragupathi, G. *Immunol. Cell Biol.* **2005**, *83*, 418–428.
34. Wan, Q.; Chen, J.; Chen, G.; Danishefsky, S. J. *J. Org. Chem.* **2006**, *71*, 8244–8249.
35. Spohn, R.; Buwitt-Beckmann, U.; Brock, R.; Jung, G.; Ulmer, A. J.; Wiesmller, K.-H. *Vaccine* **2004**, *22*, 2494–2499.
36. Cai, H.; Huang, Z-H.; Shi, L.; Zhao, Y-F.; Kunz, H.; Li, Y-M. *Chem. Eur. J.* **2011**, *17*, 6396–6406.
37. Lee, D. J.; Harris, P. W. R.; Brimble, M. A. *Org. Biomol. Chem.* **2011**, *9*, 1621–1626.
38. Dobrovolskaia, M. A.; Vogel, S. N. *Microbes Infect.* **2002**, *4*, 903–914.
39. Tang, S.; Wang, Q.; Guo, Z. *Chem. Eur. J.* **2010**, *16*, 1319–1325.
40. Lee, L. V.; Mitchell, M. L.; Huang, S-J.; Fokin, V. V.; Sharpless, B.; Wong, C.-H. *J. Am. Chem. Soc.* **2003**, *125*, 9588–9589.
41. van der Peet, P.; Gannon, C. T.; Walker, I.; Dinev, Z.; Angelin, M.; Tam, S.; Ralton, J. E.; McConville, M. J.; Williams, S. J. *ChemBioChem.* **2006**, *7*, 1384–1391.
42. Melo, E. B.; Gomes, A. S.; Carvalho, I. *Tetrahedron* **2006**, *62*, 10277–10302.
43. Rossi, L. L.; Basu, A. *Bioorg. Med. Chem. Lett.* **2005**, *15*, 3596–3599.
44. Diot, J. D.; Moreno, I. G.; Twigg, G.; Mellet, C. O.; Haupt, K.; Butters, T. D.; Kovensky, J.; Gouin, S. G. *J. Org. Chem.* **2011**, *76*, 7757–7768.
45. Perion, R.; Ferrieres, V.; Garcõa-Moreno, M. I.; Mellet, C. O.; Duval, R.; Fernández, J. M. G.; Plusquellec, D. *Tetrahedron* **2005**, *61*, 9118–9128.
46. Dedola, S.; Hughes, D. L.; Nepogodiev, S. A.; Rejzek, M.; Field, R. A. *Carbohydr. Res.* **2010**, *345*, 1123–1134.
47. Dedola, S.; Nepogodiev, S. A.; Hughes, D. L.; Field, R. A. *Acta Crystallogr. C* **2008**, *64*, o445eo446.
48. Diot, J.; Garcõa-Moreno, M. I.; Gouin, S. G.; Mellet, C. O.; Haupt, K.; Kovensky, J. *Org. Biomol. Chem.* **2009**, *7*, 357–363.
49. Zhou, Y.; Zhao, Y.; O'Boyle, K. M.; Murphy, P. V. *Bioorg. Med. Chem. Lett.* **2008**, *18*, 954–958.
50. Fátima, A.; Baptistella, L. H. B.; Pilli, R. A. *Quim. Nova* **2005**, *28*, 306–316.
51. Itzstein, M. V. *Nat. Rev. Drug Discov.* **2007**, *6*, 967–974.
52. Li, J.; Zheng, M.; Tang, W.; He, P.-L.; Zhu, W.; Li, T.; Zuo, J.-P.; Liu, H.; Jiang, H. *Bioorg. Med. Chem. Lett.* **2006**, *16*, 5009–5013.
53. Weiwer, M.; Chen, C.-C.; Kemp, M. M.; Linhardt, R. J. *Eur. J. Org. Chem.* **2009**, *16*, 2611–2628.
54. Valaperta, R.; Valsecchi, M.; Rocchetta, F.; Aureli, M.; Prioni, S.; Prinetti, A.; Chigorno, V.; Sonnino, S. J. *Neurochem.* **2007**, *100*, 708–719.
55. Chen, X. P.; Enioutina, E. Y.; Daynes, R. A. *J. Immunol.* **1997**, *158*, 3070–3080.

56. Anastasia, L.; Papini, N.; Colazzo, F.; Palazzolo, G.; Tringali, C.; Dileo, L.; Piccoli, M.; Conforti, E.; Sitzia, C.; Monti, E.; Sampaolesi, M.; Tettamanti, G.; Venerando, B. *J. Biol. Chem.* **2008**, *283*, 36265–36271.
57. Gadhoum, S. Z.; Sackstein, R. *Nat. Chem. Biol.* **2008**, *4*, 751–757.
58. Kato, K.; Shiga, K.; Yamaguchi, K.; Hata, K.; Kobayashi, T.; Miyazaki, K.; Saijo, S.; Miyagi, T. *Biochem. J.* **2006**, *394*, 647–656.
59. Azuma, Y.; Sato, H.; Higai, K.; Matsumoto, K. *Biol. Pharm. Bull.* **2007**, *30*, 1680–1684.
60. Zou, Y.; Albohy, A.; Sandbhor, M.; Cairo, C. W. *Bioorg. Med. Chem. Lett.* **2010**, *20*, 7529–7533.
61. Carvalho, I.; Campo, V. L.; Andrade, P. *Trypanosoma cruzi trans*-Sialidase. In: H. Cerecetto Meyer, M. González. Eds. *Enfermedad de Chagas: Estrategias en La búsqueda de nuevos medicamentos*, 1st edn.; Una visión Iberoamericana: Mexico, **2012**.
62. Carvalho, I.; Andrade, P.; Campo, V. L.; Guedes, P. M. M.; Sesti-Costa, R.; Silva, J. S.; Schenkman, S.; Dedola, S.; Hill, L.; Rejzek, M.; Nepogodiev, S. A.; Field, R. A. *Bioorg. Med. Chem.* **2010**, *18*, 2412–2427.
63. Campo, V. L.; Carvalho, I.; Silva, C. H. T. P.; Schenkman, S.; Hill, L.; Nepogodiev, S. A.; Field, R. A. *Chem. Sci.* **2010**, *1*, 507–514.
64. Campo, V. L.; Sesti-Costa, R.; Carneiro, Z. A.; Silva, J. S.; Schenkman, S.; Carvalho, I. *Bioorg. Med. Chem.* **2012**, *20*, 145–156.

INDEX

Click Chemistry in Glycoscience: New Developments and Strategies, First Edition.
Edited by Zbigniew J. Witczak and Roman Bielski.